Tsunami Events and Lessons Learned

Advances in Natural and Technological Hazards Research

Volume 35

For further volumes:
http://www.springer.com/series/6362

Y.A. Kontar • V. Santiago-Fandiño • T. Takahashi
Editors

Tsunami Events and Lessons Learned

Environmental and Societal Significance

Springer

Editors
Y.A. Kontar
College of Sciences
University of Findlay
Findlay, OH, USA

V. Santiago-Fandiño
Environmental Assessment Consultant
Villaviciosa, Asturias, Spain

T. Takahashi
Faculty of Safety Science
Hydrosphere Disaster Laboratory
Kansai University
Takatsuki-shi, Osaka, Japan

ISSN 1878-9897 ISSN 2213-6959 (electronic)
ISBN 978-94-007-7268-7 ISBN 978-94-007-7269-4 (eBook)
DOI 10.1007/978-94-007-7269-4
Springer Dordrecht Heidelberg New York London

Library of Congress Control Number: 2013953346

Printed on acid-free paper

Springer is part of Springer Science+Business Media (www.springer.com)

Foreword

December 26, 2004 began like any other day for the thousands of tourists enjoying the beautiful beaches of Phuket, Thailand. The sea was calm and gentle. When it slowly began to recede, the tourists happily rushed down to the shore to collect suddenly exposed shells and sea animals. Most of them had never even heard the word "tsunami" and could not have anticipated that one of the world's most destructive natural disasters was about to engulf them. By the time, the last tsunami wave had washed ashore, about 230,000 people had been killed along the coasts of 14 countries in the Indian Ocean. Citizens of about 60 countries had lost their lives. This was one of the most catastrophic and widespread natural hazards in the recorded history. The question that immediately comes to mind is: What can be done that such tragedy never happens again in future?

The devastating 2004 Sumatra tsunami generated by the M_w 9.3 megathrust earthquake was a scientific dividing line. Prior to the event, the term "tsunami" was familiar only to specialists. Within hours of the event, the entire world came to understand the power of tsunami wave forces. Thousands of new people were suddenly recruited to fight ignorance about these killing waves. They entered tsunami science and operation from different research fields bringing their experience and new ideas. Various countries from around the globe contributed major funding to tsunami research, enabling the installation of hundreds of new high-precision instruments, the development of new technology and the establishment of more modern communication systems. As a result, incredible progress has been achieved in recent years in tsunami research and operation. Tsunami warning and hazard mitigation systems have dramatically improved. The tsunami observational network of coastal tide gauges has been significantly reconstructed, upgraded and expanded. Tsunami waves began to be monitored in both the deep ocean and from space. A large number of open-ocean Deep-ocean Assessment and Reporting of Tsunamis (DART) stations have been distributed throughout the entire Pacific Ocean; DARTs are also now deployed in the Indian and Atlantic oceans. These new precise instruments have yielded thousands of coastal and hundreds of deep-water high-quality tsunami records that have enabled researchers to refute some previous misconceptions about tsunamis and to bore more deeply into the physics of tsunami generation and propagation. Modern numerical models, combined with open-ocean DART records, make it possible to predict tsunami waves for coastal sites with very high accuracy.

Despite the recent advances, tsunamis remain a major threat to coastal infrastructure and human life. Destructive tsunami events continue to kill people and create enormous damage. Several catastrophic events occurred between 2005 and 2010, including the 2005 Sumatra, 2006 Java, 2009 Samoa, 2010 Chile and 2010 Mentawai tsunamis. The Tohoku (Great East Japan) tsunami of 11 March 2011, with almost 20,000 casualties and destroyed the Fukushima nuclear power plant, was a tragic example of this chain of devastating events. We can state with considerable certainty that the number of victims would have been many times higher without tsunami mitigation programs and effective tsunami warning services in Japan and other countries.

International cooperation is a key factor in the ongoing quest to understand tsunamis and to improve our ability to predict them and to mitigate their disastrous consequences. This volume presents a unique collection of papers prepared by scientific specialists from various countries that defines the state of tsunami research and understanding at the beginning of the twenty-first century. The authors outline the problems and recent progress in tsunami assessment and prediction, numerical modeling and simulation, coastal damage and reconstruction. The volume also provides an up-to-date review on the geology of the oceanic lithosphere and geomorphology of coastal zones, chronicles important problems of after-event ecology and environmental assessment. Finally, it discusses some general questions of tsunami policy and the lessons that we learned from latest catastrophic events.

This book will be of interest to researchers, graduate and undergraduate students involved in the study of marine geology, marine geophysics, natural marine hazards, physical oceanography, seismology, carbon footprint, risk assessment and environmental science.

Russian Academy of Sciences Alexander B. Rabinovich
Institute of Ocean Sciences. Canada. Richard E. Thomson

Preface

The term ***Tsunami*** (津波) comes from the Japanese language meaning *Harbour Wave*, often confused with tidal wave, particularly by the press. The meaning has evolved becoming in general terms a large oceanic wave caused by geological activity such as earthquakes and underwater volcanic eruptions, submarine landslides or by fallen material from space such as meteorites.

Throughout history tsunamis have occurred with more or less frequency, intensity and power mainly in certain areas of the planet were the tectonic plates, faults and volcanoes are highly active i.e. the Mediterranean Sea, the Pacific Rim (or Belt of Fire) and the Indian Ocean. Tsunamis have also been recorded in the Caribbean Sea and with much less frequency in the North Atlantic.

Although these types of oceanic events often are harmless due to their small size, others may cause incalculable damage in terms of human loss and to the society; with enormous impact to the environment; local, regional or even national economy, infrastructure and lifeline services.

Coastal communities have been most vulnerable to tsunami as they directly receive the full destructive power by waves and run-up. Once the tsunami is generated reaching the shallow coastline its power is concentrated in a smaller volume causing havoc. Theoretically, the power (in terms of energy) gathered in the tsunami wave reaching Sri Lanka in 2004 (where more than 12,000 people perished) was theoretically about 1 GW per kilometer of shoreline. This energy is enough to power a city of a half a million inhabitants or the equivalent of production of one of the Fukushima Nuclear Reactors prior to destruction.

After a period relative calm of small tsunamis events since 1998 (Papua New Guinea) the definition of recent tsunami events could be considered for practical purposes as those generated from 2004 on wards, in particular as from the large Indian Ocean tsunami which hit Indonesia, India, Thailand and Sri Lanka among other countries which caused enormous social, human, economic and environmental losses to the Great East Japan Earthquake and Tsunami event of March 11, 2011. The latter likely to be the widest witnessed episode of this kind by the world population through the media and live coverage due to its power, impact and phenomenal destruction to the environment and resources, infrastructure and industry, coastal communities and services along the Sanriku coastline (and beyond). Added to this was the nuclear plant debacle of the Fukushima-Dai Ichi plant mainly as result of the tsunami hit.

There are a number of previous scientific studies and publications looking at the above events as well as others related to the nature and origin of tsunami, historical evidence and field and lab research, modelling, state of the art early warning systems, and particular episodes and their impacts. To mention a few examples amongst others are the books prepared by Kenji Satake (2005 and 2011 Springer; 2007; Barnes and Noble and 2011 World Scientific Pub. Co. Inc.); Edward Bryant (2005, Cambridge University Press and 2008, Springer); Boris Levin and Mikhail Nosov (2009, Springer), Phil Cummins, Laura Kong and K. Satake (2009, Virkhauser Verlang/Springer). On specific tsunami episodes Gail L. Karkowski and J. MacDonald (2006, Darby Creek Publishing); William W. Lace (2008, Infobase Publishing), Tad S. Murty (Taylor and Francis, 2007); Phil Cummins and Mark Leonard (AUSGeo, 2008) analyzing the tsunami in Sumatra; Douglas Fernando in Sri Lanka (2005, Verlang).

Some publications related to the social and economic impacts of recent tsunamis are available like those by Tricia Wachtendorf et al. (2006, Cambridge University Press, Research Institute), E.N. Bernard et al. (Springer 2005, 2006), The Economist Intelligence Unit (2005) and the Asian (2005) Development Bank Report about the Indian Ocean Tsunami of 2004. On the side of the environmental or Ecological impacts by tsunami not much has been written. The United Nations Environment Programme published in 2005 the Post Tsunami Environmental Assessment of the Seychelles Islands hit by the Indian Ocean Tsunami of 2004, and E. Bernard and A. Robinson include a few articles on the environmental impacts of the Tsunami in the Thailand Coast (2009, Harvard College; USA).

There as been an enormous advancement in the science and knowledge related to tsunamis, nonetheless, plenty remains to be learned in fields such as modelling, ecology, socioeconomic aspects, early warning and forecasting, hazard and risk prevention, etc. An opportunity arose to increase existing knowledge with the latest information gathered from the Asia Oceania Geosciences Society (AOGS)-AGU (WPGM) Joint Assembly in Singapore and the American Geophysics Union (AGU) meeting in San Francisco held in 2012 as well as the International Union of Geodesy and Geophysics (IUGG) GeoRisk Commission conference which was held in 2012 at Chapman University, Orange, California, USA (amongst other sources).

It is from these meetings and conferences that experts have been invited to participate in the production of the present collection by sharing their findings, experience and lessons-learned from recent tsunami events. The information included considers a broad number of fields which directly or indirectly would help to better understand, quantify, and forecast tsunami events and their consequences in order to better protect coastal water resources and ecosystems as well as the society and coastal communities affected by these oceanic events.

Although not a comprehensive book, this publication brings a series of independent chapters on a variety of topics related to recent tsunami events, including their ecological and societal significance.

The editors would like to express deep appreciation and thanks to all of the experts who kindly participated in this publication by sharing their knowledge, experience and research findings through their respective papers. Moreover, deep appreciation is also expressed to the large number of experts who kindly reviewed the papers providing most important comments and suggestions.

Contents

Chapter 1
The Tsunami and Earthquake in Miyagi Prefecture and Sanriku 2011–2012: An Overview

Vicente Santiago-Fandiño

Abstract Due to its geographical location and geology, Japan is a country susceptible to a large number of weather-related events and destructive forces such as typhoons and tsunamis, volcanic eruptions and earthquakes. The Tohoku Region has six prefectures including Iwate and Miyagi, having the latter a surface area of 7,285 km^2 with varied topography with mountainous areas and the Sanriku coastline with rich aquatic/marine and land ecosystems. As well as important economic sectors such as agriculture, fisheries, forestry, manufacturing, food processing and tourism amongst others.

In March, 2011 the Great East Japan Earthquake and Tsunami occurred with enormous environmental, economic and societal impacts heavily impacting the coastline as well as the economic and social milieu as cities and towns, industrial and agricultural facilities, farms, fishing grounds, ports, schools, hospitals, roads, and infrastructure in general were either severely damaged or washed away completely. In a period of 2 year since large efforts have been made by the local and central government as well as the society to start restoring the damage.

Although far from being exhaustive, the present paper provides an overall view of the situation towards restoration during this period in key sectors such as Agriculture and Forestry, Environment, Fisheries, and Urban Areas, including Debris Management in Miyagi prefecture and the Sanriku coastline.

Keywords Agriculture • Debris management • Environment • Fisheries • Forestry • Impact • Society • Tsunami • Urban

V. Santiago-Fandiño (✉)
Private Consultant, El Curbiellu 28 Peon, Villaviciosa 33314, Asturias, Spain
e-mail: v.santiago.f@gmail.com

Y.A. Kontar et al. (eds.), *Tsunami Events and Lessons Learned: Environmental and Societal Significance*, Advances in Natural and Technological Hazards Research 35,
DOI 10.1007/978-94-007-7269-4_1,

1.1 Introduction

Japan, a country divided into various regions, is formed by a series of islands with 35,000 km of coastline exhibiting complex morphology. Due to its geographical location, the country is susceptible to a large number of weather-related events such as winter storms, heavy rains, strong seasonal winds and ocean surges, as well as nature's more powerful and often destructive forces such as typhoons and tsunamis. Furthermore, due to the activity of Japan's complex geological environment, volcanic eruptions, earthquakes and tremors are common.

The Tohoku Region, located in the northern part of the Honshu Island, consists of Yamagata and Akita prefectures in the Sea of Japan coast while Aomori, Iwate, Miyagi and Fukushima prefectures in the Pacific coast. Miyagi Prefecture has a surface area of 7,285 km^2 and about 2.35 million inhabitants, as of 2011 (JETRO 2012). The prefecture has a varied topography with mountainous areas including three large ranges: the Zao, Funagata and Kurikoma ranges. It also features vast plains and a magnificent coastline known as Sanriku (Fig. 1.1).

Although the climate varies considerably along the prefecture's coastline, in winter and early spring it is generally snowy with low temperatures, hot and humid in summers, and cool in autumn (12.4C annual temperature and 1,254 ml yearly average precipitation; Miyagi Prefecture Government 2012a). Fog banks from the sea frequently move inland and the prefecture is even occasionally hit by typhoons.

Fig. 1.1 Miyagi prefecture and Sanriku coastline

Miyagi Prefecture is known for its beautiful natural scenery and large number of aquatic/marine and land ecosystems as well as important economic sectors such as agriculture, fisheries, forestry and livestock as well as manufacturing, food processing, electronics commerce and trade. The tourism sector, albeit on a smaller scale than the other sectors, has played an important role in Miyagi's economy, particularly in areas relating to sightseeing, natural scenery, protected areas such as forests and national parks, high-quality seafood production, local summer festivals and recreational sites.

The prefecture has been hit by a variety of catastrophic natural events over the millennia, as the first records go as far back as 869 AD and thereafter in 1611, 1896 and 1933 (Imamura and Anawat 2011). Moreover, a 1960 tsunami originating in Chile also hit this prefecture. The Great East Japan Earthquake was predicted to happen by a technical investigation committee of the Central Disaster Management Council of Japan (Kunihiko 2012). A series of large earthquakes in March 2011 topping the Japanese scale of magnitude level 7 (recognized as the largest in the country's history) was a result of the sudden release of a colossal amount of energy as the Pacific Plate re-accommodated itself off the coast of Miyagi Prefecture, displacing the ocean floor horizontally by 50 m and vertically by about 10 m (Fushiwara et al. 2011). This produced a mega-tsunami of cataclysmic proportions, with heights striking most of the east coast of Japan and having the heaviest impacts on Iwate, Miyagi and Fukushima Prefectures.

The environmental, economic and societal impacts resulting from these catastrophic events, particularly the tsunami, were immense as entire coastal cities and towns, industrial and agricultural facilities, farms, fishing grounds, ports, schools, hospitals, roads, and infrastructure in general were either severely damaged or washed away completely (Fig. 1.2). Likewise, lifeline systems for water, sanitation, electricity, fuel and food were either heavily disrupted or totally wiped out. Economically, the damage has totaled more than 16 trillion yen (more than US $200 billion), as reported by the OECD and the Japanese Government in 2012 (Gurría 2012; Saito 2012), which equaled about 3.5 % of the nation's GDP (Waldenberger and Eilker 2011).

More than 15,000 people perished when the tsunami struck and nearly 3,000 people remain missing as of September 2012 (NPAJ 2012), although some say the number is 20,000 already considering the missing as dead. A most interesting analysis of fatality rates is done by Ando et al. (2012) and ICHARM (2011). More than 1,500 people have died due to post-disaster related health problems. About 340,000 displaced people still live in 136,000 temporary housing units and other emergency sites provided by the Japanese government. About 71,000 residents have evacuated the affected areas to sites outside Iwate, Miyagi and Fukushima and many have left the region permanently (Japan Times 2012a).

In Miyagi prefecture alone, the Police Agency stated that a total of more than 85,000 buildings and houses were damaged or destroyed by the tsunami, while 390 roads were damaged in various ways; 51 landslides occurred and at least 45 dykes suffered different levels of damage severity, from partial to complete destruction. Immediately following the event. Moreover, the account for human losses were 9,530 people, and 1,359 missing (NPAJ 2013).

Fig. 1.2 Minamisanriku city hall; 2011

In March 2011, the number of evacuees rose to 321,000 who were placed in more than11,000 temporary shelters (Onodera 2012), and the relocation of these evacuees to shelters and temporary housing would cause further distress in which refugees suffered psychological stress in addition to the enormous psychological impact of the cataclysm. Soon after the disaster, a large number of the 7,515 evacuees located in 49 emergency evacuation shelters in Minamisanriku were suffering from acute stress disorders such as multiple trauma, PTSD, sleepiness, and pneumonia, in addition to other health problems related to the loss of properties and family members (Fig. 1.3) (Nishizawa 2011; Médecins sans Frontières 2011). The most affected were the elderly and people above the age of 65 as more than 90 % suffering from the above mentioned health problems died as a result of anxiety that more would happen in the near future (Vervaeck and Daniell 2012).

Children in the hit areas are also suffering from emotional stress, having developed different types of disorders including fear of water, prone to sickness and other ailments which require long-term counselor support (Japan Times 2012a). A report produced by the Miyagi government stated that the total damage caused to date by the earthquake and tsunami was US$115.54 billion, in which agriculture suffered US$6.75, livestock US$65.65, forestry US$183.81 and fisheries US$9 billion in losses. Moreover, the loss suffered by ports has totaled US$1.43 billion, coastal damage US$1.08, rivers and dams US$3.31 billion, sewage US$4.88 and roads about US$6 billion (Onodera 2012).

Fig. 1.3 An elderly man visiting the remains of the families' house has been a common scene in Minamisanriku

To deal with this enormous challenge, the Reconstruction Agency was established in February 2012 by the Japanese Government to lead reconstruction efforts, budget and coordinate policies with the central and local governments and municipalities (Reconstruction Agency of Japan 2012).

The total budget assigned by the Japanese government towards the restoration of the affected areas within the next 10 years will reach well beyond 23 trillion yen (US $300 billion), and by 2015 a total of 19 trillion yen are to be spent (Saito 2012). In Miyagi, the prefectural government developed a detailed plan in 2011 to be implemented for a period of 20 years. During the first 10 years, full recovery is expected while completing the remaining projects within the following 10 years. Three main stages have been devised: Restoration, Reconstruction and Development, which include eleven Urgent Priority Items and ten Recovery Points in a number of sectors including agriculture and forestry, marine products, tourism, disaster-resistant communities, Eco-towns, and development of human resources. Furthermore, the East Special Recovering Zones System was established to provide particular restoration incentives to various sectors including the manufacturing and fisheries industries (Miyagi Prefecture Government 2011; World Press Recovering Tohoku 2012).

Although far from being exhaustive, the present paper provides overall view of the situation as of October 2012 towards restoration in key sectors such as Agriculture and Forestry, Environment, Fisheries, and Urban Areas, including Debris Management in Miyagi prefecture and the Sanriku coastline.

1.2 Agriculture

Japan's once thriving agriculture industry has been on the decline since the end of the 1950s, from more than 6 million to 4.56 million hectares of land, due to a variety of reasons including the advancing age of farmers, migration of young adults to cities, transformation of agricultural lands (mainly rice) into residential areas and use for other alternative purposes (about 80 km^2), added lack of development incentives as well as land valuation encouraging sales. About 15,000 ha of fertile land is lost every year and the labour force has fallen to half since the tsunami hit. Moreover, about 10 % (or near 400,000 ha) of the country's farmland has not been used (Daily Yomiuri 2012a).

At its height, Miyagi prefecture's agriculture extended to about 136,600 ha in 2009 (MAFF 2010a), producing a large variety of products such as rice (Fig. 1.4), the most common product, representing 73 % of total agricultural production, followed by leguminous plants (11 %), green vegetables (9.5 %) as well as wheat and barley (2.7 %). Soya beans and fruits such as apples, plums and strawberries

Fig. 1.4 Rice agriculture along the Okawa river watershed between Miyagi and Iwate prefectures

were also very important products, Miyagi producing the country's eight-largest crop of strawberries (MIAC 2011).

By 2010, Miyagi's rice production was about 400,000 t, holding the seventh position in the nation, and soybeans totaled approximately 12,000 t (Japan's fourth-largest producer). Aside from agricultural products, the prefecture also produces high-quality pork meat and beef, the latter industry of which Miyagi held the country's eight-largest position (Onodera 2012).

One of the main problems in agriculture in Miyagi, as in the rest of Japan, is farmers reaching advanced ages, as well as emigration of the work force which has weakened agricultural production and adversely affected management of water and irrigation systems as there are no new recruits to support management activities. Furthermore, the profitability of rice paddy farming depends heavily on the cultivation of plots of land with areas smaller than one hectare and the majority of the specific monocultures along the Sanriku coast are unprofitable. In order to improve the situation and encourage community integration particularly in disaster-hit areas, new policies must be implemented to open up agriculture to fisherman, fishing industries and vice versa (Kazunuki 2011).

In terms of sustainable agriculture, a policy introduced in 2007 under the Outline of Farm Management Stabilization Program Measures to Conserve and Improve Rural Resources (Land, Water and Environment) has shown modest success. The policy aims at supporting the collaboration and pioneering of farming activities to improve irrigation as well as reducing (by up to 50 %) the use of chemical fertilizers and synthetic pesticides (MAFF 2005). Miyagi prefecture encourages this type of approach by incorporating sustainable agricultural practices under the Eco-farmer policy to ensure sound soil management in farms, reducing artificial chemical fertilizers while increasing the use of organic ones, for example, compost produced from cattle manure and straw. The Organic Agricultural Products and Special Cultivated Agricultural Products programs have been put into place for this purpose (MAFF 2005, 2007).

Other sustainable methods include cultivation rotation, the "aigamo" or duck farming, the "fuyu mizu tambo" or the flooding of rice paddies during winter like in Kabukuri-numa rice fields, which in 2005 became a RAMSAR site (Okubo 2007) and the oldest and most environmentally sustainable production system for agriculture and forestry known and as Satoyama (Fig. 1.5). This system is based upon the natural benefits of ecosystem services provided by forest, fauna, flora and human intervention, hence the unnecessary use of artificial fertilizers and pesticides. Recently this system has been losing its sustainability due to the introduction of modern technology, fertilizers and loss of specialized agricultural knowledge with the diminishing population of older farmers (Kobori and Primack 2003a, b).

As a result of the earthquake and tsunami, large areas of agriculture were badly damaged (Fig. 1.6), including those areas where measures for sustainable agriculture were undertaken. The estimated amount of damaged land accounts for more than 15,000 ha or 11 % of the prefecture's total arable land (136,300 ha). Paddy fields were most severely affected by the tsunami, with a total loss of 12,685 ha, while lands affected at higher elevations totaled more than 2,300 ha (MAFF 2010b).

Fig. 1.5 Mechanized Satoyama system in Shizugawa; Miyagi prefecture

Fig. 1.6 Sunken and badly damaged rice paddy taken over by seagull flocks as resting and nesting grounds

As a result, the restoration of land affected by the tsunami has become imperative, requiring transferral of new soil which has proved to be a big problem as there is not enough ready soil in the prefecture. Furthermore, transportation costs and rising prices of soil has further deepened the crisis and the limited budget is affecting the

Fig. 1.7 Former rice paddy restored using different soil likely meant to grow other types of crops

speed of restoration (Onodera 2012). The government of Miyagi has estimated that 20 million cubic meters of soil are required to restore more than 14,000 ha of damaged agricultural land in the lower parts of the watershed (Nakamura 2012).

To ameliorate the situation, the prefecture has undertaken efforts to restore damaged soil and change crops where necessary. Trials are underway in salt-damaged rice fields in Ishinomaki city (Hebia) with extract from Stevia plants (Stevia spp.), and have had successful results (ACN newswire 2012). Cotton, a more salt-tolerant and easier plant to cultivate, has also been introduced in pilot sites in Sendai and Natori through the Tohoku Cotton Project in a joint collaboration between the government of Australia, the prefectural government and local farmers and retailers. So far, the project has successfully yielded 1.8 ha of cotton crops (Ministry for Agriculture, Fisheries and Forestry, Australia 2012; Cotton Australia 2012; Takashige 2012). Also, the development of new salt-tolerant rice strains in certain areas is being tested (Nagano 2012; The Economist 2012). Soil desalination by flushing fields with large amounts of fresh water is also being considered to assist with the soil recuperation, however, it is only suitable for some areas where the earthquake and tsunami did not destroy or affect irrigation or drainage systems, as large amounts of deposited salt are difficult to remove. Siphoning-out seawater ponds and flooding of rice paddies during winter to wash out salts have also been proposed to improve soil conditions. Seasonal rains are also likely to be helpful in flushing out salt deposited in the soil, helping with restoration.

Some local farmers along the Sanriku coast are also exchanging rice fields for vegetable crops using shale rock and limestone as soil (Fig. 1.7), but with the inconvenience of having to add artificial fertilizers. In other cases, farmers are switching to hydroponic systems in order to grow tomatoes and soya beans (USDA 2012), the latter in particular, as it is the basis for the production of *miso* paste which is one of the most important food ingredients in the prefecture's local cuisine.

The accident of the Fukushima Dai-Ichi Nuclear Plant caused the release and spread of a large amount of radioactive material, affecting agriculture mainly in that prefecture. Nonetheless, some material fell in certain areas of Miyagi and Iwate. For example in the mountainous area of Murata town radioactive cesium was found in shiitake mushrooms cultivated on logged trees, forcing the government to request farmers to refrain from distributing them (MAFF 2012e; The Asahi Shinbum 2012). Likewise, this seems to be happening in other forest areas (hot spots) causing lasting dispersion of radioactive materials as they are gradually moving into rivers, estuaries and the ocean.

1.3 Forestry

In 2007, more than 25 million hectares (two-thirds of Japan) was covered with trees, half of which consisted of natural forests and another 40 % consisting of plantations. The forests are owned by the national and municipal governments as well as privately (MAFF 2011b). Amongst the most popular tree plantations (Fig. 1.8) are the Japanese cedar or *sugi* (*C. japonica*), cypress or *hinoki* (*C. obtusa*), red pine or *akamatsu* (*P. densiflora*) and larch or *karamatsu* (*L. kaempferi*). All of them are commonly found in Miyagi and along the Sanriku coastline (Mitsui & Company Limited 2012).

Fig. 1.8 Mixed forest and plantations along Sanriku coastal watershed

Fig. 1.9 Damaged Sugy tree plantation as result of high salinity accumulation in the soil

In the past, Japan has relied on its own forests for supplying timber materials, but since the timber pick production in 1967, it has heavily decreased by 2010. One of the reasons for the downfall is the reduction of the workforce due to aging as a large number of people dedicated to forestry are over 65 years old (MIAC 2012) while the most important reason has been the increase of cheaper imports from overseas causing the price of the wood to drop.

Forests in Miyagi presently cover 57 % (420,000 ha) of the prefecture's total surface area (Miyagi Prefecture Government 2012b), which in 2008 contained 209,000 ha of artificial forest: 96,000 of Japanese cedar, 59,000 ha of cypress and 3,000 ha of larch. About 95,000 ha are privately owned, the rest is owned by the government (MIAC 2012).

The tsunami damaged large areas of the coastal forest along the coastlines of Aomori, Iwate, Miyagi, Fukushima and Chiba Prefectures totaling 140 kilometres (MAFF 2012a). In Rikuzentakata City, Iwate Prefecture, a green belt of 4000 pine trees for coastal protection were totally devastated (Harada and Imamura 2005); an additional account has been given stating that 70,000 forest between Takata and Matsubara coastline was over run by the tsunami (Sugimura 2013). Likewise, in the cities of Ishinoamaki and Sendai local communities coastal forests were devastated (Figs. 1.9 and 1.10) as also reported in Mongabay.com (2011). These tree barriers proved to be irrelevant and instead caused large secondary damage to houses and buildings when dragged along by the tsunami. Other types of coastal forest destruction occurred in Kesennuma City, due to fire from burning spilled oil, which burned part of the adjacent Oshima forest areas (Kobayashi 2012).

Miyagi has a large number of sawmills as well as wood and wood-related industries which make products such as lumber, plywood, millwork and panels

Fig. 1.10 Pine forest washed away by the tsunami in Sendai city

Fig. 1.11 Wooden temporary housing being built in Utatsu; Miyagi prefecture

for construction (Dun and Bradstreet 2012), and the city of Ishinomaki alone contains the largest number of plywood manufacturing plants in Japan (JETRO 2012; MAFF 2011a). The earthquake and tsunami caused vast devastation to some of these industries mainly along the coastline. Nonetheless, with the support of the Forestry Agency, plywood processing companies and lumber mills are gradually returning back to business (Fairwood Partners 2012).

Forest products also played an important role in the prefecture after the March 11 events as a large number of temporary houses were built with wood originating from the prefecture (Fig. 1.11); this type of housing has proved to be more comfortable than others. To restore the lost forests, in 2011 many regional and local governments in Miyagi, Aomori, Iwate and Fukushima prefectures drafted reconstruction policies focusing on forestry and the forest related industries as well as the recuperation of coastal forests. In Miyagi prefecture, the central government has planned to build a disaster prevention coastal forest along the Sendai Plains and general coastal area (Japan Times 2012b; Sendai City 2012). Moreover, the Japanese NGO OISCA has developed a 10-year program to assist in the restoration of the prefecture's coastal forests based on the prefectural and national government plans (OISCA 2012), where the private sector, the general public, and other organizations as well as foreign countries are expected to be partners (Tokyo Marine & Nichido Fire Insurance Co. Ltd. 2012). A nursery has been already established in Natori City where seeds of nematode-resistant Japanese Black Pine and ordinary Black Pine are growing. A total of about 500,000 seedlings are expected to be raised within the next 10 years of the program.

As in the case for agriculture concerns about radiation in forest and forest products, bamboo and bamboo shoots as well as other plants grown in decaying forest trees, such as the Log-Grown Shiitake mushroom (*Lentinula edodess*), have been important in inducing the government to take measurements of radioactive cesium – and iodine – (MAFF 2012c). Radioactive cesium was found in mushrooms in the cities of Kurihara and Marumori, and the provincial Government has asked producers to voluntarily refrain from shipping contaminated products from April and June, 2012 (Miyagi Prefecture Government 2012c, d). Concerning outdoor-grown mushrooms in decaying forest logs, the Tome policy was also implemented in May, 2012 (Miyagi Prefecture Government 2012e). In the case of bamboo measurements undertaken in bamboo shoots near and around the area, tests proved to be negative as no radiation was present (Miyagi Prefecture Government 2012f).

1.4 Environment

As a prefecture very rich in nature, Miyagi has a large number of ecosystems, from high mountainous biomass to the coastline; the geology is varied and complex, particularly on the Sanriku coastline, which presents paleozoic, mesozoic and quaternary formations. This coastline stretches for about 600 km starting from the south of Aomori to Miyagi prefecture along the Pacific coast, a portion spanning for about 280 km reaches northern Miyagi prefecture with a distinctive zipper-like morphology featuring beaches, mudflats and natural wetlands, lowlands, forests, cliffs, inlets and small islands as well as many rivers and streams. The watershed has luxurious forests and tree plantations. Rice fields are also abundant due their rich fauna and flora while some of them being most important for local migratory waterfowl species (Fig. 1.12), particularly in lowlands in which the Satoyama mixed ecosystem is common. A large number of fjord-type formations known as *rias* form narrow and irregular estuaries along the coastline, hence the name of this region, *Rias Coast*.

Fig. 1.12 Commonly found egrate in Miyagi's rice paddies

Fig. 1.13 Rich biota found along the rias and bays in Sanriku

The prefecture's ocean side is generally highly dynamic, with strong storms and winds producing large waves and swells. These harsh conditions are so common, they require shoreline protection in many places to reduce erosion and abrasion. Off of Miyagi's coast, the "Oyashio" (cold) and "Kuroshio" (warm) ocean currents meet creating large turbulence as they mix, concentrating large amount of nutrients including dissolved iron from the Amur river basin as shown by the Amur-Okhotsk Project (Shiraiwa 2012; Nishioka et al. 2007, 2011; Amur-Okhotsk Project Report Series 2003–2010) and other reports, eventually this iron is used by phytoplankton creating one of the of the richest fishing areas in the world (Midori II 2012) – (Fig. 1.13).

Miyagi has a number of protected areas which include the shared Rikuchukaigan National Park (12,200 ha), stretching from Kuji in Iwate prefecture to Kesennuma, the Kesen Numa Prefectural National Park (21,079 ha), Minamisanriku-Kinkansan

Fig. 1.14 (**a**) and (**b**) Eroded rocky formations and damaged beach along the Sanriku coastline

Quasi National Park (13,902 ha), Keyosan-Mangokuura Prefectural National Park (9,933 ha) and the Matsushima Kinkansan Prefectural Park found in the south of the prefecture. They have a large portion of coastline, which are visited by a large number of tourists particularly in the summer season (National Parks of Japan 2012; Takahashi 2012; MOE 2012a; Miyagi Prefecture Government 2012g).

As a result of the catastrophic events of March 11, the ground actually sank about 80 cm or rose in a few areas due to the earthquake. Tsunami heights well beyond 10 m (Satake et al. 2012) not only produced environmental havoc in many locations but also created profound morphological and ecological alterations along the coastline, including protected areas, wetlands, mud and tidal flats, sand beaches, estuaries, as well as other type formations (Fig. 1.14a, b) and related fauna and

Fig. 1.15 Brocken cliff formation in Motoyoshi, Miyagi prefecture

flora. Examples of the changes in morphology are plenty along Sanriku coastline as newly created mud flats and marshes have been created like in Shizugawa (Katsurawa 2012) and Monue Bay (Tanaka 2012a, per com.), while some wetlands were left isolated from the estuarine system. Heavy erosion also occurred along the coastline like in Minamisanriku where a well know beach suffered large alterations becoming deeper and steeper; a ria in Minato town was totally eroded leaving only the bare rock having had all the bottom sediments removed as the tsunami run-up went deep onto the lower watershed with devastating consequences. Sea grass beds in shallow coastal areas suffered similar impacts for example in the Mangokuura Sea in Ishinomaki city (Hori 2012) whilk Eelgrass beds (*Zoostera japonica*) in Nagatsuraura bay. In Sendai, a beach was badly eroded by the tsunami's run-up and backwash, but unexpectedly within a few months after the event recovering seems to be occurring (Richmond 2012).

In the case of cliffs and rocky formations, they better stood the impact of the tsunami and earthquake at first glance, but large sections suffered breakages (Fig. 1.15), cleavage and abrasion while exposed soils were severely eroded or totally removed; most of the boulders and broken sections may have likely been washed out into the ocean by the tsunami's backwash. Studies undertaken on the Otsu Chi coast (Iwate prefecture) showed that enormous changes occurred not only along the shoreline, but also in the morphology of the ocean floor as it suffered changes in depth and some other features, most likely mirroring what has happened in other affected areas along the Sanriku Coastline (University of Tokyo 2012).

Rivers and streams along the coastline suffered enormous damage after having their courses filled with debris, sand and other materials, banks broken or eroded, depths altered and seemingly flow rates and discharge reduced. Although the lower

Fig. 1.16 Kitakami river and steel frame left by the tsunami about 10 km from river mouth

sections in the majority of them have been cleaned out within 1 year, the situation still remains critical, particularly concerning midstream sections where the embankments are still in very poor condition.

Aquatic environments including associated fauna and flora were also hit hard due to the speed, turbulence, and salinity from the incoming seawater; a myriad of ecosystems were affected: agricultural lands including rice paddy fields, forests, swamps and natural wetlands. Along the Kitakami river (Fig. 1.16) the tsunami's run-up reached almost 50 km upstream while in its estuary in Ishinomaki city flooded about 73 km^2 marshes and wetlands (Government of Japan et al. 2011).

The transformation of natural wetlands into rice paddy fields has been the main reason for the loss of these ecosystems; Miyagi has already lost about 92 % natural wetlands over the past 100 years, earning it one of the highest loss rates in the country (MOE 2011). The impact of the earthquake and tsunami have further aggravated this situation along the coastline, although new wetlands and tidal areas have been created either permanent or temporal being the latter in areas where freshwater is still flowing along spoiled rice fields. A number of international bodies including the RAMSAR Convention is encouraging the development of integrated rice and fish production system paddy fields, which would also protect and enhance biodiversity as well as ecosystem services like in the traditional Satoyama (UNU 2010, 2012). This may well help to recover some environmental equilibrium in the damaged rice paddies in disaster-hit areas.

In the future (2015), the Miyagi Prefectural Government and the Central Government of Japan have decided to build larger and longer concrete embankments as well as more sluice gates in rivers (classified as grade 1 and 2) for tsunami protection purposes (Mainichi 2012a). This decision is likely to further exacerbate the environmental impact on rivers as more ecotone will be destroyed when transformed into concrete and or cement walls hence reducing the interaction between its water, flora and fauna and the shore. Moreover, more soil and land will be required for their construction, which in turn poses another large environmental problem.

Fig. 1.17 Exposed coastline to wave embattlement in Utatsu

To restore and protect the land from tsunamis and large waves (Fig. 1.17), a plan also by the prefectural government has been drafted considering three main geographical sectors, i.e. Sanriku Area, Ishinomaki/Matsushima Area and Sendai Port/Southern Miyagi. The construction of multiple tsunami protecting barriers such as walls and dykes have been considered for the second and third sectors, however not on the Sanriku Coastline (Onodera 2012). Local authorities in Sanriku have also decided to build single dykes along coastal areas, despite the negative environmental impacts this may bring. An exception has been made in Monue Bay (Kesennuma), where the pressure from local citizens has stopped authorities from building these types of walls (Tanaka 2012a, per com.).

Considering the reconstruction and building of dykes, seawalls, and river embankments, leveling-up areas with dykes as well as raising their foundations by several meters, raising levels of sunken lands and rebuilding cities and towns, the amount of soil required will be gargantuan (Japan Times 2012c), and most certainly will cause enormous environmental problems (Fig. 1.18). The local governments of Sendai and Ishinomaki have plans to raise the height of coastal roads as an anti-tsunami measure which will require about 5 million cubic meters of soil, while elevating the level of seawalls may require 15.4 million cubic meters. Furthermore, large amounts of soil have been required to prepare land for temporary housing, while in the near future even more soil will be required to prepare areas at higher grounds in order to be urbanized for the relocation of coastal towns and cities (Nakamura 2012). Mining such large amounts of soil from other mountainous and hilly areas in the prefecture is likely to extend the damage beyond tsunami-hit areas.

Forest ecosystems and plantations will also be hit hard as construction of new housing roads, lifeline systems will certainly increase deforestation, changing the characteristics of the affected watersheds and resulting in soil compacting and quality alteration as well as modification of surface water flow and groundwater recharge.

Fig. 1.18 Leveling-up the ground in Ksennuma port

In terms of pollution, it is most likely that contamination from a number of chemicals and compounds from fertilizers and pesticides has spread throughout soil, freshwater bodies and coastal waters along the tsunami-hit coastline impacting ecosystems and related species. Similarly, a large number of industries were destroyed, including paper factories, chemicals, fertilizers, oil and food-related companies, resulting in large amounts of solvents, heavy metals, PCBs, and other chemicals being released along the coastline. This has resulted in extensive contamination of the soil, groundwater, and freshwater bodies as well as coastal waters (Maksymov 2012).

Aside from being used in agriculture, Organo-Phosphates and Pyrethroid pesticides such as Fenitrothion EC as well as Etofenprox – also a wood preservation compound – (European Union 2012) were used in 2011 for a few months in Kenssen Numa to control pests in rotten fish at frozen storage warehouses which were destroyed by the tsunami, and ended up as piled-up debris awaiting to be processed (Tabaru et al. 2011). Contamination from such debris at various sites has most likely reached the soil and aquatic ecosystems through leaching and run-off likely with environmental consequences.

Besides the direct damage inflicted by debris being carried by the tsunami, a secondary event has resulted in damage from debris storage and handling has also taken place; the large amount of debris scattered in the affected coastline needed to be concentrated in certain areas and managed to avoid health problems, therefore the clean-up had to be undertaken as quickly as possible. Due to the prevailing emergency situation and environmental impact related to the pile-up, storage and management of debris have had to be coordinated by various clean-up teams.

Pollution from leach and run-off originating from piled-up wooden debris which was exposed to rain prior to recycling processes in order to wash-off salt as well as

Fig. 1.19 (**a**) and (**b**) Polluted water reaching the soil, rivers and coastal waters

Fig. 1.20 Polluted runoff from debris reaching coastal waters

copper-chrome arsenate -CCA- (UNEP 2012) will certainly pose environmental repercussions, as it will seep into surrounding soil, reaching surface freshwater bodies, groundwater and coastal waters. Contaminated liquid will also spread pesticides from piled-up debris; a similar situation will occur with leach and run-off from other types of debris such as cars, metals and electronics, as well as other devices which will produce a variety of toxic chemicals that will leach into groundwater and adversely affect the environment (Figs. 1.19a, b and 1.20).

Fig. 1.21 Salt-resistant pine tree striving after the tsunami; Kessenuma

Forest and plantations growing along watersheds, as well as coastal forests and related ecosystems, were damaged to different degrees when they were hit by the tsunami. Of the three main tree species found in forest and plantations, the Japanese cedar, or sugi (*C. japonica*), was the most severely damaged due to its low salinity resistance while the black pine tree, or kuromatsu (*P. thunbergii*; Fig. 1.21), and the red pine tree, or akamatsu (*P. densiflora*) once again proved to be were the most resilient (Nakamura et al. 2012; Santiago-Fandiño 2011a, b). Likewise, forest soil at many locations was vastly damaged by the abrading and eroding run-up and backwash, removing the soil and leaving only bare rock, while also devastating microflora particularly in the most exposed areas. However, due to the area's characteristic heavy rain and snow, the quality of the soil in some areas has improved as salt has been washed away from many slopes and incline lowlands, allowing primary and secondary vegetation to make a comeback, growing in many damaged areas and helping to rebuild the damaged soil and ecosystems (Fig. 1.22).

As such extensive damage to coastal forestlands and ecosystems requires urgent attention, the initiative to activate and develop ecotourism as well to restore damaged natural parks, the Miyagi the prefectural government has developed the Initiative for Sanriku Reconstruction (Fukko) National Park (under discussion as of September 2012). The existing Rikuchukaigan National Park (Fig. 1.23) will become the core protected area within Fukko National Park which will include other parks both within and outside the prefecture. The Initiative fosters the creation of linkages between the forest and the ocean through Sato systems (Satoyama and Satoumi), vast forestation in large surface areas, protection and increase of biodiversity, and strengthening the cooperation between agriculture, forestry and

Fig. 1.22 Recolonization of primary vegetation in tsunami affected soils

Fig. 1.23 Sanriku Park board with sites and access information for interesting tourist spots prior to the tsunami

fisheries, amongst others. The establishment of the park will require a revision of the existing boundaries and management plans as well as conservation policies to ensure that important ecosystems are not altered by the reconstruction efforts. Moreover, research and monitoring of the effects of the earthquake and tsunami on the environment and biodiversity together with their recovery processes are to be encouraged and supported (Sanriku Reconstruction (Fukko) National Park Initiative 2012; Takahashi 2012; Katsurawa 2012; MOE 2012b).

Radiation from the Dai-Ichi nuclear plant in Fukushima has most likely had a lasting impact to various degrees on some ecosystems in Miyagi Prefecture, as during the nuclear crisis winds blew northwest carrying contaminated rain, which was deposited in many areas of Miyagi. This can be inferred by the fact that some products like mushrooms grown in forest areas and other plants in Miyagi (see agriculture section) has already been shown to have contamination with cesium (MAFF 2012b; The Asahi Shinbum 2012; Miyagi Prefecture Government 2012h). There are some unofficial accounts from sources in the Japanese government of alarming levels of radiation having reached Miyagi prefecture, contaminating some species (Simply Info 2012; Government of Japan 2012).

1.5 Fisheries

In 2010, Japan ranked as the second-largest world consumer of ocean resources (Fig. 1.24), totaling 582 million tons (World Fishing and Aquaculture 2010) while in catches ranked seventh in the same year (FAO 2010a). Furthermore, the country rated as the world's largest single importer of fish (60 %) and fishery products (54 %) in 2008 (FAO 2010b). The country is now going through a slow-down in total fisheries output, as catches from coastal fishing in 1989 dropped from almost 9 million tons to less than 4 million tons in 2011 (excluding marine fishery and aquaculture production in Iwate, Miyagi, Fukushima and Ibaraki prefectures); a similar fate has been reserved for the cases of sea-farming and offshore fisheries as

Fig. 1.24 Fishing fleet anchored in Kessenuma port

Fig. 1.25 Bidding for Skipback tuna (katsuo) in Kessenuma fishmarket

reported by the Ministry of Agriculture, Forestry and Fisheries in 2011 (MIAC 2012). Nonetheless, in terms of production, Japan holds 2.9 % of total sea-weed production in the world and appears to be the second-most important producer of nori and in 2008 it had the largest production of cultured seaweed in the world, i.e. kelp -kombu- (*Laminaria japonica*) with 4.8 million tons, eucheuma seaweed (*Kappaphycus alvarezii* and *Eucheuma* spp.) with 3.8 million tons, and wakame (*Undaria pinnatifida*) with 1.8 million tons (FAO 2010b). Eucheuma seaweed species bring large revenue as they are widely used for a variety of purposes (The SuriaLink 2012). The Tohoku Region and five other prefectures participating in fishery production accounted for about 21 % of Japan's total fishing and aquaculture production prior to the event on March 11, 2011 (Wright 2012).

Employment in the Japanese fishing industry has lately been on a steady decline for reasons such as smaller catches, increased productivity due to new technology innovations, an aging workforce and emigration to cities, to name but a few. From 1990 to 2010, the number of people employed in marine fishing decreased by 45 %, the number fishermen and fish farmers suffered a reduction in the workforce from 370,000 (1990) to 200,000 by 2010 (FAO 2010b). From a workforce of 200,000 fishermen and fisheries in Japan, 34 % are 65 years old or older therefore young fisherman from other countries are brought to work as trainees in fishing boats (Asahi Shinbum 2012).

Miyagi prefecture fisheries benefit from a convergence of the Oyashio and Kuroshio ocean currents and other rich fishing grounds and areas along the Tohoku region where a large number of fish species flourish which are commercialized (Fig. 1.25), i.e. blue-fin tuna -Kuromaguro- (*T. orientalis*), Skipback Tuna -Katsuo- (*K. pelamis*), albacore -Tombo or bin-naga- (*T. alalunga*), mackerel -Saba- (*S. japonicus*), strip jack mackerel -Aji- (*S. niphonius* and *T. japonicus*), yellow-fin

Fig. 1.26 Common local anchoring sites and ports along Sanriku coastline

tuna -Maguro- (*T. obesus*), and the Pacific saury -Sanma- (*C. saira*). Also, black marlin -Kurokajiki- (*M. nigrikans*) and the swordfish -Mekajiki- (*X. gladius*) are two of the most important fish species (Kimoto and Yokawa 2012).

The Prefecture has been a major producer of marine products; in 2008, it held one of the largest numbers of marine fishing establishments in the country, accounting for about 4,000 which were mostly individually managed. Miyagi also had 142 designated fishing ports (Fig. 1.26), again being one of the highest in Japan, of which more than half dealt with marine farming (MIAC 2011). Miyagi held the third position in processed marine products and fourth in fishing and farming before the earthquake and tsunami struck (Onodera 2012). In terms of trading and commerce, the Kesennuma fish market in this prefecture's market, aside from being one of the largest and most important centers in Japan, also ranks number one in the shark industry (Fig. 1.27), providing 90 % of the shark fin trade in Japan with 14,000 t reported in 2009. As resources have become over-exploited, the central government has developed the NPOA-Japanese Government National Plan of Action for the Conservation and Management of Sharks in 2009 in order to regulate and manage the shark fishing industry (FAO 2012).

In Miyagi, as well as in many other parts of the country, the fishing industry is going through difficult times due to international fishing regulations and an increase of imports as well as aging of the dwindling workforce dedicated to the fishing trade (Miyagi Prefecture Government 2012b). In 2008, a survey was made showing the percentage of fishermen in Miyagi, Iwate and Fukushima prefectures aged 50 years of age and below was only 27 %, and since that time the number of fishermen seems to have fallen by as much as 10 %. Furthermore, a survey after the tsunami hit showed that 30 % of the fishermen in Miyagi alone wanted to change their occupation (Daily Yomiuri 2012b). In Kesennuma Port (Miyagi prefecture), due to the lack in workforce during last part of the 80s, the fishing fleet has fallen from 64 long-line tuna fishing boats to only 18 vessels in 2010 (Daily Yomiuri 2010).

Fig. 1.27 Sharks ready for auction in Kessenuma fishmarket

The earthquake and tsunami of 2011 created havoc and paralyzed the fishing industry in the hit areas, either badly damaging or completely destroying the majority of the fishing facilities, equipment and infrastructure (Fig. 1.28). As of March 2012, the damage to the fisheries and related facilities amounted to more than 1 trillion Yen or US$12.5 billion. More than 28,000 fishing boats were destroyed, 319 harbors and an enormous amount of sea-farming facilities and products (MAFF 2012d).

In Miyagi, the situation was overwhelming as about 90 % of the fishing fleet (13,579 crafts) was lost, and many fishing-related businesses such as ship ice-makers and seafood processing companies suffered greatly (Wright 2012). Kessen Numa, Onagawa and Ishinomaki, where some of the most important companies have been established (Fig. 1.29), are still struggling to make a come-back and continuing to struggle with their important rebuilding efforts (Nikkei 2012a, b). In Ishinomaki alone, one of the largest and most important food processing companies was forced to close down due to the extensive damage and the enormous cost of rebuilding (Daily Yomiuri 2012c).

Ports damaged in Miyagi prefecture totaled 142 ports, with a total economic loss of more than US$4 billion (MAFF 2011b), excluding other damage (Santiago-Fandiño 2011a). Furthermore, large areas sank below high water, rendering many ports unusable. A calculation by the prefectural government showed that marine areas together with inland grounds below high water at or below historical high tide increased by 216 km^2 (Onodera 2012). Sinking depth range in coastal areas appeared to be from 23 cm in Natori City to 114 cm in Ishinomaki Port (MAFF 2011b), and 70 cm in Kesennuma (Tanaka 2012a, per com.).

Fig. 1.28 Ishinomaki fishmarket after the tsunami

Fig. 1.29 One of the many Ishinomaki's seafood processing companies after the tsunami hit

Large efforts on national and local levels are ongoing to rebuild the devastated industries. The Miyagi prefectural government has developed the Miyagi Prefecture Earthquake Disaster Recovery plan with a number of key points, including the "Special Reconstruction Zone for Fisheries". The objective is to consider legal and

Fig. 1.30 Kessunuma fishmarket almost in full operation

technical/technological aspects of rebuilding and further developing the fishing industry and promote the reconstruction of fishing cities by attracting the private sector. The goal is to strengthen the sector of local fisherman, (Daily Yomiuri 2012d; Government of Japan 2011; Miyagi Prefecture Government 2011); which is strongly opposed to losing their historical fishing rights to outside companies overtaking their territories (Santiago-Fandiño 2011a). The situation at the end of 2012 has become more complicated as fisherman are strongly opposing the Government (Kotani 2012, per com.).

At the local level, for example, the Kessennuma fish market, which sank about 70 cm and whose facilities were totally washed away, has almost been fully rebuilt (Fig. 1.30) has already received important loads of seasonal fish species like skipback tuna -Katsuo- (*K. pelamis*) and Pacific saury -Sanma- (*C. saira*), as well as other species such as yellow-fin tuna maguro -(*Kihada maguro*)-, black marlin -Kurokajiki- (*M. nigrikans*), swordfish -Mekajiki- (*X. gladius*) and a variety of shark species. The market is slowly regaining its activity and economic importance, although the availability of ice remains an obvious problem as the majority of ice-making factories were destroyed by the tsunami.

As of 2012, the Ishinomaki fish market, considered to be one of the most important in Japan, the port facilities and related industrial complexes are being rebuilt thanks to emergency reconstruction funds and are planned to be finished by 2015. The smaller-scale Shizugawa fish market in Minamisanriku, also having suffered from similar destruction, has been rebuilt and some seafood processing companies as well as a boat manufacturing company are already working (Fig. 1.31a–c).

At the local community level, despite the fact that there is not enough manpower to harvest seaweed as young workers prefer jobs related to reconstruction than a common 2-month harvesting contract (Daily Yomiuri 2012e), seaweed farmers have been able to install back racks and nets and have obtained their first harvest in 2012. Cooperatives have been established through personal financial contributions, changing many family-run businesses into streamlined corporation-like organizations, which aim to become more efficient, reduce costs and share equipment. An example of this new approach can be found with wakame seaweed farmers in Utatsu (Minamisanriku city). Cooperation between fishermen in Kesennuma and Ishinomaki is also taking place to rebuild the local industry, which in turn is helping to stabilize prices by looking for alternative markets, introducing new and different approaches to the way business is carried out.

Fig. 1.31 Fishmarket (**a**), food processing company (**b**) and boat factory (**c**) already in operation in Shizugawa; 2012

Fig. 1.31 (continued)

Signs of the industry recovering through the returning of fisherman are starting to appear in Ogatsu (McCurry 2012).

The Kessenuma-based NPO, "Mori wa Umi no Koibito" – "The Sea is Longing for the Forest"-, located on the hard hit shores of Moune Bay (Fig. 1.32) brings a combination of science and a social movement, supports education and raising awareness as well as scientific research through the H2O Studies Programme linking fishermen and researchers in studies in this bay (Fig. 1.33a, b). In collaboration with forest and parks authorities and local agricultural communities, voluntary reforestation efforts have taken place over the past 20 years in the upper watershed of the Okawa river (Fig. 1.4) flowing along the Murone and Yagoshi mountains, where 40,000 large variety trees have been planted (Fig. 1.34), improving the once highly polluted waters of Kessenuma bay. This in turn has benefited Moune bay where important oyster and scallop cultivation take place due to nutrients flow from the mountain side. This effort has earned to the president of the NPO the 2011 UN "Forest Heroes Awards" for Asia (Mori wa Umino Koibito 2012; UNO 2011).

Shellfish cultivated in Moune Bay (Fig. 1.35) appears to be growing faster than in previous years, which seems to be due to the improved water quality flowing from the forested mountainous areas enhancing plankton growth (although some research is being done to evaluate the tsunami effect if any). A number of universities and institutions from various parts of the country are undertaking the studies research to evaluate this possibility and prioritized the identification and analysis of the deep linkages between forest, humans and the sea. Moreover, Miyagi

Fig. 1.32 Moune bay in Kessenuma; Miyagi

is a birthplace of this social movement hence its philosophical and practical idea could be use in rebuilding a more sustainable and interactive society (Yokoyama and Hatakeyama 2012; Harino and Yatsuzuka 2012; Yamamoto et al. 2012; Nishitani et al. 2012; Yamada 2012; Masuda 2012; Tanaka 2012b).

As result of the Fukushima nuclear problem, to reduce the possibility of contaminated fish being shipped to market, radioactivity inspection is being conducted by the government in various migratory fish species. Salmon -Sake- (*Oncorhynchus keta*) is inspected in October and November, following the Fisheries Agency Inspection Protocol, as well as skipjack tuna crossing the Joban Oki fishing grounds off Fukushima's coastline as high level radiation in the ocean and seafloor has already prompted closure of many important fishing areas for flounder -Hirame- (*Paralichthys olivaceus*), sole -karei- (*Limanda sp.*), abalone -awabi- (*Haliotis sp.*) and others, further restricting their distribution (MAFF 2012e; Biello 2012).

The presence of radioactive materials is closely related to the fishing industry in Miyagi, and are particularly high in various non-migratory species which have already shown to have radioactive cesium, such as the black porgy (*Acanthopagrus schlegelii*) caught offshore Yoshida, Watari Town, and the white spotted char (*Salvelinus leucomaenis*), In Taihaku-ku, Sendai City. Fish shipment restrictions imposed by the government have been enforced since mid-2012. A similar fate has befallen Japanese flounder (*Paralichthys olivaceus*) as fisherman were called to voluntarily refrain from shipping the fish since April 24, 2012 (Miyagi Prefecture Government 2012d).

To reduce the spread and exposure of radioactive materials into the ocean and sediment, which is strongly suspected of affecting the food chain, fish resources along Tohoku and other areas south of the prefecture the company in charge of the facility (TEPCO) have started to cover 70,000 m^2 of the ocean floor with clay and cement to guard against further contamination for the next 50 years (Daily Yomiuri 2012f).

Fig. 1.33 (**a**) and (**b**) Devastated natural wetland at Moune bay and sampling studies

1.6 Urban Areas and Society

Japan is a highly-populated country with a large number of cities along its coastlines as they provide flat, ideal areas for development. Tohoku is, however, one of Japan's least populated regions, bearing only a small number of large and medium-size cities in comparison with other prefectures as populations mainly reside in the Kanto and Kansai regions. Miyagi prefecture, with 2.34 million inhabitants as of 2010, holds nine principal or main cities and twenty-two minor cities and towns including Sendai, Ishinomaki, Kessenuma (Fig. 1.36), Natori, Sendai, Shiogama, and Matsushima (considered to be one of the three most

Fig. 1.34 Reforestation sign map at the Yagoshi mountain and project coordinator; Iwate Prefecture

Fig. 1.35 Recently harvested scallops at Moune bay; Miyagi

beautiful area in Japan) as well as Shichiganama, Minamisanriku, Onagawa and Watari, located along the coastline. Serving as its prefecture's capital, Sendai city has the largest number of inhabitants with more than one million residents (City Population 2012; Regions & Cities 2012).

Miyagi prefecture's population has been decreasing due to emigration being about 7 % rate in 2008, considered as one of the highest rates in the country, while its unemployment rate was 6.4 % in 2009, representing again one of the highest in Japan (Statistics Japan 2012).

Due to the earthquake and tsunami of March 11, most of the cities and towns along the coastline were either partially damaged or devastated. Ofunato and

Fig. 1.36 Kessunma City port being reconstructed in 2102; Miyagi

Fig. 1.37 Onagawa port after the tsunami in June 2011

Rikuzen Takata in the south of Iwate Prefecture, and Minamisanriku, Onagawa (Fig. 1.37), Kesennuma and Ishinomaki in Miyagi are examples of the tsunami's devastation, as well as Sendai city, where large areas along the coastline were destroyed or badly damaged like the Sendai airport which started to operate with temporary airline counters its fully in service (Fig. 1.38). Streets, avenues and roads

Fig. 1.38 Rebuilt Sendai airport; June 2012

sank, cracked or broke due to the earthquake; most of the houses and buildings collapsed or were badly damaged and means of transportation such as local railways systems were either badly affected or totally destroyed. Water, sewage, telecommunications and electricity networks, as well as fuel storage and distribution lines completely collapsed.

Initial records of casualties immediately after the events varied as further research was undertaken; for example, Minamisanriku records stated higher figures, which were considered to be the highest casualties in the prefecture (Japanese Red Cross Society 2012). However, a report in 2012 stated that about 600 people died and almost 300 were missing, representing 5.2 % of the population, making Ishinomaki city the area hardest hit by the tsunami (Vervaeck and Daniell 2012).

With tsunami wave heights beyond 10 m, many places along the coastline in the Tohoku Region (Satake et al. 2012), especially urban centers and villages, had little possibility of survival. In the region as a whole, about 370,000 houses were damaged and about 130,000 totally destroyed in coastal cities, towns and villages including all kinds of infrastructure (Fig. 1.39a, b) adding to this damage (Sendai City 2012). The coastline actually sank, allowing for a high number of square kilometers to be flooded where the tsunami struck. One example of the sunken coastline is the flooded areas in Miyagi prefecture which equal 4.5 % of its total surface. In some cities such as Minami Sanriku, the tsunami reached up to 20 m, while in Ishinomaki plains waters ventured up to 5 km inland, flooding large areas (Building Research Institute Incorporated Administrative Agency 2012; Mori et al. 2011; EERI 2012).

After the events the inhabitants of cities and towns as well as villages further suffered distress from the lack of food and clothing shortages, medicine availability, health services and security added to the difficulties posed by the thousands of tons of debris spread everywhere, making supplies extremely difficult to distribute

Fig. 1.39 (**a**) and (**b**) Provisional water supply PVC line and submerged drainage, a commonality along Sanriku cities and towns

Fig. 1.40 Minato ward privately own shelter in Utatsu; April 2011

and relief operations could not reach the isolated stricken communities for some time in Miyagi. To help in ameliorating the problem, private emergency facilities (Fig. 1.40), city halls and other government facilities as well as community centers, were provided for evacuees as provisional shelters, while volunteer centers were set-up to support the relief efforts and establish health-checkpoints.

Fig. 1.41 Temporary housing in Utatsu; Miyagi

As damaged urban areas will require many years to rebuild, 55,000 emergency temporary housing units have been constructed in Iwate, Miyagi and Ibaraki prefectures using prefabricated materials and wood (Fig. 1.41). In Miyagi, more than 11,000 units were built, of which almost 3,000 units, or about one-quarter of the total, were built using wood as shortages of prefabricated materials limited the availability of construction supplies (Sendai City 2012).

Local government plans to reconstruct destroyed urban areas are facing problems as communities often disagree with development proposals. An example of this is the development of new housing areas in the highlands, leaving only the coastal plains for business and agricultural purposes; at the end of 2011, roughly 18,000 households in 170 districts of 12 cities and towns in Miyagi were considered for relocation to cover 951 ha of land, an enormous challenge both citizens and the government alike face (Daily Yomiuri 2012g). In Miyagi, developers are purchasing land at higher elevations at inflated prices for development purposes (Fig. 1.42), earning the prefecture a place on the list of top-ten highest appreciation rates for residential and commercial properties purchase in Japan (Mainichi 2012b). This situation will likely create a shortage of land available for relocation to be purchased by the government (with an already limited budget) and most likely will create further problems with local communities.

Changing purposes of land use in the prefecture is occurring at a fast rate as conversion of farmland for uses such as construction is increasing privately. This activity is likely related to the building of new homes in areas outside of local authorities' consideration for development, which may result in problems related to provision of lifeline services such as water, electricity, and sanitation services.

Fig. 1.42 Land reclamation in forest areas for construction purposes along Sanriku

However, the central Government of Japan seems to be encouraging applications to change purposes of land use in tsunami-hit areas. More than ten cities and towns within the prefecture have received permission by local authorities to change the purpose of land use from agriculture to residential areas, corresponding to about 2.5 times more than the same period in 2010. In Kessenuma alone, authorization for such changes has risen more than six times (Daily Yomiuri 2011).

Other problems the prefectural government and local communities face are the merging of fishing villages into larger ports while also allowing for larger industries to participate in the construction of tsunami "fence" structures near ports and cities. However, this will prevent coastal cities and towns from enjoying sea views. In Kessenuma City, actions by local citizens are being taken to force the local government to change their development plans as there has not been enough discussion (Tanaka 2012a, per com; Japan Times 2012c).

Sea fences to protect Tohoku's cities, towns and ports as well as industries were built along the coastline but were severely damaged (Fig. 1.43a, b). These structures were designed to stop high waves, storm surges, tsunamis and erosion and included dykes, walls, jetties, breakwaters, rip-rap revetments, concrete bags, tetrapods and groin lines. In northern Miyagi prefecture, dyke heights were designed after careful consideration of past tsunami records while the southern area of the prefecture has specialized their coastline fortifications development for storm surges (Kato et al. 2003; Tori and Kato 2001).

After the Chilean tsunami in 1969, further efforts to protect the coastline were undertaken, but the March 11 earthquake and tsunami proved many of the fortifications to be inefficient as the tsunami run-up was larger than previously

Fig. 1.43 (**a**) and (**b**) Sluice gate (*above*) and seawall (*below*) overpowered by the tsunami along the Sanriku coastline

calculated (Fig. 1.44). Instead of providing the protection they were designed for, their structures also became agents of destruction. One hundred and ninety kilometers of the 300 km of sea fences placed along the coastline from Iwate to Fukushima prefectures were severely damaged or destroyed (Tori and Kato 2001). River mouths or estuaries proved to be efficient means for the tsunami to hit lowlands and urban centers near the coast. A large part of the city of Ishinomaki (13 %) was flooded as also the tsunami went through the Kitakami river and more than 20,000 houses were crushed. A year after the tsunami struck, the city has mostly dried-out but the situation remains far from normal as can be observed in comparative photographic images from NASA's EO-1, ASTER and Terra satellites (NASA 2012).

Fig. 1.44 Sea fences could not match the power of the tsunami run-up along the coast of Miyagi

The city of Kamaishi in Iwate Prefecture was severely damaged by a 4.3-m high tsunami wave, despite an almost 2-km long, 63-m deep tsunami breakwater was completed in 2009 after 30 years of construction (Onishi 2011a, b); although good examples of protection exist like in Fudai City (Today 2011).

In Miyagi prefecture, a two-tier approach has been taken for to protect urban areas from tsunamis; both property development and economic activities are carefully addressed by the government. The first is based on conventional structures and reinforcements capable of dealing with frequent tsunamis occurring every 100 years while the other plans for new and improved designs, including nonstructural systems, for less frequent tsunamis, every 1,000 years. Local characteristics including the environment, finance, and sustainability are being taken into account when designing them (Imamura and Anawat 2011). A suitable approach to concrete dykes structures are green dykes (forest) as they last longer and are more economical (Tanaka 2012a, per com.).

Among the existing structures to be reinforced or redesigned for flood protection in coastal urban areas are river embankments which are found all along the coastline in Miyagi Prefecture. Due to their low heights, however, the tsunami of March 11 was able to get through and flood large areas, causing enormous devastation. In the city of Minamisanriku, a 8.7-m high embankment wall will be constructed along the four rivers flowing into the bay (Fig. 1.45). Moreover, to avoid having the city sectioned by the embankment walls, built-up areas will be

Fig. 1.45 Aerial view of the city of Minamisanriku and the rivers flowing across the city (Image taken from a picture exhibited at the Minamisanriku City Hall)

raised to almost the same height as the embankment (Mainichi 2012a). In Sendai City a coastal breakwater will be built as well as a large coastal forest (likely mixed trees species rather than monoculture) while also raising the height of the Shiogama-Watari Line prefectural road will be designed as tsunami-prevention barriers (Sendai City 2012). On the other, it is important to highlight that Sanriku has a tremendous potential for the sightseeing industry due to its well preserved and beautiful nature, but if huge concrete dykes and walls are to be built the result can be devastating for the industry and urban scenery.

For the reconstruction of various cities along Sanriku in Iwate and Miyagi prefectures, there are plans to develop them based upon energy efficiency using the latest technology development and social service policies aimed at resolving environmental and population aging problems. Rikuzentakata and Ofunato cities in Iwate prefecture are envisaged to be the first mega-solar power cities in the world (Miyata and Uemura 2012; MOFA 2012). In Miyagi prefecture, as a part of the Miyagi Recovery Plan, the local government is planning to rebuild cities based on the Eco-town concept and develop resilient communities utilizing recent technological developments for renewable energy (Onodera 2012).

Although the restoration and rebuilding of cities and towns started from the end of 2011, most of the debris clean-up has been completed, however, many concrete buildings still remain to be removed (Fig. 1.46) as well as and boats; one exception might be a 330-t fishing ship (Fig. 1.47) likely to become a part of the tsunami memorial park in Kesennuma City (Daily Yomiuri 2012h; Japan Times 2012c). Others include disagreements between the local governments and the affected local citizens. Ishinomaki City started to be rebuilt in May 2012 by setting-up special

Fig. 1.46 Removal of a tsunami-damaged Senior Home building in Minamisanriku

Fig. 1.47 The Kyotoku Maru fishing trawler idling by the Sishiorikarakuwa train station in Kessenuma City

companies under the Special Zone Reconstruction Law, which in turn will be the first disaster devastated municipality to follow this scheme in Tohoku, and will provide business and employment opportunities to local residents. Moreover, the downtown area will be designed to consider designs to allow senior-citizens to take up residence; it furthermore considers attracting health care and pharmaceutical businesses, among others (Nikkei 2012c), something that other cities may follow.

1.7 Debris Management

The massive destruction by the tsunami along the coastline created a gargantuan amount of debris of every type imaginable (Fig. 1.48). So great is the quantity of debris that there are discrepancies in its estimated weight measurement by different authorities; the range varies from more than 15.5 million tons by UNEP to 18 million tons by the Miyagi prefectural government, equaling about 23 years of waste production by the entire prefecture and from which 10.44 million tons, or about 67 %, has already been dealt with (UNEP 2012; Onodera 2012). On the other hand, recent studies have revealed that roughly around 20 % of the total debris generated by the tsunami in Tohoku has been swept into the Pacific Ocean and spread over an area 4,000 km long by 2,000 km wide; a large amount of this floating debris is expected to eventually reach the coast of North America and the Hawaiian Islands (Ozawa and Kaji 2012; Amos 2012; Simon 2012).

To clean up all the debris in and around the many cities and towns along the coastline (Fig. 1.49), a strong, decisive campaign to separate and dispose of the debris was put into action by the central and local governments. The challenges so far have been enormous, as more than 1.3 million tons of debris was generated in Kesennuma city, 56,000 t in Minami Sanriku and 6.1 million tons (the largest amount recorded) in Ishinomaki city (UNEP 2012). A financial and managerial challenge appeared as the daily workforce assigned to separate this debris by type required the work of more than 1,000 employees (Daily Yomiuri 2012g).

The local government's managerial approach and policy for dealing with the debris has been divided into various steps. Once collected from the field, they are moved to temporary storage areas where they are sorted into burnable (lumber and other organic materials), non-burnable (metals and household appliances) hazardous

Fig. 1.48 Concentrated mix debris in Kesennuma City awaiting for recycling

Fig. 1.49 Debris collection and gathering along the Pacific coastline in Sanriku

Fig. 1.50 Waste management facility in Kessenuma City (Hajikami) sorting and processing tsunami-scattered debris

materials (asbestos, PCBs, gas cylinders, etc.); damaged boats and cars go through a completely separate process. After this step, debris goes through a second separation process into one of five different holding blocks along the coastline which include cities, towns and villages such as Kesennuma (Fig. 1.50), Ishinomaki, Miyagi-Tobu, Sendai and Watari-Natori. Once finalized, the material then goes through a breakdown and/or incineration process for recycling or final disposal (Onodera 2012).

In terms of technology, the 68-ha-large Ishinomaki block holding 7 million tons of mixed debris and containing mud, structural housing materials among other debris originating from the sea, will be shredded by specialized equipment and fed into one of five large incinerators with a processing capacity of 1,500 t/day. This approach is expected to deal with all the debris within 2 years (Waste Management World 2012; Kajima Corporation 2012). As the situation has been overwhelming from the beginning, other prefectures such as Aomori, Yamagata, Tokyo and later Aichi have helped in this incinerating process. A number of cities on the other hand are already considering the use of debris for reconstruction purposes like Sendai city which is to use treated debris and as sand to increase the height of roads and building hills as protecting tsunami barriers (Sendai City 2012).

1.8 Conclusions

Miyagi prefecture and the coastline including Sanriku in particular suffered the consequences of one of the worst cataclysmic events of modern times, causing enormous environmental, social, urban and economic debacle. Central and local governments faced challenges never seen before amidst unfathomable hardships of irreparable losses to many communities.

In a most remarkable effort during the first year, most of the debris has been cleared and is ready to be processed, which due to the overwhelming amount, distribution complications and seemingly infinite types was certainly a most difficult task to achieve, and we are now left with the process of reconstruction for the future. Certainly, the problems lying before us are not easy to foresee nor easy to solve, as societal and government participation in the restoration process is bound to create countless more hurdles.

To protect the coastline, people's lives and possessions against future cataclysms of this nature, existing plans need to be revised to come up with fresh risk and crisis management assessments, vulnerability studies as well as environmental and social assessments, which are crucial for the well-being of society and nature. There is an important priority that we must first and foremost avoid what could become enormous collateral environmental damage beyond the tsunami and earthquake impact due to human error in the process and implementation of programs and projects to restore and rebuild disaster-hit areas.

Being the most extensive and well-documented event the world has yet seen, added to the extensive amount of data gathered and processed, as well as the existing experience from past events in the country, it is most likely that the large majority of the programs and actions will be successful in their objectives to prevent and mitigate future damage from earthquakes and in particular by tsunamis, albeit with limitations due to the most difficult and complex predictions of tsunami's behaviour, climate change and the rise in sea level. To that end, a large amount of scientific and social research should continue, considering modeling and forecasting, prevention and awareness, monitoring and evaluation, environmental and social impact assessment as well as risk management, to mitigate resources and human impact.

Other no less important existing issues, such as soil procurement to build tsunami barriers and leveling grounds and deforestation, are of paramount importance. Furthermore, pollution of surface water bodies, groundwater and coastal waters as well as their resources originating from atmospheric dissemination, runoff, leaching and seepage of materials being scattered from devastated industries, warehouses and facilities of various types, concentrated debris as well as the impact on the watershed due to urban and town relocation, social impact due to changes in practices such agriculture and emigration and workforce loss all need closer attention. Specific research on short, medium and long-term acute and chronic effects of the compounds as well bioaccumulation in the biota, including aquatic organisms, should be carefully assessed, particularly in the case of fisheries and sea-farming species.

Previous government policy was completely separated between land and sea, but the tsunami events demonstrated once again that those two areas are strongly linked as for example radioactive substances spread over forest area are gradually reaching rivers, estuaries and the ocean hence the government's basic policy should be more integrated when dealing with coastal areas and watersheds. Moreover, in the case of Miyagi prefecture as the coastline can be separated by its morphology i.e. the south wider and longer while highly complicated and rough in the north a different planning and management policy may be required (Tanaka 2012a, per com.).

The human component is the most important aspect to be dealt with, as personal losses may never be recovered, lifestyles and communities may have been decimated for good and health and psychological problems may last for life. Science will bring knowledge and understanding about the origin, behavior and expected damage of natural hazards such as earthquakes and tsunamis, but human nature and behavior requires more careful attention. As enormous amounts of economic resources are already in place, only through an integrated approach whereby both the human and environmental component as well as economic and social aspects comes together in harmony may we be able to move forward. Compromises will have to be undertaken by the government and the communities alike, as there is not one simple solution nor a perfect answer for the many problems facing the reconstruction and further development of the particular situations.

Future catastrophic events like the Nankai earthquake and the mega-tsunami are expected to occur along the south Pacific coast and the Cabinet Office's Central Disaster Prevention Council has forecasted that 30 prefectures could be hit at any moment. The human toll projection is already calculated to be beyond 300,000 people; flooding is expected to be beyond 1,000 km^2 covering 24 prefectures of which at least 602 km^2 may suddenly find itself under a meter of water (Japan Times 2012d).

Certainly all the knowledge and experience being gained in the aftermath of the tsunami and earthquake supported by strong scientific and social research will be most valuable to mitigate and prevent as much damage as possible in future events. Surely, a number of local and central authorities, NGOs and MPOs, research institutions, the private sector and other groups are already taking timely action on this line.

Acknowledgements The visit to the stricken areas in Japan was been possible thanks to the kind support of a number of individuals and professionals that the author highly appreciates. Special recognition is due to M. Nakayama, President and CEO of Nakayama Industries in Shiga Prefecture; Prof. M. Tanaka and N. Kimura from Kyoto University, S. Hatakeyama President of the MPO *Mori wa Umino Koibito* in Kesennuma, and K. Sakakibara in Tome City, Miyagi Prefecture. Last but not least to H. Kotani Ex-Secretary General of ILEC in Shiga Prefecture, G. Grant for English proofreading.

References

ACN newswire (2012) B and L's Stevia Laboratory demonstrates successful salt removal in brine damaged rice paddies. http://en.acnnewswire.com/press-release/english/7758/b&l's-stevia-laboratory-demonstrates-successful-salt-removal-in-brine-damaged-rice-paddies. Accessed Oct 2012

Amos J (2012) Path of tsunami debris mapped out. The BBC News. 2 Feb 2012. http://www.bbc.co.uk/news/science-environment-17122155. Accessed Oct 2012

Amur-Okhotsk Project Report Series No 1 to 6 (2003–2010) Shiraiwa T (ed) Accessed Jan 2013

Ando M, Ishida M, Hayashi Y, Mizuki C, Nishikawa Y (2012) Interviewing insights regarding the high fatality rate inflicted by the 2011 Tohoku earthquake. http://g-ever.org/en/materials/sharing/G-EVER1_Ando_AS.pdf. Accessed Feb 2013

Asahi Shinbum (2012) Japan relies on 'trainee' fishermen to bolster aging industry. http://ajw.asahi.com/article/behind_news/social_affairs/AJ201209270004. Accessed Oct 2012

Biello D (2012) Radioactive fish near Fukushima suggest ongoing contamination. Scientific American (on line). 25 Oct 2012. https://www.scientificamerican.com/article.cfm?id=fukushima-fish-remain-radioactive-suggesting-ongoing-radionuclide-release. Accessed Oct 2012

Building Research Institute Incorporated Administrative Agency (2012) Summary of the field survey and research on "The 2011 off the Pacific coast of Tohoku Earthquake" -the Great East Japan Earthquake- . Japan. http://www.kenken.go.jp/english/contents/topics/20110311/0311summaryreport.html. Accessed Oct 2012

City Population (2012) Japan. Miyagi. http://www.citypopulation.de/Japan-Miyagi.html. Accessed Oct 2012

Cotton Australia (2012) Cotton Australia assists tsunami affected rice growers. Thursday, 14 June 2012. http://cottonaustralia.com.au/news/article/cotton-australia-assists-tsumani-affected-rice-growers

Daily Yomiuri (2010) Kesennuma fishermen band together. 15 Mar 2010. http://www.yomiuri.co.jp/dy/national/T120314004789.htm. Accessed Oct 2012

Daily Yomiuri (2011) Farmland conversion up after disaster. http://www.yomiuri.co.jp/dy/national/T111106003219.htm. Accessed Nov , 2011

Daily Yomiuri (2012a) Revitalizing Japan- creative use of land/don't let abandoned farmland remain fallow. May 2012. http://www.yomiuri.co.jp/dy/national/T120508005672.htm. Accessed Oct 2012

Daily Yomiuri (2012b) One year after the disaster/fishermen shifting to incorporation. 5 Mar 2012. http://www.yomiuri.co.jp/dy/national/T120304004408.htm. Accessed Oct 2012

Daily Yomiuri (2012c) One year after the disaster/20% disaster-hit business still closed. 2 Mar 2012. http://www.yomiuri.co.jp/dy/national/T120301006659.htm. Accessed Oct 2012

Daily Yomiuri (2012d) Programme creates more jobs in quake-hit areas (2012). 26 Apr 2012. http://recoveringtohoku.wordpress.com/category/livelihood/fishing/. Accessed Oct 2012

Daily Yomiuri (2012e) Workers shortage leaves seaweed farmers in lurch. 14 Mar 2012. http://www.yomiuri.co.jp/dy/national/T120313004757.htm. Accessed Oct 2012

Daily Yomiuri (2012f) Seabed near nuke plant to be paved. 23 Feb 2012. http://www.yomiuri.co.jp/dy/national/T120222005379.htm. Accessed Oct 2012

Daily Yomiuri (2012g) Reconstruction agency faces urgent problems. 12 Feb 2012. http://www.yomiuri.co.jp/dy/national/T120211003511.htm. Accessed Oct 2012

Daily Yomiuri (2012h) Emergency repairs stalled for 35,000 houses. 15 Feb 2012. http://www.yomiuri.co.jp/dy/national/T120214005440.htm. Accessed Oct 2012

Dun and Bradstreet (2012) Companies located in Japan-Miyagi in the forest products (2012) http://dnb.alacrastore.com/research/d-and-b/Japan_Miyagi/Forest+Products_Lumber,+Plywood,+Millwork,+and+Wood+Panels. Accessed Oct 2012

EERI (2012) Tohoku Tsunami Joint Research Group. Learning from earthquakes The Japan Tohoku Tsunami of March 11, 2011. Special earthquake report; November 2011. http://www.eqclearinghouse.org/2011-03-11-sendai/files/2011/11/Japan-eq-report-tsunami2.pdf. Accessed Oct 2012

European Union (2012) Directive 98/8/EC concerning the placing of biocidal products on the market. Inclusion of active substances in Annex I or IA to Directive 98/8/EC. CIRCAB. Assessment report. Etofenprox. Product-type 8 wood preservatives. https://circabc.europa.eu/faces/jsp/extension/wai/navigation/container.jsp. Accessed Oct 2012

Fairwood Partners (2012) The allocation of JPY 5.3 billion first supplementary budget has been decided to rebuild tsunami affected factories. Japan Forestry Investigation Committee: June 6, 2011: http://www.fairwood.jp/eng/info/indst_news/JFN_2011.html. Accessed Oct 2012

FAO (2010a) Yearbook; fisheries and aquaculture statistics. ftp://ftp.fao.org/FI/CDrom/CD_yearbook_2010/booklet/ba0058t.pdf. Accessed Oct 2012

FAO (2010b) The state of world fisheries and aquaculture. http://www.fao.org/docrep/013/i1820e/i1820e.pdf. Accessed Oct 2012

FAO (2012) Fisheries and Aquaculture Department. International plan of action for the conservation and management of sharks. http://www.fao.org/fishery/ipoa-sharks/npoa/en. Accessed Oct 2012

Fushiwara T, Kodaira S, No T, Kaiho Y, Takahashi N, Kaneda Y (2011) The 2011 Tohoku-Oki earthquake: displacement reaching the trench axis. Science 6060(334):1240

Government of Japan (2011) Resolution reconstruction design council in response to the Great East Japan Earthquake- main opinions expressed in the deliberations thus far. May 2011. http://www.cas.go.jp/jp/fukkou/english/pdf/main_opinions.pdf. Accessed Oct 2012

Government of Japan (2012) WSPEEDI-II records (in Japanese). http://radioactivity.mext.go.jp/ja/contents/5000/4813/24/1305748_0316_08.pdf. Accessed Oct 2012

Government of Japan, GFDRR, The World Bank (2011) Structural measures against tsunamis. Knowledge Note 1-1. http://wbi.worldbank.org/wbi/Data/wbi/wbicms/files/drupal-acquia/wbi/drm_kn1-1.pdf. Accessed Oct 2012

Gurría A (2012) An amazing come-back: working together for a strong recovery and sustainable growth in Japan. OECD. 14 Mar 2012. http://www.oecd.org/about/secretary-general/anamazingcome-backworkingtogetherforastrongrecoveryandsustainablegrowthinjapan.htm. Accessed 28 Oct 2012

Harada K, Imamura F (2005) Effects of coastal forest on tsunami hazard mitigation -a preliminary investigation. In: Satake K (ed) Tsunamis: case studies and recent developments. Springer, New York, pp 279–292

Harino H, Yatsuzuka E (2012) Current status of chemical substances in Kesennuma and Moune bay after the earthquake. Aquabiology 34(6):531–537

Hori M (2012) Effects of coastal seascape diversity on seagrass-associated fish production. FRA-Japan. http://wwz.ifremer.fr/biarritz_2011/content/download/57385/798385/file/Mazakazu_Hori.pdf. Accessed Oct 2012

ICHARM - International Centre for Water Hazard Risk Management (UNESCO) and CTI Engineering Co. Ltd. Japan (2011) Meta and longitudinal analysis of high death rates of some particular municipalities in GEJET. http://www.irdrinternational.org/wp-content/uploads/2012/05/Tadashi%20Nakasu.pdf. Accessed Feb 2013

Imamura F, Anawat S (2011) Damage due to the 2011 Tohoku earthquake tsunami and its lessons for future mitigation. In: Proceedings of the international symposium on engineering lessons learned from the 2011 Great East Japan Earthquake, 1–4 Mar 2012, Tokyo. http://www.jaee.gr.jp/event/seminar2012/eqsympo/pdf/papers/118.pdf. Accessed 28 Oct 2012

Japan Times (2012a) Tohoku long way from healing 18 months on. Wednesday, 12 Sept 2012. http://www.japantimes.co.jp/text/nn20120912a2.html. Accessed Oct 2012

Japan Times (2012b) Aichi to build quake debris incinerator. 19 Mar 2012. http://www.japantimes.co.jp/text/nn20120319a2.html. Accessed Oct 2012

Japan Times (2012c) Recovering Tohoku. 2012. Tsunami-hit structures eyed as memorials locals split on need to rebuild or remember that fateful day. 25 July 2013. http://recoveringtohoku.wordpress.com/category/miyagi/kesennuma. Accessed Oct 2012

Japan Times (2012d) Nankai quake projected toll radically raised. 30 Aug 2012. http://www.japantimes.co.jp/text/nn20120830a1.html. Accessed Oct 2012

Japanese Red Cross Society (2012) Japan: earthquake and tsunami. http://www.jrc.or.jp/vcms_lf/kokusai010212.pdf. Accessed Oct 2012

JETRO - Japan External Trade Organization (2012) Investing in Japan, Miyagi. http://www.jetro.go.jp/en/invest/region/miyagi/. Accessed 28 Oct 2012

Kajima Corporation (2012) Annual report. Removing and processing debris in damaged areas. http://www.kajima.co.jp/ir/annual/2012_pre/pdf/ar2012pre_p08-09.pdf. Accessed Oct 2012

Kato F, Inagaki S, Fukuhama M (2003) Wave force on coastal dike due to tsunami. http://www.pwri.go.jp/eng/ujnr/joint/37/paper/13kato.pdf. Accessed Oct 2012

Katsurawa H (2012) Rule of protected areas for recovery from natural disaster. Ministry of the Environment. http://cmsdata.iucn.org/downloads/1_3_mr_hirokikatsuragawa.pdf. Accessed Oct 2012

Sugimura K (2013) Tsunami "miracle" pine tree was younger than thought. The Asahi Shinbum. Feb 28 2013. http://ajw.asahi.com/article/0311disaster/quake_tsunami/AJ201302280060. Accessed Mar 2013

Kazunuki O (2011) Towards universally accepted agriculture (Part II). New agriculture entries for boosting managerial strength. (7) August-September 2011. JapanEchoWeb. http://www.japanechoweb.jp/jew0710/. Accessed Oct 2012

Kimoto A, Yokawa K (2012) Review of size data for blue marlin caught by Japanese fisheries in the Pacific Ocean since 1970s. Western and Central Pacific Fisheries Commission; Scientific Committee Eight Regular Session, Japan. 2012. WCPFC-SC8-AR/CCM-09. isc.ac.affrc.go.jp/pdf/BILL/ISC12BILLWG-1_WP09.pdf. Accessed Oct 2012

Kobayashi K (2012) Damage caused by tsunami and fire in Iwate and Miyagi prefectures. Tokyo University, Global COE Programme. http://gcoe.moritalab.com/cms/cms_e/wp-content/uploads/2012/02/7016220-e5b08fe69e97e38090e4bfaee6ada3e6b888e38091_english-1.pdf. Accessed Oct 2012

Kobori H, Primack RB (2003a) Participatory conservation approaches for Satoyama, the traditional forest and agricultural landscape of Japan. Ambio 32(4):307–311

Kobori H, Primack RB (2003b) Conservation for Satoyama, the traditional landscape of Japan. Arnorldia 62(4). The Arnold Arboretum of Harvard University. http://arnoldia.arboretum.harvard.edu/pdf/articles/2003-62-4-conservation-for-satoyama-the-traditional-landscape-of-japan.pdf. Accessed Oct 2012

Kotani H (2012) International Lake Environment Committee Foundation (ILEC), per comm.

Kunihiko S (2012) The giant tsunami had been forecasted, but not been included in disaster design. In: Japan Geoscience Union meeting, Chiba, 2012. http://www2.jpgu.org/meeting/2012/session/PDF_all/U-03/U03_O_e.pdf. Accessed 28 Oct 2012

MAFF - Ministry of Agriculture, Forestry and Fisheries (2005) Annual report on food, agriculture and rural areas in Japan –Summary- FY2005. http://www.maff.go.jp/e/pdf/fy2005_rep.pdf. Accessed Oct 2012

MAFF - Ministry of Agriculture, Forestry and Fisheries (2007) Basic principles for promoting organic farming in Japan. Summary. MAFF Update. http://www.maff.go.jp/e/maffud/2007/pdf/no672.pdf. Accessed Oct 2012

MAFF - Ministry of Agriculture, Forestry and Fisheries (2010a) The 85th statistical yearbook. http://www.maff.go.jp/e/tokei/kikaku/nenji_e/85nenji/index.html. Accessed Oct 2012

MAFF - Ministry of Agriculture, Forestry and Fisheries (2010b) Annual report. Special feature: Great East Japan Earthquake. http://www.maff.go.jp/e/annual_report/2010/pdf/e_feature.pdf. Accessed Oct 2012

MAFF - Ministry of Agriculture, Forestry and Fisheries (2011a) The damages caused by the Great East Japan Earthquake and actions taken by the Ministry of Agriculture, Forestry and Fisheries. http://www.maff.go.jp/e/quake/press_110725-2.html and http://www.maff.go.jp/e/quake/press_110406-1.html. Accessed Oct 2012 and http://www.maff.go.jp/e/quake/press_110406-1.html. Accessed Oct 2012

MAFF - Ministry of Agriculture, Forestry and Fisheries (2011b) Forest land and forest resources. Chapter III forest management. Japan. http://www.rinya.maff.go.jp/j/kikaku/hakusyo/23hakusyo/pdf/23_e.pdf. Accessed Oct 2012, and xls. IV crops. https://docs.google.com/viewer?url=http%3A%2F%2Fwww.maff.go.jp%2Fe%2Ftokei%2Fkikaku%2Fnenji_e%2F86nenji%2Fother%2Fn189_191.xls. Accessed Oct 2012

MAFF (2012a) Annual report on forest and forestry in Japan. Fiscal year 2012. Summary. http://www.rinya.maff.go.jp/j/kikaku/hakusyo/24hakusyo/pdf/h24summary.pdf.

MAFF - Ministry of Agriculture, Forestry and Fisheries (2012b) Results of the measurements of radioactivity in agricultural, forestry and fishery products in Miyagi prefecture. Press Release. 4 Apr 2012. http://www.jfa.maff.go.jp/j/kakou/kensa/pdf/miyagi240404e.pdf. Accessed Oct 2012

MAFF - Ministry of Agriculture, Forestry and Fisheries (2012c) N189_191 xls. IV crops. https://docs.google.com/viewer?url=http%3A%2F%2Fwww.maff.go.jp%2Fe%2Ftokei%2Fkikaku%2Fnenji_e%2F86nenji%2Fother%2Fn189_191.xls. Accessed Oct 2012

MAFF - Ministry of Agriculture, Forestry and Fisheries (2012d) Fisheries Agency. Results of the inspection on radioactivity materials in fisheries products. http://www.jfa.maff.go.jp/e/inspection/. Accessed Oct 2012

MAFF - Ministry of Agriculture, Forestry and Fisheries (2012e) Fisheries Agency. White paper of fisheries. Damaged related to the fishing industry (2). http://www.jfa.maff.go.jp/j/kikaku/wpaper/h23_h/trend/1/t1_1_1_2.html. Accessed Oct 2012

Mainichi (2012a) Miyagi prefecture to build embankments along rivers that carried tsunami inland. 4/1/12. Recovering Tohoku. http://recoveringtohoku.wordpress.com/2012/04/01/miyagi-pref-to-build-embankments-along-rivers-that-carried-tsunami-inland-mainichi-4112. Accessed Oct 2012

Mainichi (2012b) Recovering Tohoku Developers inflate land prices in upland parts of tsunami-hit areas. 20 Sept 2012. http://recoveringtohoku.wordpress.com/2012/09/20/developers-inflate-land-prices-in-upland-parts-of-tsunami-hit-areas-mainichi-92012/. Accessed Oct 2012

Maksymov L (2012) Environment related problems caused by Tohoku earthquake and Fukushima nuclear disaster. http://www.jeef.or.jp/yelp/2012/02/environment-related-problems-caused-by-tohoku-earthquake-and-fukushima-nuclear-disaster. Accessed Oct 2012

Masuda R (2012) Underwater visual census of fish assemblage in Moune bay, Kesennuma. Aquabiology 34(6):562–569

McCurry J (2012) After the tsunami: fishermen return to the sea to rejuvenate economy. The Guardian 8 Mar 2012. http://www.guardian.co.uk/world/2012/mar/08/tsunami-fishermen. Accessed Oct 2012

Médecins sans Frontières (2011) Japan: MSF supports psychologists working in disasters' aftermath. 24 Mar 2011. http://www.doctorswithoutborders.org/news/article.cfm?id=5126&cat=field-news. Accessed Oct 2012

MIAC - Ministry of Internal Affairs and Communications (2011) Chapter 7 agriculture, forestry and fisheries. www.stat.go.jp/english/data/nenkan/1431-07.htm. Accessed Oct 2012

MIAC - Ministry of Internal Affairs and Communications (2012) Chapter 5; overview of agriculture, forestry and fisheries. Statistics Bureau. http://www.stat.go.jp/english/data/handbook/c05cont.htm#cha5_3. Accessed Oct 2012

Midori II (2012) Phytoplankton bloom in the spring adjacent to the sea around Japan. Midori II. Earth view. ftp://ftp.eorc.jaxa.jp/cdroms/EORC-071/e_view/12/pdf/12.pdf. Accessed Oct 2012

Minister for Agriculture, Fisheries and Forestry, Australia (2012) Australia provides relief to Japanese farmers. 13 June 2012 http://www.daff.gov.au/ludwig/media_office/media_releases/media_releases/2012/june/relief-to-japanese-farmers. Accessed Oct 2012

Mitsui & Company Limited (2012) Forest. Trees cultivated in our forest. http://www.mitsui.com/jp/en/csr/contribution/forest/about/wood.html. Accessed Oct 2012

Miyagi Prefecture Government (2011) Miyagi prefecture earthquake disaster recovery plan -summary-. http://mozism.files.wordpress.com/2012/01/miyagi-prefecture-earthquake-disaster-recovery-plan-english.pdf. Accessed Oct 2012

Miyagi Prefecture Government (2012a) Data. http://www.myu.ac.jp/english/guide/gu03.html. Accessed 28 Oct 2012

Miyagi Prefecture Government (2012b) Profile of Miyagi; industries in Miyagi. Forestry. http://www.pref.miyagi.jp/english/profile/index.htm. Accessed Oct 2012

Miyagi Prefecture Government (2012c) Press release; 29 June. Results of the measurements of radioactivity in agricultural, forestry and fishery products in Miyagi Prefecture. http://www.jfa.maff.go.jp/e/inspection/pdf/120629_miyagi_en.pdf

Miyagi Prefecture Government (2012d) Press release. 25 April. http://www.jfa.maff.go.jp/e/inspection/pdf/120425_miyagi_en.pdf. Accessed Oct 2012

Miyagi Prefecture Government (2012e) The situation of request for shipment retrain and others. As of 11 May 2012. http://www.maff.go.jp/e/quake/pdf/eng_press_may11.pdf

Miyagi Prefecture Government (2012f) Results for detection of radiation levels in Sendai forest products. June 2012. http://www.city.sendai.jp/language/__icsFiles/afieldfile/2012/07/23/120723en.fore.pdf

Miyagi Prefecture Government (2012g) Natural parks. http://www.pref.miyagi.jp/kankou/EN/sightseeing/natural_park/00_index.htm. Accessed Oct 2012

Miyagi Prefecture Government (2012h) Radiation information site Miyagi. Measurements results by local municipalities. http://www.r-info-miyagi.jp/r-info/en/food/#. Accessed Oct 2012

Miyata H, Uemura K (2012) Rebuilding Eco "Future Cities" in Tohoku. Forum for East Japan Smart City Project. 5 Apr 2012. www.accj.or.jp/en/.../92-rebuilding-eco-future-cities-in-tohoku. Accessed Oct 2012

MOE - Ministry of the Environment of Japan (2011) Successful cases on sustainable rice paddy farming and practices and wetland conservation in Asia. 2011. http://www.wetlands.org/_strp/WGAgandwet/Library/Report%20Takashima%20meeting-Successful%20Cases%20on%20Sustainable%20Rice%20Paddy%20Farming%20Practices%20and%20Wetland%20Conservation%20in%20Asia_final.pdf. Accessed Oct 2012

MOE - Ministry of the Environment of Japan (2012a) General overview of area figures for natural parks by prefecture. 2012. http://www.env.go.jp/en/nature/nps/park/doc/files/np_2.pdf. http://www.env.go.jp/en/nature/nps/park. Accessed Oct 2012

MOE - Ministry of the Environment of Japan (2012b) Forest, rivers, sea and satoyama. Reconnecting with nature and the future. Green reconstruction: creating a new national park. 2012. http://www.env.go.jp/jishin/park-sanriku/images/sanriku_fukkou_project_eng.pdf. Accessed Oct 2012

MOFA - Ministry of Foreign Affairs (2012) Summary of the results of the international energy seminar on smart community proposals for reconstructing disaster-affected areas of Japan. 2 Mar 2012. http://www.mofa.go.jp/policy/economy/energy/seminar120302.html. Accessed Oct 2012

Mongabay.com (2011) Tsunami: before and after photos of Central Ishinomaki. 15 Mar 2011. http://news.mongabay.com/2011/0315-satellite_pictures_japan_google.html. Accessed Oct 2012

Mori N, Takahashi T, Yasuda T, Yanagisawa H (2011) Survey of 2011 Tohoku earthquake tsunami inundation and run-up. Geophys Res Lett 38:LOOG14. doi:10.1029/2011gl04920

Mori wa Umino Koibito (2012) http://www.mori-umi.org/base14.html. Accessed Oct 2012

Nagano Y (2012) Rice farmers seek to save their crops from salt. The New York Times, Global Business; 22 Apr 2012. http://www.nytimes.com/2012/04/23/business/global/rice-farmers-seek-to-save-their-crops-from-salt.html?pagewanted=all&_r=0. Accessed Oct 2012

Nakamura N (2012) Soil shortage a major problem in Tohoku disaster areas. The Asahi Shinbum/ Asia and Japan Watch. 4 May 2012. http://ajw.asahi.com/article/0311disaster/AJ201205040069. Accessed Oct 2012

Nakamura K, Aikawaa T, Ichiharaa Y, Mizuta N (2012) Detection of the pinewood nematode and its insect vector in the tsunami-damaged trees of Pinus thunbergii and P. densiflora. http://hyoka.nenv.k.u-tokyo.ac.jp/alien2012/alien_abst/p14.pdf

NASA (2012) Wreckage and recovery in Ishinomaki, Japan. Nasa Earth Observatory. 13 Mar 2012. http://earthobservatory.nasa.gov/IOTD/view.php?id=77352. Accessed Oct 2012

National Parks of Japan (2012) http://www.env.go.jp/en/nature/nps/park/parks/index_2.html, and http://www.unepwcmc.org/medialibrary/2012/04/13/7b66605b/Protected%20Areas%20in%20Japan%20-%20Keisuke%20Takahashi.pdf. Accessed Oct 2012

Nikkei (2012a) S'tomo, Mitsui to help disaster-hit seafood firms. Nikkei.Com June 30 2012. http://e.nikkei.com/e/fr/tnks/Nni20120629D2906A06.htm. Accessed Oct 2012

Nikkei (2012b) Fishery industry in quake-hit Sanriku area rebuilding (2012). Nikkei.com. August 6 2012. http://e.nikkei.com/e/fr/tnks/Nni20120806D03HH091.htm. Accessed Oct 2012

Nikkei (2012c) Recovering Tohoku (2012). Ishinomaki moves first on city rebuilding scheme. Nikkei April 24 2012. http://recoveringtohoku.wordpress.com/2012/04/24/articles-tuesday-april-24-2012-ishinomaki-moves-first-on-city-rebuilding-scheme-nikkei-42512/. Accessed Oct 2012

Nishioka J, Ono T, Saito H, Nakatsuka T, Takeda S, Yoshimura T, Suzuki K, Kuma K, Nakabayashi S, Tsumune D, Mitsudera H, Johnson WK, Tsuda A (2007) J Geophys Res 112:C10012. doi:10.1029/2006JC004055

Nishioka J, Ono T, Sakaoka K, Yoshimura T (2011) Oceanic iron supply mechanisms which support the spring diatom bloom in the Oyashio region, western subarctic Pacific. J Geophys Res 116: C02021. doi:10.1029/2010JC006321

Nishitani G, Yamamoto M, Natsuike M, Liu D, Yoshinaga I (2012) Dynamics of phytoplankton in Kesennuma bay and Moune bay after the disaster 3.11. Aquabiology 34(6):545–555

Nishizawa M (2011) Disaster relief report -the 2011 off the Pacific coast of Tohoku Earthquake-Minami-Sanriku City in Miyagi prefecture. http://www.sgim.org/userfiles/file/AM12%20Handouts/IPA%20Japan.pdf. Accessed Oct 2012

NPAJ, National Police Agency of Japan (2012) Damage situation and police countermeasures associated with 2011 Tohoku District- off the Pacific Ocean Earthquake; 17 Oct 2012. Accessed Oct 2012

NPAJ, National Police Agency of Japan (2013) Damage situation and police countermeasures associated with 2011Tohoku district - off the Pacific Ocean earthquake. http://www.npa.go.jp/archive/keibi/biki/index_e.htm/higaijokyo_e.pdf. Accessed 28 Feb 2013

OISCA (2012) Tohoku (in Japanese). http://www.oisca.org/project/disaster/index.html#touhoku. Accessed Oct 2012

Okubo E (2007) Sustainable agriculture in Miyagi. Agriculture, Forestry and Fisheries Department; Miyagi Prefectural Government. http://jaef.la.coocan.jp/ifaj/eng/document/tua/slide/4-2.pdf. Accessed Oct 2012

Onishi N (2011a) Japan's seawalls were little security against tsunamis. The New York Times. 13 Mar 2011. http://www.nytimes.com/2011/03/14/world/asia/14seawalls.html?pagewanted=all&_r=0. Accessed Oct 2012

Onishi N (2011b) In Japan seawall offered a false sense of security. The New York Times. 2011. http://www.nytimes.com/2011/04/02/world/asia/02wall.html?pagewanted=all. Accessed Oct 2012

Onodera Y (2012) Miyagi prefecture's crisis management system-based on the Great East Japan Earthquake experience. http://csis.org/event/miyagi-prefectures-crisis-management-system-based-great-east-japan-earthquake-experience. Accessed Oct 2012

Ozawa R, Kaji E (2012) Japan, US monitor drifting tsunami debris; The Daily Yomiuri Online. February 14, 2012. http://www.yomiuri.co.jp/dy/national/T120213005317.htm. Accessed Oct 2012

Reconstruction Agency of Japan (2012) Role of reconstruction agency. http://www.reconstruction.go.jp/english/topics/about-us.html. Accessed Oct 2012

Regions & Cities (2012) Tohoku Region. http://web-japan.org/region/tohoku.html. Accessed Oct 2012

Richmond B (2012) Erosion, deposition and landscape change on the Sendai coastal plain, Japan, resulting from the March 11, 2011 Tohoku-Oki tsunami. Sediment Geol 282:27–39, Elsevier. August 2012. http://www.sciencedirect.com/science/article/pii/S003707381200231X. Accessed Oct 2012

Saito T (2012) Reconstruction design global forum. 22 Mar 2012. http://www.reconstruction.go.jp/english/topics/final%20G_Forum20120322_DCCS_Saito_ENG.pdf. Accessed Oct 2012

Sanriku Reconstruction (Fukko) National Park Initiative (2012) Re-formation of natural parks in the disaster-affected area. http://www.env.go.jp/en/nature/nps/sanriku-fukko/initiative.pdf. Accessed Oct 2012

Santiago-Fandiño V (2011) Impacts of the earthquake and tsunami in the coast of Japan –Miyagi prefecture, Sanriku coastline. LOICZ-INPRINT 2001/3. http://www.loicz.org/imperia/md/content/loicz/snapshot/Inprint_2011_3_Tsunami_in_the_Coast_of_Japan.pdf and http://www.loicz.org/imperia/md/content/loicz/print/newsletter/Inprint_2011_3_online72.pdf. Accessed Oct 2012

Santiago-Fandiño V (2011) Newsletter of the IWA Specialist Group on watershed & river basin management. The tsunami in the Northeast Coast of Japan- the case of Minamisanriku, Miyagi prefecture. W&RBM News – July 2011 pp 9–12. http://www.iwahq.org/contentsuite/upload/iwa/all/Specialist%20groups/Specialist%20groups/Watershed%20and%20River%20Basin%20Management/SG%20Newsletters%20and%20Reports/Watershed%20%26%20RBM%20NewsletterJune2011.pdf. Accessed Oct 2012

Satake K et al (2012) Tsunami heights along the Pacific Coast of Northern Honshu from the 2011 Tohoku and past great earthquakes. AOGS-AGU joint assembly meeting. Singapore, 2012.

Sendai City (2012) Disaster reconstruction plan; digest version (2012) Tsunami reduction and housing reconstruction project. http://www.city.sendai.jp/shinsai/shinsaihukkokentou/pdf/keikakushiryou/plan%20English.pdf. Accessed Oct 2012

Shiraiwa T (2012) Giant fish-breathing forest: a new environmental system linking continental watershed with open water. In: Takayuki M, Shiraiwa T (eds) The dilema of boundaries. Towards a new concept of catchment, Global environment studies. Springer, Tokyo

Simon R (2012) Tsunami debris from Japan poses costly coastal cleanup emergency. Los Angeles Times. 14 Aug 2012. http://articles.latimes.com/2012/aug/14/nation/la-na-nn-tsunami-debris-cleanup-20120814. Accessed Oct 2012

Simply Info (2012) The Fukushima project. 2012. Contamination rampant in Miyagi prefecture. July 2012. http://www.simplyinfo.org/?p=6953. Accessed Oct 2012

Statistics Japan (2012) Prefecture comparisons. http://stats-japan.com/t/categ/50008. Accessed Oct 2012

Tabaru Y, Kanno K, Kawabata T, Ishikawa Y, Tanaka K, Hirao M, Kumon K, Watanabe M (2011) Outbreak of blow flies in the tsunami affected areas and their control. Med Entomol Zool 63 (1):71–83, 2012

Takahashi K (2012) Protected areas based on laws under jurisdiction of Ministry of the Environment. http://www.unep-wcmc.org/medialibrary/2012/04/18/7b66605b/Protected%20Areas%20in%20Japan%20-%20Keisuke%20Takahashi.pdf. Accessed Oct 2012

Takashige H (2012) Tohoku cotton goods sold for first time since tsunami. The Asahi Simbun/Asia Japan Watch. 25 June 2012. http://ajw.asahi.com/article/economy/business/AJ201206250047. Accessed Oct 2012

Tanaka M (2012a) International Institute for Advance Studies. Kyoto University. Personal Communication

Tanaka M (2012b) The aim and future perspective of "Kesennuma-Moune investigation" on effects and recovery after the 3/11 disaster conducted in Kesennuma and Moune bay, Miyagi prefecture, Japan. Aquabiology 34(6):515–523

The Asahi Shinbum/Asia (2012) Tougher food safety standards of radiation levels creates hardship for many farmers. 07 Apr 2012. http://ajw.asahi.com/article/behind_news/social_affairs/AJ201204070058. Accessed Oct 2012

The Economist (2012) Salt-tolerant rice; nuclear powered crops. 5 May 2102. http://www.economist.com/node/21554169. Accessed Oct 2012

The SuriaLink Seaplants Handbook (2012) http://surialink.seaplant.net/HANDBOOK/Genera/reds/Eucheuma/Eucheuma.htm. Accessed Nov 2012

Today (2011) How Fudai Village defy the tsunami. 16 May 2011. World 28. http://imcmsimages.mediacorp.sg/cmsfileserver/documents/006/pdf/20110516/1605wnp028.pdf. Accessed Oct 2012

Tokyo Marine & Nichido Fire Insurance Co., Ltd (2012) Participation. In: Ten-Year Restoration Project of the Coastal Forests of Tohoku Region. 24 Feb 2012. http://www.tokiomarine-nichido.co.jp/en/new/pdf/120224.pdf. Accessed Oct 2012

Tori K, Kato F (2001) Risk assessment on storm surge flood. http://www.pwri.go.jp/eng/ujnr/joint/34/paper/83kato.pdf. Accessed Oct 2012

UNEP (2012) Managing post-disaster debris: the Japan experience. Report of the International Expert Mission to Japan. 47 pp. http://www.unep.org/pdf/UNEP_Japan_post-tsunami_debris.pdf. Accessed Oct 2012

University of Tokyo Atmosphere and Ocean Research Institute (2012) Survey on the physico-chemical environment and plankton of Otsuchi http://www.kaiyo-gakkai.jp/sinsai_eng/2011/09/a-survey-on-the-physicochemical-environment-and-plankton-of-otsuchi-bay-performed-by-the-university.html. Accessed Oct 2012

UNO (2011) UN international year of forest 2011. http://www.un.org/en/events/iyof2011/forests-for-people/awards-and-contests/award-winners. Accessed Oct 2012

UNU - United Nations University (2010) Satoyama initiative. http://satoyama-initiative.org/en. Accessed Oct 2012

UNU - United Nations University (2012) Biodiversity and livelihoods; the Satoyama concept and practice. http://satoyama-initiative.org/wp-content/uploads/2011/09/biodiversity_booklet_en_web.pdf. Accessed Oct 2012

USDA (2012) Foreign agricultural service; trip report; Japan agricultural situation. http://www.pecad.fas.usda.gov/highlights/2012/08/Japantrip/. Accessed Oct 2012

Vervaeck A, Daniell J (2012) Japan – 366 days after the quake… 19000 lives lost, 1.2 million buildings damaged, $574 billion. In: Earthquake report. March 2012. http://earthquake-report.com/2012/03/10/japan-366-days-after-the-quake-19000-lives-lost-1-2-million-buildings-damaged-574-billion. Accessed Oct 2012

Waldenberger F, Eilker J (2011) The economic impact of the Tohoku earthquake. In: L'Ecole des Hautes Etuden En Sciences Sociales. http://ffj.ehess.fr/index/article/283/the-economic-impact-of-the-tohoku-earthquake.html. Accessed Oct 2012

Waste Management World (2012) 7m tons of tsunami debris to be shredded for incineration in Japan. http://www.waste-management-world.com/index/display/article-display/7327388476/articles/waste-management-world/waste-to-energy/2012/08/7m_Tonnes_of_Tsunami_Debris_to_be_Shredded_for_Incineration_in_Japan.html. Accessed Oct 2012

World Fishing and Aquaculture (2010) China, Japan, US lead fish consumption. http://www.worldfishing.net/news101/report-china,-japan,-us-lead-fish-consumption. Accessed Oct 2012

World Press Recovering Tohoku (2012) Miyagi, Iwate special zones approved, Yomiuri. 8 Feb 2012. http://recoveringtohoku.wordpress.com/2012/02/08/miyagi-iwate-special-zones-approved-yomiuri-2812. Accessed Oct 2012

Wright H (2012) Tohoku fisheries fight back from 3/1. In: The Japan Times. Thursday 23 Aug 2012. http://www.japantimes.co.jp/text/fl20120909x1.html

Yamada Y (2012) Seasonal changes of zooplankton community structure in Moune-Kesennuma inlet and adjacent waters. Aquabiology 34(6):556–561

Yamamoto M, Yokoyama K, Yoshinaga I (2012) Changes of heavy metals and oil spilled in Kesennuma bay and Moune bay after the disaster 3.11. Aquabiology 34(6):538–544

Yokoyama K, Hatakeyama M (2012) Tsunami disaster in Moune bay and recovered tidal flat. Aquabiology 34(6):524–530

Chapter 2
Assessment and Modeling of Dispersal Contamination Incoming with Submarine Groundwater Discharge (SGD) in Tsunami Affected Coastal Areas

Y.A. Kontar, K.A. Korotenko, and V. Santiago-Fandiño

Abstract The submarine groundwater discharge (SGD) transports a significant amount of various contaminants into the coastal zone especially in tsunami affected areas. An assessment of the impact of intruded pollutants in the coastal ecosystems requires understanding the fate of the pollutants and processes of their dispersal in ambient waters. In this paper, we proposed a methodology for SGD data collection and data assessment, using different methods, technology, techniques and instruments as well as the 3-D coupled ocean circulation/particle-tracking model for assessment and predicting the transport and dispersal of pollution-containing SGD into a coastal environment. Among the proposed methods to use for data collection and the SGD assessment primary attention was paid to geophysical, hydrologic, remote sensing and hydro-geologic measurements, using natural radiotracers, measurements by seepage meters and benthic chambers, biogeochemical and biological measurements. Also, several new modeling approaches were considered in particular those which use the particle-tracking model. The particle-tracking model takes currents and turbulent diffusivities predetermined by the ocean circulation model and uses the Lagrangian approach to predict the motion of individual droplets, the sum of which constitutes a contaminant plume which is the result of discharge of contaminant-rich submarine groundwater. Presently, we limited our simulations to elucidate the effect of tides on the SGD/nitrate plume formation. The model predicts behaviour of a nitrate plume, its shape and variation during a tidal cycle in the shallow waters. The model can be used to predict

Y.A. Kontar (✉)
College of Sciences, The University of Findlay, Findlay, OH, USA

Federal GEOS Funding, Inc, Findlay, Ohio, USA
e-mail: ykontar@findlay.edu

K.A. Korotenko
P.P. Shirshov Institute of Oceanology, Russian Academy of Sciences, Moscow, Russia

V. Santiago-Fandiño
United Nations Organization (retired), Private Consultant, El Curbiellu 28 Peon, 33314 Villaviciosa, Asturias, Spain

Y.A. Kontar et al. (eds.), *Tsunami Events and Lessons Learned: Environmental and Societal Significance*, Advances in Natural and Technological Hazards Research 35, DOI 10.1007/978-94-007-7269-4_2,

contamination of coastal waters with various pollutants incoming with SGD in the aftermath of a tsunami when impact of the latter on aquifers can be significant.

Keywords Tsunami • Coastal zone contamination • Submarine groundwater discharge • Assessment • Evaluation • Geophysical and hydrologic measurements • Natural radiotracers • Seepage meters • Benthic chambers • Biogeochemical and biological measurements • Eulerian-Lagrangian model

2.1 Introduction

It has been recognized and discussed at the Session OS07-17 entitled Groundwater/Surface Water Exchange in Tsunami Affected Areas of Japan – Ecological and Societal Significance during the Asia Oceania Geosciences Society (AOGS) and the American Geophysical Union (AGU) Western Pacific Geophysics Meeting (WPGM) Joint Assembly being held on the 13–17 of August 2012 at Sentosa Island in Singapore, that **submarine groundwater discharge** (hereinafter SGD) is a potentially significant contributor of hazardous pollution to the coastal ocean in tsunami affected coastal areas (Kontar 2011; Kontar et al. 2012a). While tsunamis cause direct loss of life and infrastructure destruction, there is a need to study their longer-term hazardous effects on SGD. Availability of clean groundwater is key to society's survival (Williamson et al. 2009) and SGD is an important pathway for solute and energy transport, including freshwater, nutrients, trace metals, bacteria, and other pollutants.

Although not as obvious as river discharge, groundwater can also discharge directly into the coastal ocean (Biddanda and Cotner 2002; Biddanda et al. 2009; Boehm et al. 2005; Bokuniewicz et al. 2004, 2008; Burnett et al. 2002, 2003, 2006; Cable et al. 2002; Church 1996; IOC 2004; Kontar 2007a, b; Kontar 2008a; Kontar and Burnett 1999; Kontar and Zektser 1999; Kontar and Ozorovich 2006; Kontar et al. 2002a, b, 2006, 2010, 2012; Lin et al. 2010, 2011; Lobkovsky et al. 2003; Moore 2006; Ozorovich and Kontar, 2007; Paytan et al. 2006; Povinec et al. 2006, 2008, 2012; Slomp and Van Cappellen 2004; Stieglitz et al. 2008; Swarzenski et al. 2002, 2006a, b, Swarzenski et al. 2007a, b, c, 2009; Taniguchi et al. 2006, 2008; Zavialov et al. 2012). Like surface water, groundwater flows down gradient. Therefore, groundwater flows directly into the coastal ocean wherever a coastal aquifer is connected to the sea. Furthermore, artesian aquifers can extend for considerable distances from shore, underneath the continental shelf with discharge to the coastal ocean at their points of outcrop. In some cases, these deeper aquifers may have fractures or other breaches in the overlying confining layers, allowing groundwater to flow into the sea. Figure 2.1 schematically illustrates shallow and deep aquifers and processes associates with SGD.

SGD may be a significant pathway for diffuse pollution to enter the coastal zone (Fig. 2.2) where coastal aquifers have become contaminated by septic systems or other pollution sources under influence of tsunami waves (Kontar 2007b; Kontar 2008a; Kontar et al. 2012a, b). Although SGD contribution in the ocean's water

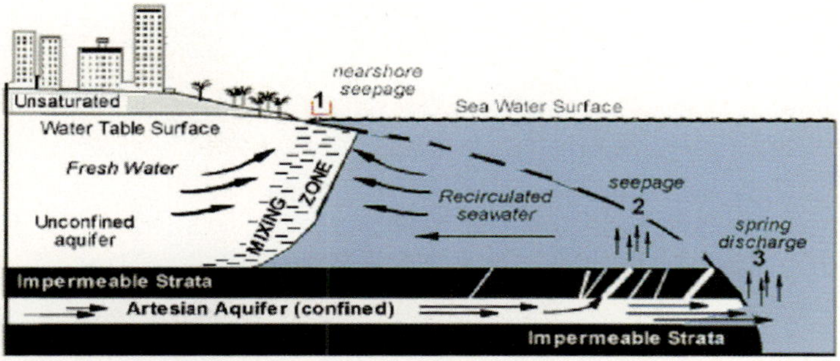

Fig. 2.1 Diagrammatic view of the relationships between coastal aquifers, seawater, and groundwater discharge. Three types of submarine groundwater discharge are illustrated: (1) Nearshore seepage; (2) Offshore seepage; and (3) Submarine springs (Modified from Burnett et al. 2003)

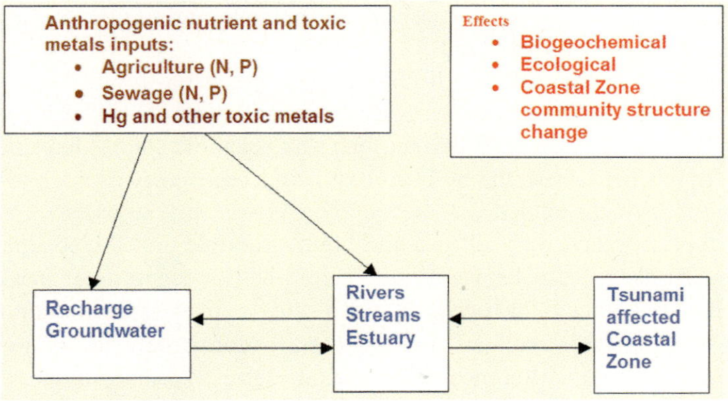

Fig. 2.2 Diagram of groundwater – coastal zone water interaction in tsunami affected areas

budget is generally small, it represents an important source of nutrients and other contaminants that affect the ecology of the estuaries and coastal fresh water bodies (Andersen et al. 2007; Black et al. 2009; Boehm et al. 2005; Chen and Wang 1999; Charette 2007; Kontar and Burnett 1999; Kontar et al. 2002a, b; Kroeger and Charette 2008; Leote et al. 2008; Li et al. 1999; Niencheski et al. 2007; Nowicki et al. 1999; Paytan et al. 2006; Portonoy et al. 1998; Santos et al. 2009; Simmons 1992; Slomp and Van Cappellen 2004; Smith et al. 2002; Spiteri et al. 2008; Swarzenski et al. 2006b, 2007a; Uchiyama et al. 2000; Weiskel and Howes 1992; Winter et al. 1998). There are reports that SGD may also be an important source of alkalinity and carbon to shelf waters (Lin et al. 2010; Wang et al. 2008, 2009).

Submarine groundwater coming from confined and unconfined aquifers into the coastal ocean carries dissolved contaminants, concentration and type of which considerably varied from region to region. Multiple studies (Boehm et al. 2005; Paytan et al. 2006; Swarzenski et al. 2007a) indicate that the concentration of groundwater contaminants increases with increasing housing density and agricultural

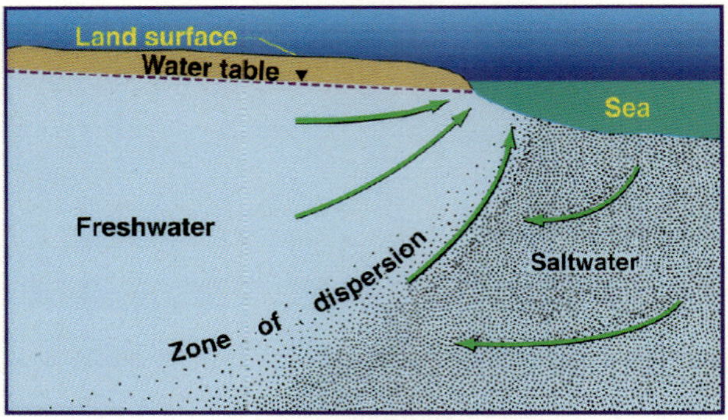

Fig. 2.3 Diagrammatic view of hydrodynamic interface or zone of dispersion between fresh groundwater and saline sea water (Modified from Burnett et al. 2006)

activity. Expanding residential and commercial near-shore development is leading to increased nutrient inputs to groundwater that eventually migrate into to coastal waters. Several-decades long research shows that nitrogen inputs via non-point sources over large coastline areas cause decline of ecological health and may support harmful algal blooms (Lin et al. 2011).

Investigations of the offshore discharge of groundwater, for many years, were largely motivated by water resource related issues. As was noted by Andersen et al. (2007), there were at least two reasons why scientific studies have developed so slowly in this field. First, the SGD process is inherently very difficult to measure and monitor, which tended to discourage serious investigations. Nearshore seepages typically have very diffuse and highly variable unit fluxes although the cumulative discharge can be very significant when it occurs over a wide area. Second, SGD is a process that occurs across a land-sea interface that spans different scientific disciplines as well as environments. Unfortunately, there are distinct cultural and structural differences that separate terrestrial and marine scientists. Literally, hydrologists and coastal oceanographers are looking at the same problem from different ends (Fig. 2.3).

For the last decade, intensive field investigations of coastal environment pollution associated with SGD were undertaken. It was found out that aquifers are significant sources of trace elements and other chemical constituents to the coastal waters where on land chemicals as NH_4, SO_4, NO_3, PO_4 and SiO_2, hydrocarbons, trace metals (Ni, Cu, Cd), NaOH, fertilizers, and sewage were. Uchiyama et al. (2000) also reported the exploratory discovery of SGD by using salinity for the first time in the coastal area and pointed out that SGD may occur near the river estuary. Recently Spiteri et al. (2008) and Leote et al. (2008) have shown that submarine groundwaters are also clearly indicated by tracers of oxygen, strontium isotopes and barium content. They noted that the concentration of many contaminants in groundwater is typically several times higher than in seawater. Usually with distance from seepage the concentration of contaminants decreased rapidly. As observations of nitrate (NO_3) plumes near shallow seepages by Santos et al. (2009) and Lee et al. (2009) show, the flux of nitrate towards the marine environment was highly variable even over a short distance from the coast. In addition, greater dispersion near the seepage face due to movement of water masses

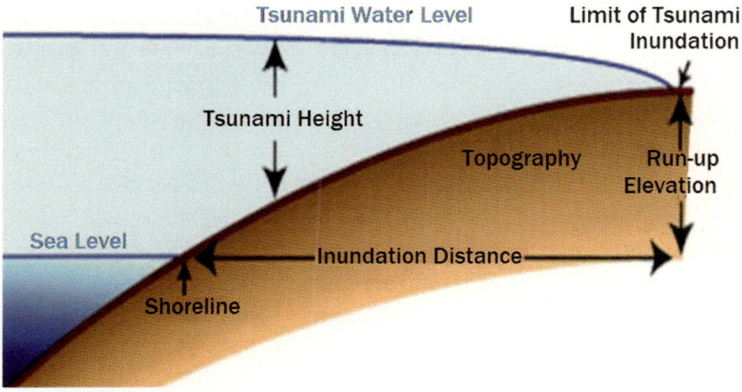

Fig. 2.4 Diagrammatic view of tsunami run-up elevation which rapidly changed the hydraulic gradient of coastal aquifers and created saltwater intrusion replacing process of SGD

in response to tides and to seasonal cycles of recharge to the fresh aquifer. Outside an aquifer, nitrate diminished substantially in concentration so that horizontal and vertical sizes of a detectable inclined nitrate plume did not exceed 6 and 4 m, respectively (Wang et al. 2008).

In the coastal ocean, processes of spreading and fate of contaminants coming from SGD aquifers are governed by ocean dynamics and depend on many factors such as seepage zone, discharge rate and type of contaminants. Mathematically, predicting such processes is very a complicated task that requires developing numerical models for reproducing/predicting the ocean circulation and transport of contaminants. Until now, numerical modeling directed to the solution of problems associated with pollution inputs from SGD was mostly focused on modeling the transport and biogeochemical processes (Kontar 2008a; IOC 2004) inside aquifers. Some works have been dedicated to analysis and modeling an influence of tides on oscillation of groundwater discharge rate (Kontar et al. 2012a, b; Taniguchi et al. 2006; Kontar and Ozorovich 2006). By theoretical modelling for SGD and the associated chemical transfer to the ocean, Burnett et al. (2003) have demonstrated that wave setup and tidal pumping may be the processes largely responsible for the high rate of SGD and, in turn, a considerable intensification of the transport of chemicals to the ocean.

Despite a wide spectrum of models and approaches that were built for the solution of various problems associated with contamination of coastal zone due to SGDs, yet, no attempt has been made on the development of a complex approach for predicting the regional circulation and its effect on the transport and mixing of pollutants coming into seawater from submarine seepages. Therefore one of the objectives of this publication is to fill this gap and propose a pollution transport model coupled with an ocean circulation model. We will focus on the shelf zone affected by tsunami (Fig. 2.4); however, based on this generalized approach, the model can be adapted easily for any region of interest and considered not only SGD process, but also process of seawater intrusion in the coastal aquifers.

As reported recently, after the process of seawater intrusion ends, SGD transports a significant amount of various contaminants into the coastal zone in tsunami affected areas (Kontar et al. 2012a, b). We expect that increased SGD of pollution due to tsunami impact may be particularly important for increasingly populated coastal zones.

Detailed knowledge of hazardous SGD is currently limited to a few well studied locations (Kontar et al. 2012a), and to our knowledge there has been no multi-disciplinary comprehensive SGD study to date in tsunami or storm affected areas, to quantify hazardous SGD, the associated nutrient and trace metal fluxes, and their impacts on ecosystem health/aquatic community indicators. A multi-disciplinary comprehensive SGD study could significantly advance: (i) the next generation of field sampling; (ii) predictive model simulations; (iii) mathematical tools; and (iv) laboratory, remote sensing, and social science techniques, of groundwater-seawater interaction processes in tsunami affected coastal zones.

2.2 Development of Methodology of the SGD Data Collection and Data Assessment

We propose a methodology which combines field studies and laboratory analyses to quantify SGD nutrient and select pollutant fluxes to the tsunami affected key coastal zone locations (Kontar 2007c, 2008b). We propose to identify the hydro geological and biogeochemical controls of the fluxes, and assess their impact on the water quality and ecological health of the coastal zones affected by tsunami to assess the effects of tsunami induced changes in groundwater nutrient and trace metals discharge on the cycles of C, N and P, and food webs in the tsunami affected coastal zones.

Through using this methodology the anthropogenic changes in SGD and the coastal water cycle can be elucidated. The specific goals of using this methodology could be:

1. To assess the relative importance of the terrestrial (hydraulic gradient, seasonal recharge cycle), coastal zone conditions (tsunami wave setup and pumping, water level differences, wind, tide), and geologic factors that influence fluid exchange across the shelf at three contrasting sites.
2. To evaluate the fluxes of nutrients and trace metals within the study sites and their influence on the dreissenid mussel distribution.
3. To compare the impacts of nutrient inputs at one relatively pristine site and two sites with heavy anthropogenic activity on the benthic invertebrate community as an index of overall ecological health.
4. To place the obtained data, results/interpretations of the study into the broader context of science and management for further modeling and prediction of potential implications related to dreissenid mussel dispersal and colonization and its influence on food webs.

5. To achieve an easily-measured set of indicators that can be implemented in the future to gauge changes in the coastal ecosystem, and specifically in dispersal and colonization of dreissenid mussel.

The primary focus in this kind of research and field work could be field data collection for assessment of SGD and associated constituent loading to the coastal zone and assess the resulting benthic ecosystem response under influence of tsunami. Since many SGD drivers are known to have temporal variations, we propose to deconvolute the influence of these forces by measuring the frequency of the relevant controlling parameters and the ensuing groundwater seepage responses. Numerical tools can be calibrated using observations for predictive purposes (e.g., simulations of future tsunamis and their effects). Such tools can be incorporated in future projects as part of virtual reality platforms, in order to allow coastal zone managers and policymakers to formulate and compare potential scenarios.

2.3 Possible Field Research Plan and Methods

2.3.1 General Strategy

Tsunami-induced processes may have resulted in exposure of aquifers to contaminated surface waters, initiating desorption of elements and increasing reactivity and bioavailability of organic matter, leading to higher rates of decomposition and release of P and N as well as other constituents. We expect that increased nutrient loads from SGD due to tsunami impact may be important in northeastern Japan, because of high recharge rates, fractured dolomite terrains conducive to rapid flow, and recent large-scale changes in land-use (e.g., urbanization, fertilizer application).

The first step of the plan of an assessment of tsunami's influence on the spatial distribution of SGD could be fulfilled by performing a qualitative radon survey along the coastline. This tool has the ability to reveal areas where SGD is more prevalent than others to allow researchers to target the areas for more in-depth studies. At least three tentative study sites for the fieldwork could be selected, but the results of the radon survey may yield more interesting areas for the quantitative studies. The survey could be done twice during two seasons, with the exception of the mussel sampling, which will occur once during the reproductive season. In order to evaluate the afore-mentioned driving forces that control SGD and associated constituent loading, time-series measurements at three contrasting sites (selected based on the survey and existing hydrogeology data) could be conducted to measure all parameters relevant to characterization of the driving forces of SGD at each of the study sites. These methodologies are now well established (Povinec et al. 2006, 2008, 2012) so that sound assessments could be made. Since different drivers operate on different time scales, deconvolution of various forcing functions could be performed by defining these cycles.

Table 2.1 Measurement parameters, methods, and purpose to the quantification of SGD and benthic nutrient fluxes in tsunami affected areas

Measurement parameter	Method, PIs	Purpose	Overall product
^{222}Rn	RAD7 (Durridge Co.); LSC	SGD Tracer	Quantitative understanding of SGD rates
Direct seepage flux	Manual and EM seepage meters	Direct measurement of SGD	
Chemical constituents: nutrients, trace metals	Flow injection autoanalyzer ICP-MS	Chemical loading; source identification	Chemical loading rates
SGD/surface water interface	TEM, GPR, multi-electrode resistivity	Mapping and qualitative SGD assessment	Qualitative understanding of hydrodynamics, chemical and biological changes
Basic water quality: temperature, dissolved oxygen, conductivity, pH, Eh, chlorophyll, phycocyanin	In-situ probes and sondes/well sampling	Qualitative SGD assessments	
Biogeochemical and benthic metabolism measurements	Changes in tracers and dissolved oxygen within benthic metabolism chambers	Benthic metabolism assessment	Metabolism inside and outside of SGD
Biological measurements and benthic imagery	10 cm square cores	Identification of organisms	Understanding effects of SGD inputs on benthos
	Underwater camera	Community composition	

Field experiments to investigate the relative importance of SGD drivers could employ geochemical tracers, including radon (Burnett et al. 2003, 2006), together with manual and automatic seepage meters (Taniguchi et al. 2006, 2008, Povinec et al. 2008, 2012; Bokuniewicz et al. 2008; Lobkovsky et al. 2003).

Biogeochemical and various hydrogeological and geophysical measurements could be made in both surface and pore waters (specifically for nutrients and trace metals). The work could enhance understanding and predictive capability of groundwater discharge to near-shore waters (Table 2.1). A summary listing of the parameters is presented below.

Parameter to be measured:

(a) Location of groundwater/surface water interface using geophysical methods (TEM and GPR) and monitoring wells at the focused study stations.
(b) ^{222}Rn, as a groundwater tracer using RAD7, Durridge Co. in survey and time series modes.
(c) Chemical constituents associated with groundwater seepage, such as nutrients (flow injection autoanalyzer) and trace metals (ICP-MS).

(d) Meteorological parameters (wind speed/direction, temperature, precipitation, etc. – from existing weather station networks).
(e) Variations of sea level and tsunami wave's height – (from multiple water level recorders).
(f) Seepage flux at the three study sites (multiple seepage meters).
(g) Basic water quality characteristics, temperature, salinity, pH, dissolved oxygen, redox conditions in aquifers and pore waters via piezometers at the focused study sites.
(h) Rates of benthic metabolism (production and respiration), oxygen demand and sedimentary organic carbon content.
(i) Measurement of adult dreissenid density and larval settling rates at the focused SGD study sites.

Specifically, we could study SGD and related processes caused by tsunamis using field, laboratory, and numerical modeling studies to quantify groundwater, nutrient and pollutant fluxes to the coastal zone at several sites, one primary and several more which can be used for comparison. Sites can be selected based on high tsunami impact, results of earlier work on geology, and the presence of a sand aquifer that rapidly transmits groundwater and dissolved fluxes to the ocean (Kontar et al. 2012a, b).

2.3.2 Study Sites

It could be reasonable to conduct a detailed study at three contrasting sites. Tentatively, one of the sites can be located for example within a park, is relatively pristine, and influenced by rapid surface water exchange. The tentative second site could be located close to a tailing pond and might be affected by different pollutants including heavy metals. The tentative third site could be located close to the discharge point of a water treatment facility or a sewage treatment plant, and so is representative of a heavily affected site from human activity. There, water circulation (especially in the summer) is somewhat hindered and water residence times are generally long. At such a site, dissolved nutrients generally could be highest.

Depending on results from the geophysical and tracer survey the study sites may be modified or additional sites added. Specifically, if the survey indicates "hot spots" of groundwater seepage as indicated by geophysics and high radon activities further investigations of these sites and their evaluation and locations as potential sites for more focused study in the future. At the minimum several additional samples can be collected at the selected locations to reinforce any results from the survey.

It would be also useful to examine the poorly known processes of tsunami impact on coastal groundwater and provide new understanding of SGD in coastal systems, by using improved assessment technologies such as high-resolution process models based on hydrologic, geologic, and chemical data (some newly acquired), combined with tsunami inundation simulation results. We can also examine the influence of nutrients and pollutants on the coastal zone food webs by monitoring aquatic communities, such as diatoms, using both in situ sampling and remote sensing in sites with high/low anthropogenic influence on SGD inputs. From this integrated assessment, we could develop adaptive groundwater management programs for local communities that will address hazard in tsunami-affected areas. These could be integrated into adaptive policy models to adjust regulatory stringency based on the severity of the pollution risks forecasted for different tsunami/source scenarios (serious game concept).

2.3.3 Methods

2.3.3.1 Geophysical, Hydrologic and Hydrogeologic Measurements of SGD

At the beginning of fieldwork we can conduct measurements using a new geophysical system (Kontar and Ozorovich 2006; Ozorovich and Kontar 2007) based on the transient electromagnetic method (TEM) sounding technology (Fig. 2.5) to study the movement and position of the groundwater/surface water interface relative to the wind-wave pumping range and examine possible changes in these patterns.

Results of geo-electromagnetic surveys and 3D mapping of the spatial distribution of the groundwater/surface water interface will help us select the best position for the study transects. Also we will examine the coastal subbottom geologic formations using multi-electrode (multi-channel) resistivity geophysical measurements (Stieglitz et al. 2008; Swarzenski et al. 2006a, 2007a, c, 2009), a ground penetrating radar (GPR) technology and high resolution seismic profiling methodology (Ismail et al. 2011). Hydraulic gradient and conductivity measurements will be made in existing monitor wells and piezometers. Water levels will be monitored continuously in several of the piezometers using pressure transducers (corrections for barometric pressure fluctuations will be applied). We will also use seepage meters as a direct, physical measurement of the discharging groundwater at each of the selected focused study sites.

Hydraulic gradient and water-quality measurements could be made in monitoring wells. We could use seepage meters as a direct, physical measurement of SGD.

Continuous Waterborne Electrical Imaging (CWEI)

CWEI is a promising technology for the field evaluating surface water/groundwater exchange and SGD (Fig. 2.6), where a string of electrodes is pulled along the water surface with continuous recording (Cohen et al. 2010; Day-Lewis et al. 2006).

This technique can be effectively applied to map SGD and lithological heterogeneity. The coastal zone sediments are ideally suited for imaging with induced polarization (IP) as relatively modest changes in grain size result in measurable field-scale IP polarization signals. Using this technique it is possible to survey multiple shore-parallel lines to build a pseudo 3D image of lithology and SGD.

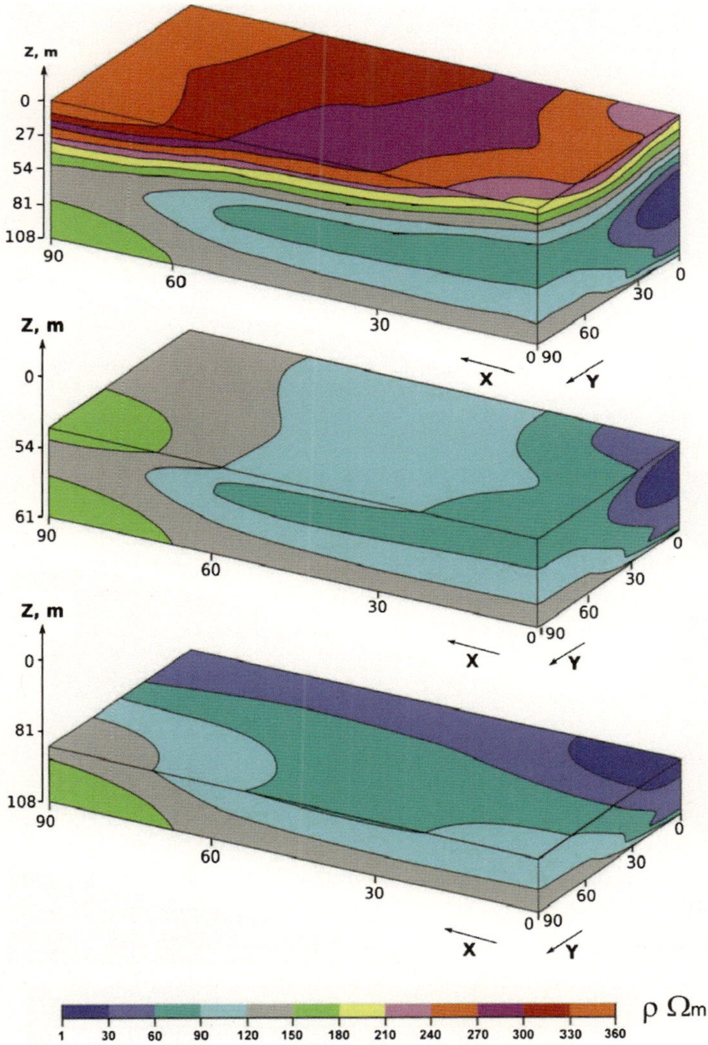

Fig. 2.5 Example of processed the transient electromagnetic method (TEM) sounding technology 3D image of the spatial distribution of the formation resistivity with depth in the study area from diverse coastal settings to demonstrate the spatial/temporal dynamics of the groundwater/surface water interface (Modified from Kontar and Ozorovich 2006)

2.3.3.2 Fiber Optic Distributed Temperature Sensing (FO-DTS)

(FO-DTS) is a new technology, exploiting temperature-dependent backscatter mechanisms (Henderson et al. 2009; Selker et al. 2006a, b) initially used in fire detection and industrial process monitoring, but recently for monitoring surface-water/groundwater exchanges in coastal zones (Tyler et al. 2009) by identifying

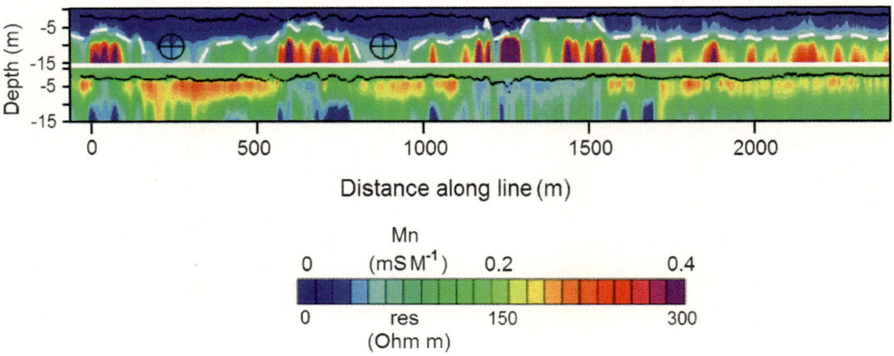

Fig. 2.6 Induced polarization (mS/M) and resistivity (Ohm m) from CWEI (Modified from Day-Lewis et al. 2006). The inversion used the known water layer (*black line*) and resistivity as a hard constraint. Thickness of a coarse grained, permeable upper unit and an underlying fine-grained confining unit are obvious in the IP image (*white line*)

temperature plumes of colder groundwater (4 °C). Since simultaneous measurements (in the thousands) are made for a section of cable (based on time-of-flight calculation), FO-DTS can generate spatially rich maps showing SGD.

2.3.3.3 Hydrogeological Modeling

Geological, geochemical, biogeochemical, and temperature data could be used in process-based models of the hydrogeologic system, forced or informed as appropriate by results of tsunami impact (inundation) models. Forward models with detailed stratigraphy and time-varying boundary conditions could be used to simulate transients in groundwater flow, and solute and reactive transport (Clement et al. 1998; Burnett et al. 2006). We could build on an earlier modeling approach (Burnett et al. 2006) that used solute transport models of offshore freshwater anomalies to show that SGD can vary significantly with climate (Church 1996).

Similar modeling shows that SGD and solute transport are greatly affected by silt connectivity (Coering and Cline 1970; Hunter et al. 1993). This emphasizes the need for good stratigraphic control. We will adapt our continental shelf and kyr-to-myr-scale model for ocean-shoreline and decadal-to-seasonal-scale processes. The footprint facilitates higher temporal and spatial resolution. We could generate a hydrostratigraphic cross-section based on regional geology, geophysical surveys, and other existing subsurface data. Sensitivity studies on model parameters (stratigraphy, permeability) could be performed. Model results could be compared with radon data to validate modeled SGD.

Also we could evaluate SGD sensitivity to various forcing functions, including time-varying inundation/ocean levels and variable rainfall infiltration. This could help define key driving forces and system response to ongoing climate change that will allow better isolating tsunami impact on SGD. Lastly, we could couple solute and heat transports with groundwater flow, to understand migration of fluids, nutrients, chemicals, and energy (e.g., Andersen et al. 2007). Modeling reactions could also provide better constraints on the physico-chemical processes and flow paths in the coastal zone system. These models could also be validated with field data and used to identify which environmental changes (i.e., anthropogenic loading, climatic changes) could have the larger control on the evolution of the groundwater/surface water system in the tsunami affected coastal zones.

2.3.3.4 Remote Sensing

Remote sensing observations could be used as an indicator of SGD and to gauge changes in the coastal zone ecosystem and water quality. Time-series from remote sensing satellite data sets could be used to obtain: (1) water clarity; (2) estimates of Chlorophyll (*Chl*), DOC, and suspended minerals (SM); (3) maps of *Cladophora* extent and biomass to ~15 m depth; and (4) HAB maps. The data could have spatial resolution of several meters to 1 km based on satellite source. Water clarity time series could be generated using Landsat data. Water visibility could be estimated on an area as small as 30 m.

Also we could generate yearly water visibility values dating back to the early 1970s with readily available historical satellite imagery, for each of the SGD work sites. We could develop and test an algorithm to estimate *Chl*, DOC and SM (CPA's) from MODIS and SeaWiFS satellite data (Grimaldi et al. 2011a, b). Although the resolution is ~1 km, they have been shown useful in assessing river DOC inputs. We could investigate the applicability of using the MERIS satellite sensor in CPA retrieval routines which is of interest because of its finer spatial resolution, (300 m vs 1 km for MODIS), and similar spectral band sets to MODIS and SeaWIFS. Chlorophyll estimate time series are a critical input into primary productivity models. Suspended mineral estimates are useful to separate storm events from local effects on SGD. Monthly estimates of *Chl*, DOC and SM for a 100 km area centered on each of the SGD work sites could be derived back to 1998 (SeaWiFS launch). Bottom feature maps will be generated over each site to see if there is a correlation between SGD and bottom features.

A model could be specifically developed to estimate primary productivity for the SGD study coastal zone. The model could use satellite derived *Chl*, water depth, and optical properties of the water column. For each work site, we could calculate primary productivity on a monthly basis, back to 1998. A MODIS based algorithm has been already developed to detect HABs (Grimaldi et al. 2011a, b). This algorithm could be used on selected MODIS data back to 2002 (its launch) to map any HABs in the vicinity of the focus sites. The MERIS sensor will be investigated in this capacity as well in order to derive higher spatial resolution maps.

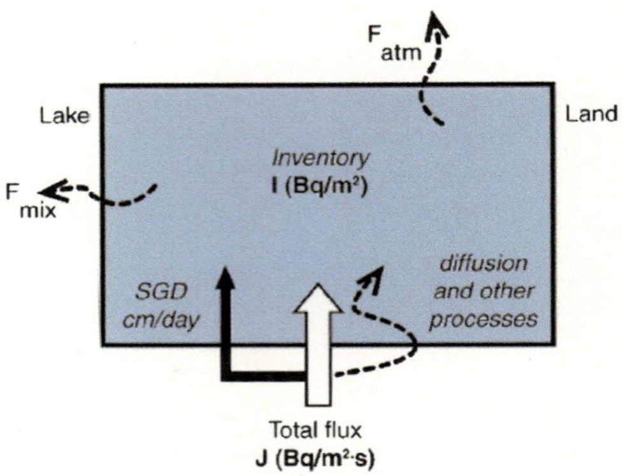

Fig. 2.7 Conceptual model of continuous radon measurements for estimating SGD. The inventory refers to the total amount of excess ^{222}Rn per unit area (Modified from Burnett and Dulaiova 2003; Burnett et al. 2006). It is possible to complement the Rn work with geophysical survey to select the right locations for the seepage meters arrays (transects)

2.3.3.5 Natural Radiotracer

Temporal/spatial patterns and magnitudes of SGD could be obtained using natural tracer ^{222}R. An advantage of this natural tracer is that the coastal water column integrates the tracer coming into the system via various groundwater pathways. Smaller-scale variations, which are not of regional interest, are smoothed out. This tracer approach is ideal for addressing the large spatial heterogeneity that is invariably associated with groundwater flow paths (Burnett et al. 2002, 2003, 2006; Swarzenski et al. 2002, 2006a, 2009). We can recommend using an automated radon monitor (RAD7, Durridge Co.) for continuous measurements over long time periods. The change in radon inventories over time could be used to estimate fluxes into and out of the coastal system. Assuming that benthic fluxes of radon are driven mainly by groundwater (pore water) advection, one can convert calculated ^{222}Rn fluxes to water fluxes by dividing by the radon concentration of the advecting fluid (Fig. 2.7).

The high temporal resolution of radon measurements provides the ability to examine the dynamics of groundwater seepage inputs (Burnett et al. 2006; Paytan et al. 2006). Temperature and biogeochemical parameters (nutrients, trace metals, pH, and redox conditions) could be monitored in parallel with the natural radioactive tracer (Swarzenski et al. 2009).

2.3.3.6 Assessment of SGD Processes and Landform Evolution upon Tsunami Impact

We could assess the spatial distribution of SGD by performing joint qualitative geophysical, remote sensing and radon surveys (Fig. 2.8) and a preliminary screening of SGD water quality, using a Trident subsurface seepage monitor and temporary drive-point piezometers. These tools will reveal areas where SGD is more prevalent so we can target these areas for monitoring, characterization, and modeling studies. The radon surveys could be done during spring and late summer. To evaluate the forces that control SGD and associated constituent loading, time-series measurements could be made at several work sites. These sites may be refined based on the survey and

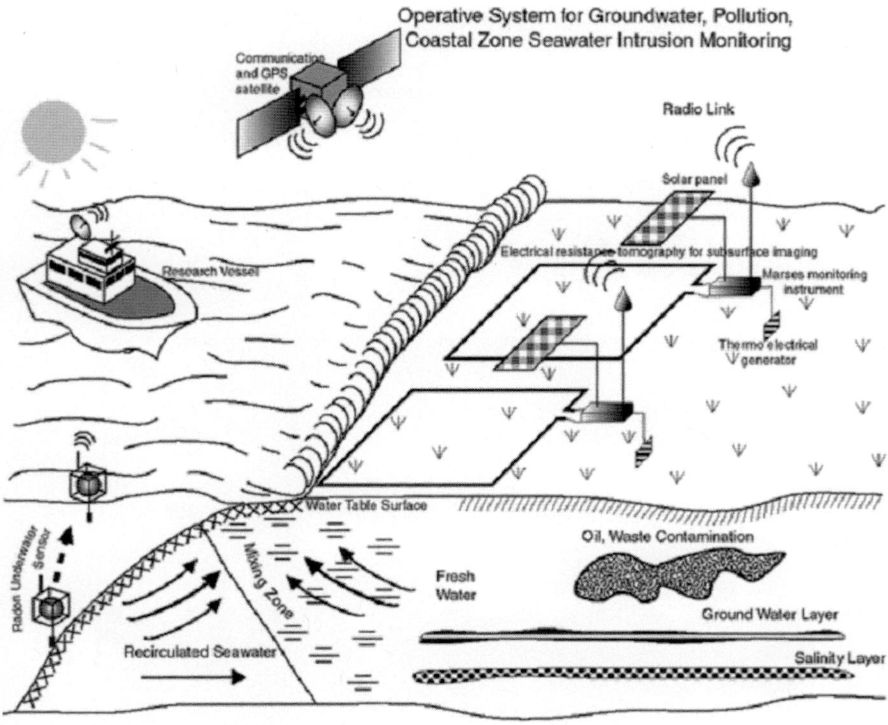

Fig. 2.8 Operational system for assessing SGD and coastal aquifers using jointly geophysics, remote sensing and tracer measurements (Modified from Lobkovsky et al. 2003)

existing hydrogeology data. We could measure all parameters (e.g., rainfall, aquifer water level, discharge fluxes, lake and wave levels, and geochemistry) relevant to characterization of the driving forces. The methodology of jointly using geophysics, remote sensing and tracer measurements are now well established (Lobkovsky et al. 2003) so that sound assessments could be made.

Since different drivers operate on different time scales, deconvolution of various forcing functions could be performed. Field experiments could also employ ^{210}Pb as a tracer together with manual and automatic seepage meters (Boehm et al. 2005; Swarzenski et al. 2009).

2.3.4 Measurements by Seepage Meters and Benthic Chambers

Measurements of SGD rates into surface water bodies can be made using manual seepage meters (Figs. 2.9 and 2.10). The seepage meter drum forms an open bottom chamber that is inserted into the sediment (Lee 1977; Burnett et al. 2006). Water seeping through the sediment displaces water trapped in the chamber forcing it up through the port into a plastic bag.

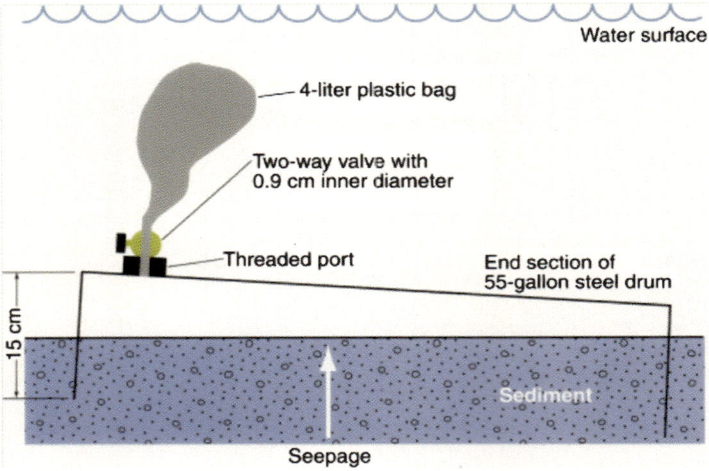

Fig. 2.9 Diagram of a manual seepage meter (Modified from Burnett et al. 2006)

Fig. 2.10 Measurement and sampling SGD using seepage meters during the low tide period

We could use seepage meters designed to be resistant to waves (Swarzenski et al. 2009). The bags will be isolated from wave action and will have large valves to minimize fluid resistance. Automatic electromagnetic seepage meters (EM) could be also used (Taniguchi et al. 2006, 2008). SGD estimates from seepage meters could complement estimates from groundwater modeling (Burnett et al. 2006). Porewater geochemistry is important for SGD study, however, sediment porewater sampling in sand is problematic.

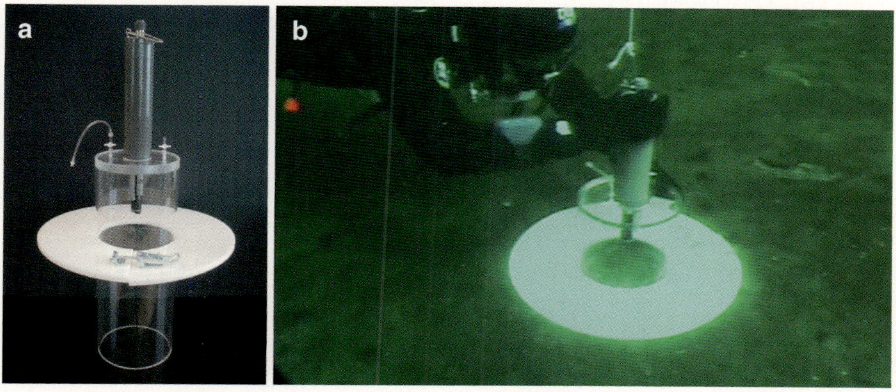

Fig. 2.11 (**a**) Benthic metabolism chamber; (**b**), Diver deployment of benthic metabolism chamber (Modified from Biddanda et al. 2009)

The change in volume of water in the bag over a measured time interval provides the flux measurement and a sample of groundwater (Lee 1977; Burnett et al. 2006).

Also for SGD measurements we could use a drive point piezometer (Burnett et al. 2006; Povinec et al. 2012) with a short (2 cm) screened interval. The piezometer could be driven to progressively greater depths to develop a geochemical profile. Piezometer development could be done concurrently with the geophysical surveying and modeling.

A benthic metabolism chamber (Fig. 2.11) fitted with sounds to measure a number of variables in the benthic layer, including: temperature, conductivity, salinity, dissolved oxygen, pH, turbidity, depth, and benthic biological productivity (production and respiration) at the different sites along transects also can be used (Biddanda and Cotner 2003; Biddanda et al. 2009).

2.3.4.1 Biogeochemical Measurements

Because SGD can be an important mechanism for nutrient delivery to the coastal waters, we recommend to construct a detailed nutrient budget for the selected contrasting sites of the coastal zone within the tsunami affected area. Potential sources of nutrients to this area include surface water inflow, atmospheric deposition, and diffusion across the sediment/water interface, and groundwater seepage. It is possible to follow the general approach of (Biddanda and Cotner 2003; Biddanda et al. 2009) for assessing SGD-associated nutrient contributions to coastal waters. Nutrient concentrations could be measured in groundwater samples collected from the seepage face from an array of piezometers at both sites, and along transect of seepage meters offshore. It is also possible to sample at various points within the coastal aquifer to assess possible input and removal processes. For example, nutrients and trace metals associated with SGD may be significantly modified by redox-dependent transformations, mobilization, and removal processes in the mixing zone. Total N, nitrate, nitrite, ammonia and DON, inorganic and total phosphate, silica, and DOC can be measured via a flow injection autoanalyzer and a total C analyzer and a suite of trace metals (Cd, Ba, Pb, Cr, Fe, Mn, Cu, V, Co, Ni, Zn, and U) could be measured by ICP-MS.

2.3.4.2 Biological Measurements

Mussels could be sampled for density of adults and larval settling rates at each of the sampling sites. Density of adults can be measured by collecting 10 cm square cores selected at random from a 1 m transect frame place randomly within a mussel colony as described in detail in (Biddanda and Cotner 2003). In addition it is possible to measure larval settling rates using microscope slide racks suspended at 3 and 5 m depths (Biddanda et al. 2009). Mussel's larvae prefer to settle on hard substrates, and microscope slides present a convenient surface to collect larvae and then examine on a microscope. Slides could be collected from each site at regular intervals throughout the reproductive season. Ancillary data to be collected could include groundwater seepage rate, direction and rate of near-bottom current, temperature, Secchi depth, water hardness, adjacent substrate type, and identification of other organisms settling on slides. An appropriate diversity index and biotic index can be calculated to compare how SGD from the impacted and non-impacted site may be affecting their water quality and ecological health. Results could also be compared to the data set of benthic invertebrates collected before tsunami. These studies can be supported by underwater photography that could be used to chronicle the changes in benthic community composition and differences in the mussel population densities under influence of tsunami on SGD.

The SGD assessment methodology could use an innovative approach to synthesize food webs and dreissenid mussel dispersal and colonization in tsunami affected areas under influence of contaminated SGD (fluxes and pathways) through fusing (combining) different types of data (biological, biogeochemical, ecological, radiological, geophysical, hydrological, remote sensing and hydrological) which in the future could be used in biogeochemical numerical models.

Reliable estimates of the dreissenid mussel dispersal and colonization and SGD are difficult to obtain by traditional methods due to limited data availability and large uncertainty of the data and models. Since these data are measured in different units ("apples and oranges") and no single-objective metric (distance between data and estimate as well as distance between data and models) can be introduced, our approach can quantify data uncertainty in terms of fuzzy sets and search for the optimal estimate of the dreissenid mussel dispersal and SGD through multi-objective optimization. This approach is not traditional assimilation data procedure.

Model results (independently obtained with and without data assimilation) could be used as independent prior knowledge on the dreissenid mussel dispersal and SGD in tsunami affected areas and fused with other data series. This innovative approach (a) provides more accurate representation of data uncertainty than traditional mean and auto-covariances because its specification requires considerably less data; (b) does not require a priori knowledge on statistical weights to fuse data of different types. Any statistical weights are subjectively given that introduces arbitrariness in the estimate; (c) allows for fusing incomparable data, i.e. data or pieces of same data which contradict one to another due to different spatial resolution or biases; (d) allows utilizing non-traditional information such as digital video films, expert opinions etc.; (e) it is preferable due to factors, such as low cost, high reliability, and ease of use.

This innovative approach can also be combined with de-trended fluctuation and multi-fractal analyses to quantify essential statistical features (correlations, scaling and non-linearity) in time series of biological, hydrological and biogeochemical characteristics to understand underlying dynamic processes and find significant correlations between tsunami, SGD, anthropogenic forcing, and dreissenid mussel dispersal and colonization.

Biogeochemical and various remote sensing, hydrogeological and geophysical measurements (see details above) could be made in surface/pore waters (specifically nutrients and trace metals). The availability of groundwater dissolved organic carbon (DOC) could be assayed (Kontar et al. 2012a, b). Biological responses (benthic algal abundance, sediment bacterial abundance, and bacterial activity) to SGD could be measured as well. The field work and SGD experiments could enhance our understanding and predictive capability of SGD in nearshore waters of tsunami affected areas (see Table 2.1).

2.4 Development of New Modelling Approach

For the numerical simulations, we chose the coastal zone of south eastern Taiwan (Fig. 2.12), where areas with SGD are likely to be widespread off the shore. For the last decade, intensive field investigations of coastal environment pollution associated with SGD were undertaken in coastal zone of Taiwan including its southwestern part. It was found that aquifers are significant sources of trace elements and other chemical constituents to the southwestern coastal waters of Taiwan where inland chemicals as NH_4, SO_4, NO_3, PO_4 and SiO_2, hydrocarbons, trace metals (Ni, Cu, Cd), NaOH, fertilizers, and sewage (Wang et al. 2009). Recent observations has discovered SGD phenomena in area east of Gaoping (submarine) Canyon at depth of 30 m (Zavialov et al. 2012).

The marine environment in this area is rather unique in terms of geomorphology. A major feature of the latter is Gaoping Canyon, belonging to the river extension type canyon, and Siaoliuciou Island. The canyon is well aligned with the Gaoping River on land, runs across the continental shelf, continues its course south westward onto the continental slope, and terminates at a depth of about 3,000 m. The head of the canyon, cutting the continental shelf, is characterized by high and steep walls and complex cross-sectional morphology (Korotenko et al. 2012). As field

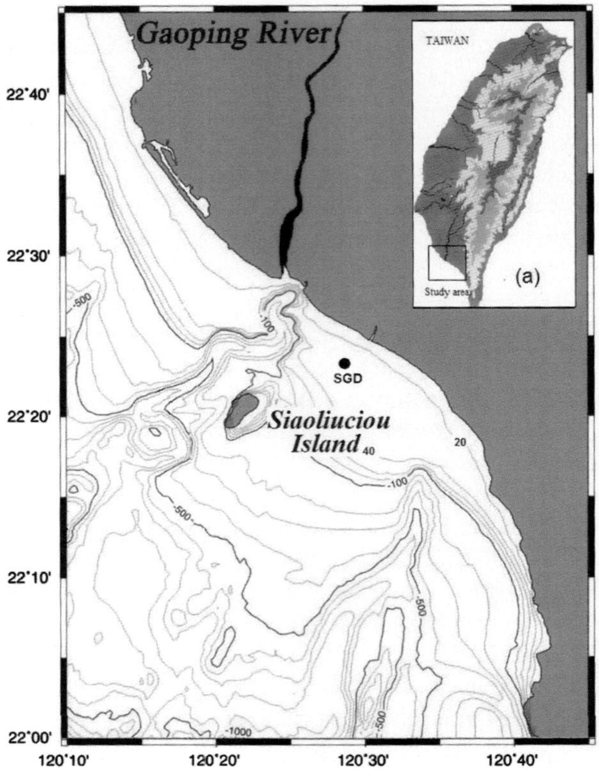

Fig. 2.12 Map of the Pingtung coastal zone of the southwestern Taiwan showing the bathymetry off southwestern Taiwan and the shallow SGD well marked by the solid circle (Modified from Korotenko et al. 2012)

observations exhibited (Wang et al. 2008, Lee et al. 2009), such peculiarities in geomorphology have a great effect on coastal circulation, mixing processes and devastating results caused by tsunamis due to the effect of wave shoaling when surface waves entering shallower water increase in wave height.

Regions of southwestern and eastern Taiwan are potential tsunami affected areas due to active seismic zones with frequent earthquake events. A recent earthquake with magnitude 5.5 occurred on 20 January 2013, 75 km east from Suao (northwestern Taiwan), approximately 3 months before, a 4.5-magnitude earthquake occurred near Tungkang (southwestern Taiwan). A powerful 6.0-magnitude earthquake happened in 1950 when significant wave height reached 10 m. Fortunately there was no deadly tsunami in the last 100 years in Taiwan, but the potential threat is ticking. That is why raised public awareness and education is the most urgent task to minimize possible tsunami disaster damage and to understand possible contamination of coastal waters by various potential sources including SGDs triggered by tsunami.

In this work, using the Lagrangian particle tracking model coupled to a hydrodynamic model we attempt to elucidate an effect of SGD carrying contaminants on the coastal water contamination and its dependence on external forcing.

2.4.1 Coupled Model Description

Modeling the transport and dispersal of contaminant-containing SGD from a coastal aquifer into the adjacent shallow marine environment affected by tsunami presents a real challenge to create a model. Difficulties are associated with many factors such as the hydrostatic instability of bottom fresh water discharging into the saline environment, uncertainties of the SGD rate estimate, seepage location(s) as well as difficulties predicting coastal currents induced by variable forcing (tsunami waves, wind, tides, etc.). All these factors are necessary to take into account in deciding on the choice of a particular numerical model and approach. Our approach to the solution of the transport and dispersal of contaminant-containing fresh groundwater is based on the Lagrangian particle tracking method (LPTM) that is significantly more effective than any other approaches based on finite-difference models since LPTM describes the advective transport with a high accuracy. This is important for realistically reproducing the movement of a plume of contaminants. In addition, the Lagrangian method is better suited to modeling the fate and dispersal of contaminants, when the attribution of fate properties to a particular contaminant is required.

As any other similar coupled models, our consists of a hydrodynamic model embedded into a particle transport model (PTM). PTM takes currents and turbulent diffusivities predetermined by the hydrodynamic model and uses LPTM to predict the motion of individual particles, the sum of which constitutes a contaminant plume. The basic concept of the proposed model is similar to that presented earlier (Korotenko 1999; Korotenko et al. 2000, 2004, 2010; Sentchev and Korotenko 2004, 2005, 2007; Korotenko and Senchev 2008). Therefore, herein, we shortly discuss only those extensions/improvements of the model that are made for the present specific problem associated with SGD into a coastal zone.

Generally, the procedure of predicting behavior and fate of a contaminant plume is divided into two parts: (i) predetermination of currents, turbulent diffusivities with the hydrodynamic model; and (ii) applying the predetermined three-dimensional motions to individual particles, the aggregate of which constitutes the plume. The model thus simulates the advection and turbulent diffusion of particles. The physicochemical decay and processes modifying the structure and properties of a contaminant plume are simulated with a use of specific algorithms.

2.4.2 Hydrodynamic Module

To predict circulation, we chose the sigma-coordinate Princeton ocean model (POM) (Blumberg and Mellor 1987) and adapted it for the abovementioned area of south-eastern Taiwan. The POM has a resolution of 1/60 Deg along longitude (Δx) and latitude (Δy). The calculation domain, 39 $\times$ 72 cells, covers the region of interest. Over the vertical, the model has 31 uneven σ-levels, which are distributed to provide the maximal resolutions near the surface and the bottom. The calculation of time steps for the external, *DTE*, and for the internal, *DTI*, modes were chosen from the CFL criterion of stability and were equal to 6 s and 120 s, respectively.

For the momentum fluxes in the near-bottom layer $(1 + \sigma_{kb-1})H/z_0$, we used the quadratic dependence on the velocity: $K_M(\partial\overline{u}_i/\partial\sigma)/D = C_d\left[\overline{u}_1^2 + \overline{u}_2^2\right]^{1/2}(\overline{u}_i)$ for $\sigma \rightarrow -1$, where $H(x, y)$ is depth, $D=H + \eta$, $\eta(x, y)$ is the elevation of the free surface, K_M is the coefficient of turbulent viscosity, z_0 is the roughness parameter, $\overline{u}_i$ $(i = 1, 2)$ are the horizontal components of the mean current velocity, and drag coefficient, C_d is defined as $C_d = MAX\left[\frac{\kappa^2}{[\ln\{(1+\sigma_{kb-1})H/z_0\}]^2},\ 0.0025\right]$, where $k = 0.4$ is the Karman constant.

Open boundary conditions. At the open (western and southern) boundaries of the calculation domain, two types of boundary conditions for the temperature and salinity were used: the condition for the inflow and that for the outflow. In the case of the outflow outside the calculation domain, the values for T and S at the corresponding open boundary were setup. In the case of the inflow into the calculation domain, the radiation equation $\partial(T,S)/\partial t + \overline{u}_n\partial(T,S)/\partial n$ was solved (Korotenko and Senchev 2004). The index n, here, represents the coordinate normal to the open boundary. The components of the mean current and the turbulent diffusivities were output to the particle transport model at every time step during the simulations.

Forcing. Since the effect of wave shoaling is extremely difficult to parameterise, in this study we limited our simulations to the elucidation of effects of tidal current on spreading contaminants incoming with SGD. Note that effect of wind forcing was studied recently for favourable seasonal winds by Korotenko et al. (2012). For the tidal forcing, we analysed the ratio between amplitude of tidal oscillations of coastal currents along south western Taiwan induced by the astronomical semidiurnal, M_2, S_2 and diurnal, K_1, and O_1 constituents. They showed that the semidiurnal, M_2, constituent was the major one in energy spectra of tidal oscillation. Thus, for the first order approximation, we can simplify tidal forcing by a use of the dominant constituent M_2 only and, for tuning the circulation module, we utilized current velocity measurements conducted by Zavialov et al. (2012) near by the SGD well. We applied a radiation condition based on the long gravity-wave speed, to determine the boundary elevation and a gravity-wave radiation condition to specify the normal component of the depth-averaged current at the model open boundaries.

POM initialisation. The hydrodynamic model is spun up from rest and with temperature and salinity taken from the ocean climatological database (Levitus 2009) with adjusting T, S- profiles, in the water column over the SGD.

2.4.3 Particle Tracking Model

In the Lagrangian particle tracking method, an algorithm for updating particle coordinates (Eq. 2.1) is the following: at every time step, particles are moved in the 3-dimensional Cartesian reference frame by an advective displacement added to a diffusive jump:

$$(\Delta x_i)_{j,k} = \overline{u}_{i,j}\,\Delta t + (\eta_i)_{j,k} \quad (i = 1-3;\; j = 1,2....,N_t;\; k = 1,2,\ldots,N_p) \qquad (2.1)$$

The displacements, $(\Delta x_i)_{j,k}$, are defined as sum of a deterministic displacement caused by mean velocities, $\overline{u}_{i,j}$, and a random displacement, $(\eta_i)_{j,k}$, due to fluctuations of velocity. Here N_t is the number of time steps, Δt is the time step and N_p is the total of particles released during a numerical experiment.

The advective movement within a grid cell is determined by linear interpolation of the velocity values from the 8 nodes of the grid cell, and computation of the displacement vector is a product of the interpolated velocity vector and the time step Δt. Diffusive jumps of particles (random displacement due to sub-grid fluctuations of velocity) along horizontal (i = 1, 2) and vertical (i = 3) axes are determined differently. For the horizontal axes, we used so-called 'naïve random walk' (NRW) scheme, i.e., $\eta_i = \gamma_i(2K_{i,j}\Delta t)^{1/2}$ to simulate diffusive jumps. The random vector, γ_i, normally distributed with an averaged value of zero and unit standard deviation is converted later to yield the Gaussian distribution with zero mean and unit standard deviation. Coefficients $K_{i,j}$ represent time dependent diffusivities along the i- axis. Such simple scheme for the lateral transport and dispersal of particles is feasible due to a weak variation of horizontal diffusivities, $K_{i,j}$, along the correspondent axes.

Unlike $K_{i,j}$, profiles of the vertical diffusivity, $K_{3,j}$, usually exhibits significant variations in coastal waters where current and density structures are formed under tidal and wind-driven circulation (Korotenko et al. 2013), and, often, under strong influence of freshwater input (Korotenko 1999; Sentchev and Korotenko 2005). Such combined forcing leads to the formation of non-uniform vertical diffusivity profiles that, in case of the use of the NRW scheme, can form artificial particle accumulation zones in layers with weak vertical mixing. To avoid this effect, we employed so-called a 'consistent random walk' (CRW) approach (Hunter et al 1993; Visser 1997; Sentchev and Korotenko 2005) in order to obtain vertical particle displacements correctly. For this, we applied the formula $\eta_i = K_3'(z)\,\Delta t + \gamma_3[2K_3(z^*)\Delta t]^{1/2}$, adopted from Visser (1997). As seen, this formula includes deterministic and diffusive (or random) components. The deterministic component causes a net displacement of the centre of mass of the neutrally buoyant particles toward increasing diffusivity at a rate K_3' (a local gradient of K_3 in the vertical direction), thus allowing avoidance of the artificial particle accumulation within layers of low vertical diffusivity. The diffusion coefficient K in the CRW model is estimated from the diffusivity profile at a vertical coordinate z^* shifted from the particle coordinate z by a small distance $0.5K_3'\Delta t$. Note that the CRW model should be used for simulating horizontal displacements too. However, according to our assessment, the largest horizontal diffusivity gradient, in the coastal zone studied, is the order of 10^{-3} m s^{-1} (cf. $K'_{3\max} \approx 10^{-1}$ m s^{-1}) so that the effect of $K'_{1,2}$ on horizontal distribution of particles is negligible for a given model grid spacing and time step (see below).

As was mentioned above, the horizontal and vertical diffusion coefficients, $K_{1,2} = K_H$ and $K_3 = K_V$ as well as the mean current velocity components, $\overline{u}_{i,j}(x, y, z, t)$, and density, $\rho(x, y, z, t)$, are provided by the POM. The horizontal diffusivities are computed with a use of Smagorinsky formula and the vertical diffusivity is obtained from the level–2.5 turbulence closure scheme (Mellor and Yamada 1982). Thus, 3-dimensional parameters available from the POM are used as forcing in the particle tracking model, which estimates particle coordinates at every time step Δt that can be equal to (or longer than) the internal time step, *DTI*, of POM. The time steps is chosen to be 360 s that is long enough but prevents particle jumping more than one grid cell, and, thus, guarantees an accurate estimate of particle displacement in each of three directions.

It should be noted that, based on the random walk concept, particle tracking models coupled with the hydrodynamic models have certain structural features associated with the ratio between model time step, *DTI* (Δt), and the Lagrangian timescale, T_{Lag}. For the vertical motion, the external timescale for small-scale turbulence in the ocean, T_{turb}, is of the order 1–10 sec, so that $\Delta t \geq DTI >> T_{turb}$. It means that vertical velocity fluctuations and displacements of a Lagrangian particle at the time lag Δt are not correlated we can use Eq. 2.1 for predicting vertical ($i = 3$) particle coordinates. In contrast, for horizontal currents in the ocean, the typical value of the Lagrangian timescale T_{Lag} is 1–3 days, so that $\Delta t << T_{lag}$ and the random horizontal velocities and displacements of a Lagrangian particle within the time lag Δt are correlated. For this reason, the horizontal motion ($i = 1, 2$) of the particle is described by equations

$$\overline{u}_{i,j+1} = \overline{u}_{i,j}\left(1 - \Delta t / T_{lag}\right) + \sigma_a A_i \Delta t, \tag{2.2}$$

$$x_{i,j+1} = x_{i,j} + \overline{u}_i\left(x_{i,j}, z_i\right)\Delta t + \left(\overline{u}_{i,j+1} + \overline{u}_{i,j}\right)\Delta t/2, \tag{2.3}$$

where σ_a is the root mean square random acceleration, A_i is the normally distributed random vector value with zero mean and unit variance, $x_{i,j}$ and $x_{i,j+1}$ are the horizontal radius-vector of the particle at moments t and $t + \Delta t$, respectively. Note that (2) is a finite difference analogue of the Langevin equation. Note that, from POM, we have explicitly only one parameter to describe horizontal random motion of a Lagrangian particle, namely the horizontal diffusivity, K_H, while Eqs. 2.2 and 2.3 include two parameters, T_{Lag} and σ_a. However, taking into account that POM uses the Smagorinsky formula for the horizontal diffusivity, K_H:

$$K_H = C_H \Delta x \Delta y \left[(\partial \overline{u}_1 / \partial x_1)^2 + 0.5(\partial \overline{u}_1 / \partial x_1 + \partial \overline{u}_2 / \partial x_2)^2 + (\partial \overline{u}_2 / \partial x_2)^2\right]^{1/2}$$

where C_H is a constant, we can infer the Lagrangian timescale as $T_{Lag} = \Delta x \Delta y / K_H$ and the random acceleration as $\sigma_a = (2(K_H)^3/\Delta t)^{1/2}/\Delta x \Delta y$ [26] and thus close the system of Eqs. 2.2 and 2.3.

Definition of the source. A very important step for the model setup is to set correctly the near-seepage zone processes and the discharge rate of contaminant-rich SGD. Submarine groundwater coming into the seawater from coastal aquifers carries dissolved contaminants, type and properties of which depend on regional factors. Contaminants may be considered as passive and either conservative or nonconservative tracers but, being dissolved in fresh groundwater, such tracers would have vertical velocity due to the buoyancy force. The magnitude of the rise velocity depends on difference between densities of SGD and adjacent saline water.

In the particle tracking approach applied, we presume that submarine groundwater discharging from an aquifer breaks up into fresh water droplets with diameter, d, defined by empirical formula

$$d = 9.52\nu^{2/3}/\left(g^{2/3}(1-\rho_0/\rho)^{1/3}\right),$$

where ρ_0 is density of SG droplets, ρ and ν is density and viscosity of ambient water, respectively and g is the acceleration of gravity. The terminal velocity of the freshwater droplet can be estimated as $w = [8gd(1 - \rho_0/\rho)/3]^{1/2}$. Thus, the magnitude of rise velocity grows with the droplet diameter and the ratio ρ_0/ρ. For $d = 1$ mm, the rise velocity is about 0.25 m s^{-1} and thus in shallow waters (~ 30 m), SG droplets may appear at the sea surface in a several minutes after their release.

In the work, we applied a 'stationary' source of particles, i.e., a certain number of 'particles-in-SGW' (cluster), N_j, was released every Δt seconds at the position (X_0,Y_0) of seepage at the bottom. Amount of particles in a cluster N_j is set to prescribe the initial concentration $C_0 = (N_j/\Delta V_0)$ in the cell- volume $\Delta V_0 = \Delta x \Delta y \Delta z$.

Two types of particles were used in numerical experiments: (i) conservative particles that mimicked 'long-life' tracers or contaminants (as ^{222}Rn, ^{226}Ra, ^{3}H, ^{4}He) without significant attenuation in their concentration due to physico-chemical reactions; and (ii) nonconservative particles that emulate 'short-life' contaminants experiencing a relatively rapid decay in result of their own properties and reactions. As an example of a nonconservative contaminant we will take nitrate (NO_3) that, as recognized now, is the major contaminant of SGD. According to measurements by Zavialov et al (2012), we can set the initial concentration of nitrate as ($C_0 =$ 0.48 mg L^{-1}.), The reduction of nitrate due to denitrification and other reactions can be described by a first-order degradation process with an exponential decrease in concentration $C = C_0 \exp(-\ln 2/\lambda)$, where λ is 'half-life' parameter. In the particle tracking approach, effects of degradation can be treated in a similar way as radioactive decay using e-folding times $T_e = 1/\lambda$ for the probability of removal of particle in a time step. According to research by Coering and Cline (1970) the process of denitrification in seawater spans about 100 h and thus, in the model, we can set T_e to be 30 h and use procedure of randomise half-life parameter as was described in Korotenko et al. (2004, 2010).

Particle release. To simulate a continuous source, particles are launched every *DTI* (180 s) time step coordinates of the source could be chosen at the site of SGD. In the numerical experiments with nitrate, every time step $\Delta t = DTI$, we launched 15480 particles to fit to the initial concentration of 0.48 mg/L^{-1}. We limited our simulation by short-term effects of simulated currents on the development of particle plumes. Therefore in numerical experiments with tidal forcing, duration of simulations was chosen to be 72 h to study behaviour of a nitrate plume in oscillating current and its effect of nitrate concentration variability.

2.5 Results of Modelling and Discussion

The groundwater discharging into the sea often contains nitrate with high concentrations. However, denitrification and dilution of nitrate apparently reduces its concentration. As measurements indicated by Coering and Cline (1970), the attenuation of nitrate concentration in the seawater proceeds quite rapidly. Of interest here is to estimate scales of the nitrate contamination of adjacent seawater nearby the SG well as well study an influence of tidal circulation on the distribution of nitrate in the impacted zone. For this, we forced the hydrodynamic model with dominant constituent M_2 and tune the model to adjust computed current velocity to that obtained in the field observations (Zavialov et al. 2012).

Figure 2.13 shows the integral[1] nitrate concentration $C(x, y)_H$ formed under tidal circulation. As seen, the area occupied by particles is stretched along the coast. An asymmetry in distribution of particles caused by the northwestward residual current is clearly seen. To show a detailed structure of the nitrate plume, in Fig. 2.14, we present its successive phases developed in tidal current for periods of flood, slack water and ebb.

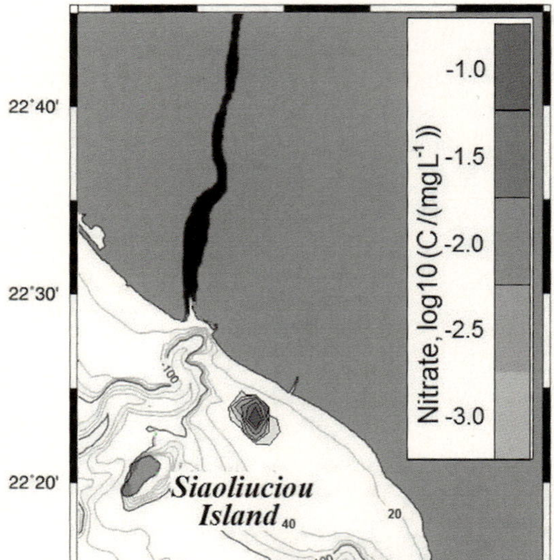

Fig. 2.13 An integral concentration of nitrate formed by tidal circulation after 24 h of discharge (After Korotenko et al. 2012)

---

[1] $C(x, y)_H = (N/\Delta V)$, where N-number of particles, $\Delta V = \Delta x \cdot \Delta y \cdot H$ and $H(x, y)$ is depth.

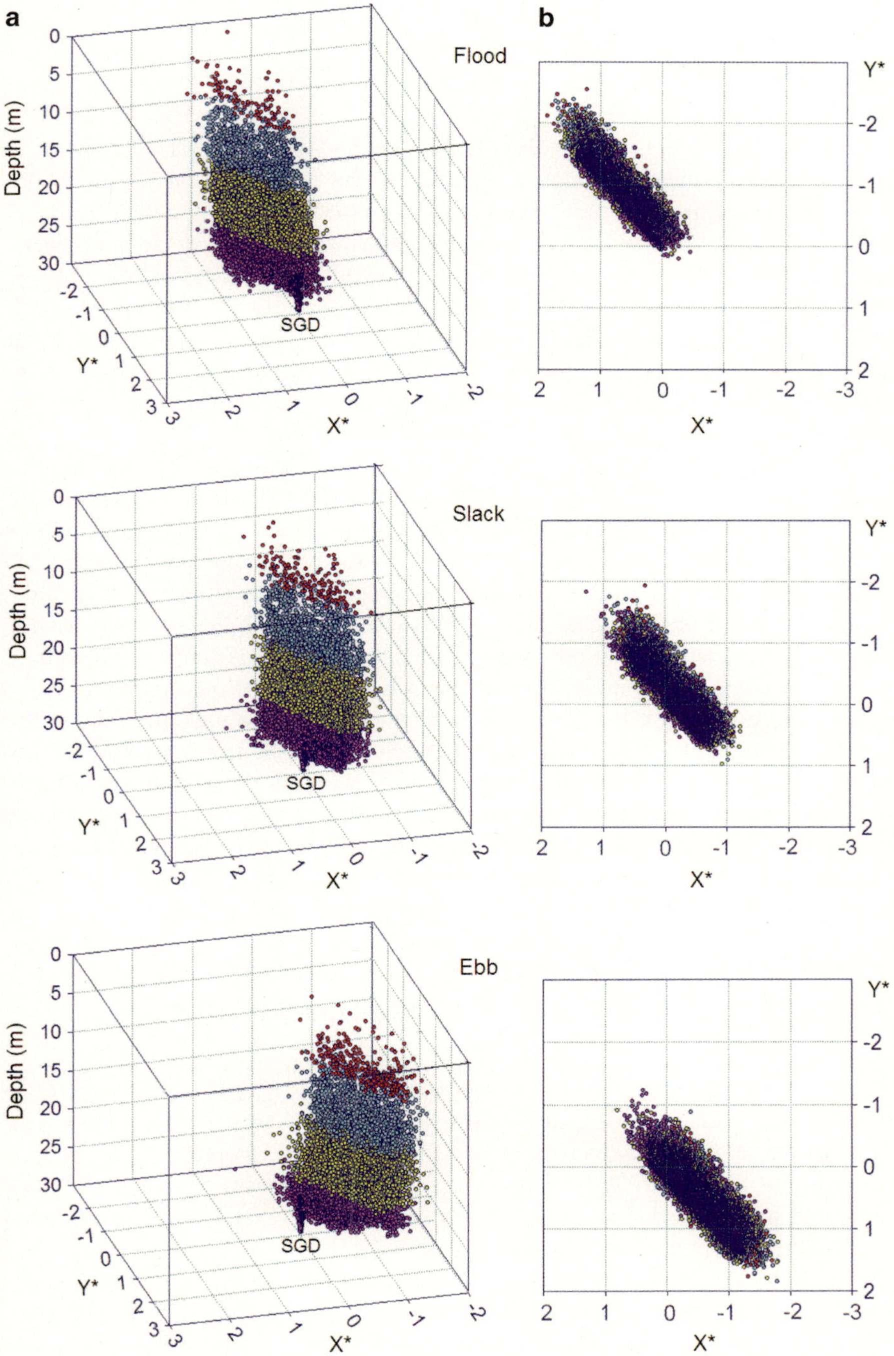

Fig. 2.14 Successive phases of the 3-D structure (panel a) and planar structure (panel b) of the nitrate plume in tidal current for periods of peak flood, slack water and peak ebb. Dimensionless latitudinal X* and meridional Y* coordinates are given in number of grid cells from the SGD source, i.e., $X^* = X - X_0/\Delta x$ and $Y^* = Y - Y_0/\Delta y$, where (X,Y) is position of a particle, (X_0, Y_0) is position of the source. Particle colour denotes the vertical range of particle position: *red*, *blue*, *yellow* and *magenta* correspond to 0–7.5, 7.6–15, 15.1–22.5 and 22.6–30 m, respectively

2.6 Conclusion

An assessment of SGD-originated contamination impact on ecosystems requires in tsunami affected areas a complex approach to the problem including fieldwork sampling, experimental study and numerical modeling. Until now, numerical modeling focused on the solution of problems associated with contaminant distribution inside aquifers and only a few works focused on the processes near seepage zones. In the paper, we presented a new methodology which combined fieldwork and data collection with the 3-D coupled circulation/particle transport model that could predict coastal circulation and its effect on the transport and dispersal of particles mimicking conservative and nonconservative contaminants discharging in seawater as the result of seepage of submarine groundwater.

We focused on assessment of SGD using a new modelling approach to investigate a nitrate plume formation as a result of contaminated SGD in the coastal zone of Taiwan. Given the dynamics of SGD in tsunami affected areas it is extremely complicated and exhibits non-linear effects of the interaction of tsunami, tidal and wind/wave-induced currents with bottom topography, and development of a sophisticated model for predicting all these effects is an extremely difficult task. That is why we limited our simulations to elucidate the effect of tides on the SGD/nitrate plume formation. The simulations have shown behaviour of the plume, its shape and variation during a tidal cycle. For discharge rate chosen, nitrate could be detected in area of a few miles around the SGD well. It is worthwhile to note that, tsunami aftermath, impact on aquifers in shallow waters and, in turn, on marine environment due to contaminating groundwater appears to be much longer than impact during the event, hence, the developed model, as it is, could be used to estimate consequences after tsunamis.

2.6.1 *Future Work: Critical Issues and the Way Forward*

As any other, our model requires further improvements that particularly associate with realistic tsunami forcing and correct initialisation as well as the consideration of various contaminants incoming with SGD, their specific properties and reactions (Kontar and Korotenko 2013). Also we are looking forward to develop a more efficient and relatively inexpensive new generation of the in-situ and remote sensing techniques to detect and describe SGD (Kontar et al. 2012b).

Using the 2011 Tohoku-oki tsunami in Japan as a case study, we plan to conduct field work to assess SGD in tsunami affected areas. The main objective of this study is to better understand, quantify, forecast, and protect water resources in tsunami affected coastal ecosystems. The interdisciplinary project team regroups investigators from 3 US and 14 international research institutions (Australia, England, France, Italy, Japan, New Zealand, Spain and Russia). Building on recent research results in tsunami and groundwater flow modeling, we plan to use new sampling techniques and observations, secured via the project and some international SGD programs to advance the study of SGD in tsunami affected coastal areas (Kontar et al. 2012a). This will be accomplished by developing and applying the next generation of field sampling, predictive model simulations, laboratory and remote sensing techniques, to studying, forecasting and mitigating (in future events) tsunami/groundwater-seawater interaction processes and their socio-economic impact.

Acknowledgements Dr. Kontar would like to thank the members of the AGU Natural Hazards Focus Group, IUGG GeoRisk Commissions and the International Commission for Groundwater-Seawater Interactions for their useful suggestions and discussion of this study. Dr. Korotenko would like to acknowledge that this work was supported in part by the bilateral Taiwanese-Russian Project "Monitoring, Assessment, and Management Implications of Submarine Groundwater Discharge in Taiwan" between P. P. Shirshov Institute of Oceanology, Russian Academy of Sciences and Tainan Hydraulics Laboratory, National Cheng Kung University, Taiwan. Both authors are grateful to the anonymous reviewer for useful comments which helped us improve the manuscript.

References

Andersen MS, Baron L, Gudbjerg J et al (2007) Discharge of nitrate-containing groundwater into a coastal marine environment. J Hydrol 336:98–114. doi:10.1016/j.jhydrol.2006.12.023

Biddanda BA, Cotner JB (2002) Love handles in aquatic ecosystems: role of dissolved organic carbon drawdown resuspended sediments and terrigenous inputs in the carbon balance of Lake Michigan. Ecosystem 5:431–445

Biddanda BA, Cotner JB (2003) Enhancement of dissolved organic matter bioavailability by sunlight and its role in the carbon cycle of Lakes Superior and Michigan. J Gt Lakes Res 29:228–241

Biddanda BA, Nold SC, Ruberg SA (2009) Great lakes sinkholes: a microbiogeochemical frontier. EOS 90(8):61–62. doi:10.1029/2009EO080001

Black FJ, Paytan A, Knee KL, de Sieyes NR, Ganguli PM, Gray E, Flegal AR (2009) Submarine groundwater discharge of total mercury and monomethyl mercury to central California coastal waters. Environ Sci Technol. doi:10.1021/es900539c

Blumberg F, Mellor GL (1987) A description of a three-dimensional hydrodynamic model of New York harbor region. J Hydrol Eng 125:799–816

Boehm AB, Paytan A, Shellenbarger GG, Davis KA (2005) Composition and flux of groundwater from a California beach aquifer: implications for nutrient supply to the surf zone. Con Shelf Res 26(2):269–282. doi:10.1016/j.csr.2005.11.008

Bokuniewicz H, Kontar E, Rodrigues M, Klein DA (2004) Submarine Groundwater Discharge (SGD) patterns through a fractured rock: a case study in the Ubatuba coastal area, Brazil. AAS Rev Asoc Arg Sed 11(1):9–16

Bokuniewicz H, Taniguchi M, Ishitoibi T, Charette M, Allen M, Kontar E (2008) Direct measurements of Submarine Groundwater Discharge (SGD) over a fractured rock aquifer in Flamengo Bay Brazil. J Est Coast Shelf Sci 76(3):466–472. doi:10.1016/j.ecss.2007.07.047

Burnett WC, Dulaiova H (2003) Estimating the dynamics of groundwater input into the coastal zone via continuous radon-222 measurements. J Environ Radioact 69:21–35

Burnett WC, Chanton J, Christoff J, Kontar EA, Krupa S, Lambert M, Moore W, O'Rourke D, Paulsen R, Smith C, Smith L, Taniguchi M (2002) Assessing methodologies for measuring groundwater discharge to the ocean. EOS 83(11):117–123. doi:10.1029/2002EO000069

Burnett WC, Chanton JP, Kontar EA (eds) (2003) Submarine groundwater discharge. The Netherlands Biogeochemy, vol 66. Kluwer, The Netherlands, p 202

Burnett WC, Aggarwal P, Aureli A, Bokuniewicz H, Cable JE, Charette MA, Kontar EA, Krupa S, Kulkarni KM, Loveless A, Moore WS, Oberdorfer JA, Oliveira J, Ozyurt N, Povinec PP, Privitera AMG, Rajar R, Ramessur RT, Scholten J, Stieglitz T, Taniguchi M, Turneret JV (2006) Quantifying submarine groundwater discharge in the coastal zone via multiple methods. Sci Total Environ 367:498–543

Cable JE, Corbett DR, Walsh MM (2002) Phosphate uptake in coastal limestone aquifers: a fresh look at wastewater management limnology. Oceanogr Bull 11:1–4

Charette MA (2007) Hydrologic forcing of submarine groundwater discharge: insight from a seasonal study of radium isotopes in a groundwater-dominated salt marsh estuary. Limnol Oceanogr 52(1):230–239

Chen C-TA, Wang S-L (1999) Carbon alkalinity and nutrient budget on the East China Sea continental shelf. J Geophys Res 104(C9):20675–20869

Church TM (1996) An underground route for the water cycle. Nature 380:579–580

Clement TP, Sun Y, Hooker BS, Petersen JN (1998) Modeling multi-species reactive transport in groundwater aquifers. Ground Water Monit Remediat J 18(2):79–92

Coering JJ, Cline JD (1970) A note on denitrification in seawater. Limnol Oceanogr 15:306–309

Cohen D, Person M, Wang P, Gable CW, Hutchinson D, Marksamer A, Dugan B, Kooi H, Groen K, Lizarralde D, Evans RL, Day-Lewis RD, Lane JW Jr (2010) Origin and extent of fresh paleowaters beneath the Atlantic continental shelf. New Engl Ground Water 48 (1):143–158. doi:10.1111/j.1745-6584.2009.00627.x

Day-Lewis FD, White EA, Johnson CD, Lane JW Jr (2006) Continuous resistivity profiling to delineate submarine groundwater discharge-examples and limitations. Lead Edge 25(6):724–728

Grimaldi CSL, Casciello D, Coviello I, Lacava T, Pergola N, Tramutoli V (2011a) An improved RST approach for timely alert and near real time monitoring of oil spill disasters by using AVHRR data. Nat Hazards Earth Syst Sci 11:1281–1291

Grimaldi CSL, Coviello I, Lacava T, Pergola N, Tramutoli V (2011b) A new RST-based approach for continuous oil spill detection in TIR range: the case of the deepwater horizon platform in the Gulf of Mexico. In: Liu Y, MacFadyen A, Ji Z-G, Weisberg RH (eds) Monitoring and modeling the deepwater horizon oil spill: a record-breaking enterprise, geophysical monograph series, vol 195. AGU/Geopress, Washington, DC, pp 19–31

Henderson R, Day-Lewis GFD, Harvey CF (2009) Investigation of aquifer-estuary interaction using wavelet analysis of fiber-optic temperature data. Geophys Res Lett 36:403–424

Hunter J, Craig C, Phillips H (1993) On the use of random-walk models with spatially-variable diffusivity. J Comp Phys 106:366–376. doi:10.1016/S0021- 9991(83)71114-9

IOC (2004) Intergovernmental oceanographic commission submarine groundwater discharge: management implications measurements effects. IOC Manuals and Guides. Paris (UNESCO), France, 44

Ismail A, Kontar Y, Smith E, Phillips A, Stumpf A (2011) Misleading interpretation of shallow seismic methods: three case studies from MASW, P-wave reflection and S-wave reflection surveys. In: Proceedings of SAGEEP, March 29–April 2, 2009, Fort Worth, CD-ROM edition

Kontar EA (2007a) Development of operational system for monitoring and studying groundwater discharge and seawater intrusion in coastal zones. IAEA-TECDOC-1595, pp 125–138

Kontar Y (2007b) Groundwater-seawater interactions in tsunami affected areas: solutions and applications. In: A new focus on groundwater-seawater interactions vol 312, IAHS, UK, pp 19–27

Kontar YA (2007c) Development of new approaches of coastal geo-hazards observation and warning systems. EOS, OSG23b-03, AGU fall meeting 2007, San Francisco, 10–14 Dec 2007

Kontar YA (2008a) Assessment of groundwater – surface water interaction in tsunami affected areas. In: From the watershed to the global ocean, AGU ASLO TOS Ocean science meeting, Orlando, 2–7 Mar 2008

Kontar Y (2008b) Hazards in the coastal zones. Symposium PEH-01: hazards: minimizing risk maximizing awareness. In: Beer T (ed) IYPE hazards megasymposium: The 33rd international geological congress, Oslo, 6–14 Aug 2008

Kontar Y (2011) Haiti earthquake aftermath: urgent action needed to improve scientific communication in the Caribbean region. UNAVCO COCONET workshop: community science, station siting, and capacity building, San Juan, 3–4 Feb 2011

Kontar Y (2012) International studies of hazardous groundwater/surface water exchange in the volcanic eruption and tsunami affected areas of Kamchatka. EOS, NH009-02, Control Number 1464091, AGU Fall Meeting 2007, San Francisco, 3–7 Dec 2012

Kontar EA, Burnett WC (1999) Study of groundwater discharge to the coastal zone and evaluation of potential earthquakes. In: Proceedings: international conference on marine environment the past present and future; National Sun Yat-Sen University, Kaohsiung, 26–28 Jan 1999

Kontar YA, Korotenko KA (2013) Assessing hazardous contamination incoming with submarine groundwater discharge in tsunami affected coastal areas. EOS, NH-02, Control Number 1667672, AGU Meeting of the Americas, Cancun, 14–17 May 2013

Kontar EA, Ozorovich YR (2006) Geo-electromagnetic survey of fresh/salt water interface in the coastal Southeastern Sicily. Cont Shelf Res 26(7):843–851. doi:10.1016/j.csr.2005.12.012

Kontar EA, Zektser IS (1999) Submarine discharge and its effect on oceanic processes in the coastal zone. Water Resour 26(4):459

Kontar EA, Burnett WC, Povinec PP (2002a) Submarine groundwater discharge and its influence on hydrological trends in the Mediterranean sea proceedings of the CIESM workshop: tracking long term hydrological change in the Mediterranean sea, Monaco, 22–24 Apr2002, pp 109–114

Kontar EA, Ozorovich YR, Salokhiddinov A, Azhigaliyev YB (2002b) Study of groundwater-seawater interactions in the Aral Sea basin. In: Proceedings of the international conference on low-lying coastal areas hydrology and integrated coastal zone management, Bremerhaven, 9–12 Sept 2002, pp 225–230

Kontar EA, Lobkovsky LI, Korotenko KA (2006) On risk assessment and sustainable industrial development of shelf zones. In: Ismail-Zadeh AT (ed) Recent geodynamics georisk and sustainable development in the Black Sea to Caspian Sea region, vol 825. American Institute of Physics, Melville, pp 84–94

Kontar EA, Ozorovich YR, Salokhiddinnov AT (2010) Hazards in the coastal zones related to groundwater-seawater interactions. In: Beer T (ed) Geophysical hazards-minimizing risk maximizing awareness: international year of planet earth. Springer, Science + Business Media, pp 179–194

Kontar YA, Swarzenski YP, Paytan A et al (2012a) Groundwater/surface water exchange in tsunami affected areas in Japan. OS07-A025 AOGS-AGU (WPGM) Joint assembly, Singapore, 13–17 Aug 2012

Kontar YA, Korotenko KA, Santiago-Fandiño V, Tamoyuki T (2012b) Assessment and prediction the transport and dispersal of contaminants incoming with submarine groundwater discharge in tsunami affected coastal areas in Japan. In: Extreme natural hazards and their impacts. IUGG SRC Conference, Chapman University, December 8–11, 2012, pp 121–122

Korotenko KA (1999) Matter transport in meso-scale oceanic fronts of river discharge type. J Mar Syst 24(1):85–95. doi:10.1016/S0924-7963(99)00080-9, 10.1016/S0924-7963(99)00080-9#doilink

Korotenko KA (2003) Warfare chemicals dumped in the Baltic Sea: modeling transport processes of pollution resulting from possible leakages. Oceanology 43:11–23

Korotenko KA, Senchev AV (2004) A on the formation of anomalies in the ichthyoplankton concentration field along the French coast in the eastern English channel. Oceanology 44(5):644–653

Korotenko KA, Senchev AV (2008) Effects of particle migration on the features of their transport by tidal currents in a region of freshwater influence. Oceanology 48:622–633. doi:10.1134/S0001437008050020

Korotenko KA, Mamedov RM, Mooers CNK (2000) Prediction of the dispersal of oil transport in the Caspian Sea resulting from a continuous release. Spill Sci Technol Bull 5/6:323–339. doi:10.1016/S1353- 2561 (01)00050-0, 10.1016/S1353-2561(01)00050-0

Korotenko KA, Senchev AV, Schmitt FG, Jouanneau N (2013) Variability of turbulent quantities in the tidal bottom boundary layer: case study in the eastern English Channel. Coast Shelf Res 58:21–31, http://dx.doi.org/10.1016/j.csr.2013.03.001

Korotenko KA, Mamedov RM, Kontar AE, Korotenko LA (2004) Particle tracking method in the approach for prediction of oil slick transport in the sea: modelling oil pollution resulting from river input. J Mar Syst 48:159–170. doi:10.1016/j.jmarsys.2003.11.023

Korotenko KA, Bowman MJ, Dietrich DE (2010) High-resolution model for predicting the transport and dispersal of oil plumes resulting from accidental discharges in the black sea, recent advances in numerical ocean modeling and prediction. Terr Atmos Ocean Sci 21:123–136. doi:10.3319/TAO.2009.04.24.01(IWNOP)

Korotenko KA, Zavialov PO, Kao R-C, Ding C-F (2012) Model for predicting the transport and dispersal of contaminants incoming with submarine groundwater: case study for the south-western Taiwan coastal zone. Open J Mar Sci 2:70–83. doi:10.4236/ojms.2012.22010

Kroeger KD, Charette MA (2008) Nitrogen biogeochemistry of submarine groundwater discharge. Limnol Oceanogr 53(3):1025–1039

Lee DR (1977) A device for measuring seepage flux in lakes and estuaries. Limnol Oceanogr 22:140–147

Lee I-H, Lien R-C, Liu JT, Chuang W-S, Xu J (2009) Turbulent mixing and internal tides in Gaoping (Kaoping) submarine canyon. J Mar Syst 76:383–396. doi:10.1016/j.jmarsys.2007.12.011

Leote C, Ibanhez JS, Rocha C (2008) Submarine groundwater discharge as a nitrogen source to the Ria Formosa studied with seepage meters. Biogeochemistry 88:185–194. doi:10.1007/s10533-008-9204-9

Levitus S (2009) World ocean atlas/NOAA Atlas. U.S Department of Commerce NOAA/NODC, Silver Spring, MD, USA

Li L, Barry DA, Stagnitti F, Parlange J-Y (1999) Submariane groundwater discharge and associated chemical input to a coastal sea. Water Resour Res 35:3253–3259. doi:10.1029/1999WR900189

Lin I-T, Wang C-H, You C-F, Lin S, Huang K-F, Chen Y-G (2010) Deep submarine groundwater discharge indicated by tracers of oxygen strontium isotopes and barium content in the Pingtung coastal zone of southern Taiwan. Mar Chem 122:51–58. doi:10.1016/j.marchem.2010.08.007

Lin I-T, Wang C-H, Lin S, Chen Y-G (2011) Groundwater – seawater interactions off the coast of southern Taiwan: evidence from environmental isotopes. J Asian Earth Sci 41:250–262. doi:10.1016/j.jseaes.2011.03.001

Lobkovsky LI, Kontar E, Garagash I, Ozorovich Y (2003) Monitors and methods for investigation of submarine landslides seawater intrusion and contaminated groundwater discharge as coastal hazards. In: Beer T, Ismail-Zadeh A (eds) Risk science and sustainability: science for reduction of risk and sustainable development of society, Kluwer Academic Publishers, Kluwer, pp 191–207

Mellor GL, Yamada T (1982) Development of a turbulence closure model for geophysical fluid problems. Rev Geophys Space Phys 20:851–875. doi:10.1029/RG020i004p00851

Moore WS (2006) The role of submarine groundwater discharge in coastal biogeochemistry. J Geochem Explor 88(1–3):389–393

Niencheski LFH, Windom HL, Moore WS, Jahnke RA (2007) Submarine groundwater discharge of nutrients to the ocean along a coastal lagoon barrier Southern Brazil. Mar Chem 106:546–561. doi:10.1016/j.marchem.2007.06.004

Nowicki BL, Requintina E, Vankeuren D, Portnoy J (1999) The role of sediment denitrification in reducing groundwater-derived nitrate inputs to nauset marsh estuary Cape Cod Massachusetts. Estuar Coast 22:245–259. doi:10.2307/1352981

Ozorovich YR, Kontar Y (2007) Possibilities of geophysical survey for groundwater contamination and subsurface pollution determination and monitoring in the coastal zone. In: Sanford W, Langevin C, Polemio M, Povinec P (eds) A new focus on groundwater-seawater interactions. IAHS Publication 312, Oxfordshire, pp 93–99, Red books. ISBN 978-1-901502-04-6

Paytan A, Shellenbarger GG, Street JH, Gonneea ME, Davis K, Young MB, Moore WS (2006) Submarine groundwater discharge: an important source of new inorganic nitrogen to coral reef ecosystems. Limnol Oceanogr 51:343–348

Portonoy JW, Nowicki BL, Roman CT, Urish DW (1998) The discharge of nitrate contaminated groundwater from developed shoreline to marsh-fringed estuary. Water Resourc Res 34:3095–3104. doi:10.1029/98WR02167

Povinec PP, Aggarwal PK, Burnett W, Kontar E, Kulkarni K, Moore W, Rajar R, Taniguchi M, Comanducci J-F, Cusimano G, Dulaiova H, Gatto L, Hauser S, Levy-Palomo I, Ozorovich Y, Privitera AMG, Schiavo M (2006) Characterization of Submarine Groundwater Discharge (SGD) offshore South-Eastern Sicily – IAEA-UNESCO SGD collaboration. J Environ Radioact 89(1):81–101. doi:10.1016/j.jenvrad.2006.03.008

Povinec PP, Bokuniewicz H, Burnett WC, Cable J, Charette M, Comanducci JF, Kontar EA, Moore WS, Oberdorfer JA, de Oliveira J, Peterson R, Stieglitz T, Taniguchi M (2008) Isotope tracing of submarine groundwater discharge offshore Ubatuba, Brazil: results of the IAEA – UNESCO SGD project. J Environ Radioact 99(10):1596–1610. doi:10.1016/j.jenvrad.2008.06.010

Povinec PP, Burnett WC, Beck A, Bokuniewicz H, Charette M, Gonneea ME, Groening M, Ishitobi T, Kontar E, Kwong LW, Marie DEP, Moore WS, Oberdorfer JA, Peterson R, Ramessur R, Rapaglia J, Stieglitz T, Top Z (2012) Isotopic geophysical and biogeochemical investigation of submarine groundwater discharge: IAEA-UNESCO intercomparison exercise at Mauritius island. J Environ Radioact 104:24–45

Santos R, Burnett WC, Dittmar T, Suryaputra IGNA, Chanton J (2009) Tidal pumping drives nutrient and dissolved organic matter dynamics in a Gulf of Mexico subterranean estuary. Geochim Cosmochim Acta 73:1325–1339. doi:10.1016/j.gca.2008.11.029

Selker JS, Thévenaz L, Huwald H, Mallet A, Luxemburg W, van de Giesen N, Stejskal M, Zeman J, Westhoff M, Parlange MB (2006a) Distributed fiber-optic temperature sensing for hydrologic systems. Water Resour Res 42:117–184

Selker JS, Van De Giesen N, Westhoff M, Luxemburg W, Parlange MB (2006b) Fiber optics opens window on stream dynamics. Geophys Res Lett 33:64–66

Sentchev A, Korotenko KA (2004) Stratification and tidal current effects on larval transport in the eastern English channel: observations and 3D modeling. Environ Fluid Mech 4:305–331. doi:10.1023/B:EFMC.0000024246.39646.1d

Sentchev A, Korotenko KA (2005) Dispersion processes and transport pattern in the ROFI system of the eastern English channel derived from a particle-tracking model. Cont Shelf Res 25:2294–2308. doi:10.1016/j.csr.2005.09.003

Sentchev A, Korotenko KA (2007) Modeling distribution of flounder larvae in the eastern English channel: sensitivity to physical forcing and biological behavior. Mar Ecol Program Ser 347:233–245. doi:10.3354/meps06981

Simmons GM Jr (1992) Importance of Submarine Groundwater Discharge (SGWD) and seawater cycling to material flux across sediment/water interfaces in marine environments. Mar Ecol Program Ser 84:173–184

Slomp CP, Van Cappellen P (2004) Nutrient inputs to the coastal ocean through submarine groundwater discharge: controls and potential impact. J Hydrol 295:64–86. doi:10.1016/j.jhydrol.2004.02.018

Smith C, Smith L, Taniguchi M (2002) Assessing methodologies for measuring groundwater discharge to the ocean. EOS 83(11):117–123

Spiteri C, Slomp CP, Tuncay K, Meile C (2008) Modeling biogeochemical processes in subterranean estuaries: effect of flow dynamics and redox conditions on submarine groundwater discharge of nutrients. Water Resour Res 44(2):W02430. doi:10.1029/2007WR006071

Stieglitz T, Bokuniewicz H, Burnett W, Cable J, Charette M, Kontar E, Martin J, Moore W, Oberdorfer J, Peterson R, Povinec P, de Olivera J, Taniguchi M (2008) Submarine groundwater discharge from a fractured rock aquifer on the Ubatuba coastline Brazil: an overview of the IAEA-UNESCO project. EOS, transactions American geophysical union (AGU) 89(23), Western pacific geophysics meeting, Cairns, 29 July–1 August 2008, Suppl Abstract H25B-05 B-05

Swarzenski PW, Reich CD, Spechler RM, Moore WS (2002) Using multiple geochemical tracers to characterize the hydrogeology of the submarine spring off crescent beach, Florida. Chem Geol 179:187–202

Swarzenski PW, Burnett WC, Weinstein Y (2006a) Combined time-series resistivity and geochemical tracer techniques to examine submarine groundwater discharge at dor beach, Israel. Geophys Res Let 33:L24405

Swarzenski PW, Orem WG, McPherson BF, Baskaran M, Wan Y (2006b) Biogeochemical transport in the Loxahatchee river estuary: the role of submarine groundwater discharge. Mar Chem 101:248–265

Swarzenski PW, Kruse S, Reich C, Swarzenski W (2007c) Multi-channel resistivity investigations of the fresh water/saltwater interface: a new tool to study an old problem. In: A new focus on groundwater – seawater interactions, vol 312. IAHS Publ, pp 100–108

Swarzenski P, Simonds F, Paulson T, Kruse S, Reich C (2007b) A geochemical and geophysical examination of submarine groundwater discharge and associated nutrient loading estimates into lynch cove hood canal, WA. Environ Sci Technol 41:7022–7029

Swarzenski PW, Reich C, Kroeger K, Baskaran M (2007c) Ra and Rn isotopes as natural tracers of submarine groundwater discharge in Tampa Bay, FL. Mar Chem 104:69–84

Swarzenski PW, Dellapenna TM, Reich CD, Noll CJ (2009) Examining submarine ground-water discharge into simpson bay, Alaska with electrical resistivity and radon. EEGS/SAGEE, Ft Worth

Taniguchi M, Burnett W, Duevalova H, Kontar E, Povinec P, Moore W (2006) Submarine groundwater discharge measured by seepage meters in Sicilian coastal waters. Cont Shelf Res 26(7):835–842

Taniguchi M, Burnett W, Dulaiova H, Siringan F, Foronda J, Wattaykorn G, Rungsupa S, Kontar Y, Ishitobi T (2008) Groundwater discharge as an important land-sea pathway into Manila Bay Philippines. J Coast Res 24(1A):15–24

Tyler SW, Selker JS, Hausner MB, Hatch CE, Torgersen T, Thodal CE, Schladow SG (2009) Environmental temperature sensing using Raman spectra DTS fiber-optic methods. Water Resour Res 45:91–99

Uchiyama Y, Nadaoka K, Peter RP, Adachi K, Yagi H (2000) Submarine groundwater discharge into the sea and associated nutrient transport in a sandy beach. Water Resour Res 36:1467–1479. doi:10.1029/2000WR900029

Visser A (1997) Using random walk models to simulate the vertical distribution of particles in a turbulent water column. Mar Ecol Program Ser 158:275–281. doi:10.3354/meps15827

Wang YH, Lee IH, Liu JT (2008) Observation of internal tidal currents in the Kaoping canyon off southwestern Taiwan. Estuar Coast Shelf Sci 80:153–16. doi:10.1016/j.ecss.2008.07.016

Wang R-M, You C-F, Chu H-Y, Hung J-J (2009) Seasonal variability of dissolved major and trace elements in the Gaoping (Kaoping) river estuary southwestern Taiwan. J Mar Syst 76:444–456. doi:10.1016/j.jmarsys.2007.11.012

Weiskel PK, Howes BL (1992) Differential transport of sewage-derived nitrogen and phosphorus through a coastal watershed. Environ Sci Technol 26:352–360. doi:10.1021/es00026a017

Williamson CE, Soros DW, Schindler JE (2009) Sentinels of change. Science 323:887–888

Winter TC, Harvey JW, Franke OL, Alley WM (1998) Ground water and surface water – a single resource, vol 1139. U.S Geological Survey Circular, 79 p

Zavialov PO, Kao R-C, Kremenetskiy VV, Peresypkin VI, Ding C-F, Hsu R-T, Kopelevich OV, Korotenko KA, Wu Y-S, Chen P (2012) Evidence for submarine groundwater discharge on the southwestern shelf of Taiwan. Coast Shelf Res 4:18–25. doi:10.1016/j.csr.2011.11.010

Chapter 3
Tsunami Inundation Modeling of the 2011 Tohoku Earthquake Using Three-Dimensional Building Data for Sendai, Miyagi Prefecture, Japan

Toshitaka Baba, Narumi Takahashi, Yoshiyuki Kaneda, Yasuyuki Inazawa, and Mariko Kikkojin

Abstract In conventional modeling of tsunami inundation, based on nonlinear shallow water theory in a finite-difference scheme, the effect of buildings and structures is represented by a bottom friction parameter rather than by three-dimensional (3D) building shapes. But large, strong buildings should offer direct protection against an incoming tsunami, like seawalls. In this study, therefore, we incorporated 3D building data obtained from lidar measurements in modeling the tsunami from the 2011 Tohoku earthquake at the port of Sendai, Miyagi Prefecture, Japan, and compared the results from conventional modeling based on a digital elevation model. In the model incorporating 3D building data, the maximum inundation height was greater than in the conventional model at the front of coastal buildings and structures and smaller behind them. High-velocity currents appeared in the corridors between these buildings, and the tsunami inundation area was smaller in the residential zone because of the obstacles that buildings presented to the tsunami. These results mean that solid buildings and structures have a significant influence on the propagation of tsunamis on land. The effects of the 3D shapes of buildings and structures should be further investigated for detailed tsunami hazard assessments in urban areas.

Keywords 2011 Tohoku Tsunami • Tsunami modeling • 3D Building data

---

T. Baba (✉) • N. Takahashi • Y. Kaneda
Japan Agency for Marine-Earth Science and Technology (JAMSTEC), Yokohama Institute for Earth Sciences, 3173-25 Showa-machi, Kanazawa-ku, Yokohama, Kanagawa 236-0001, Japan
e-mail: babat@jamstec.go.jp

Y. Inazawa • M. Kikkojin
Geospatial Information Authority of Japan (GSI), 1 Kitasato, Tsukuba, Ibaraki 305-0811, Japan

Y.A. Kontar et al. (eds.), *Tsunami Events and Lessons Learned: Environmental and Societal Significance*, Advances in Natural and Technological Hazards Research 35,
DOI 10.1007/978-94-007-7269-4_3,

3.1 Introduction

The 2011 Tohoku earthquake (M 9.0) occurred on 11 March near the Japan Trench, where the Pacific plate subducts westward with respect to the North America plate at a rate of approximately 8–9 cm/year (DeMets et al. 2010). Finite fault models of this earthquake have been proposed based on seismic wave data (e.g., Ammon et al. 2011; Yagi and Fukahata 2011), crustal displacement data (e.g., Suito et al. 2011; Ito et al. 2011), and tsunami data (e.g., Satake et al. 2013). The dimensions of the rupture zone have been estimated to be about 400 $\times$ 200 km, and the largest slip appears to have exceeded 30 m on the plate interface, although there is some disagreement among fault models. It was the fourth greatest earthquake in the world during the last century, and it caused catastrophic damage in Japan. The Pacific coast of northeastern Japan was widely and deeply inundated by the tsunami. The Japan Meteorological Agency (JMA) immediately issued early tsunami warnings that prompted residents living in the tsunami hazard area to take steps to evacuate. Although the warnings saved many lives, the earthquake and tsunami were responsible for 15,870 fatalities and 2,814 missing (National Police Agency 2012).

The first step to mitigate tsunami disasters is a tsunami hazard map, which indicates in advance the expected inundation area from various possible tsunamis. This map can be used to construct appropriate evacuation routes to elevated locations, to select locations and candidates for tsunami evacuation buildings, and to aid in estimation of human and economic losses. Construction of these maps begins with possible scenarios for generating tsunamis based on seismological, geological, and historical evidence. Next, tsunami inundation modeling is carried out based on the earthquake scenarios. Modeling usually relies on nonlinear shallow water equations along with accurate bathymetric and topographic data. Three-dimensional (3D) shapes of breakwaters and seawalls are directly incorporated in tsunami computations because of their effects on tsunami propagation, but the shapes of buildings and structures farther inland usually cannot be considered owing to limits of spatial resolution and lack of comprehensive building data. Land areas are represented instead by a bottom friction parameter, which varies in space to represent land-use types such as building lots, agricultural land, and mountainous areas.

Lidar measurements are being carried out along the Japanese coast by the Geospatial Information Authority of Japan (GSI), which collects reflections from the ground surface plus reflections from elevated surfaces such as building roofs, roads, bridges, and the tops of trees, with high spatial resolution. Large, strong buildings may be able to interact with a tsunami like a seawall rather than be subsumed in a uniform bottom friction parameter. We infer that incorporating 3D shapes of buildings and structures obtained by lidar measurements may lead to improved modeling of tsunami inundation. In this study, therefore, the 3D building data derived from lidar measurements were embedded as topographic highs in a tsunami inundation model to investigate the effect of structure shapes. This chapter presents the results obtained with and without the 3D building data. We also discuss the advantages of using 3D building data for tsunami damage mitigation.

3.2 Construction of 3D Building Data

Since the 2011 tsunami disaster, GSI has accelerated its ongoing program of airborne lidar surveys in coastal regions for disaster mitigation purposes. For compiling our 3D building data file, defined as data on ground heights of buildings and structures distributed in a gridded pattern, we used the airborne lidar data obtained by GSI.

In addition to ground surface elevations, airborne lidar provides locations of points on the tops of buildings, structures, trees, and so on in plane rectangular 3D coordinates. These points are randomly distributed in space at intervals from 1 to 2 m. The elevated objects such as buildings and trees are filtered from the original lidar data to obtain a point dataset of ground elevations, usually called a digital elevation model (DEM) (Fig. 3.1). The DEM may then be subtracted from the original lidar data to create a point dataset of the heights of objects above ground level.

This initial dataset contains all above-ground objects, including objects such as trees and bridges that do not obstruct tsunamis. To extract only buildings and structures that would affect tsunami inundation, we made use of the government's Fundamental Geospatial Data collection, which contains outlines of buildings and structures. We were able to use the four-part classification of building types (solid building, common building, solid wall-less building, and common wall-less building) to restrict ourselves to those (solid building and common building) that could provide tsunami protection. Figure 3.2 shows an example of this step in our compilation of 3D building data.

The resulting building height data were randomly distributed in the plane rectangular coordinates. We next converted this coordinate system to latitude and longitude as required by the tsunami code we used, and then sampled the points into a grid at intervals of 2/9 arcsec, consistent with the finest grid in the tsunami simulation.

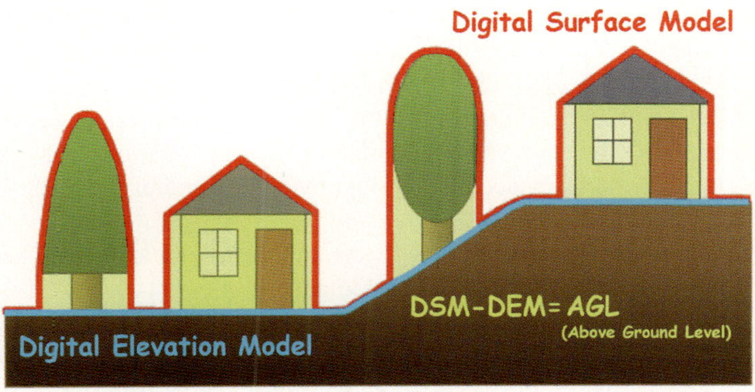

Fig. 3.1 Concepts of digital elevation model (*DEM*), digital surface model (*DSM*), and derivation of "*above ground level*" (*AGL*) data

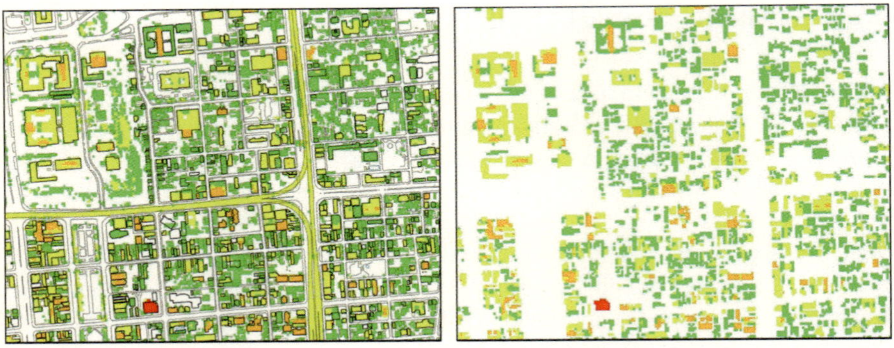

Fig. 3.2 AGL data (*left*) and 3D building data extracted from AGL data (*right*)

3.3 Tsunami Inundation Modeling

3.3.1 Computational Code and Nesting Scheme

To model the propagation and inundation of the 2011 Tohoku tsunami, we were going to use the URSGA code of Jakeman et al. (2010). This model uses a variable nested-grid scheme which allows the spatial resolution of the study region to be easily increased. It was originally based on the uniform finite-difference scheme of Satake (2002) and solves either the linear or nonlinear shallow water equations. However, a large number of grids with high spatial resolution is needed in order to assess effect of 3D shape of buildings and structures on tsunami inundation. In our case, the total number of the grids needed was going to be about 20 million. This would make the computation time too long if a serial code like URSGA is used. We therefore parallelized the URSGA code by using OpenMP and MPI in order to speed up the tsunami inundation modeling. In addition, several improvements were also made to how the nesting grids communicate. A function compatible with a GIS viewer was also developed to create snap shots of wave height distribution. Finally we also rewrote the code into Fortran90 from the original version written in C and renamed it "JAGURS". JAGURS stands for the JAMSTEC parallelized the code developed by Geoscience Australia and URS Corporation which uses Satake's kernel. Figure 3.3 shows a result of a parallel performance evaluation test for the JAGURS. It shows that we successfully upgraded the speed of computation by the parallelizing the code.

Five nested grids as shown in Fig. 3.4 were defined for our analysis. The coarsest grid represents the entire computational domain (34–43°N, 140–147.5°E), including the tsunami source and the target area of Sendai, Miyagi Prefecture (Fig. 3.4a). The bathymetry in the grid was made using a combination of the M7000 map series provided by the Marine Information Research Center, Japan Hydrographic Association; the Tohoku bathymetric grid from JAMSTEC (Kido et al. 2011); and GEBCO data (British Oceanographic Data Centre 2010), which was interpolated

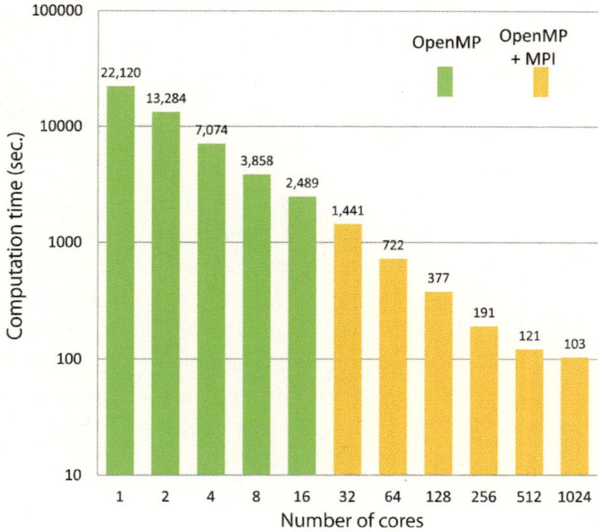

Fig. 3.3 Parallel performance test for the JAGURS code

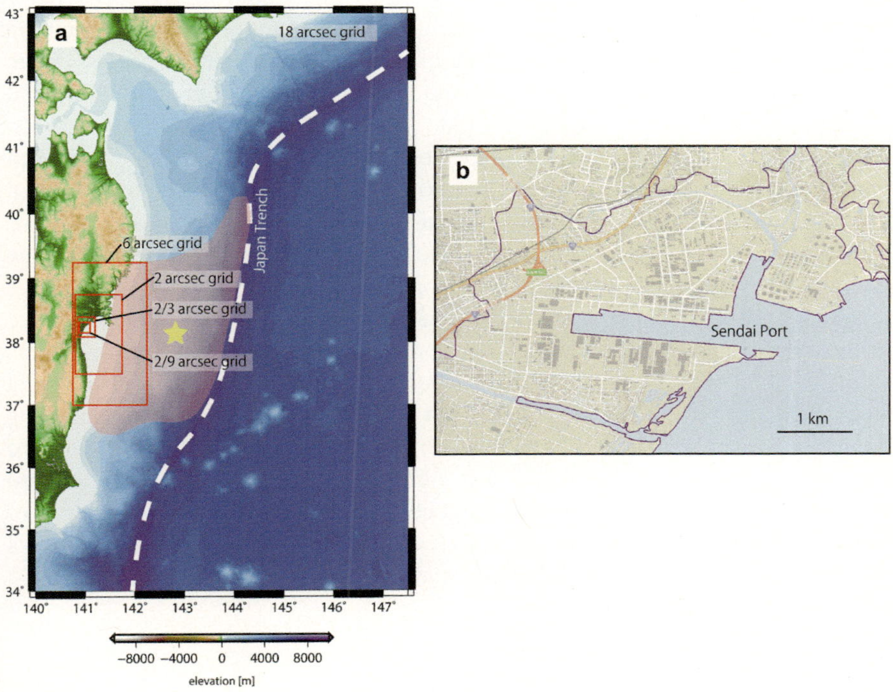

Fig. 3.4 (**a**) Bathymetry and topography of the computational domain (18″ grid spacing) and outlines of nested grids. *Star* indicates epicenter of the 2011 Tohoku earthquake determined by the U.S. Geological Survey. *Colored zone* is area of large slip in the fault model of Satake et al. (2013). (**b**) Target region of this study within the 2/9 arcsec grid. *Purple line* encloses the inundation area of the 2011 Tohoku earthquake surveyed by GSI (2011)

to 18 arcsec intervals. The M7000 series is a set of digital bathymetric contours made by combining the basic map of the coastal waters of Japan and other bathymetric information. The Tohoku bathymetric grid includes all results from JAMSTEC's multi-narrow beam surveys conducted in the Japan Trench. GEBCO provides global bathymetry datasets for the world's oceans with a spatial resolution of 30 arcsec. These datasets were also subsampled and interpolated, respectively, to make grids with spacings of 6, 2, 2/3, and 2/9 arcsec for the nesting scheme. The finest grid includes the area around the port of Sendai (Fig. 3.4b), where the inundation height was measured as 6–7 m by the Tohoku Tsunami Joint Survey Group (2011). Its spacing of 2/9 arcsec is almost equal to 5 m spacing.

For the land area, we resampled the GSI data to make topographic grids. The 50-m interval topographic data assembled by GSI covering all of Japan was used for the topography in grids of 18, 6, and 2 arcsec spacing. The 5-m interval topographic data provided by GSI was resampled and interpolated for the 2/3 arcsec and 2/9 arcsec grids. These topographic grids were mated to the bathymetric grids to yield seamless bathymetric–topographic grids for the entire region. The shape of the coastline, which is important in tsunami modeling, relied on the GSI topographic data. The shapes of tsunami defense facilities such as seawalls and breakwaters larger than 7.5 m were included as topography in the finest DEM data. The resulting grid is referred to here as the "DEM model." The 3D building data created from the lidar data was added to the 2/9 arcsec DEM model to obtain the "incorporated model," representing the total elevation including building and structure heights.

To assess the effect of the 3D shapes of buildings and structures on tsunami propagation, we also ran a simulation using a heterogeneous bottom friction parameter determined by land-use type using the DEM model instead of the incorporated model. We also added a function into the JAGURS to incorporate spatially heterogeneous bottom friction. For the Manning roughness, we used the segmentalized land-use data from the digital national land information website (Ministry of Land, Infrastructure, Transport and Tourism 2006). The roughness value was 0.02 s $m^{-1/3}$ in field, orchard, river, and lake areas, 0.03 s $m^{-1/3}$ in forest areas, 0.04 s $m^{-1/3}$ in residential areas, and 0.025 s $m^{-1/3}$ in other areas including the sea. However, this dataset has a relatively poor spatial resolution of 100 m. We accordingly interpolated it to fit the 2/9 arcsec (~5 m) grid. The poor resolution also meant that we could not resolve roads and buildings in residential areas in terms of bottom friction. For the finest grid scale, we assigned a Manning roughness of 0.04 s $m^{-1/3}$ to residential roads and buildings, and for the other four grids, we adopted a uniform value of 0.025 s $m^{-1/3}$.

3.4 Tsunami Computation and Results

Our tsunami model generates tsunamis by seafloor crustal deformation due to an earthquake. For the 2011 tsunami computation, we first calculated the crustal deformation of the fault model provided by Tohoku University (version 1.1)

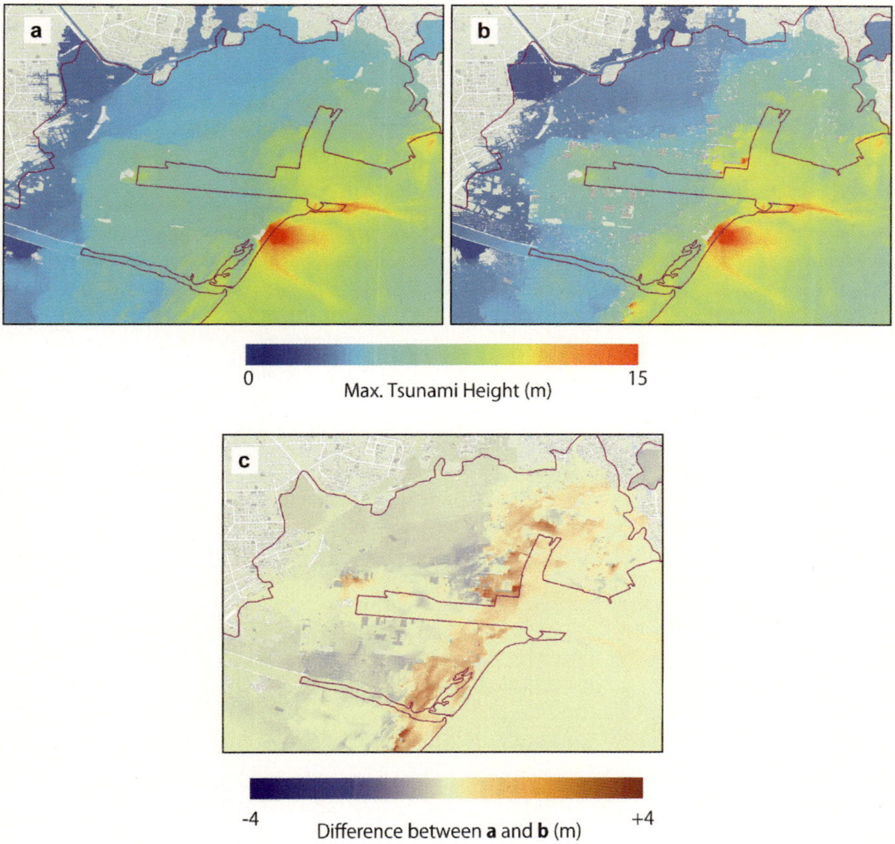

Fig. 3.5 (**a**) Map of maximum tsunami and inundation height computed with the DEM model. (**b**) Map of maximum tsunami and inundation height obtained with the incorporated model. (**c**) Difference between the DEM and incorporated models, with positive values indicating greater height in the incorporated model

(Imamura et al. 2011) by using the method of Okada (1985). The sea surface was deformed according to the seafloor crustal deformation with a finite rise time that was assumed to be 60 s. The topography was also modified by the same amounts at the same time. The time step used in the computation was 0.05 s in order to satisfy the stability condition of the finite difference scheme. We calculated tsunami propagation for 3 h after the Tohoku earthquake. To simulate tsunami wave propagation, we solved linear shallow water equations in the coarsest (18 arcsec) grid to save computation time and ensure stability. Nonlinear shallow water equations were applied for the four inner grids.

Figure 3.5 shows results of the tsunami analysis in the study region around Sendai. On average, the incorporated model yielded an increase in tsunami height of 2–3 m in front of coastal buildings and structures over the DEM model. For the maximum increase, the incorporated model yielded inundation heights at the front

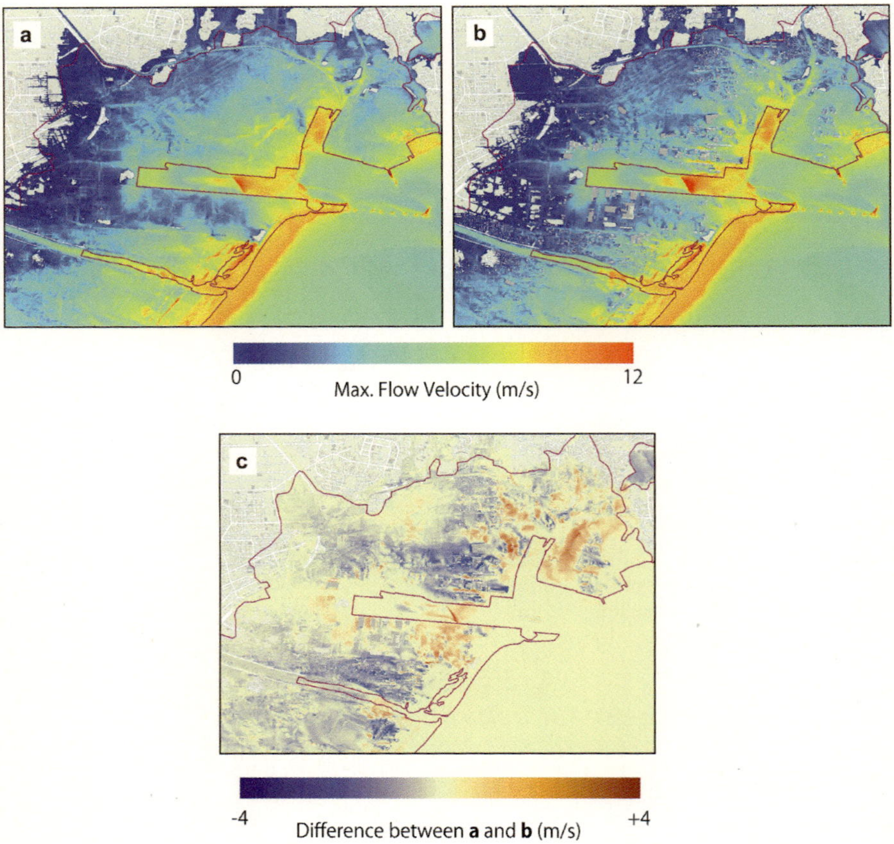

Fig. 3.6 (**a**) Map of maximum flow velocity computed with the DEM model. (**b**) Map of maximum flow velocity obtained with the incorporated model. (**c**) Difference between the DEM and incorporated models, with positive values indicating greater flow velocity in the incorporated model

of coastal buildings near the bay mouth that were locally as much as 9.3 m higher than those in the DEM model. But we would need detail investigations on the maximum increase of 9.3 m because the area the great increase appears is much localized. It may be an abnormal value caused by instabilities during the tsunami computation. Behind coastal buildings, inundation heights were smaller with the incorporated model than with the DEM model. The incorporated model yielded a smaller tsunami inundation area in the residential zone.

Similarly, the modeled flow velocity in the area between the buildings (Fig. 3.6) was as much as 4.75 m/s greater in the incorporated model than in the DEM model. On average, the flow velocity increased by 2–3 m/s in corridors parallel to the advance of the tsunami near the coast, a difference that gradually decreased with distance from the sea. Lower flow velocities were seen in areas directly behind buildings.

3.5 Discussion

Figures 3.5 and 3.6 show that in our simulations that incorporate the 3D shapes of buildings and structures, the tsunami height was greater in front of buildings, the flow velocity was greater in building corridors parallel to the direction of tsunami travel near the coast, and the tsunami threat was weakened behind buildings. This result indicates that buildings and structures near the coast, if they are resilient enough against the force of a tsunami, may complement seawalls and breakwaters to decrease tsunami damage in the hinterland. We conclude that incorporating 3D shapes of buildings and structures is useful in tsunami inundation modeling, especially for urban areas.

Using 3D building data in tsunami inundation modeling may also be advantageous for tsunami hazard mitigation. The tsunami hazard maps provided by the incorporated model would be useful for selecting tsunami evacuation buildings because they would account for tsunami amplification by the buildings themselves. The modeled amplification of flow in the corridors between buildings would be significant in planning tsunami evacuation routes and in planning mitigation measures to strengthen buildings. The response of residents to a tsunami may be more realistically forecasted on the basis of more accurate accounts of tsunami behavior in the urban setting. And simulations that incorporate 3D building data may be helpful in designing a tsunami resilient city in which large buildings can be designed and oriented to resist tsunami invasion into the residential hinterland.

This paper has presented the effect of using the 3D shapes of buildings and structures, derived from lidar data, in tsunami inundation modeling by incorporating these objects as topographic highs in a finite-difference scheme solving nonlinear long wave equations. This incorporated model yielded tsunami characteristics on land that differ from those in the conventional DEM model. We propose that 3D shapes of buildings and structures should be investigated in more detail for tsunami hazard assessments in urban areas. For example, we treated them as intact solid bodies in our calculation, but after the 2011 Tohoku earthquake, many coastal buildings were destroyed by the tsunami. How do we account for destruction of buildings and the loss of their effectiveness as a barrier in tsunami modeling? The spatial resolution we applied, which was 2/9 arcsec (about 5 m) in the finest grid, is still too coarse to represent narrow alleys in residential districts, which may result in underestimating tsunami inundation areas because these small tsunami corridors are erased in the interpolation process used to make the incorporated model. A more desirable spatial resolution, at a meter scale or finer, would impose severe computational demands. Because the tsunami flow around buildings is likely to be very complicated, it may be better to apply the governing equations with a high degree of accuracy rather than rely on two-dimensional nonlinear shallow water theory. We are investigating these issues using the very large dataset from the 2011 Tohoku earthquake for mitigation of future tsunami damage.

Acknowledgments Dr. Hong Kie Thio, Dr. Phil Cummins, and Dr. David Burbidge kindly provided us with the URSGA tsunami code. We also thank the editor, Vicente Santiago-Fandiño and an anonymous reviewer. Some Figures were made using Generic Mapping Tools (Wessel and Smith 1998) and ArcGIS.

References

Ammon CJ, Lay T, Kanamori H, Cleveland M (2011) A rupture model of the 2011 off the Pacific coast of Tohoku earthquake. Earth Planets Space 63:693–696

British Oceanographic Data Centre, GEBCO (General Bathymetric Chart of the Oceans) (2010) http://www.gebco.net/data_and_products/gridded_bathymetry_data

DeMets C, Gordon RG, Argus DF (2010) Geologically current plate motions. Geophys J Int 181:1–80

Geospatial Information Authority of Japan (2011) Tsunami inundation map of the 2011 Tohoku-oki earthquake on a scale of 1 to 100000 (in Japanese). http://www.gsi.go.jp/kikaku/kikaku60003.html

Imamura F, Koshimura S, Murashima Y, Akita Y, Nitta Y (2011) Implementation of the tsunami simulation for the 2011 Tohoku earthquake (in Japanese). http://www.tsunami.civil.tohoku.ac.jp/hokusai3/J/events/tohoku_2011/model/dcrc_ver1.1_111107.pdf

Ito T, Ozawa K, Watanabe T, Sagiya T (2011) Slip distribution of the 2011 off the Pacific coast of Tohoku earthquake inferred from geodetic data. Earth Planets Space 63:627–630

Jakeman JD, Nielsen OM, VanPutten K, Mleczeko R, Burbidge D, Horspool N (2010) Towards spatially distributed quantitative assessment of tsunami inundation models. Ocean Dyn 60:1115. doi:10.1007/s10236-010-0312-4

Kido Y, Fujiwara T, Sataki T, Kinoshita M, Kodaira S, Sano M, Ichiyama Y, Hanafusa Y, Tsuboi S (2011) Bathymetric feature around Japan Trench obtained by JAMSTEC multi narrow beam survey, MIS036-P58 (in Japanese). Japan Geoscience Union Meeting, 2011

Ministry of Land, Infrastructure, Transport and Tourism (2006) Segmentalized land utilization data. http://nlftp.mlit.go.jp/ksj/jpgis/datalist/KsjTmplt-L03-b.html

National Police Agency (2012) Damage caused by the 2011 off the Tohoku Pacific coast earthquake (in Japanese). http://www.npa.go.jp/archive/keibi/biki/higaijyokyo.pdf

Okada Y (1985) Surface deformation due to shear and tensile faults in a half-space. Bull Seismol Soc Am 75:1135–1154

Satake K (2002) Tsunamis. In: Lee WHK, Kanamori H, Jennings PC, Kisslinger C (eds) International handbook of earthquake and engineering seismology, vol 81A. Academic, Amsterdam, pp 437–451

Satake K, Fujii Y, Harada T, Namegaya Y (2013) Time and space distribution of coseismic slip of the 2011 Tohoku earthquake as inferred from tsunami waveform data. Bull Seismol Soc Am 103:1473, accepted

Suito H, Nishimura T, Tobita M, Imakiire T, Ozawa S (2011) Interplate fault slip along the Japan Trench before the occurrence of the 2011 off the Pacific coast of Tohoku earthquake as inferred from GPS data. Earth Planets Space 63:615–619

Tohoku Tsunami Joint Survey Group (2011) The 2011 off the Pacific coast of Tohoku earthquake tsunami information. http://www.coastal.jp/tsunami2011

Wessel P, Smith WHF (1998) New, improved version of generic mapping tools released. EOS Trans Am Geophys Union 79:579

Yagi Y, Fukahata Y (2011) Rupture process of the 2011 Tohoku-oki earthquake and absolute elastic strain release. Geophys Res Lett 38(19):L19307. doi:10.1029/011GL048701

Chapter 4
Numerical Simulation of Coastal Sediment Transport by the 2011 Tohoku-Oki Earthquake Tsunami

Daisuke Sugawara and Tomoyuki Takahashi

Abstract Erosion and deposition by the 2011 Tohoku-oki tsunami in the coastal areas of the Sendai Plain were numerically investigated using the tsunami sediment transport model (STM). The simulation suggested that much of the sediment deposited on land originated from the beach and dune, whereas the contribution from sea-bottom sediments was quite minor. Erosion observed at the backs of engineering structures such as reinforced concrete (RC) dikes was reproduced well in the simulation. Simulated deposition and erosion inside the nearby coastal forest and farmlands were generally consistent with the observed data. Further investigation of the model parameters and implementation of dynamic change in the roughness coefficient and destruction of the structures are required to simulate detailed local variability of erosion and deposition and of the thin sand layer in the inland areas.

Keywords Numerical modeling • Sediment transport • Tsunami deposit

4.1 Introduction

Tsunami deposits have been widely used for identifying past large-scale tsunamis and assessing the magnitudes of paleotsunami events. Recent studies have attempted to estimate tsunami height and flow speed based on sedimentary data, such as the thickness and grain size of deposits. Tsunami erosion and deposition can be calculated by the sediment transport model (STM) developed by Takahashi et al. (2000). The model simulates bathymetric and topographic changes under given flux and bottom shear conditions. The STM has been validated using modern examples prior

D. Sugawara (✉)
International Research Institute of Disaster Science, Tohoku University, Sendai, Japan
e-mail: sugawara@irides.tohoku.ac.jp

T. Takahashi
Faculty of Safety Science, Kansai University, Takatsuki, Japan

Y.A. Kontar et al. (eds.), *Tsunami Events and Lessons Learned: Environmental and Societal Significance*, Advances in Natural and Technological Hazards Research 35,
DOI 10.1007/978-94-007-7269-4_4, © Springer Science+Business Media Dordrecht 2014

to being applied to paleotsunami events. For example, Gusman et al. (2012) recently applied the model to sedimentation caused by the 2004 Indian Ocean tsunami at Lhok Nga, Banda Aceh, Indonesia. They showed that the model reasonably reproduced the thickness distribution of the sand layer found in the field and suggested that the STM can be applied to numerical investigation of the source process of large tsunamis. If this is so, the magnitudes of the paleotsunami events can be quantified based on paleotsunami deposits. The 2011 Tohoku-oki tsunami offers a valuable opportunity for validation of the STM. An abundance of sedimentary data, as well as measurements, video footage and eyewitness accounts, are available for the tsunami.

However, recent studies of the Tohoku-oki tsunami deposits from the coastal plains have revealed a number of specific sedimentological features. The distribution of the sandy deposit was typically limited to within 57–76 % of the inundation distance (e.g., Abe et al. 2012). Beach sand and dune soil have been estimated to be the main source of the tsunami deposits found in the zone up to 1 km from the shoreline, and only a small contribution of marine sediments has been detected based on grain-size and paleontological analysis (Szczuciński et al. 2012). Engineering structures, such as reinforced concrete (RC) dikes, are considered to have played an important role in sediment transport. Marked scouring and erosion at the backs of such structures were commonly observed in the study area (e.g., Richmond et al. 2012). Some of the RC dikes installed at the coast were severely damaged and collapsed. Hydraulic jump on the back of the structures is attributed to the intensive erosion of sediments and loss of the foundations of the structures (Tappin et al. 2012). It is likely that natural and anthropogenic coastal topographic features may have contributed to the limited distribution of the sand and the faint marine signature in the tsunami deposits.

In the present study, a numerical simulation of sediment transport caused by the 2011 Tohoku-oki earthquake tsunami was performed using the STM to reproduce and investigate the change of the coastal geomorphology of the Sendai Plain in northeast Japan (Fig. 4.1a, b). The performance of the model was validated using an available high-resolution digital elevation model (DEM) and field data on erosion and deposition caused by the tsunami.

4.2 Materials and Method for Simulation

4.2.1 Topography Data for Simulation

The computational domain from the open ocean to the Sendai Plain was covered by a nesting grid system. The tsunami generation and propagation was calculated on the coarser grids with spatial resolutions (dx) of 405, 135, 45 and 15 m. In the study area, a pre-tsunami DEM with dx = 5 m is available (Fig. 4.1b) and was used for the calculation of the STM. Elevations of the engineering structures with paved surfaces, such as roadways and RC dikes, are well reproduced in the DEM. The elevation was surveyed by the Geospatial Information Authority of Japan in

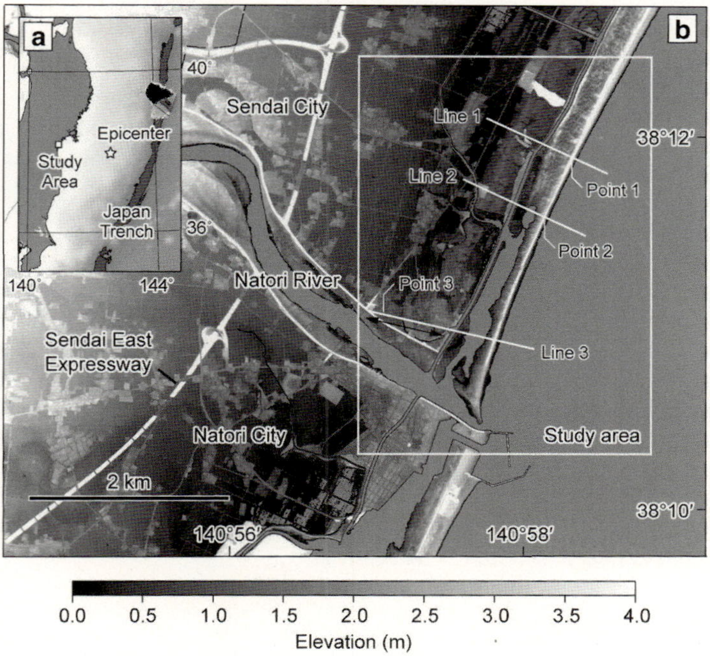

Fig. 4.1 Location map of the study area (**a**) and pre-tsunami topography of Sendai Plain and the study area (**b**)

the winter season of 2005–2006, nearly 5 years before the Tohoku-oki tsunami. Some recent topographic changes, such as a slight retreat of beach, implementation of new roadways and land development, which have occurred since 2006, are not incorporated in the DEM.

Manning's roughness coefficient, n, is an important parameter in the calculation of flow speed and bottom shear. A land use map was developed based on the aerial photographs (Fig. 4.5). A value for Manning's roughness coefficient was assigned to each grid cell on the basis of the land use map and findings by Kotani et al. (1998). In the simulation, engineering structures with paved surfaces are distinguished using the land use map. Although some of the structures were severely damaged and collapsed, they were not the main source of the tsunami deposits. The paved structures were not directly eroded by tsunami flow but rather collapsed due to erosion of their foundations. Some of the RC dikes might have been damaged directly by the impact of the tsunami wave. The STM in its present form cannot treat such dynamic and complex processes as the destruction of these structures. Thus, the paved surfaces were assumed to be intact during the computation. Erosion of the sediments does not extend below the elevations of these features. In addition, surfaces with muddy sediments, such as dry fields and rice paddies, were also assumed to be intact during the computation. This assumption was made because the present study was focused on the transport of sand. As noted in Sect. 4.2.2, the surficial sediments of these farmlands were in fact eroded. However, they did not contribute to the thickness of sand erosion and deposition.

Fig. 4.2 Erosion at the back of the reinforced concrete dike at Point 1 in Fig. 4.1b

4.2.2 Observed Data for Validation

A post-tsunami DEM with $dx = 5$ m is available for the study area. The survey was carried out immediately after the tsunami (during 19–23 March 2011), although most of the lowland was still flooded at that time. The vertical precision of the pre- and post-tsunami DEMs is around $\pm$ 15 cm. The post-tsunami DEM may include the elevation change due to the accumulation of debris. The data was processed by eye, and the debris greater than 1 m in height and 10–20 m in length was removed. Coseismic subsidence of 20–30 cm was observed in the study area (Ozawa et al. 2011). Because the Sendai Plain is alluvial in origin, the surficial sediments of the study area are composed of unconsolidated sediments of natural levees, beach ridges and back marshes (Matsumoto 1985). Local subsidence due to liquefaction was observed at many locations in the study area. A correction was made for this local subsidence based on flat areas with compacted or paved surfaces, such as schoolyards outside of the tsunami inundation area, which serve as reference levels for tsunami erosion and deposition. Taking into account the local subsidence that occurred, coastal elevation change between the pre- and post-tsunami DEMs was assumed to be an approximation of tsunami-induced topographic change. Detailed comparison of the deposition and erosion base in the DEMs is difficult, particularly if the erosion is shallow and/or deposition is thin, because of the limit of the precision of the survey and the correction for local subsidence.

Topographic changes and damage to engineering structures were observed in the field on 5 June 2011 (Points 1 and 2 in Fig. 4.1b).

In Sendai, RC dikes with typical heights of 5–6 m were installed on the sand dunes. At Point 1, the RC dike was still present. Severe erosion was observed at the back of the RC dike, and the depth of the erosion was up to 2–3 m at this location (Fig. 4.2). At Point 2, the dike was totally destroyed (Fig. 4.3). A considerable number of tetrapods and concrete casts were scattered on the beach and transported

Fig. 4.3 Collapsed RC dike and erosion of the sand bar at Point 2 in Fig. 4.1b. Concrete casts that comprised the dike were scattered and buried by sand

inland and even buried by beach sand. Although the pre-tsunami coastline extended further south of this location, the post-tsunami coastline was terminated due to the massive erosion of the beach.

The tsunami deposit was investigated on 26 May 2011 at 16 pits along the inland part of Line 3 (Fig. 4.1b). The observation pits were arranged at 50-m intervals. The area was originally used for farmland, such as dry fields and rice paddies. The tsunami deposit was composed of a coarse- to medium-grained sand layer, capped by a thin (approximately 1 cm) mud layer (Fig. 4.4). The thickness of the sand layer varies significantly from nearly 0 to 15 cm. The sand layer was clearly in contact with the underlying muddy soil and included many rip-up clasts entrained from the soil. This suggests that severe erosion might have taken place prior to the deposition of the sand layer (Fig. 4.5).

4.2.3 Method of Calculation

The STM was coupled with a tsunami propagation and inundation model based on nonlinear shallow-water theory (TUNAMI-N2; Goto et al. 1997; IOC/IUGG TIME Project). The simulation reproduced offshore tsunami generation, propagation to the coast and inundation for three model hours. Source models for the Tohoku-oki tsunami have been investigated in a number of studies. A composite fault model proposed by Japan Nuclear Energy and Safety Organization (JNES 2011) was used for the tsunami source. The model was developed based on the inversion analysis of tsunami records from tide gauges and GPS buoys, taking into account the tsunami heights measured near the coast and the observed coseismic subsidence on land. Crustal deformation by the fault movement was calculated using the method developed by Okada (1985). The subsidence calculated from the fault model approximates the observed data well (JNES 2011).

Fig. 4.4 Tsunami deposit found at Point 3 in Fig. 4.1b. The deposit was 12 cm thick and composed mainly of medium-grained sand capped by a thin mud layer

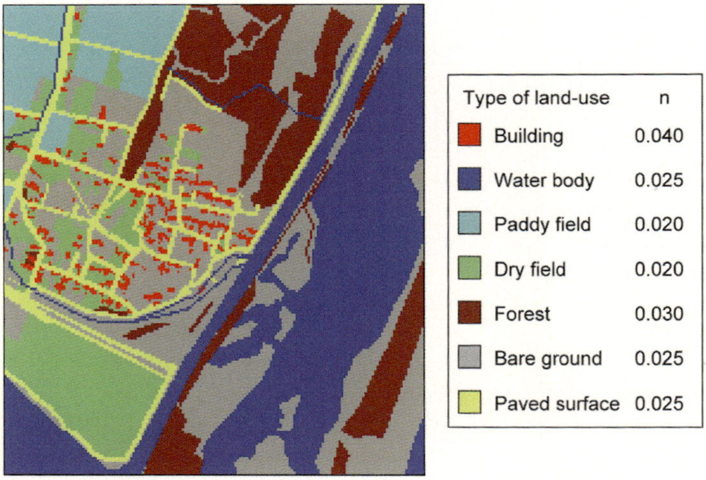

Fig. 4.5 Example of the land use map used in the present study. *n*: Manning's roughness coefficient. Buildings, paddy fields, dry fields and paved surfaces were assumed to be intact during the computation

The STM assumes two distinct layers of bed load and suspended load and assumes exchange between the two layers. Details of the model are given by Takahashi et al. (2000) and Gusman et al. (2012). Two key parameters control the behavior of the STM, namely, the grain size (d) and the saturated concentration of the suspended load (C_s). Previous studies documented that most beach and dune sands from the Sendai Plain are composed of fine- to medium-grained sand with a mean diameter of 2 phi (0.25 mm) (e.g., Matsumoto 1985). A uniform grain size of

$d = 0.25$ mm was assumed in the present study. Takahashi et al. (2011) found that the coefficients for the pick-up rate equations for the bed and suspended loads depend on the grain size. The grain-size-dependent parameters for the pick-up rate equations were chosen based on Takahashi et al.'s description of their findings. The saturated concentration of the suspended load (C_s) was introduced to the STM to maintain numerical stability during extreme flow conditions, such as at the bore front. In a previous study, Gusman et al. (2012) used a value of $C_s = 1$ % in their calculations and obtained good agreement between simulated and observed results. Although there is still some uncertainty associated with the physical model and reasonable determination of the parameter (Takahashi 2012), a value of $C_s = 1$ % was assumed in the present study.

4.3 Results and Discussion

4.3.1 General Topographic Change

Figure 4.6 shows a comparison of the simulated (A) and observed (B; the post-tsunami DEM) topography of the study area. Note that the simulation shows the

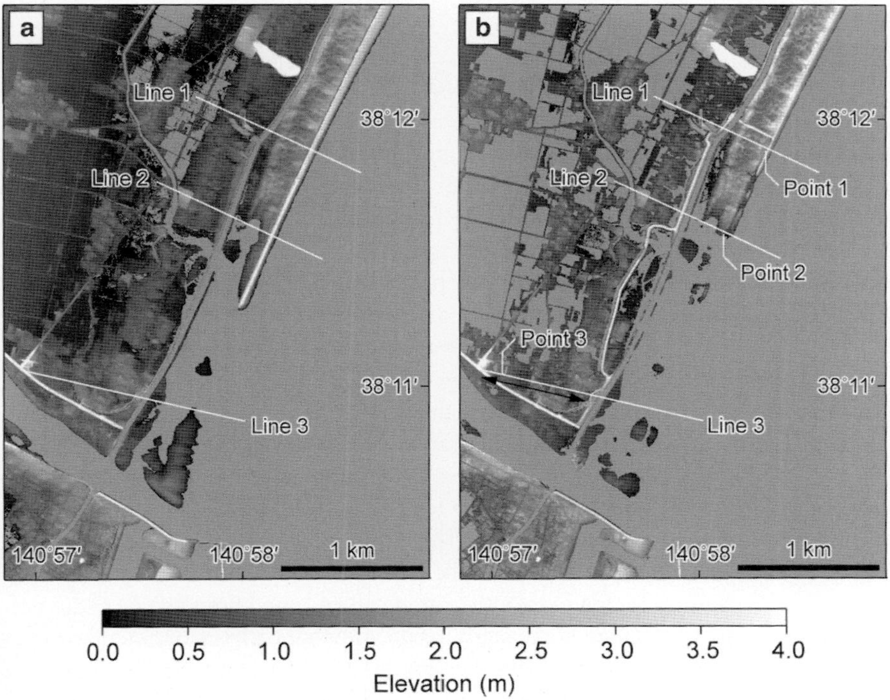

Fig. 4.6 Comparison of (**a**) simulated and (**b**) observed elevations. Note that elevations of the lowlands were not measured because of the flood water resulting from the tsunami. Surfaces covered by water are colored a plain gray

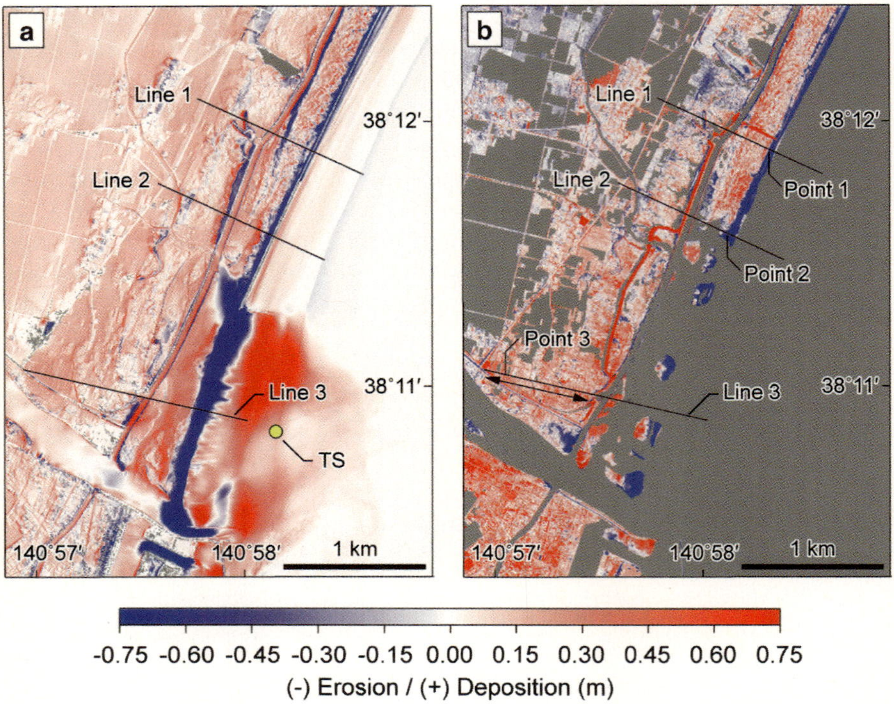

Fig. 4.7 Comparison of (**a**) simulated and (**b**) observed elevation changes. The line with arrows at both ends in (**b**) shows the location of the field survey of tsunami deposits

topography 3 h after the earthquake (nearly 2 h after the tsunami attack), and the observation was made at least 8 days after the tsunami. Subsequent tsunami waves more than 3 h later and normal wind waves might have changed the topography.

Before the tsunami, a sand bar and a lagoon (Ido-ura) were situated in the north of the mouth of the Natori River (Fig. 4.1b). The sand bar was not protected by RC dikes, except for the northern part. The post-tsunami DEM shows that the sand bar was severely eroded and the lagoon was connected directly to the open sea (Fig. 4.6b). In general, the simulation reproduced the massive erosion of the sand bar well, although the remaining area at the south was greater in the simulation. A significant difference between the simulation and observation can be observed at the north section of the sand bar (Figs. 4.6a, b), where an RC dike was installed (Fig. 4.3). In fact, the dike collapsed due to the tsunami, and this may have caused the massive erosion of the sand bar. The erosion in the simulation was more moderate than that observed because in the simulation the RC dike remained intact.

Figure 4.7 shows a comparison of the simulated (A) and observed (B) bathymetric and topographic changes in the study area. Note that bathymetric change is not included in the observed data. Both the simulation results and the observed data showed that the erosion on land was greatest near the beach and on the sand bar.

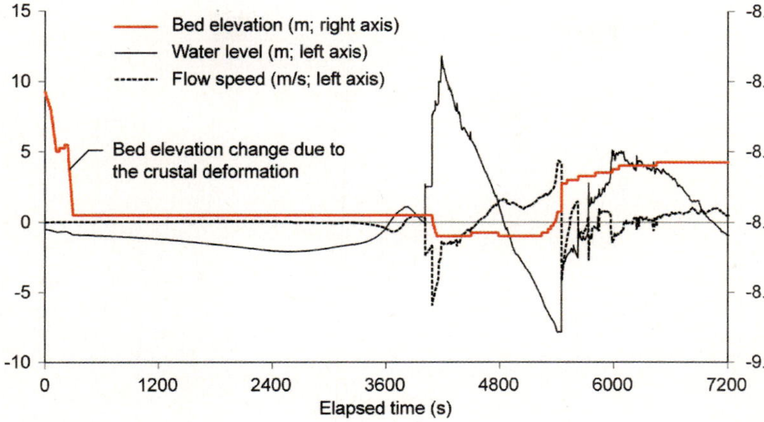

Fig. 4.8 Time series of the bed elevation, water level and flow speed at 500 m offshore of the sand bar (TS in Fig. 4.7). Note that the positive value of flow speed indicates eastward (seaward) flow direction, and the negative value corresponds to westward (landward) flow direction

Figure 4.8 shows a time series of bed elevation, water level and flow speed at 500 m offshore of the sand bar. At this location, the sea bottom was eroded by the first positive wave. The timing of the erosion is associated with the peak of flow speed at 4,100 s, rather than the maximum of the water level, which was delayed nearly 100 s. The bed level was recovered in the last 200 s of the negative wave, and even become higher than the original elevation. The simulation results suggest that most of the sediments that comprised the sand bar were washed out and deposited within 1 km offshore of the coastline, and some of the sediments were deposited in the lagoon. This implies that the total erosion of the sand bar is primarily explained by backflow. In the simulation, minor erosion of the sea bottom occurred approximately 500 m offshore of Lines 1 and 2 (Fig. 4.7a). This suggests that a quite small amount of marine sediment was transported and deposited onshore, which is consistent with findings from recent research on the region near the study area (e.g., Szczuciński et al. 2012).

4.3.2 Erosion at the Backs of RC Dikes

Comparisons of the elevations and their tsunami-induced changes along Lines 1 and 2 (Fig. 4.1b) are shown in Figs. 4.9 and 4.10. The elevation of the RC dike on Line 1 was approximately 4 m (Fig. 4.9a). According to the topographic change estimated from the pre- and post-tsunami DEMs, the erosion at the back of the dike reached more than 1.5 m (Fig. 4.9b); meanwhile, the depth of the erosion was estimated to be up to 2–3 m in field observations (Fig. 4.2). The simulation estimated erosion up to 2.5 m at this location. The difference in these results can

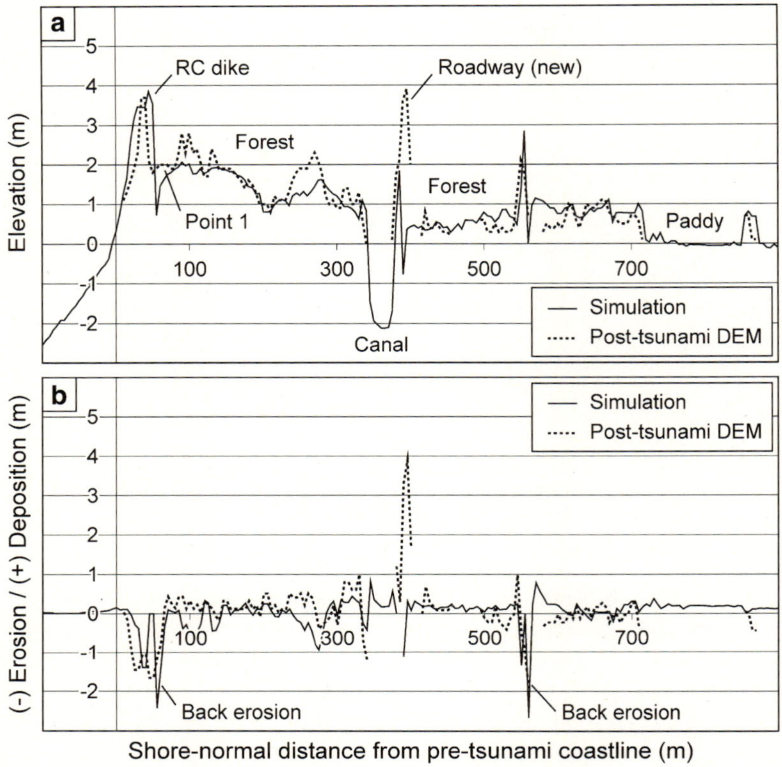

Fig. 4.9 Comparison of simulated and observed (**a**) elevation and (**b**) erosion and deposition along Line 1 in Fig. 4.1b

be attributed to a numbers of factors, such as the parameters applied to the tsunami and sediment transport models, post-tsunami reworking of sediments and precision of the surveying. Nonetheless, the simulation generally reproduced the deep excavation at the backs of the dikes.

The differences in elevation and its tsunami-induced changes are significant at locations where an RC dike collapsed. The elevation of the dike on Line 2 was 5.5 m (Fig. 4.10a), and the erosion at the back of the dike predicted in the simulation approached 3.5 m (Fig. 4.10b). In fact, the RC dike was severely damaged and the sand bar was flattened along Line 2 (Fig. 4.3). Compared to the RC dike along Line 1, it is obvious that deeper erosion occurred at the back of the higher RC dike along Line 2. This may have led to loss of foundation support and total collapse of the dike, such as described by Tappin et al. (2012). In Fig. 4.10, the differences in the simulated and observed elevations (A) and their change (B) are significant at least 300 m from the coastline. This suggests that the protective effect of the dike would have extended to this distance if the dike could have withstood the tsunami impact.

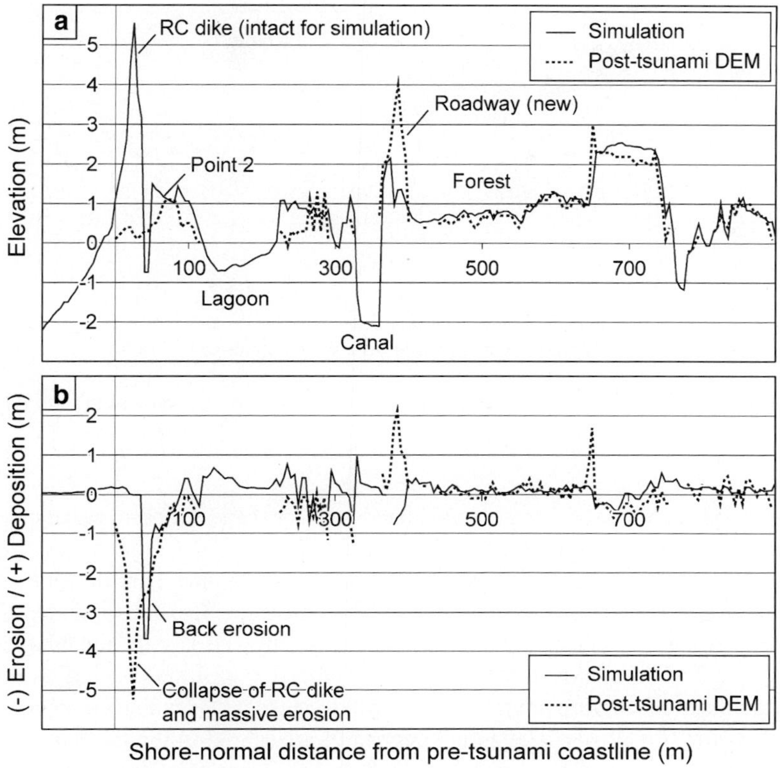

Fig. 4.10 Comparison of simulated and observed (**a**) elevation and (**b**) erosion and deposition along Line 2 in Fig. 4.1b

4.3.3 Detailed Comparison of Erosion and Deposition

In the study area, coastal forests of pine trees covered the zones from the backs of the dikes up to 600 m from the beach. Both erosion and deposition can be observed in the simulation and observation (Fig. 4.7), and the profiles of the elevation and its changes are generally consistent between the simulation and the observation (Figs. 4.9 and 4.10). However, the distribution patterns of the elevation changes are slightly different. For example, the difference in the elevation along Line 1 is significant at approximately 100 and 250 m from the coastline (Fig. 4.9). The simulation indicated that erosion up to 0.5–1.0 m occurred, whereas the observation showed that deposition dominated at these locations. This discrepancy occurred because a uniform value of Manning's roughness coefficient of $n = 0.040$ was assumed in the simulation for the coastal forest. In fact, the age, height and density of the trees varies from place to place. In addition, the trees in this area were broken or uprooted and flattened during the tsunami attack. Actual frictional resistance to flow and the capacity for sediment transport would have varied in terms of time and space. To reproduce in detail

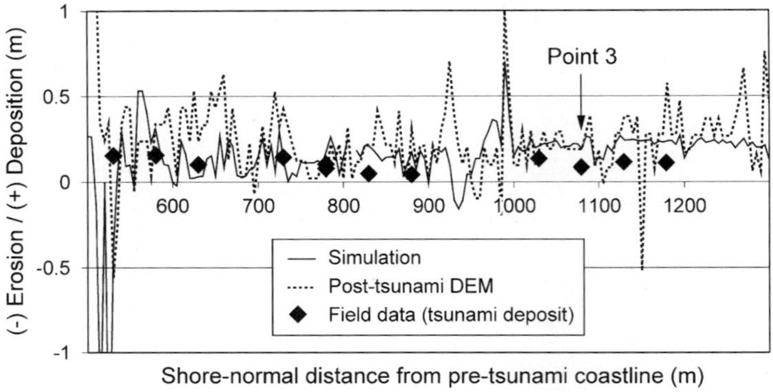

Fig. 4.11 Comparison of simulated and observed deposition and erosion along Line 3 in Fig. 4.1b

the topographic change of a densely vegetated area, local and temporal variation of the roughness coefficient should be included in the STM.

Figure 4.11 shows the erosion and deposition along the part of Line 3 that extends from 500 to 1,200 m inland of the coastline. Erosion and deposition estimated from the pre- and post-tsunami DEMs showed significant fluctuations in this section. This may be attributable to the precision of the surveying, as well as to the inclusion of tsunami debris. It is difficult to make a reasonable comparison for this line using the DEMs. The actual deposition measured in the field approached 15 cm and was thinnest approximately 800–1,000 m from the coastline. The thickness of deposition according to the simulation was comparable to that measured in the field data for some locations; however, the simulation generally overestimated the measured values.

In particular, in the section 1,000–1,200 m from the coastline, where the measured deposition was approximately 10 cm, the simulated deposition exceeded 20 cm, twice the field-measured thickness. As mentioned in Sect. 4.3.2, the difference can be explained by a numbers of factors, although post-tsunami reworking of sediments would not have taken place because the sand layer was capped by a mud layer deposited from suspension during the stagnation period of the tsunami flooding. Although the thickness of the tsunami deposition found after the Tohoku-oki tsunami approached 30 cm near the beach, it suddenly decreased to 15 cm at a distance 1 km inland of the coastline. At the landward limit of the distribution, the thickness of the sand layer thinned from a few centimeters to a few millimeters (Abe et al. 2012). Takahashi (2012) noted that the concentration of suspended load estimated by the STM is sensitive to the flow speed derived from the tsunami inundation model and the parameterization of the saturation of the suspended load. Simulation of tsunami deposition in inland areas may require additional research on the model parameters involved, such as Manning's roughness coefficient, n, the saturated concentration of the suspended load, C_s, and grain-size-dependent coefficients of the pick-up rate equations.

4.4 Conclusions

We conclude that the tsunami sediment transport model (STM) can be applied to the erosion and deposition by the Tohoku-oki tsunami. The simulated erosion on the sand bar and at the backs of the RC dikes was consistent with observations, and the eroded sediments could have been the main source of the tsunami sediments deposited inland. To simulate deposition and erosion in coastal forests, spatial and temporal change of the roughness coefficient needs to be implemented in the STM. Application of the STM to the formation of thin tsunami-deposited layers (less than 30 cm) inland requires further research on the model parameters.

Acknowledgments We acknowledge the Tohoku Regional Bureau, Ministry of Land, Infrastructure and Transport, for providing the post-tsunami digital elevation model of the study area. We wish to thank Mr. Shunji Iwama, for his efforts in making the land use map. The research was financially supported by and carried out as a part of the Strategic Program for High-Performance Computing Infrastructure sponsored by the Ministry of Education, Culture, Sports, Science and Technology in Japan.

References

Abe T, Goto K, Sugawara D (2012) Relationship between the maximum extent of tsunami sand and the inundation limit of the 2011 Tohoku-Oki tsunami on the Sendai Plain. Sed Geol 282:142–150. doi:10.1016/j.sedgeo.2012.05.0049

Goto C, Ogawa Y, Shuto N, Imamura F (1997) Numerical method of tsunami simulation with the leap-frog scheme. IUGG/IOC TIME Project, IOC manuals and guides, UNESCO, 35

Gusman AR, Tanioka Y, Takahashi T (2012) Numerical experiment and a case study of sediment transport simulation of the 2004 Indian Ocean tsunami in Lhok Nga, Banda Aceh, Indonesia. Earth Planets Space 64:817–827

Japan Nuclear Energy Safety Organization (2011) Cross-check analysis of the simulations of the 2011 Tohoku-Oki earthquake tsunami by the electric power company. http://www.nsr.go.jp/archive/nisa/shingikai/800/26/003/3-4.pdf. Retrieved on 2 Nov 2012

Kotani M, Imamura F, Shuto N (1998) Tsunami run-up simulation and damage estimation by using GIS. Proceedings of coastal engineering. JSCE 45:356–360

Matsumoto H (1985) Beach ridge ranges and the Holocene sea-level fluctuations on alluvial coastal plains, Northeast Japan. Sci Rep 35:15–46, Tohoku Univ., 7th series (geography)

Okada Y (1985) Surface deformation due to shear and tensile faults in halfspace. Bull Seismol Soc Am 75(4):1135–1154

Ozawa S, Nishimura T, Suito H, Kobayashi T, Tobita M, Imakiire T (2011) Coseismic and postseismic slip of the 2011 magnitude-9 Tohoku-Oki earthquake. Nature 475:373–376

Richmond B, Szczuciński W, Chagué-Goff C, Goto K, Sugawara D, Witter R, Tappin DR, Jaffe B, Fujino S, Nishimura Y, Goff J (2012) Erosion, deposition and landscape change on the Sendai coastal plain, Japan, resulting from the March 11, 2011 Tohoku-Oki tsunami. Sed Geol 282:27–39. doi:10.1016/j.sedgeo.2012.08.005

Szczuciński W, Kokocińsk M, Rzeszewski M, Chagué-Goff C, Cachão M, Goto K, Sugawara D (2012) Sediment sources and sedimentation processes of 2011 Tohoku-Oki tsunami deposits on the Sendai Plain, Japan – insights from diatoms, nannoliths and grain size distribution. Sed Geol 282:40–56. doi:10.1016/j.sedgeo.2012.07.019

Takahashi T (2012) Numerical modeling of sediment transport due to tsunamis and its problem. J Sed Soc Jpn 71(2):149–155

Takahashi T, Shuto N, Imamura F, Asai D (2000) Modeling sediment transport due to tsunamis with exchange rate between bed load layer and suspended load layer. Proc Int Conf Coas Eng 2000, Vol. 2, ASCE:1508–1519

Takahashi T, Kurokawa T, Fujita M, Shimada H (2011) Hydraulic experiment on sediment transport due to tsunamis with various sand grain size. J JSCE (J Seismol Coas Eng) 67:231–235, Ser. B2 (Coastal Engineering)

Tappin DR, Evans HM, Jordan CJ, Richmond B, Sugawara D, Goto K (2012) Coastal changes in the Sendai area from the impact of the 2011 Tohoku-Oki tsunami: interpretations of time series satellite images, helicopter-borne video footage and field observations. Sed Geol 282:151–174. doi:10.1016/j.sedgeo.2012.09.011

Chapter 5
NOAA's Historical Tsunami Event Database, Raw and Processed Water Level Data, and Model Output Relevant to the 11 March 2011 Tohoku, Japan Earthquake and Tsunami

Paula Dunbar, Marie Eblé, George Mungov, Heather McCullough, and Erica Harris

Abstract On 11 March 2011, a magnitude 9.0 *Mw* earthquake occurred near the east coast of Honshu Island, Japan. The earthquake generated a tsunami with wave heights up to 40 m triggering a response from the National Oceanic and Atmospheric Administration (NOAA). NOAA is the lead federal agency responsible for tsunami warnings issuance, technology innovation, research, and mitigation for the United States. These activities are matrixed across several closely integrated bureaus, including the National Weather Service National Data Buoy Center, Tsunami Warning Centers, and UNESCO/IOC – NOAA International Tsunami Information Center; the National Ocean Service Center for Operational Oceanographic Products and Services, the Pacific Marine Environmental Laboratory; and the National Environmental Satellite, Data, and Information Service National Geophysical Data Center (NGDC). Each of these bureaus work together to improve tsunami forecasting and thereby protect lives. Immediately following the March 2011 Japan earthquake and tsunami, all NOAA offices were involved in complimentary tsunami activities. In this paper, we describe the status of data associated with this tsunami event, review the processing and availability of tide gauge and tsunameter data, including Deep-ocean Assessment and Reporting of Tsunamis (DART®) and discuss the uses of these data in tsunami models. The basic data in the NGDC historical event databases include: date, time, event location, magnitude of the phenomenon, and socio-economic information such as the total number of

P. Dunbar (✉) • H. McCullough
NOAA, NESDIS, National Geophysical Data Center, Boulder, CO 80305-3328, USA
e-mail: Paula.Dunbar@noaa.gov

M. Eblé
NOAA, OAR, Pacific Marine Environmental Laboratory, 7600 Sand Point Way NE, Seattle, WA 98115, USA

G. Mungov • E. Harris
NOAA, NESDIS, National Geophysical Data Center, Boulder, CO 80305-3328, USA

CIRES, University of Colorado at Boulder, 216 UCB, Boulder, CO 80309-0216, USA

Y.A. Kontar et al. (eds.), *Tsunami Events and Lessons Learned: Environmental and Societal Significance*, Advances in Natural and Technological Hazards Research 35,
DOI 10.1007/978-94-007-7269-4_5,

fatalities and dollar damage estimates. The tsunami database includes additional information on runups (locations where tsunami waves were observed by eyewitnesses, post-tsunami field surveys, tide gauges, or deep ocean sensors). Therefore, an introduction and summary of the effects of the earthquake and tsunami is also included.

Keywords Damage • Earthquake • Runup • Tide gauge • Tsunameter • Tsunami • 2011 Tohoku Japan Tsunami

5.1 Introduction

The National Oceanic and Atmospheric Administration (NOAA) is the lead federal agency responsible for tsunami warnings issuance, technology innovation, research, and mitigation for the United States. These activities are matrixed across several closely integrated bureaus:

- National Weather Service (NWS) line offices have several responsibilities. The NWS Global Telecommunication System (GTS) provides real-time data flow of all observation data including Deep-ocean Assessment and Reporting of Tsunamis (DART®) observations from the network managed by the NWS National Data Buoy Center (NDBC). The NWS Tsunami Warning Centers (TWC) operationally forecast tsunami impact along U.S. coastlines and issue warnings. The International Tsunami Information Center (ITIC) is a United Nations Educational, Scientific and Cultural Organization (UNESCO)/Intergovernmental Oceanographic Commission (IOC) – NOAA partnership that assists Member States in establishing warning systems and serves as a clearinghouse for educational and preparedness materials.
- National Ocean Service (NOS) Center for Operational Oceanographic Products and Services (CO-OPS) maintains the U.S. coastal sea level tide-gauge network and distributes all data, including 1-min, for operations and research.
- Office of Atmospheric Research (OAR) Pacific Marine Environmental Laboratory (PMEL) works to address technology obsolescence and conducts tsunami research to improve observation and forecasting technologies. PMEL designed and developed the DART® system that is now the international tsunameter standard.
- National Environmental Satellite, Data, and Information Service (NESDIS) National Geophysical Data Center (NGDC) maintains the global tsunami archive, providing integrated access to tsunami event observational, instrumental, and socio-economic impact data. This includes the historical tsunami and significant earthquake databases, imagery, raw and processed coastal tide gauge and DART® data relevant to a tsunami event. NGDC also hosts the World Data Service (WDS) for Geophysics.

Each of the above mentioned bureaus work together to improve tsunami forecasting and thereby protect lives. CO-OPS tide gauge data and NDBC DART® and other tsunameter data are transmitted to the TWCs in real-time to verify tsunami arrival time and amplitude at ocean and coastal locations. NGDC processed tide gauge data are used to calibrate forecast models and determine tidal characteristics for tsunami isolation in any given signal. NDBC DART® and international tsunameter observations are used in real-time to estimate a tsunami source via inversion techniques and post-event to improve both the inversion process and the models. The work of NGDC to minimize background noise and all tidal contribution to the signal is essential to the forecasting effort.

Immediately following the March 2011 Japan earthquake and tsunami, all NOAA offices were involved in complimentary tsunami activities. In this paper, we describe the status of data associated with this tsunami event, review the processing and availability of tide gauge and tsunameter data, and discuss the uses of these data in tsunami models. The basic data in the NGDC historical event databases include: date, time, event location, magnitude of the phenomenon (tsunami or earthquake magnitude and/or intensity), and socio-economic information such as the total number of fatalities, injuries, houses damaged or destroyed, and dollar damage estimates. The tsunami database includes additional information on runups (locations where tsunami waves were observed by eyewitnesses, post-tsunami field surveys, tide gauges, or deep ocean sensors). Therefore, an introduction and summary of the effects of the earthquake and tsunami follows.

5.2 11 March 2011 Tohoku, Japan Earthquake and Tsunami

On 11 March 2011, at 05:46:24 UTC, a magnitude *Mw* 9.0 earthquake occurred at 38.297° N, 142.373° E, depth 29 km, near the east coast of Honshu Island, Japan (USGS 1971 to present). The earthquake generated a tsunami with wave heights up to 40 m in the near field at Aneyoshi Bay south of Miyako City (Mori et al. 2011). The tsunami, not the earthquake, was responsible for the majority of casualties and a large portion of the physical damage and loss of life (PEER/EERI/GEER/Tsunami Field Investigation Team 2011). The National Police Agency of Japan (2012) reported that as of 3 October 2012 there were 15,870 deaths, 2,813 missing and presumed deaths, and 6,111 injuries in 20 Japanese prefectures. More than 90 % of the deaths were due to drowning and 65 % of the dead were older than 60 years of age (UNESCO 2012). The earthquake and tsunami also caused a nuclear disaster with an International Atomic Energy Agency (IAEA) rating of 7 that included equipment failures, explosions, fires, nuclear core meltdowns, and radiation release (Strickland 2011). The damage costs resulting from the earthquake and tsunami in Japan are estimated at 16.9 trillion yen (approximately $220 billion) (UNESCO 2012). A review of the NGDC historical tsunami databases reveals that this is the deadliest tsunami since the 2004 magnitude 9.1 Sumatra earthquake and tsunami that caused nearly 230,000 deaths and $10 billion in damage and the most costly tsunami that has ever occurred.

5.2.1 Damage in Japan from the 2011 Tohoku Event

Mori et al. (2011) determined from field surveys that the tsunami affected a 2,000 km stretch of Japan's Pacific coast and inundated more than 400 km^2 of land. The earthquake and tsunami resulted in the complete collapse and wash out of 129,549 houses and partially damaged 265,781 houses (NPA of Japan 2012). After the earthquake, there were 345 fires in 12 prefectures, including some cases where the tsunami triggered the fire (Mimura et al. 2011). The damage to infrastructure was also widespread including damage to 4,200 areas along roads, 77 bridges, and 29 parts of the railway system (NPA of Japan 2012). Lifeline infrastructure was also damaged (Mimura et al. 2011).

This event caused explosions and leaks in three reactors at the Fukushima I (Daiichi) Nuclear Power station, located approximately 150 km from the epicenter. When the earthquake occurred the reactors that were in operation at Daiichi shut down as they were designed. Although the local power grid had been knocked out, the plant had its own generators. Approximately 1 h after the earthquake, the generators shut down as a result of the tsunami waves (Sanger and Wald 2011). The Tokyo Electric Power Company (TEPCO) estimated the tsunami at 10 m at Daiichi and 12 m at Daiini (World Nuclear News 2011). The Fukushima power plants were required to withstand a certain height from tsunamis: 5.7 m at Daiichi and 5.2 m at Daiini (World Nuclear News 2011). As of 30 September 2011 Japan lifted evacuation advisories for an area spanning five towns and cities around the nuclear power plant (Tabuchi 2011). A 12-mile exclusion zone will remain in place around the nuclear power plant and the worst contaminated areas close to the plant are likely to remain uninhabitable for decades. On 16 December the Government of Japan announced that all reactors at the Fukushima Daiichi Nuclear Power Station had been brought into a condition equivalent to "cold shutdown" (IAEA 2011).

In contrast to Fukushima, according to the IAEA, there was no damage to the Onagawa Nuclear Power Plant, located closest to the epicenter (IAEA 2012, the final report will be available in 2013). The plant experienced very high levels of ground shaking and some flooding from the tsunami, but was able to shut down safely (IAEA 2012). According to the Tohoku field survey there were runup heights of almost 13 m near the power plant (Mori et al. 2011).

5.2.2 Damage Outside of Japan from the 2011 Tohoku Event

Loss of life outside of Japan was minimal due to warnings and evacuations; but there were two confirmed distant deaths, one in California (Wilson et al. 2012) and one in Indonesia (The Jakarta Globe 2011). Financial damage outside of the Japan source region was quite extensive. Greater than $30 million damage occurred on the Hawaiian islands of Oahu, the big island of Hawaii, Maui, and Kauai (Abercrombie 2011). Moderate to strong currents were observed in all coastal harbors and marinas in California, leaving at least two dozen marine facilities with damage totaling over $50 million (Wilson et al. 2012). A relatively small but noteworthy amount of damage was inflicted on French Polynesia (Reymond et al. 2012) and the Galapagos Islands (USGS 1971 to present). Houses were destroyed in Pisco, Peru, and several buildings were destroyed at Dichato, Chile (USGS 1971 to present).

5.3 NGDC Tsunami Data Archive for the 2011 Tohoku Event

As of 3 October 2012, NGDC/WDS has collected maximum tsunami amplitudes from 289 tide gauge observations, 34 tsunameter, primarily DART®, station observations, and more than 5,400 eyewitness reports and post-tsunami field survey measurements. These observations were added to the NGDC historical tsunami database and were overlaid onto the calculated tsunami travel time map shown in Fig. 5.1.

5.3.1 Post-Tsunami Field Survey Data

After the 2011 tsunami, researchers throughout Japan conducted tsunami surveys along a 2,000 km stretch of the Japanese coast (Mori et al. 2011). These data were added to the NGDC database and represent the largest tsunami survey dataset ever collected for one event. The dataset includes more than 5,200 locations (31.5–43.7° N) with runup heights (elevation at maximum inundation), maximum water level, maximum horizontal distance of inland flooding or inundation, modeled estimated arrival time of the largest wave, and reliability. These heights were measured from sea level excluding the astronomical tide. Inundation distance and run-up heights "were measured within a few centimeters accuracy from watermarks on building, trees, and walls by using a laser range finders, a real-time kinematic (RTK) GPS receiver with a cellular transmitter, and total stations. Run-up height was determined from the maximum landward extent of debris and seawater marks" (Mori et al. 2011). The maximum runup height was 40 m at Ofunato. Figure 5.2 shows plots of the total number of deaths, runup or maximum water height, and maximum inundation by

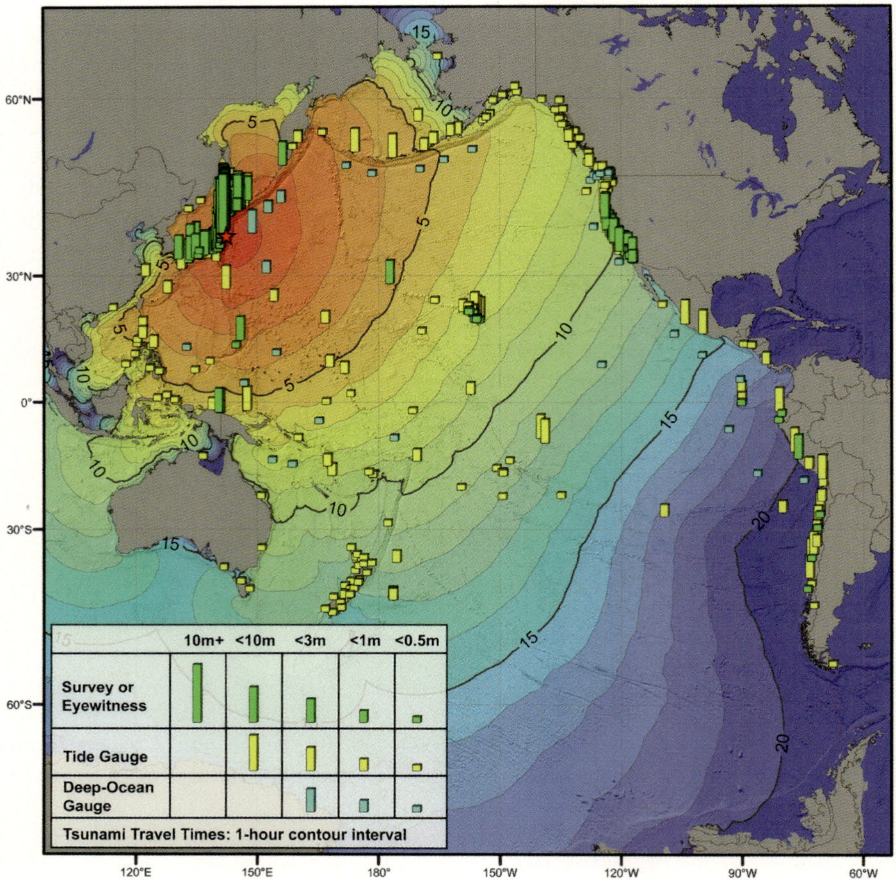

Fig. 5.1 Plot of the 11 March 2011 magnitude 9.0 Japan earthquake epicenter; water heights from tide gauges, DART® stations, field surveys, and eyewitness accounts that observed the tsunami wave; overlaid onto the calculated tsunami travel time map (contour intervals are in hours) (Source: NGDC/World Data Service for Geophysics)

latitude from 35.5° to 42° N. The highest values in all three plots are in the region from 38° to 40° N, near and just north of the epicenter at 38.3°N. These data will be very useful for understanding and modeling tsunami generation, propagation, and inundation of dry land, particularly in the near field relative to tsunami generation.

The NGDC database also includes field survey data for Indonesia (BMKG 2011), French Polynesia (Reymond et al. 2012), Mexico (Jonathan et al. 2011), and the west coast of the U.S. (Wilson et al. 2012). Additional field survey data will be added as it becomes available.

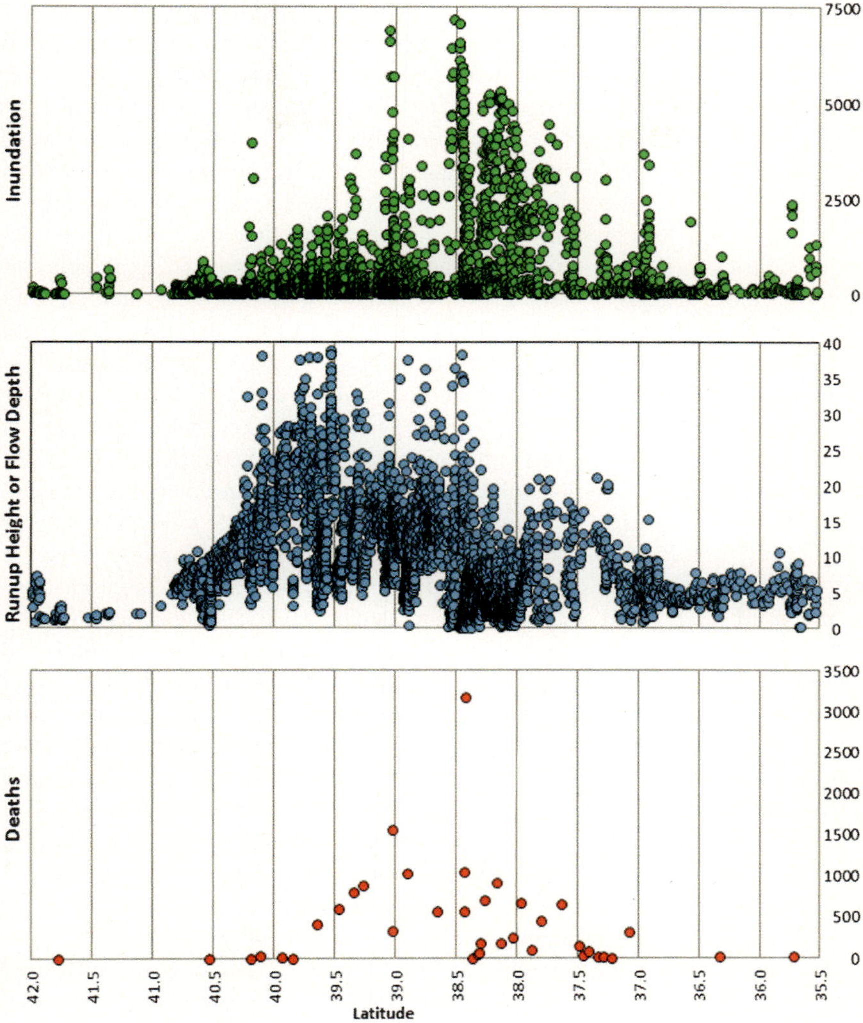

Fig. 5.2 Latitude-based graphs of total deaths for selected principalities and prefectures, run up or maximum water height, and maximum inundation distance from the 11 March 2011 Tohoku, Japan tsunami. The epicenter is at 38.3°N (Source: NGDC historical tsunami event database compiled from UNESCO/IOC 2011; Mori et al. 2011)

5.3.2 Bottom Pressure Recorder (BPR) Data

The tsunameter system is the primary component of the tsunami observational network, with each system sited in the deep-ocean near potential tsunami sources. As a tsunami wave propagates across the open ocean, DART® and internationally equivalent tsunameter systems register the passage of a tsunami and transmit observations to NOAA's two Tsunami Warning Centers and warning centers around the world. These observations are used to identify the tsunami source, a refinement of the initial earthquake parameters including magnitude that are based purely on seismic information (Percival et al. 2011). The refinement is accomplished by inversion, or fit, of observations at tsunameter stations to modeled tsunami wave forms from a precomputed propagation database. The inversion algorithm is used to assess the parameters of the tsunami source based on tsunameter observations. Source parameters revised to represent the tsunami source are then used to improve the forecast of tsunami arrival time, wave amplitude, and inundation or flooding at specific far-field coastal locations (Titov 2009). Both near-field and far-field tsunameter data are used post-event to further research tsunami wave propagation and the energy dissipation along its path, and to improve near-field forecasting.

Each tsunameter consists of an anchored seafloor bottom pressure recorder (BPR) and a companion moored surface communication buoy. Pressure is converted to sea surface height and transmitted via Iridium satellite at 6-h intervals for standard monitoring. High-resolution 15-s integrations are internally recorded on a memory card by each BPR unit. These data are available either remotely or upon recovery of a BPR unit during station service operations. Presently, there are two tsunameter technologies in widespread use: The United States and internationally deployed DART® II and the Easy-to-Deploy (ETD) system deployed by international partners. Both technologies incorporate the same electronics and sampling schemes, but the compact ETD system is deployed as a disposable unit with an operational life in excess of 4 years. Internally recorded 15-s ETD data are currently only accessible remotely via satellite in 1-h time segments. Engineering efforts are underway to support data uploads spanning 1–2-weeks. ETD systems require significantly less expenses for deployment and operations. A more detailed description of the DART®, and therefore the tsunameter system along with a history of tsunami observations can be found in Mofjield (2009). Additional information is provided on the web site of the NOAA/PMEL National Center for Tsunami (http://nctr.pmel.noaa.gov).

As of this writing, the current Pacific Ocean DART® system consists of 43 stations: 33 are owned and operated by US/NOAA, 6 by Australia, 2 by Russia, 1 by Chile, and 1 by Ecuador (set up September 2012). The 2011 Tohoku event was observed and recorded on all of these stations, except 6 NOAA, 2 Australian, and the 1 station that was not yet deployed, for a total of 34 records. Of these, 18 are 15-s high-resolution records from retrieved BPR units and 16 are real-time records from ETD systems or from BPRs that have not been serviced.

Tsunami waves as high as 10 cm were observed at deep-ocean stations located almost 16,000 km away from the earthquake source. At DART® station 21413, located 1,278 km southeast of Tokyo and approximately 1,280 km east-southeast of the epicenter, the tsunami height was 1 m. The largest wave height ever recorded in the deep ocean was observed to be 1.8 m (peak-to-trough) at DART® station 21418 located 833 km northeast of Tokyo and approximately 527 km east of the epicenter. The actual tsunami height at this station is expected to be larger since the returned observation is a 1-min average over four base frequency samples, but will likely remain unknown since this BPR was lost during recovery operations. A specific feature of this event is that the tsunami propagates in the deep ocean as one, two, or three large waves and then exhibits a strong and rapid decay with a long tail of waves having significantly lower amplitudes. Using satellite altimeter data Song et al. (2012) show that the amplification of the first tsunami waves is caused by the merging of the tsunami front when it travels along ocean ridges and seamount chains in the Pacific Ocean resulting in the doubling of the wave height and destructive potential in certain directions.

The 2011 Tohoku tsunami was recorded by a number of Japanese cable BPRs on observational networks operated by the Japan Agency for Marine-Earth Science and Technology (JAMSTEC), Earthquake Research Institute (ERI), and Japan Meteorological Agency (JMA), and at Canadian BPRs from the NEPTUNE-Canada and VENUS geophysical networks. These data are available from the corresponding agencies.

Bottom pressure recorder data from all reporting tsunameters and ETDs for the 2011 Tohoku event were added to the NGDC water-level database following data processing and quality control. The addition of these real-time and retrospective Tohoku data, brings the NGDC event database holdings, as of September 2012, to 147 BPR records from the Pacific Ocean, Caribbean Seas, Gulf of Mexico, and Atlantic Ocean. This includes 127 high-resolution 15-s time series recovered and processed retrospectively and 20 real-time records. Tsunami events that occurred after 2006 are well represented in the database due to increased hazard awareness and observational technology following the 2004 Sumatra tsunami. Longer periods of observations exist for a finite number of stations occupied for research beginning in the 1980s prior to establishment of the real-time sensing network. Three stations south of the Alaska-Aleutian Archipelago and two stations off of the U.S. coast of Oregon State have provided observations starting in 2002. In addition, DART® 46406 (51406), located in the Central Pacific 5,370 km southeast of Hawaii has been occupied since 2003. All available real-time and high-resolution 15-s tsunameter data for the 2011 Tohoku event are available on the NGDC website (http://ngdc.noaa.gov/hazard). Additional data are available upon request to haz.info@ngdc.noaa.gov.

5.3.3 Tide Gauge Data

Maximum tsunami wave heights for the 2011Tohoku event were measured in Japan by a "huge tsunami gauge" network consisting of water pressure sensors that are installed at wharves to measure extremely large tsunamis with amplitudes beyond the measuring range of the regular tide gauges. In general, this tsunami was observed at more than 60 Japanese tide stations situated on all Japanese islands according to Koizumi (2011).

The 2011 Tohoku tsunami was observed at tide gauges located in 30 Pacific Rim countries, in Antarctica, and on the West coast of the Atlantic Ocean at Arraial do Cabo, Brazil.

NGDC has processed 1-min time-series US tide gauge data that contain the 2011 Tohoku tsunami signal: 11 records from the US Pacific Islands, 7 records from the Alaska region, and 13 records from the US West coast. Including these Tohoku data, as of September 2012, NGDC has collected 1-min data from a total of 188 tide-gauge stations: US Pacific islands (11 records), Alaska region (31 records) and US West coast (35 records), US East coast (59 records), Caribbean Seas (16 records) and from the US Gulf Coast (38 records). The period with 1-min observations on most of these stations is from 01 January 2008 to the present. These data could be used to investigate tsunamis and other extreme events requiring high-resolution records. These data are available upon request to haz.info@ngdc.noaa.gov.

5.4 Model Output and Products

The NOAA tsunami forecasting system combines observations and models to produce timely and accurate predictions for guidance. Typical model output includes tsunami arrival time, wave amplitude, and flooding forecasts at model grid points corresponding to specific locations selected because of population, infrastructure, or financial interest. Additionally, current predictions are output but validation of these predictions is hampered due to the lack of observations.

The NOAA tsunami forecast system utilizes the Method of Splitting Tsunamis (MOST) model (Titov and Synolakis 1997, 1998), for both real-time forecasting and for time-insensitive hazard assessment and research, to simulate the three processes of tsunami evolution: generation, propagation, and flooding of typically dry land. Digital Elevation Models (DEMs) constructed by NGDC using the best available topography and bathymetry data, a propagation database consisting of a continuous and discrete set of fault segments along subduction zones around the world, and community specific models, are all central to the MOST methodology. Figure 5.3 shows the propagation forecast of maximum amplitudes calculated with the NOAA tsunami forecast model for the 2011 Tohoku, Japan tsunami.

The central components of the NOAA tsunami forecast system were developed with warning operations in mind, but each are resources that can be further leveraged to conduct a suite of activities ranging from basic research to hazard assessment and emergency planning at all government levels. Digital Elevation Models (DEMs), a database of precomputed tsunami propagation, and 75 community-specific forecast models for the U.S. coastal States and territories are all elements of the system. Overlapping DEMs constructed by NGDC along the coasts of Washington, Oregon, and California, DEMs encompassing the Hawaiian Islands, and those along finite

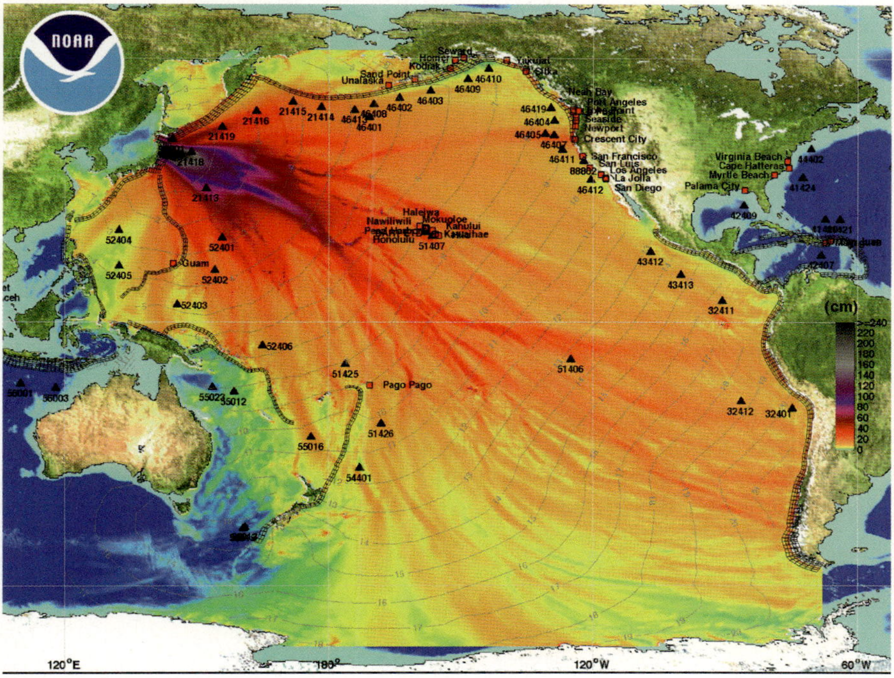

Fig. 5.3 The propagation forecast of maximum amplitude calculated by the NOAA forecast model for the 11 March 2011 Tohoku, Japan tsunami. The colors correspond with maximum amplitudes in the range 0–240 [cm] indicated by the *color bar*. – *Filled colors* show maximum computed tsunami amplitude in cm during 24 h of wave propagation. *Black contours* show computed tsunami arrival time

regions of Alaska and Pacific Island coastlines are all publically available on the NGDC website. The MOST propagation database contains pre-computed tsunami propagation results from each discrete segment and a set of tsunami waveforms across the Pacific Basin over all grid points for tsunamis generated by earthquakes with a fault rupture over each segment. The propagation database is utilized by researchers and state emergency managers to conduct hazard assessments within their areas of responsibility to improve evacuation maps and update building zone maps. The total of 75 community-specific forecast models developed for operational flooding forecasting also provide the opportunity to investigate and plan for currents in ports and harbors as well as conduct hazard assessments on a fine resolution scale. Tsunami event-specific wave height and arrival times as compared with observations within the context of research are available at: http://nctr.pmel.noaa.gov. Operationally-determined model results are available on the two NWS TWC websites.

5.5 Water Level Data Processing

The NGDC water level database focuses on tsunamis and other extreme events, not on the features of the hydrological regime. This means that "event" records should be cleaned of all other components to the extent possible. Investigation of the extreme events requires special attention to all registered high-frequency variations and any integration over smaller time intervals will mask important "extreme features." At the same time we try to avoid excessive pre- and post- processing as this could smooth or entirely remove all extreme characteristics from the record. A detailed description of tsunameter data processing and subsequent data quality control is provided by Mungov et al. (2012). The same processing techniques are applied to coastal 1-min tide-gauge data, where additional attention is paid to specific instrumental issues such as spikes and inverted observations. A modified despiking procedure is based on the application of two-level median filtering while the threshold limits are calibrated against observational data (Goring and Nikora 2002). All segments that are known to include a tsunami signal are visually, statistically, and spectrally inspected for quality assurance. This practice provides confidence that even the small details of the tsunami signal are not altered. For this reason we do not apply any automatic data processing.

To remove the tides we use a customized version of the tidal package of the Institute of Ocean Sciences (IOS, Sidney, BC, Canada) developed by Foreman et al. (2009). Details of the application are described in Mungov et al. (2012).

5.5.1 Post-Processing to Extract "Clean" Tsunami Signals and Their Spectral Characteristics

The "clean" tsunami signal is extracted by applying bandpass filtering of the residuals after the tidal analysis following Rabinovich et al. (2011) and Eblé et al. (2011). Each time band-passed filter parameters are calibrated against the data and the signal-to-noise ratio is investigated to ensure that we did not alter the tsunami record. In some cases, we prefer to leave some background noise or small spikes rather than distort or smooth the tsunami signal and corrupt the tsunami arrival times. Plots of tsunami signals obtained from DART® stations for the 2011 Tohoku, Japan tsunami are shown in Fig. 5.4.

5.6 Data Access

NGDC (http://www.ngdc.noaa.gov/hazard) and PMEL (http://nctr.pmel.noaa.gov/) are committed to delivering data and information in a manner that enables integration and supports multiple uses of data. Most of the data in the NGDC archive are in

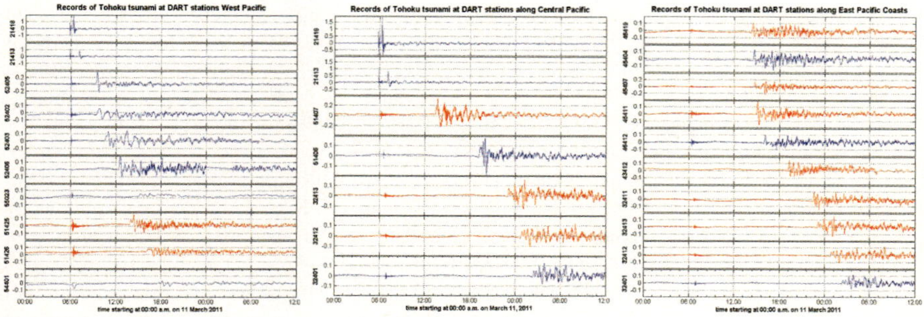

Fig. 5.4 Tsunami signals obtained from DART® stations for the 11 March 2011 Tohoku, Japan tsunami

the public domain and may be used without restriction, although users should cite data sources. These data can be searched from web maps or form interfaces by attribute (date, location, magnitude, etc.) and displayed as tables, reports, interactive maps, and imagery. For example, the maximum amplitudes from all of the tide gauge and DART® stations that observed the 2011 Tohoku tsunami can be found by searching the NGDC tsunami runup database forms interface. Additional details of the NGDC Tsunami Data Archive system are described by Dunbar et al. (2008).

The 2011 Tohoku tide gauge and DART® time-series data are not yet integrated with the runup database, but NGDC does disseminate summarized information for all recent and significant tsunami events. These web pages provide access to all available processed and archived 15-s high-resolution records and real-time data from all DART® stations where these specific events were recorded. The information provided has been verified prior to posting on the "DART® data" pages. These pages include a description of the earthquake event with links to the NGDC tsunami event and runups pages, the NDBC DART® system page, and to the Tsunami Travel Time Software used for development of tsunami travel time maps. On the tsunami travel time maps, all DART® systems reporting real-time and archived data are indicated.

5.7 Conclusion

In conclusion, NOAA covers a full suite of tsunami observations, data transmission, forecasting, post-event data processing, and data dissemination to a wide group of users. This large effort is made possible by each bureau of NOAA specializing in different areas of expertise, while simultaneously integrating their resources to create meaningful products for stakeholders. By continually examining each aspect in this process, NOAA will lead the ongoing effort to increase our understanding and ability to forecast the impact of future tsunami events, thereby allowing for science-based sound management and planning of U.S. coastal communities.

Acknowledgments The authors wish to acknowledge the partner agencies whose continued collaboration ensures the success of the United States' efforts to provide real-time tsunami warning to coastal communities during a tsunami event. NOAA's two Tsunami Warning Centers, the National Ocean Service, the National Data Buoy Center, the UNESCO/IOC – NOAA International Tsunami Information Center, the Pacific Marine Environmental Laboratory, and the National Geophysical Data Center each provide critical expertise. We also wish to acknowledge the invaluable work of the National Tsunami Hazard Mitigation Program (NTHMP). The NTHMP is a partnership between NOAA, the United States Geological Survey, the Federal Emergency Management Agency, the National Science Foundation, and the 28 U.S. Coastal States, Territories, and Commonwealths. The NTHMP is designed to reduce the impact of tsunamis through hazard assessment, warning guidance, and mitigation. The authors also wish to acknowledge Dr. Alexander Rabinovich, of the Institute of Ocean Sciences, Canada, Russian Academy of Sciences, for his valuable comments on the manuscript.

References

Abercrombie N (Governor of Hawaii) (2011) State to seek federal assistance in tsunami recovery for damage estimated at $30.6 Million, for immediate release: March 24, 2011. http://www.scd.hawaii.gov/documents/News%20Release%20Tsunami%20Damage%20Est.pdf. Accessed 25 Oct 2012

Badan Meteorologi Klimatologi Dan Geofisika (BMKG) (2011) Earthquake survey report and tsunami Honshu – Japan, 11 Mar 2011

Dunbar PK, Stroker KJ, Brocko VR, Varner JD, McLean SJ, Taylor LA, Eakins BW, Carignan KS, Warnken RR (2008) Long-term tsunami data archive supports tsunami forecast, warning, research, and mitigation. Pure Appl Geophys 165:2275–2291. doi:10.1007/s00024-008-0419-4

Eblé M, Titov V, Dendo D, Moore C, Mungov G, Bouchard R (2011) Signal-to-noise ratio and the isolation of the 11 March 2011 Tohoku tsunami in deep-ocean tsunameter records. In: Proceedings of oceans' 11 MTS/IEEE, Kona, IEEE, Piscataway, 19–22 Sept 2011, No. 6107288

Foreman MGG, Cherniawsky J, Ballantyne VA (2009) Versatile harmonic tidal analysis: improvements and applications. J Atmos Ocean Technol 26:806–817. doi:10.1175/2008JTECHO615.1

Goring DG, Nikora VI (2002) Despiking acoustic doppler velocimeter data. J Hydraul Eng 128:117–126

International Atomic Energy Agency (IAEA) (2011) Fukushima Daiichi status report 22 December 2011. http://www.iaea.org/newscenter/focus/fukushima/statusreport221211.pdf. Accessed 11 Oct 2012

International Atomic Energy Agency (IAEA) (2012) IAEA expert team concludes mission to Onagawa NPP, IAEA Press releases. http://www.iaea.org/newscenter/pressreleases/2012/prn201220.html. Accessed 11 Oct 2012

Jonathan MP, Roy PD, Sanchez-Zavala JL, Srinivasalu S, Macias-Romo MC, Lakshumanan C (2011) Field survey report on the 11th March 2011 tsunami in Pacific coast of Mexico. Nat Hazards 58:859–864. doi:10.1007/s11069-011-9871-z

Koizumi T (2011) National Report, Japan Meteorological Agency (JMA). Submitted to the international co-ordination group for the tsunami warning system in the Pacific, XXIV Session, Beijing, 24–27 May 2011

Mimura N, Yasuhara K, Kawagoe S, Yokoki H, Kazama S (2011) Damage from the great east Japan earthquake and tsunami – a quick report. Mitig Adapt Strat Glob Chang 16:803–818

Mofjeld HO (2009) Tsunami measurements. In: Robinson A, Bernard E (eds) The sea. Tsunamis, vol 15. Harvard University Press, Cambridge, MA, pp 201–235

Mori N, Takahashi T, Yasuda T, Yanagisawa H (2011) Survey of 2011 Tohoku earthquake tsunami inundation and run-up. Geophys Res Lett 38:L00G14, 10.1029/2011GL049210

Mungov G, Eblé M, Bouchard R (2012) DART® tsunameter retrospective and real-time data: a reflection on 10 years of processing in support of tsunami research and operations. Pure Appl Geophys. doi:10.1007/s00024-012-0477-5

National Police Agency (NPA) (2012) Damage situation and police countermeasures associated with 2011 Tohoku district – off the Pacific Ocean earthquake. http://www.npa.go.jp/archive/keibi/biki/higaijokyo_e.pdf. Accessed 3 Oct 2012

PEER/EERI/GEER/Tsunami Field Investigation Team (2011) Tohoku Pacific Ocean earthquake and tsunami quick observations from the PEER/EERI/GEER/Tsunami Field Investigation Team. http://peer.berkeley.edu/news/wp-content/uploads/2011/04/Tohoku-short-interim-report.pdf. Accessed 30 Sept 2011

Percival D, Denbo DW, Eblé MC, Gica E, Mofjeld HO, Spillane MC, Tang L, Titov VV (2011) Extraction of tsunami source coefficients via inversion of DART® buoy data. Nat Hazards 58:567–590. doi:10.1007/s11069-010-9688-1

Rabinovich AB, Stroker K, Thomson R, Davis E (2011) DARTs and CORK in cascadia basin: high-resolution observations of the 2004 Sumatra tsunami in the northeast Pacific. Geophys Res Lett 38:L08607. doi:10.1029/2011GL047026

Reymond D, Hyvernaud O, Okal EA (2012) The 2010 and 2011 tsunamis in French Polynesia: operational aspects and field surveys. Pure Appl Geophys. doi:10.1007/s00024-012-0485-5

Sanger DE, Wald M (2011) Radioactive releases in Japan could last months, experts say. New York Times. http://www.nytimes.com/2011/03/14/world/asia/japan-fukushima-nuclear-reactor.html. Accessed 25 Oct 2012

Song YT, Fukumori I, Shum CK, Yi Y (2012) Merging tsunamis of the 2011 Tohoku-Oki earthquake detected over the open ocean. Geophys Res Lett 39:L05606. doi:10.1029/2011GL050767

Strickland E (2011) Explainer: what went wrong in Japan's nuclear reactors. IEEE spectrum. http://spectrum.ieee.org/tech-talk/energy/nuclear/explainer-what-went-wrong-in-japans-nuclear-reactors. Accessed 25 Oct 2012

Tabuchi H (2011) Japan lifts evacuation advisories near nuclear plant. New York Times. http://www.nytimes.com/2011/10/01/world/asia/japan-lifts-evacuation-advisories-near-damaged-nuclear-plant.html. Accessed 25 Oct 2012

The Jakarta Globe (2011) Japan tsunami strikes Indonesia, one confirmed dead. http://www.thejakartaglobe.com/home/japan-tsunami-strikes-indonesiaone-confirmed-dead/428545. Accessed 25 Oct 2012

Titov VV (2009) Tsunami measurements. In: Robinson A, Bernard E (eds) The sea. Tsunamis, vol 15. Harvard University Press, Cambridge, MA, pp 371–400

Titov VV, Synolakis CE (1997) Extreme inundation flows during the Hokkaido-Nansei-Oki tsunami. Geophys Res Lett 24:1315–1318. doi:10.1029/97GL01128

Titov VV, Synolakis CE (1998) Numerical modeling of tidal wave run up. J Waterw Port Coast Ocean Eng 124:157–171

UNESCO/IOC (2011) Bulletin No. 29 – Casualties for the earthquake and tsunami of 11 March, 2011 as of 30 September 2011

UNESCO/IOC (2012) Summary statement from the Japan – UNESCO – UNU Symposium. In: The great east Japan tsunami on 11 March 2011 and tsunami warning systems: policy perspectives, Tokyo, 16–17 Feb 2012. United Nations Educational, Scientific and Cultural Organization (UNESCO), Paris

United States Geological Survey (USGS) (1971 to present) Preliminary determination of epicenters, a weekly and monthly publication. US Department of the Interior, Geological Survey

Wilson RI, Admire AR, Borrero JC, Dengler LA, Legg MR, Lynett P, McCrink TP, Miller KM, Ritchie A, Sterling K, Whitmore PM (2012) Observations and impacts from the 2010 Chilean and 2011 Japanese tsunamis in California (USA). Pure Appl Geophys. doi:10.1007/s00024-012-0527-z

World Nuclear News (2011) Stabilisation at Fukushima Daiichi. http://www.world-nuclear-news.org/RS_Stabilisation_at_Fukushima_Daiichi_2003111.html. Accessed 25 Oct 2012

Chapter 6
Tsunami Simulations in the Western Pacific Ocean and East China Sea from the Great Earthquakes Along the Nankai-Suruga Trough

Tomoya Harada and Kenji Satake

Abstract Tsunamis from great earthquakes can travel long distances and still cause damage when they strike, as was demonstrated by the March 11, 2011 Tohoku earthquake (*Mw* 9.1) which attacked a number of Pacific Ocean coasts. Along the Nankai-Suruga trough, great interplate earthquakes (M>8) have recurred historically, and the occurrence of another great earthquake can be expected in the near future. Great earthquakes along the Nankai-Suruga trough also generate large tsunamis in the western Pacific Ocean and East China Sea. In this study, we carried out tsunami numerical simulations in the western Pacific Ocean and East China Sea using existing static fault models of Tokai and Nankai earthquakes since AD 1498. Tsunami propagations were computed by use of non-linear long-wave equations, including the Corioli's force using the GEBCO 1-min bathymetry data, for a 24-h period after the earthquakes. The results of our simulations show the following: Tsunami heights in the studied oceans depend primarily on the slip amounts on the Nankai fault and are insensitive to the Tokai fault. Tsunamis from the Nankai earthquakes enter the East China Sea through the Tokara Strait, but the tsunami heights along the east coast of China were less than 1.0 m. Delayed slip on faults significantly affects the tsunami heights in the ocean areas studied. Tsunami heights computed using the 1707 and 1854 Nankai fault models were higher than those obtained from computations of other events. The tsunami behaviors computed in our study may correspond to the water disturbances recorded in Chinese historical documents on the dates of the 1707 Ho'ei and 1854 Ansei-Nankai earthquakes. However, the simulations show that tsunami heights along the east coast of China remain small when the assumed 1605 narrow-width fault model is used.

Keywords East China Sea • Maximum tsunami height (MTH) distribution • Tokai and Nankai earthquakes • Tsunami numerical simulation • Western Pacific Ocean

---

T. Harada (✉) • K. Satake
Earthquake Research Institute, The University of Tokyo, Tokyo, Japan
e-mail: haratomo@eri.u-tokyo.ac.jp

Y.A. Kontar et al. (eds.), *Tsunami Events and Lessons Learned: Environmental and Societal Significance*, Advances in Natural and Technological Hazards Research 35,
DOI 10.1007/978-94-007-7269-4_6,

6.1 Introduction

Massive tsunamis from the March 11, 2011 off the Pacific coast of Tohoku earthquake (*Mw* 9.1; Global CMT) caused ruinous damage, including 15,880 fatalities and 2,694 missing (National Police Agency, as of 27 February 2013), to the Tohoku, Kanto, and Hokkaido districts in Japan. The tsunami propagated throughout the Pacific Ocean striking oceanic islands and trans-Pacific coastlines. Shortly after the earthquake, tsunami warnings and advisories effective throughout the Pacific Ocean were issued by the Pacific Tsunami Warning Center (PTWC) in Hawaii, along with other national warning centers. The largest tsunami heights outside of Japan were recorded at 2.49 m on a tide gauge in the US state of California. Two other tide gauges on the west coast of the US recorded heights in excess of 2 m (EERI 2011). Similar tsunami heights were recorded at four locations in Chile, as well as in the Galapagos and Hawaiian Islands (EERI 2011). Those trans-Pacific tsunamis seriously damaged many coastal areas in the US, Chile, and on the Galapagos Islands. One life was lost in northern California, and another person was killed in Jayapura, Indonesia – about 8,000 and 4,500 km away from the source region, respectively (Heidarzadeh and Satake 2013).

Similarly the Pacific coasts of Japan have been repeatedly struck by trans-Pacific tsunamis from major earthquakes at other circum-Pacific subduction zones. The large tsunami from the May 22 (UTC), 1960 Chile earthquake (*Mw* 9.5; Kanamori 1977) impacted Japan's entire Pacific coasts about 22 h after the earthquake, killing 142 people and causing significant damage to the Tohoku and Kanto coasts. Trans-Pacific tsunami from the February 27 (UTC), 2010 Chilean event (*Mw* 8.8; Global CMT) also caused property damage when they reached Japan's coasts (Fritz et al. 2011). Furthermore, Japanese historical documents show damage and flooding occurred at seven sites along nearly 1,000 km of Japan's Pacific coast due to the tsunamis from the M9-class interplate earthquake on January 26 (UTC), AD 1700 at the Cascadia subduction zone along the Pacific coast of North America (Satake et al. 2003).

Far-field tsunamis originating in the western Pacific Ocean also have reached the Pacific coasts of Japan. On January 3, 2009, tsunamis from the *Mw* 7.7 and *Mw* 7.4 (Global CMT) earthquakes that occurred off Indonesian part of New Guinea, which marks the southernmost boundary of the western Pacific Ocean (Fig. 6.1), propagated throughout the entire western Pacific Ocean (Fujii et al. 2011). The Japan Meteorological Agency (JMA) issued tsunami advisories for the Japanese coast, and the actual tsunamis were recorded on Japanese tide gauges. At the time of Indonesia's Irian Jaya Earthquake on February 17, 1996 (*Mw* 8.2; Harvard CMT), which took place about 500 km east of the 2009 events (Fig. 6.1), the JMA also issued tsunami warnings for the Japanese coast.

The subduction zone along the Nankai-Suruga trough in southwest Japan coincides with the northern border of the western Pacific Ocean (Fig. 6.1). In this subduction zone, great interplate earthquakes associated with the subduction of the Philippine Sea Plate (convergence rate approximately 5–6 cm/year: Amur plate fixed; DeMets et al. 2010) have repeatedly occurred at intervals of 100–200 years

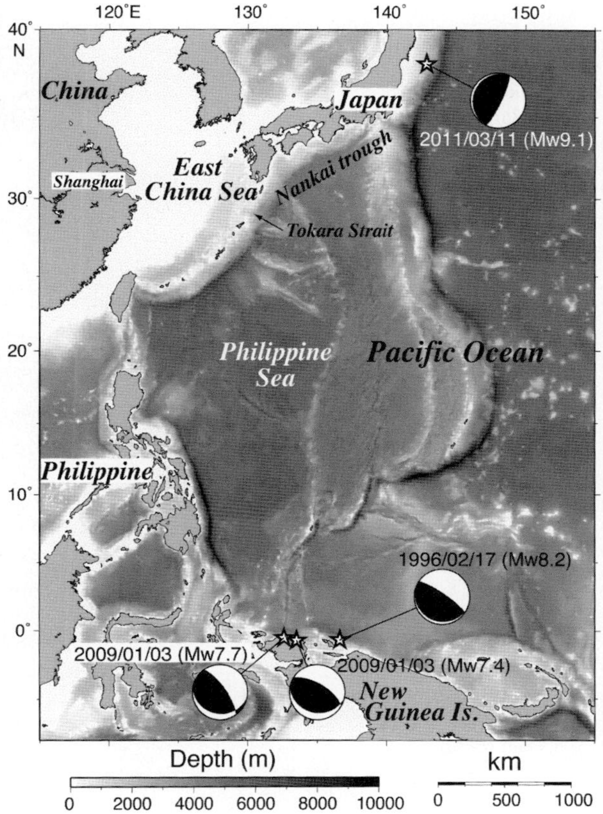

Fig. 6.1 Bathymetric map of the area in which the tsunami numerical simulations were performed. Submarine topography is shown by gray-scale based on GEBCO's 1-min (approximately 1,800 m) bathymetric data. *Stars* show epicenters of tsunamigenic earthquakes that affected the Japanese coast. Mechanism solutions and moment magnitudes are based on Global CMT Project

since AD 684 (Ando 1975; Ishibashi 2004; Fig. 6.2). These great earthquakes occurred as pairs of M~8 events, one in the eastern half (Tokai earthquake) and the other in the western half (Nankai earthquake), as was the case for the 1854 Ansei earthquakes. However, they sometimes occur as a single giant event, such as the 1707 Ho'ei earthquake. The latter had been considered the greatest earthquake in Japanese history prior to the 2011 Tohoku earthquake.

There have been nine paired Tokai and Nankai earthquakes recorded in the period from AD 684 to 1946, and destructive tsunamis from these events have struck the Pacific coasts of southwest Japan. At the same time, those tsunamis propagated across the ocean and can be presumed to have struck the coasts of Southeast Asia and Oceania. Because about 70 years have passed since the previous series of earthquakes, a great or giant interplate earthquake is expected to occur in the near future. The tsunamis generated by such an earthquake will spread across the western Pacific Ocean, and can thus be expected to damage coastal areas in Southeast Asian and Oceania countries. There has been a controversy among Japanese scientists as to whether China's Shanghai coasts and its environs along the East China Sea coast had been impacted by tsunamis from previous great

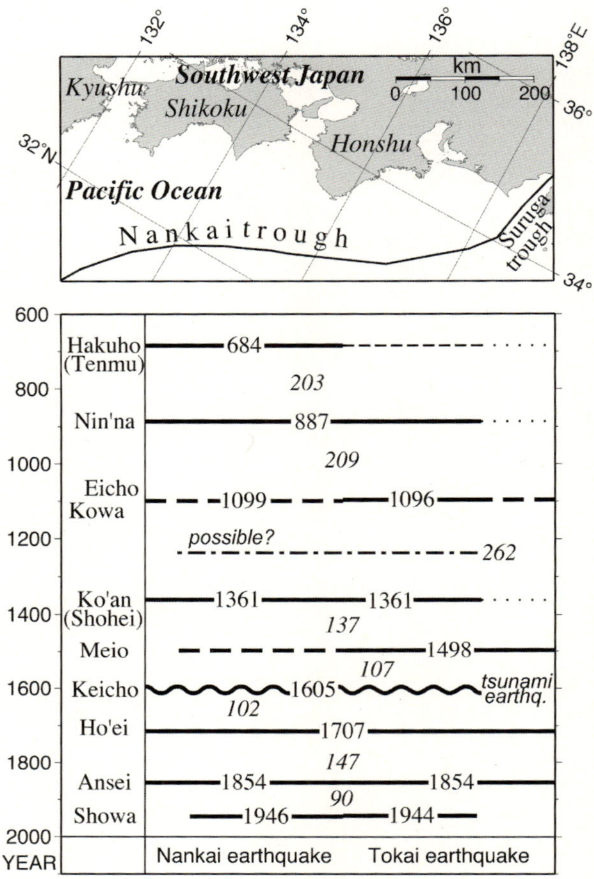

Fig. 6.2 Space-time distribution of great interplate Nankai and Tokai earthquakes along the Nankai-Suruga trough (Modified from Fig. 11 of Ishibashi 2004). *Roman* and *italic numerals* indicate earthquake occurrence years and time intervals in years between two successive series, respectively. *Thick solid*, *thick broken*, *thin broken*, and *thin dotted lines* show certain, probable, possible, and uncertain rupture zones, respectively. *Chain line* indicates a candidate for great earthquake inferred from Japanese historical documents.

earthquakes along the Nankai-Suruga trough (Fig. 6.2). Chinese historical documents report significant water disturbances in rivers, canals, ponds, and wells around Shanghai in 1498, 1707, and 1854 (Xie and Tsai 1985, 1987). The dates in 1707 and 1854 coincide with those of the Nankai earthquakes, and the date (June 30) in 1498 coincides with that of a large earthquake in southwest Japan. Tsuji and Ueda (1997) interpreted these water disturbances as being caused by large tsunamis and associated the 1498 earthquake with the Nankai earthquake. In contrast, Utsu (1988) and Ishibashi (1998) did not consider these water disturbances to have resulted from tsunamis, but instead identified them as seiches induced by long-period seismic waves. At this time, opinions remain mixed as to whether the June 30, 1498 event was a Nankai earthquake.

In this study, in order to investigate the tsunami propagation behaviors and to evaluate the coastal heights in the western Pacific Ocean and East China Sea for the great earthquakes along the Nankai-Suruga trough, we carried out various numerical simulations of tsunami propagations in those areas using existing static fault models for previous great earthquakes along the Nankai-Suruga trough.

6.2 Tsunami Numerical Simulations

6.2.1 *Fault Models for Past Great Earthquakes Along the Nankai-Suruga Trough*

Several seismologists have proposed static fault models for previous Tokai and Nankai earthquakes, including historic events. Ando (1975) developed static fault models for the 1944 and 1946 earthquakes from the aftershock areas, crustal deformation, and tsunami data on tide gauge records. He also proposed static fault models for the 1707 Ho'ei, 1854 Ansei-Tokai, and 1854 Ansei-Nankai earthquakes by comparing information on vertical crustal deformation and seismic intensities of the 1707, the 1854 Tokai, and 1854 Nankai events contained in historical documents (Kawasumi 1950; Omori 1913) with the recorded crustal deformation and seismic intensities of the 1944 and 1946 events. Slip amounts on the 1707 fault planes were twice as large as those of 1854 Tokai and Nankai events. Ando (1982) developed a revised static fault model for the 1946 Nankai earthquake.

Aida (1981a, b) produced a static fault model for the earthquakes in 1707 and 1854 by comparing tsunami heights along the Pacific coasts in southwestern Japan based on historical documents (Hatori 1974, 1978, 1980, 1981). The 1707 fault model has a significantly large slip (13.9 m) on the southwestern-most fault plane. Aida (1979, 1981b) also proposed static fault models for the 1944 and 1946 events based on the tsunami waveforms recorded on tide gauges.

An'naka et al. (2003) assumed four fault planes along the Nankai-Suruga trough. From comparisons of computed tsunami heights with tsunami heights based on the historical documents (e.g. Hatori 1988), they estimated the slip amounts on the four faults during the events since AD 1498. Based on examinations of the average convergence rate and the great earthquake recurrence intervals, they assumed that the four faults had the same geometries and slip angles, and that the slip amount on the each fault did not exceed 10 m.

6.2.2 *Fault Models for the Tsunami Numerical Simulation*

In this study, in order to examine tsunami behaviors and to evaluate coastal tsunami heights in the western Pacific Ocean and East China Sea for previous great earthquakes along the Nankai-Suruga trough, we first performed tsunami simulations by using the proposed static fault models of the Tokai and Nankai earthquakes since AD 1498. We used An'naka's (2003) static fault models of the 1498 Tokai (1498T) and 1605 tsunami earthquakes (hereinafter referred to as 1498T$_{\text{(An'naka)}}$ and 1605$_{\text{(An'naka)}}$. As for the 1707, 1854 Tokai (1854T), 1854 Nankai (1854N), 1944 Tokai (1944T), and 1946 Nankai (1946N) earthquakes (Fig. 6.2), we used static fault models by Ando (1975, 1982) (hereinafter referred to as 1707$_{\text{(Ando)}}$, 1854T$_{\text{(Ando)}}$,

1854N(Ando), 1944T(Ando), and 1946N(Ando), respectively). Those by Aida (1979, 1981a, b) are referred to hereinafter as 1707(Aida), 1854T(Aida), 1854N(Aida), 1944T(Aida), and 1946N(Aida), respectively, while those by An'naka et al. (2003) are hereinafter referred to as 1707(An'naka), 1854T(An'naka), 1854N(An'naka), 1944T(An'naka), and 1946N(An'naka), respectively.

While various modern and detailed source models have been proposed for the most recent 1944 Tokai and 1946 Nankai earthquakes (e.g. Baba et al. 2006), we decided to use the simple models developed by the abovementioned three researchers to examine the effects of model differences on the tsunamis occurring in the oceans studied because they use the same methodology for the most recent and older earthquakes along the Nankai-Suruga trough.

Next, in order to examine the effect of occurrence pattern of the Tokai and Nankai earthquake (simultaneous or separate occurrence) on the tsunami heights, we divided the 1707(Ando), 1707(Aida), and 1707(An'naka) models into Nankai and Tokai segments. Hereinafter, by using the name of each fault segment in Fig. 6.3, we designate the Nankai events as 1707(Ando)-AB, 1707(Aida)-ABC, and 1707(An'naka)-AB, while the Tokai segment events will be referred to as 1707(Ando)-C, 1707(Aida)-DE, and 1707(An'naka)-CD.

We then carried out the tsunami simulation and compared the coastal tsunami heights in the oceans studied. In addition, we assumed a simultaneous rupture for the 1854 Tokai and Nankai earthquakes (hereinafter referred to as 1854TN(Ando), 1854TN(Aida), and 1854TN(An'naka)) and then compared the computed coastal tsunami heights for the simultaneous models, 1854 Tokai models, and 1854 Nankai models.

6.2.3 Method of the Tsunami Numerical Simulation

Tsunami propagations were computed by the finite-difference method of the non-liner long-wave equations with Corioli's force (Satake 1995) in the area of 115–155°E and 8°S to 40°N. Because water depths in the East China Sea are shallower than 200 m, tsunami propagation in that area is likely to be affected by ocean bottom fiction. Figure 6.1 shows a bathymetry map of the tsunami computation area. One-minute gridded bathymetry data provided by the General Bathymetric Chart of the Oceans (GEBCO) were used. A Manning's roughness coefficient of 0.025 $m^{-1/3}$ s was assumed for the friction and a computation time step of 3 s is used to satisfy the stability condition of the finite-difference method.

The initial conditions of tsunami propagation were ocean bottom deformation due to earthquake faulting, which were computed by Okada's (1992) program. For the boundary conditions, a total reflection on the coast and an open boundary to the exterior of the computational area were used. The tsunami propagations were computed for 24 h after the events using various static models. The tsunami arrival times, based on each fault model shown in Fig. 6.3, were calculated using Geoware tsunami travel-time software (TTT ver. 3.1).

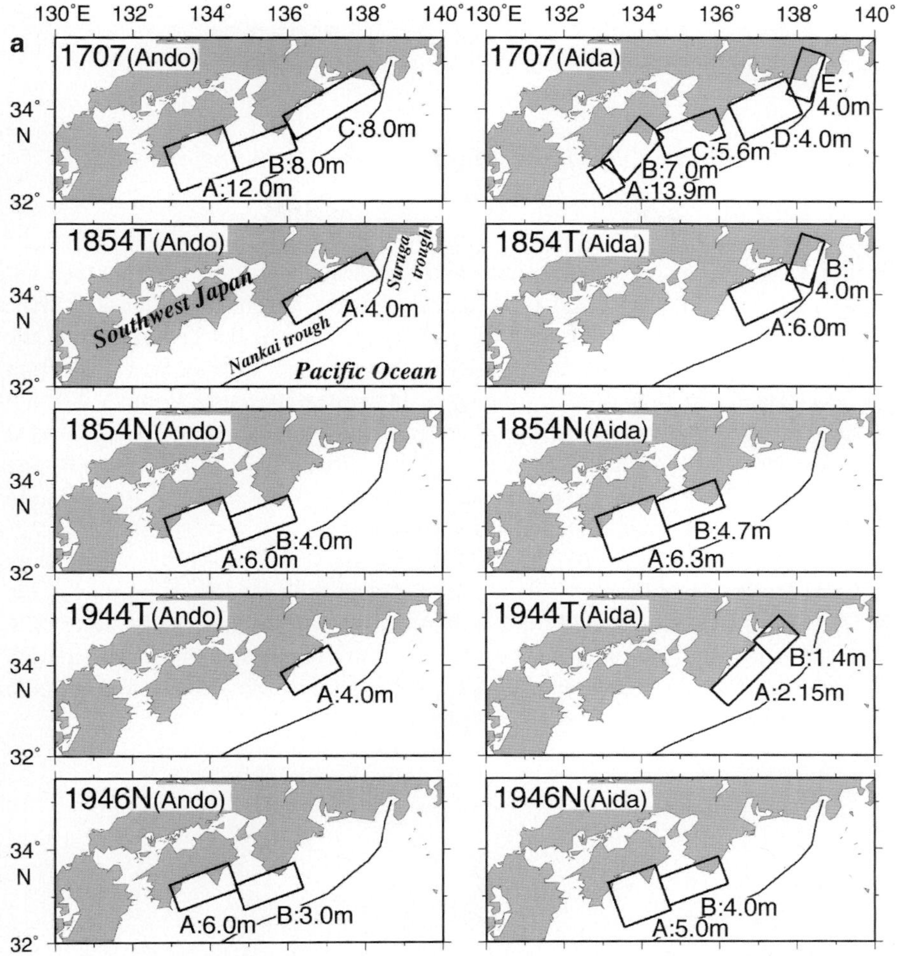

b An'naka's (2003) fault models

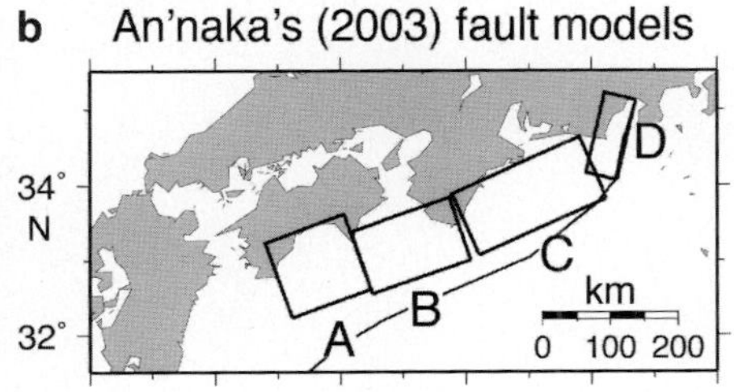

	Slip amount (m)			
	A	**B**	**C**	**D**
1498T(An'naka)	0.0	0.0	5.3	7.0
1605(An'naka)	9.7	6.0	7.0	0.0
1707(An'naka)	9.2	5.6	7.0	5.6
1854T(An'naka)	0.0	0.0	5.5	5.27
1854N(An'naka)	8.7	4.8	0.0	0.0
1944T(An'naka)	0.0	0.0	4.25	0.0
1946N(An'naka)	4.3	4.9	0.0	0.0

Fig. 6.3 (**a**) Static fault models of the 1707 Ho'ei, 1854 Ansei-Tokai (1854T), 1854 Ansei-Nankai (1854N), 1944 Tonankai (1944T), and 1946 Nankai earthquakes (1946N) used in the tsunami simulations. *Rectangles* represent horizontal projection of fault planes. *Alphabetical characters* beside each *rectangle* provide the names of each fault plane and the slip amount on each fault plane, respectively. The 1707(Ando), 1854T(Ando), 1854N(Ando), and 1944T(Ando) models are after Ando (1975). The 1946N(Ando) model is after Ando (1982). The 1707(Aida) model is after Aida (1981a, b). The 1854N(Aida) and 1946N(Aida) models are after Aida (1981b). The 1854T(Aida) model is after Aida (1981a). The 1944T(Aida) model is after Aida (1979). (**b**) Static fault models of great earthquakes along the Nankai-Suruga trough since AD 1498 after An'naka et al. (2003) were used in the present tsunami simulation. 1498T refers to the 1498 Meio-Tokai Earthquake. *Rectangles* represent horizontal fault plane projections. The table shows slip amounts on the fault planes for each event. *Alphabetical characters* beside each *rectangle* provide the name of each fault plane

6.3 Results

6.3.1 Maximum Tsunami Height (MTH) Distributions in the Computation Area

Figure 6.4 shows the maximum heights and arrival times of tsunamis from the fault models shown in Fig. 6.3. As can be seen in the figure, tsunamis from the 1605, 1707, 1854 Nankai, and 1946 Nankai earthquakes arrived along the east coast of China, the Taiwan coasts, and the northern Philippines coasts about 1 h earlier than those from the 1854 Tokai and 1944 Tokai earthquakes. In contrast, tsunamis from all the fault models arrived along the north coast of the New Guinea and the Indonesian coasts at approximately the same time. Tsunamis spread through the entire western Pacific Ocean within 7 or 8 h after the earthquakes, while they arrive at the northern margin of the East China Sea about 18 h after the earthquakes.

The maximum tsunami heights (MTHs) for the 1605 and 1707 events were estimated to be generally higher than those for other fault models. In particular, the MTH distribution for 1707(Ando) model showed the largest tsunami. The areas with high MTHs for the 1605 and 1707 events were much larger than the combined high MTH areas from the Tokai and Nankai events because the 1605 and 1707 slip amounts were larger than those of other events.

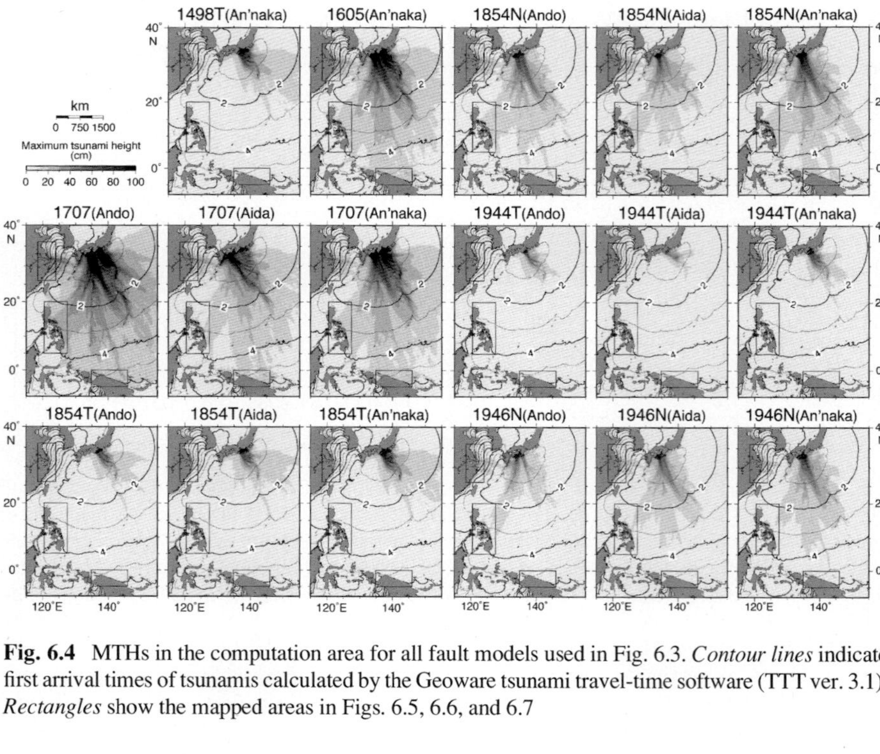

Fig. 6.4 MTHs in the computation area for all fault models used in Fig. 6.3. *Contour lines* indicate first arrival times of tsunamis calculated by the Geoware tsunami travel-time software (TTT ver. 3.1). *Rectangles* show the mapped areas in Figs. 6.5, 6.6, and 6.7

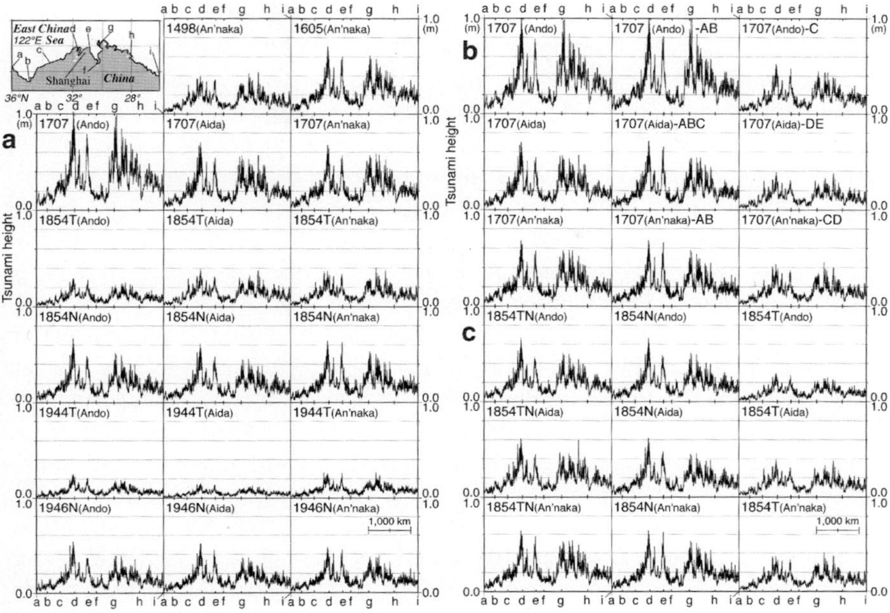

Fig. 6.5 Comparison of MTHs along the east coast of China (*thick line* in the map). (**a**) Comparison of all the fault models in Fig. 6.3. (**b**) Comparison for different models of the 1707 Ho'ei earthquake. (**c**) Comparison for different models of the 1854 Tokai and Nankai earthquakes. *Alphabetical characters* on the upper and lower side of the frames refer to map points along the coast

Tsunamis normally enter the East China Sea through the Tokara Strait (Fig. 6.1) and propagate west-northwest (the general direction of Shanghai, see Fig. 6.1). Relatively high MTHs for the 1854 and 1946 Nankai earthquakes were found to be widely distributed in the western Pacific Ocean. The 1854 and 1946 Nankai tsunamis also show the same propagation patterns in the East China Sea. Unlike the 1605, 1707, and Nankai earthquakes, tsunamis from the Tokai earthquakes spread toward southeast and rarely propagate into the western Pacific Ocean. The MTHs around Taiwan were not notably high in any of the models used. With the exception of the 1707 event, different models for the same event produced similar MTH distributions.

6.3.2 MTH Distributions Along the East Coast of China

Figure 6.5a shows the MTH distributions along the east coast of China for all the fault models shown in Fig. 6.3. The distributions have common characteristics caused by the seafloor and coast topography. For example, the MTHs gradually increase from point *a* to *c*, then suddenly increase from point *c* to *d*, and the highest peak is formed around the river mouth of the Yangtze (Chang) River near Shanghai, with another peak at around *e*. The MTHs around Hangzhou Bay (from *e* to *g*) were relatively low. The MTHs around *g* formed the second highest peak and gradually decreased from *g* to *i*.

The MTHs for the 1605 and 1707 events were found to be much higher than those for the other fault models. The 1707(Ando) MTHs were the highest and reached up to 1.0 m around *d* and *g*. The MTH distributions for 1707(Aida) and 1707(An'naka), with the highest peak of 0.7 m, were similar to each other. The MTHs for the 1605(An'naka) were similar to those for 1707(An'naka), since the slip amount on fault planes A, B, and C were the same or similar (Fig. 6.3b). The highest peaks of the MTHs for the three 1854 N models exceeded 0.6 m. The MTHs for 1854N(An'naka) and 1707(An'naka) were almost the same because they have similar slip amounts on fault planes A and B (Fig. 6.3b). The MTHs around *d* for the 1946 Nankai event exceeded 0.5 m. The MTHs for the 1498, 1854, and 1944 Tokai earthquakes were low because their tsunamis mainly propagate toward the southeast (Fig. 6.4).

Figure 6.5b shows comparisons among the MTH distributions for the Nankai segments, the Tokai segments, and all the segments of the 1707 fault models. The Figure clearly shows that the MTHs for 1707(Ando), 1707(Aida), and 1707(An'naka) were almost the same as those for 1707(Ando)-AB, 1707(Aida)-ABC, and 1707(An'naka)-AB, respectively. Therefore, while the MTHs in this region were strongly controlled by the slip amounts in the Nankai region, they were relatively unaffected by the slip amounts in the Tokai region. The MTHs were increased not by an extension of faults to the Tokai region, but by the enlarged slip amounts on the faults in the Nankai region.

Figure 6.5c shows comparisons of MTH distributions for the assumed simultaneous rupture models of the 1854 Tokai and Nankai faults, those for the 1854 Nankai fault models, and the 1854 Tokai fault models. As can be seen in the Figure, the MTHs for 1854TN(Ando), 1854TN(Aida), and 1854TN(An'naka) were almost the same as those for 1854N(Ando), 1854N(Ando), and 1854N(An'naka), respectively, which again confirmed that the MTHs in this region were strongly controlled by the slip amounts in the Nankai region.

6.3.3 MTH Distributions Along the East Coasts of the Philippine Islands

Figure 6.6a shows the MTH distributions for all the fault models (Fig. 6.3) along the east coast of the Philippine Islands. As can be seen in the Figure, the MTH distribution has three peaks (from *j* to *k*, *m* to *o*, and *p* to *q*) and the peak from *m* to *o* is the highest. The MTHs for the 1605 and 1707 events exceeded 1.0 m, and were much higher than those for the other events. It can also be seen that the MTH distribution for 1605(An'naka) and 1707(An'naka) were almost the same, while those for 1854N(An'naka) and for 1707(An'naka) were similar due to the same

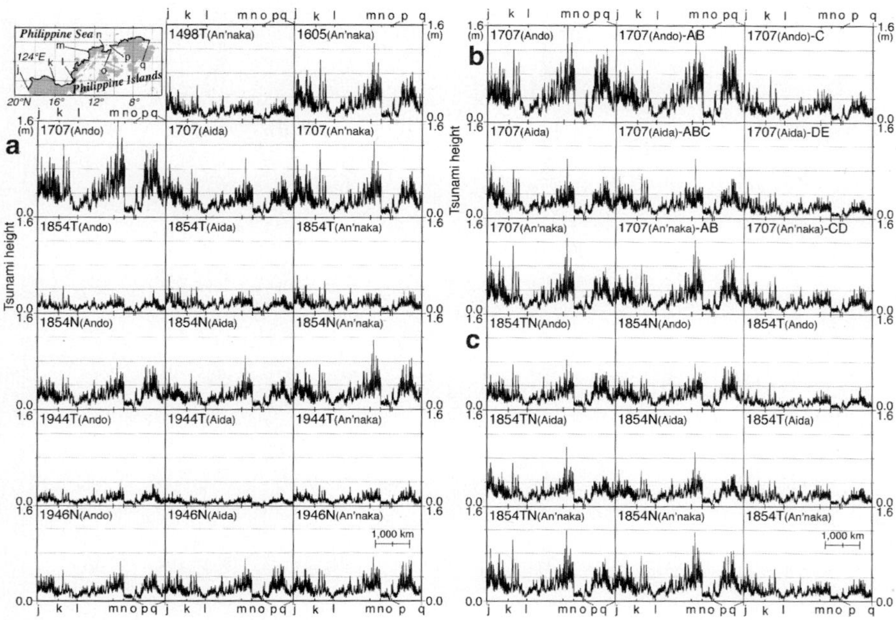

Fig. 6.6 Comparisons of MTHs along the east coast of the Philippine Islands (*thick line* in the map). (**a**) Comparison of all the fault models in Fig. 6.3. (**b**) Comparison for different models of the 1707 Ho'ei earthquake. (**c**) Comparison for different models of the 1854 Tokai and Nankai earthquakes. *Alphabetical characters* on the upper and lower side of the frames refer to map points along the coast

reason mentioned in Sect. 6.3.2. The MTHs for the 1854 and 1946 Nankai events were relatively high (maximum MTHs were from 0.8 to 1.2 m) while those for the 1498, 1854, and 1944 Tokai events were less than 0.4 m. The MTHs calculated from the three models of the 1707 event show significant differences. The variations among the different models for the other earthquakes were smaller, but more visible compared to those along the east coast of China.

Figure 6.6b shows comparisons among the MTH distributions for the Nankai segments, the Tokai segments, and all the segments of the 1707 fault models. The Figure clearly shows that the MTHs in this region also depend significantly on the slip amounts in the Nankai region, or that the MTHs increase by the enlarged slip amounts in the Nankai region.

Figure 6.6c shows comparisons of the MTH distributions for the assumed simultaneous rupture models of the 1854 Tokai and Nankai fault models, along with those for the 1854 Nankai and 1854 Tokai fault models. As can be seen in the Figure, the MTHs for 1854TN(Ando), 1854TN(Aida), and 1854TN(An'naka) were the almost same as those for 1854N(Ando), 1854N(Ando), and 1854N(An'naka), respectively. This indicates that MTH distributions in this region were also controlled by the slip amounts in the Nankai region.

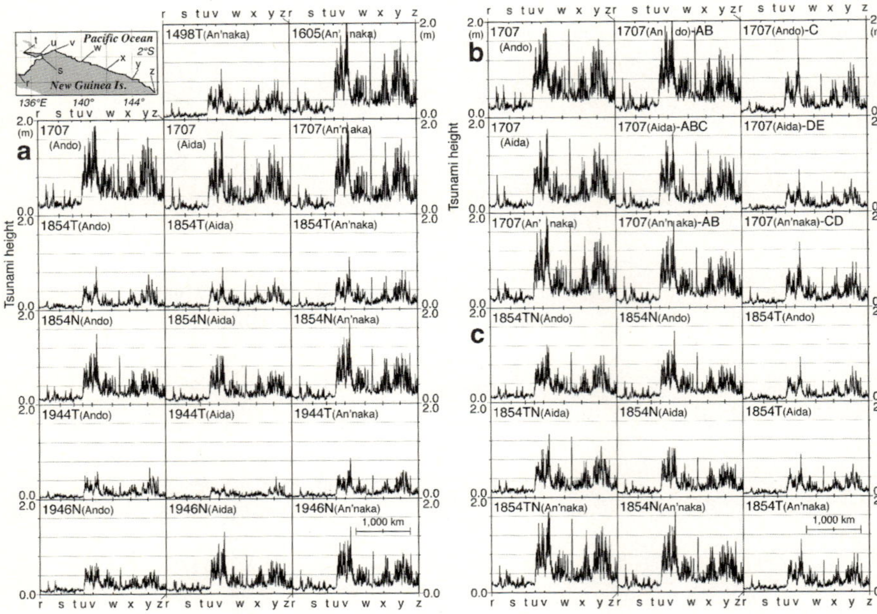

Fig. 6.7 Comparisons of MTHs along the north coast of New Guinea (*thick line* in the map). (**a**) Comparison of all the fault models in Fig. 6.3. (**b**) Comparison for different models of the 1707 Ho'ei earthquake. (**c**) Comparison for different models of the 1854 Tokai and Nankai earthquakes. *Alphabetical characters* on the upper and lower side of the frames refer to map points along the coast

6.3.4 MTH Distributions Along the North Coast of New Guinea Island

Figure 6.7a shows the MTH distributions for all the fault models (Fig. 6.3) along the north coast of New Guinea Island. While MTHs from point *r* to point *u* were low, they became significantly higher from point *u* to point *z*. The MTHs show two peaks: one is from point *t* to *w*, which provides the highest MTHs, and the other is from *x* to *z*. The MTHs exceeded 1.8 m for the 1605 and 1707 events, and were much higher than the other events. The MTH distribution for the 1605(An'naka) and 1707(An'naka) models were nearly identical for same reason mentioned in Sects. 6.3.2 and 6.3.3. However, the MTHs from *x* to *z* for 1707(An'naka) were approximately 1.5 times as high as those for 1854N(An'naka), while the MTHs from *r* to *x* for 1707(An'naka) were similar to those for 1854N(An'naka). This is probably due to the slip amounts of 1707(An'naka), since the coast from *x* to *z* is located to the east of New Guinea Island, unlike the China or Philippine coasts. In addition, the MTHs for the 1854 and 1946 Nankai earthquakes were relatively high while those for the 1498, 1854, and 1944 Tokai earthquakes were low. The 1707 MTH distribution for all three fault models were relatively similar to each other, even though the difference in MTHs among the fault models for the other events were considerably larger.

Figure 6.7b shows comparisons among the MTH distributions for the Nankai segments, the Tokai segments, and all the segments of the 1707 fault models. As can be seen in the Figure, the MTHs from point *r* to *x* for the 1707 events were similar to those for the Nankai segments, thus indicating a control of the slip amounts in the Nankai region. However, the MTHs from *x* to *z* for the 1707 events were about 1.5 times as high as those of the Nankai faulting of the 1707 models. This is believed to be caused by the superposition of tsunamis from the Tokai and Nankai events, because the coast from *x* to *z* is located the east of the New Guinea Island.

Figure 6.7c shows comparisons of the MTH distributions for the assumed simultaneous rupture models of the 1854 Tokai and Nankai faults, for the 1854 Nankai models, and for the 1854 Tokai models. As can be seen in the Figure, the MTH distributions from *r* to *x* for the simultaneous rupture model were almost the same as those for the 1854 Nankai events, and the MTHs from *x* to *z* for the simultaneous rupture model were about 1.5 times as high as those for the 1854 Nankai events, primarily due to the superposition of the 1854 Tokai tsunami.

6.4 Discussion

6.4.1 Effect of Slip Delay of Fault Planes

The abovementioned simulations revealed that MTHs in the western Pacific Ocean, except for the eastern New Guinea Island, and eastern China Sea coasts were strongly dependent on the slip amounts in the Nankai region, and that the extension of the fault plane to the Tokai region had insignificant effects on the MTH in the oceans studied. However, according to the scaling law (Kanamori and Anderson 1975), the slip amount correlates with the fault area. Thus, we can conclude that the fault extension affects the MTHs indirectly.

Tsunamis from a great earthquake are also affected by the delayed slip on the fault. The 2011 Tohoku earthquake experienced the largest slip near the trench at around 38.5°N, whereas the measured tsunami height distribution shows a peak near Miyako at around 39.6°N (Tsuji et al. 2012). Satake et al. (2013) showed that the computed coastal tsunami heights for a delayed slip, which were estimated by multiple time-window tsunami inversion, reproduced the observed inundation heights along the Sanriku coast.

Imai et al. (2010) examined the effect of delayed slip of the 1707 Ho'ei earthquake on tsunami heights on the Pacific coasts in southwest Japan by assuming various delays to the rupture initiation for each fault of 1707(An'naka). Their results showed that the average tsunami heights on the coasts reach highest when the rupture starting on fault plane C is followed by the rupture on fault plane D within 21 min, on fault plane B within 15 min, and on fault plane A within 15 min. In this study, in order to examine the behavior of tsunami from the delayed slip model of the 1707 event in the western Pacific Ocean and East China Sea, we

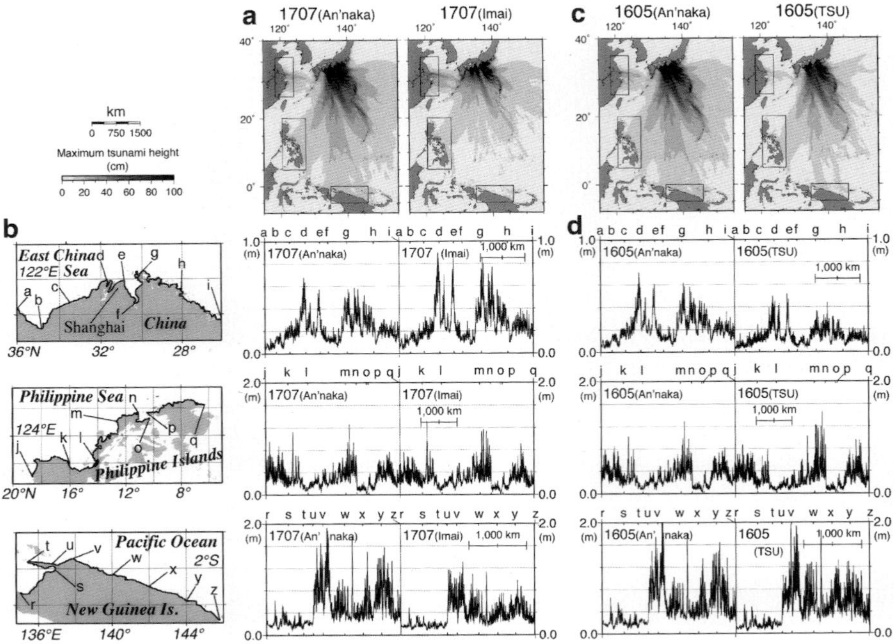

Fig. 6.8 (**a**) MTHs in the computation area for the simultaneous (An'naka) and delayed (Imai) rupture models of the 1707 earthquake. (**b**) Comparisons of MTHs for 1707(An'naka) and 1707(Imai) along the east coast of China, north coast of New Guinea, and the east coast of the Philippine Islands (*thick lines* in the left-side maps). (**c**) MTHs for the wide (An'naka) and narrow (TSU) fault models of the 1605 tsunami earthquake. (**d**) Comparisons of MTHs for 1605(An'naka) and 1605(TSU) along the same coasts in Fig. 6.8 (**b**). *Rectangles* in the maps in Fig. 6.8 (**a**) and (**c**) show the areas of the maps in Fig. 6.8 (**b**). *Alphabetical characters* on the upper and lower side of the frames refer to map points along the coast

carried out the tsunami simulations using the delayed 1707 model described by Imai et al. (2010). Hereinafter, this is called the 1707(Imai) model.

Figure 6.8a shows the MTH distributions for 1707(An'naka) and 1707(Imai) in the computation area. As can be seen in the Figure, tsunamis from 1707(Imai) were lower than those for 1707(An'naka) in the western Pacific Ocean, whereas MTHs for the two models were similar in the East China Sea. Figure 6.8b shows comparisons between the MTHs for 1707(An'naka) and for 1707(Imai) along the east coast of China, the east coasts of the Philippine Islands, and the north coast of New Guinea Island, respectively. The MTH distribution for 1707(Imai) was approximately 1.2 times higher than that for 1707(An'naka) along the east coast of China, while the MTHs for 1707(Imai) were approximately 78 % the height of those for 1707(An'naka) along the north coast of New Guinea Island. Along the east coast of the Philippine Islands, MTHs for the two models were almost the same. Therefore, it was confirmed that the delayed slip of great earthquakes along the Nankai-Suruga trough affects tsunami heights along the coasts in the western Pacific Ocean and East China Sea in different ways.

6.4.2 Water Disturbance Around Shanghai

There has been controversy among Japanese scientists as to whether Shanghai and its environs on the East China Sea coast were affected by large tsunamis due to the great earthquakes along the Nankai-Suruga trough (Utsu 1988; Tsuji and Ueda 1997; Ishibashi 1998). The tsunami simulations in this study clarified that the tsunamis from these earthquakes hit the east coast of China and that the MTHs around Shanghai were the highest. The simulations also clarified that the MTHs for the 1707 and 1854 Nankai earthquakes were higher than those for the Tokai earthquakes. These results imply that the documented water disturbances around Shanghai were caused by tsunamis, because they occurred on the exact dates of the 1707 and 1854 Nankai earthquakes. However, the MTHs along the east coast of China were limited, and the highest MTH for 1707(Ando) did not exceed 1.0 m, even though the tsunami computation was made on a coarse grid.

Tsunami computation using more accurate bathymetry data around Shanghai would make the MTHs along the east coast of China higher than those in the present simulation. However, in this study, we could not definitely conclude that the documented water disturbances were caused by the tsunamis from the great earthquakes along the Nankai-Suruga trough.

The MTH distribution for 1605(An'naka) was found to be close to that of 1707(An'naka). However, no water disturbances were reported on the date of the 1605 event in historical Chinese documents. It should be noted, however, that the quantity and quality of those documents may not have been uniform throughout Chinese history, and it is possible that the actual 1605 tsunami simply went unreported. Alternatively, the 1605(An'naka) model may not provide a good approximation of the 1605 tsunami earthquake.

A tsunami earthquake is a relatively slow rupture on the shallow plate interface near the trench axis (Kanamori 1972). Thus, tsunami earthquake static fault models need to have a narrow fault plane near the trench axis. Because the 1605(An'naka) model has slip amounts on wide fault planes (Fig. 6.3b), we applied an alternative model for the 1605 tsunami earthquake (hereinafter 1605(TSU)). In this model, the width of fault planes were narrowed to 50 km and the slip amounts were adjusted so that the MTHs along the Pacific coasts of southwest Japan for 1605(TSU) correspond to those for 1605(An'naka). The slip amounts of 1605(TSU) were as follows: A: 9.2 m, B: 5.3 m, and C: 7.0 m.

Figure 6.8c shows the MTH distributions for 1605(An'naka) and 1605(TSU) in the computation area. In the East China Sea, the MTHs for 1605(TSU) seem to be lower than those for 1605(An'naka). Figure 6.8d shows comparisons between the MTHs for 1605(An'naka) and those for 1605(TSU) along the east coasts of China, the Philippine Islands, and the north coast of New Guinea Island. As can be seen in the Figure, the MTHs for 1605(TSU) closely resemble those for 1605(An'naka) along the east coast of the Philippine Islands and the north coast of New Guinea Island. In contrast, along the east coast of China, the MTHs for 1605(TSU) were approximately 60 % of those for 1605(An'naka). Because the 1605(TSU) tsunami wavelength was relatively short due to its narrow fault plane, the attenuation of tsunamis from that event were larger than from 1605(An'naka) in the East China Sea where the water depth is shallower than 200 m. Thus, the actual 1605 tsunami around Shanghai probably lacked sufficient energy to cause excessive water disturbances.

6.5 Conclusions

In order to examine the behavior of tsunamis in the western Pacific Ocean and East China Sea from previous great earthquakes along the Nankai-Suruga trough, we carried out tsunami numerical simulations using previously proposed static fault models. The simulations showed following results: (1) Tsunami heights in the areas studied, except for the eastern New Guinea coast, strongly depend on the slip amount of the Nankai fault, even though they are generally insensitive to the Tokai fault. (2) The most energetic tsunamis from Nankai earthquakes enter the East China Sea through Tokara Strait and propagate in the direction toward Shanghai. Those waves have caused the largest tsunami height levels recorded around Shanghai. (3) The maximum tsunami heights for a 1707 delayed slip model were approximately 1.2 times higher than those for the instantaneous rupture model along the east coast of China, but were about 0.8 times as high along north coast of New Guinea Island. The difference shows that the tsunami heights were controlled by the delay pattern in a complex way. (4) Computed tsunami heights for the 1707 and 1854 Nankai fault models were higher than those for the other events, which implies that the water disturbances recorded in Chinese historical documents on the dates of the 1707 Ho'ei and 1854 Ansei-Nankai earthquakes may have been caused by tsunamis. Lack of 1605 tsunami records may be because it was a tsunami earthquake. (5) Tsunami heights along the eastern New Guinea coasts were influenced by the superposition of the Tokai and Nankai faulting.

Responding to the unexpected occurrence of the 2011 Tohoku Earthquake, the Cabinet Office of Japanese government has assumed that a simultaneous rupture of the Tokai and Nankai faults would result in M9.1 earthquakes along the Nankai-Suruga trough. Tsunami heights for those earthquake scenarios exceed 10 m on the coasts of 11 prefectures in Japan, with a maximum tsunami height of 34.4 m predicted on the Pacific coast of Shikoku Island in Southwest Japan. In the future, it will also be necessary to evaluate tsunami heights for such earthquake scenarios impacting the coasts in the western Pacific Ocean and East China Sea.

Acknowledgments We would like to express our deep thanks to the Reviewer of this paper and Dr. Vicente Santiago-Fandiño, Coordinator of the Springer Monograph, and Dr. Herman Fritz for reading this manuscript and providing valuable comments. This study was supported by the "Evaluation and disaster prevention research for the coming Tokai, Tonankai and Nankai earthquakes", a project of Japan's Ministry of Education, Culture, Sports, Science and Technology (MEXT). Figures were generated using generic mapping tools (Wessel and Smith 1998).

References

Aida I (1979) A source model of the tsunami accompanying the Tonankai earthquake of 1944. Bull Earthq Res Inst Univ Tokyo 54:329–341 (in Japanese with English abstract)

Aida I (1981a) Numerical experiments of historical tsunamis generated off the coast of the Tokaido District. Bull Earthq Res Inst Univ Tokyo 56:367–390 (in Japanese with English abstract)

Aida I (1981b) Numerical experiments for the tsunamis generated off the coast of the Nankaido District. Bull Earthq Res Inst Univ Tokyo 56:713–730 (in Japanese with English abstract)

An'naka T, Inagaki K, Tanaka H, Yanagisawa K (2003) Characteristics of great earthquakes along the Nankai trough based on numerical tsunami simulation. J Earthq Eng JSCE 27:307, http://www.jsce.or.jp/library/open/proc/maglist2/00578/2003/mg03.htm. Accessed 30 Aug 2012 (in Japanese)

Ando M (1975) Source mechanisms and tectonic significance of historical earthquakes along the Nankai Trough, Japan. Tectonophysics 27:119–140

Ando M (1982) A fault model of the 1946 Nankaido earthquake derived from tsunami data. Phys Earth Planet Inter 28:320–336

Baba T, Cummins PR, Hori T, Kaneda Y (2006) High precision slip distribution of the 1944 Tonankai earthquake inferred from tsunami waveforms: possible slip on a splay fault. Tectonophysics 426:119–134

DeMets C, Gordon RG, Argus DF (2010) Geologically current plate motions. Geophys J Int 181 (1):1–80. doi:10.1111/j.1365-246X.2009.04491.x

Earthquake Engineering Research Institute (2011) EERI Report on the performance of engineered structures in the Mw 9.0 Tohoku, Earthquake of March 11, 2011. http://www.eqclearinghouse.org/2011-03-11-sendai/files/2011/03/Japan-eq-report-Buildings-medrez.pdf. Accessed 30 Aug 2012

Fritz HM, Petroff CM, Catalán P, Cienfuegos R, Winckler P, Kalligeris N, Weiss R, Barrientos SE, Meneses G, Valderas-Bermejo C, Ebeling C, Papadopoulos A, Contreras M, Almar R, Dominguez JC, Synolakis CE (2011) Field survey of the 27 February 2010 Chile tsunami. Pure Appl Geophys 168(11):1989–2010. doi:10.1007/s00024-011-0283-5

Fujii Y, Satake K, Nishimae Y (2011) Observation and modeling of the January 2009 West Papua, Indonesia tsunami. Pure Appl Geophys 168:1089–1100

Hatori T (1974) Sources of large tsunamis in Southwest Japan. Zisin 2nd 27(1):10–24 (in Japanese with English abstract)

Hatori T (1978) Monuments of the Nankaido tsunamis of 1605, 1707 and 1854 in the Shikoku District: behavior of historical tsunamis and their comparison with the 1946 Nankaido tsunami. Bull Earthq Res Inst Univ Tokyo 53:413–445 (in Japanese with English abstract)

Hatori T (1980) Field investigation of the Nankaido tsunamis in 1707 and 1854 along the Osaka and Wakayama coasts, west Kii Peninsula. Bull Earthq Res Inst Univ Tokyo 55:505–535 (in Japanese with English abstract)

Hatori T (1981) Field investigation of the Nankaido tsunamis in 1707 and 1854 along the southwest coast of Shikoku. Bull Res Inst Tokyo 56:547–570

Hatori T (1988) Tsunami behaviors in the Seto Inland Sea and bungo channel caused by the Nankaido earthquakes in 1707, 1854, and 1946. Historical Earthquakes 4:37–46 (in Japanese)

Heidarzadeh M, Satake K (2013) Waveform and spectral analyses of the 2011 Japan tsunami records on tide gauge and DART stations across the Pacific Ocean. Pure Appl Geophys 170:1275–1293

Imai K, Satake K, Furumura T (2010) Amplification of tsunami heights by delayed rupture of great earthquakes along the Nankai Trough. Earth Planets Space 62:427–432

Ishibashi K (1998) Was the earthquake of July 9, 1498 in Southwest Japan really the great Nankai earthquake? Abstracts of the 1998 Japan Earth and Planetary Science Joint Meeting p. 313 (in Japanese)

Ishibashi K (2004) Status of historical seismology in Japan. Ann Geophys 47(2/3):339–368

Kanamori H (1972) Mechanism of tsunami earthquake. Phys Earth Planet Inter 6:346–359

Kanamori H (1977) The energy release in great earthquakes. J Geophys Res 82:2981–2987

Kanamori H, Anderson DL (1975) Theoretical basis of some empirical relations in seismology. Bull Seismol Soc Am 65(5):1073–1095

Kawasumi H (1950) Crustal deformations as deduced from mareographic data. Comm For the Synoptic Dev Of the Shikoku District Tokyo

Okada Y (1992) Internal deformation due to shear and tensile faults in a half-space. Bull Seism Soc Am 82:1018–1040

Omori F (1913) An account of the destructive earthquakes in Japan. Publ of the Earthquake Invest Comm, vol 68B. Earthquake Invest Comm, Tokyo

Satake K (1995) Linear and nonlinear computations of the 1992 Nicaragua earthquake tsunami. Pure Appl Geophys 144(3–4):455–470

Satake K, Wang KL, Atwater BF (2003) Fault slip and seismic moment of the 1700 Cascadia earthquake inferred from Japanese tsunami descriptions. J Geophys Res 108:2535. doi:10.1029/2003JB002521

Satake K, Fujii Y, Harada T, Namegaya Y (2013) Time and space distribution of coseismic Slip of the 2011 Tohoku earthquake as inferred from tsunami waveform data. Bull Seism Soc Am 103(4):1473–1492

Tsuji Y, Ueda K (1997) Proof of the existence of the 1498 Meio Nankai earthquake and the date of its occurrence Abstracts of the 1997 Japan Earth and Planetary Science Joint Meeting p 169 (in Japanese)

Tsuji Y, Satake K, Ishibe T, Kusumoto S, Harada T, Nishiyama A, Kim HY, Ueno T, Murotani S, Oki S, Sugimoto M, Tomari J, Heidarzadeh M, Watada S, Imai K, Choi BH, Yoon SB, Bae JS, Kim KO, Kim HW (2012) Field surveys of Tsunami heights from the 2011 off the Pacific Coast of Tohoku, Japan earthquake. Bull Earthq Res Inst Univ Tokyo 86:29–279, in Japanese with English abstract

Utsu T (1988) Historical materials of China related to earthquakes in Japan -Western Japan earthquake of June 31, 1498 and others-. Zisin 2nd 41(4):613–614 (in Japanese with English abstract)

Wessel P, Smith WHF (1998) New improved version of the generic mapping tools released. Eos Trans AGU 79:579

Xie YB, Tsai MB (eds) (1985) Compilation of historical materials of Chinese earthquakes, vol 2. Science Press, Beijing (in Chinese)

Xie YB, Tsai MB (eds) (1987) Compilation of historical materials of Chinese earthquakes, vol 3. Science Press, Beijing (in Chinese)

Chapter 7
Impacts of Tsunami Events on Ecosystem Services Provided by Benthic Macro-Invertebrate Assemblages of Marine Coastal Zones

Gwynne S. Rife

Abstract Significant environmental and societal impacts have ensued from the tsunami events during the last decade particularly those of the Indian Ocean (2004) and Pacific Coast of Japan (2011). Ecosystem services provided by marine inter- and subtidal benthic macroinvertebrate assemblages are tied to the changes in physical, chemical, and hydrological short and long term alterations to their habitats. Globally, benthic macroinvertebrate assemblages can be categorized to examine ecosystems services provided by these highly productive coastal areas. In addition to ongoing coastal human activity related threats to these areas, the disturbances to these assemblages immediately after a Tsunami event are currently a focus of research. Quantifying the impacts across the subunit of macroinvertebrate benthos is a necessary function for proper assessment of impacts to ecosystem services provided by coastal zones. The current knowledge base and predicted recovery timeframes, in addition to the need for further investigation of long term environmental societal factors are important globally.

Keywords Ecosystem services • Tsunami event • Macroinvertebrate assemblages • Seagrass • Coral reef • Macroalgae • Macrobenthic • Coastal marine communities

7.1 Introduction

Tsunami events and related earthquakes trigger toxic land run off, changes in the hydrology, alteration to the topography, and increased sedimentation that have an immediate and devastating negative impact on the coastal macrobenthos that inhabit near shore marine waters. Global coastal zones are the most productive and highly used regions and support fisheries and myriad other human activities and

G.S. Rife (✉)
University of Findlay, 1000 N Main St., Findlay, OH 45840, USA
e-mail: rife@findlay.edu

Y.A. Kontar et al. (eds.), *Tsunami Events and Lessons Learned: Environmental and Societal Significance*, Advances in Natural and Technological Hazards Research 35,
DOI 10.1007/978-94-007-7269-4_7,

Fig. 7.1 Coastal habitats of a variety support macroinvertebrate assemblages that provide important ecosystem services and are at risk following a tsunami event, they can be categorized by the faunal or physical structural base. Vascular plant based (*VP*) for example sea grass beds of a variety of species (**a**, photo credit creative commons NOAA stock photo); Unconsolidated sediments (*US*) of sandy to fine substrates (**b**, photo credit G. Rife); Macro-algae habitats (*MA*) including nearshore beds and kelp forests (**c**, photo credits G. Rife, **d**, photo credit Creative commons); Hard substrate habitats (*HS*) including rocky shorelines (**e**, photo credit G. Rife) and Coral Reef (*CR*) and other invertebrate reef based habitats (**f**, photo credit *G*, Rife)

impacts after Tsunami events are only beginning to be a focus of attention but the scientific community from multi-disciplinary research (Kontar et al. 2012). Societal impacts and related loss of associated ecosystem services are interrelated concerns when environmental changes negatively impact the biota of near shore marine waters that provide them. The near shore biota provide both the structural diversity and trophic base for these ecosystem services, and the macroinvertebrates communities in many cases are the foundation for these services (see Fig. 7.1). Macroinvertebrate assemblages that make up the near shore biota occur across coastal habitat types. Assemblages in major biotopes can be categorized into five major categories: (1) vascular plant dominated (VP), (2) macroalgae/protista dominated (MA), (3) unconsolidated substrate dominated (US), (4) hard substrate dominated (HB), and (5) reef species dominated (RS) as indicated in Table 7.1. The macrobiota that provide the trophic base for macroinvertebrate assemblages may be intertidal or subtidal, tropic or temperate, and have either a direct source of primary producers or subsist on suspended or

In depth Coastal and Marine Ecological Classification Standards (CMECS) are currently being vetted to categorize biotopes based upon water column, geoform, substrate and biotic components in nearshore waters of the Atlantic Coasts (Madden and Goodin 2007; 2008) but are already being applied globally outside of North American Atlantic waters (Gandomi et al. 2003). The sub-ecosystems are described

Table 7.1 Categories of habitats that support coastal marine benthic macroinvertebrate assemblages and their location in the marine coastal zone, dominate climate zone, and nutrient base

Habitat type	Intertidal/ Subtidal	Temperate/ Temporal	Nutrient base
Sea grass bed (VP)	Subtidal	Temperate to tropical	Primary productivity
Salt marsh (VP)	Intertidal	Temperate	Primary productivity
Tidal mangrove (VP)	Intertidal	Tropical	Primary productivity
Kelp forest (MA)	Subtidal	Temperate	Primary productivity
Calcareous algae bed (MA)	Intertidal	Tropical	Primary productivity
Mud flat (US)	Intertidal/ Subtidal	Temperate/ Tropical	Suspended organics and infauna
Sandy bottom (US)	Intertidal/ Subtidal	Temperate/ Tropical	Suspended organics and infauna
Cobble/Boulder (HS)	Intertidal/ Subtidal	Temperate/ Tropical	Suspended organics and infauna and epifauna
Rocky shoreline (HS)	Intertidal	Temperate/ Tropical	Suspended organics and epifauna
Human created (HS)	Intertidal/ Subtidal	Temperate/ Tropical	Suspended organics and epifauna
Coral reef (RS)	Subtidal	Tropical	Primary productivity, suspended organics and infauna and epifauna

in terms of macrobiota for the identified biotopes, with the majority being named by the dominant macroinvertebrate faunal species (see Table 7.2).

The biogeographic and aquatic setting for these coastal habitats will likely continue to be defined in a framework that will be useful for continued global comparisons of macroinvertebrate assemblages for economic valuation, prediction of recovery times, and quantification of losses resulting from a Tsunami event based on the coastal marine biotopes that are impacted.

Ecosystem Services provided by coastal macroinvertebrates span across a continuum of direct to indirect benefits. Marine ecosystem services that directly provide benefit include food, medicine, recreation, support of fisheries, and storm protection. Other ecosystem services are less tangible, such as the habitat's role in absorbing carbon from the atmosphere – a positive effect on our global climate. In addition to the economic supports coastal areas provide, human attitudes, beliefs, behaviors, customs, and traditions are often associated with the surrounding nature and environmental quality. Ecosystem Services provided by marine coastal zones are generally classified by four categories identified most commonly as Provisional, Supporting, Regulating, and Cultural Services (Millennium Ecosystem Assessment 2005). Provisioning services include food, water, and products such as building materials from mangrove and coral reef, and pharmaceutical compounds derived from marine algae and invertebrates. Supporting services include soil formation, primary productivity, and nutrient cycling, coastal habitats such as seagrass beds and mangroves are important nursery areas for the young stages of fishes and invertebrates that support coastal communities and commercial and recreational fisheries. Regulating services include regulation of climate; natural hazards such as

Table 7.2 Marine coastal macrobiotic assemblages as determined to comprise the benthic component for CMECS standards (After Madden et al. 2009)

Category of macrobenthic community	Examples of sub-units identified by CMECS	Ecosystem services provided	Direct/ Indirect	Supporting literature
Vascular plant dominated (VP)	Seagrass bed, tidal mangrove, brackish tidal aquatic vegetation	**Provisioning services** Provides building materials, areas for fisheries and associated industries **Supporting services** Soil formation, primary productivity, and nutrient cycling; nursery areas for the young stages of fishes and invertebrates **Regulating services** Capturing and filtering sediments and organic wastes in transit from inland regions to the ocean	Direct and indirect	Batker et al. (2010); Alongi (2002); Cebrian (2002); Dahdouh-Guebas et al. (2005); Quarto and Suryadiputra (2005); de Graaf et al. (1998); Ewel et al. (1998); Fast and Menasveta (2003); Lacerda and Abrao (1984)
Macro-algae dominated (MA)	Kelp forest Calcareous algal bed Canopy-forming algal bed Coralline/Crustose algal bed	**Provisioning services** Pharmaceutical compounds derived from marine algae and invertebrates **Regulating services** Capturing and filtering sediments and organic wastes in transit from inland regions to the ocean	Indirect	Vásquez (1993)
Unconsolidated sediment dominated (US)	Tunneling megafauna Burrowing anemones Bivalve bed Other non-molluscan invertebrate bed	**Provisioning services** Pharmaceutical compounds derived from marine algae and invertebrates **Regulating services** Capturing and filtering sediments and organic wastes in transit from inland regions to the ocean; sediment stabilization Primary production of benthic algae, high levels of secondary production and great diversity in benthic animals, provide forage for crabs, finfish and shorebirds	Indirect	Bolam et al. (2002); Lenihan and Micheli 2001 (review paper)

Hard substrate dominated (HS)	Mineral/Wood boring fauna Diverse colonizers Attached tube-building fauna Mobile crustaceans and gastrodpods on hard or mixed substrates Sessile/attached molluscs and/or non molluscan invertebrate communities	**Provisioning services** Pharmaceutical compounds derived from marine algae and invertebrates **Regulating services** Capturing and filtering sediments and organic wastes in transit from inland regions to the ocean; hard substrate for attached animals, provides finfish, crustacean and shorebird forage. Filters suspended material from the water for improved water quality Sediment stabilization erosion control via wave reduction High levels of secondary production and great diversity in benthic animals, forage for crabs, finfish and shorebirds	Indirect	Fox et al. (2003); Sasaki and Shepherd (2001)
Reef species dominated (RS)	Branching/columnar/foliose/plate/table coral reef Encrusting coral reef Massive coral reef Shallow molluscan dominated Mesophotic reef	**Provisioning services** Provides building materials, areas for fisheries and associated industries Pharmaceutical compounds derived from marine algae and invertebrates **Supporting services** Soil formation, photosynthesis and nutrient cycling **Cultural services** Scuba diving and other nature-based tourism	Direct and indirect	Rudi and Siregar (2005); Bellwood et al. (2004); Berg et al. (1998); Bryant et al. (1998); Burke and Maidens (2004); Burke et al. (2002); Halfpenny (2002); Hawkins et al. (1999); Fernando et al. (2005)

Fig. 7.2 Cultural ecosystems services of a variety provided by macroinvertebrate communities near the coasts include those tied to the culture and traditions of coastal peoples in many developing nations by supporting local small scale fisheries (**a**), recreational and aesthetic services across the globe as a source of natural interest and exploration for people of all ages (**b**), scientific and sociological endeavors (**c**), and ecotourism opportunities like scuba diving and sport fishing (**d**)

floods, disease, wastes, and water quality, coastal wetlands play an important role in water quality regulation by capturing and filtering sediments and organic wastes in transit from inland regions to the ocean. On a global scale, fixation of atmospheric carbon by oceanic algae and its eventual deposition in deep water represents an important part of the global carbon cycle and thus influences climate trends. Cultural services include recreational, aesthetic, and spiritual benefits derived from nature (see Fig. 7.2). Coastal tourism is the fastest-growing sector of the global tourism industry (Hall 2001), and is a major part of the economies of many small island developing nations. Moreover, the cultures and traditions of many coastal peoples are intimately tied to the marine ecosystems on which they depend.

Coastal marine ecosystem services are also provided directly, through human use or experience of the service or indirectly, via impacts of supporting and regulating services on other services and environments.

Macroinvertebrate assemblages form the basis for the majority of the coastal marine services as illustrated by the biotopes that are defined by the species that characterize the biotic components.

Changes in the local coastal marine environments following perturbations are myriad and occur in both the short term and long term spatial and temporal realms (Vásquez (1993); Constable (1999); Johnston and Roberts (2009); http://www.sciencedirect.com/science/journal/02697491; Schiel et al. (2004); Peterson et al. (1996)). Changes to these environments initially impact the resident macroinvertebrate assemblages and the ecosystem services they provide in a variety of ways. The majority of the current threats identified to these communities is heightened after a Tsunami event and are suggested to be altered long term for certain near shore biotopes (See Table 7.3).

Delineating the impacts of Tsunami event on coastal marine benthic invertebrate assemblages is closely related with the literature from other natural hazards such as hurricane and earthquake events (Batker et al. (2010); Jaramillo et al. (2012); Castilla et al. (2010). To examine the global effects that result in terms of the macrobenthic assemblages, one needs to characterize each major habitat type and synthesize current findings with related environmental disturbance known impacts.

7.2 Vascular Plant Dominated Habitat (VP)

Sea grass beds, dominated by rooted flowering aquatic grasses are significant refugia for macroinvertebrate assemblages and are dominated by turtle grass species in the tropical zones (*Thalassia* spp, *Halodule* spp, *Syringodium* spp., etc.), and *Posidonia* spp, *Ruppia* spp, *and Zostera* spp. in the more temperate waters (Short et al. 2007). These habitats not only stabilize and protect the shorelines, but support a diverse array of invertebrates that support the higher order consumers and thus, support various fisheries. Additionally they provide a complex structural habitat that serves as a nursery area for many commercially important species.

Perturbations to sea grass beads, and impacts of tsunami events have indicated that seagrass beds are resilient to perturbations but macroinvertebrate diversity of major taxonomic groups are tied to density of vegetation (Whanpetch et al. 2010; Nakaoka et al. 2007).

Salt Marshes of the temperate and tropic areas and related vegetated coastal habitats are well known to of high value as a nursery grounds, land run off filters, and high diversity of macroinvertebrate species significant to both commercial and sport fishing activities. Tsunami events load toxic land run off, scour vegetative areas, and deposit debris that compromises the health of these habitats and thus the macroinvertebrate assemblages (Kearney et al. 1995).

Mangrove habitats also known as mangels, are a group of coastal tropical halophytes that provide structural complexity and protect the shoreline by stabilizing sediments. Tsunami impacts have been examined for some habitats, and it appears mangroves may never fully recover from events that result in the extirpation of these halophytes (Alongi 2002; Dahdouh-Guebas et al. 2005). Loss of the mangroves mean loss of the ecosystem services they provide in addition to losing the associated macroinvertebrate fauna.

Table 7.3 Threats and potential for heightened effects to macroinvertebrate near shore communities after a Tsunami event

Identified threats to coastal marine macro-invertebrate communities	Mechanisms of impact	Potentially heightened by a Tsunami event
Toxic substances	Organochlorine compounds, heavy metals, organic tin compounds, organophosphates, polycyclic aromatic hydrocarbons, synthetic detergents, and surfactants	Yes – reach the oceans either directly (because the pollutants originate in coastal area), or indirectly through river systems or the atmosphere. In some cases they are released as a result of ocean dumping
Organic pollution	Excessive input of organic water and/or nutrients, or to a deterioration in the natural cleansing power	Yes – more pronounced in bays and other enclosed or semi-enclosed waters
Introduction of debris	Either direct dumping or in-direct introduction of waste materials	Yes – more significant adjacent to urban areas
Nutrient depletion	Over development and urbanization resulting in depletion of key nutrients and the indirect impact to decreased productivity and/or fertility	No – not directly impacted due to Tsunami event
Radioactive contamination	Above-ground nuclear tests conducted in years past constitute the principal source of such pollutants. Nuclear-powered ships, discharges by land-based nuclear facilities, and ocean dumping (including illegal dumping) are major sources of marine radioactive contamination	Yes – particularly in the case of facilities begin breached by earthquake activity
Depletion of resources vital to preservation	Land reclamation operations, embankment reinforcement projects, and other physical alterations to shallow-water environments have directly as well as indirectly contributed to the loss of seaweed beds, tidal marshes, coral reefs, man-grove forests	Yes – marine nutrient imbalances as well as degeneration of the natural resilience or cleansing ability of marine ecosystems
Public awareness	Lack of understanding of the aquatic habitats and biotic interaction and their role is goods and services such as assuring human populations opportunities for closer contact with the natural world	Yes – but perhaps in a positive manner if the event increases awareness and understanding

(continued)

Table 7.3 (continued)

Identified threats to coastal marine macro-invertebrate communities	Mechanisms of impact	Potentially heightened by a Tsunami event
Biotic disruptions	Many non-native wildlife species have penetrated marine ecosystems simply because they were attached to ship hulls or concealed in ship ballast water	Yes – a potential for introduction of previously un established species that have the potential to effect the biotic balance
Thermal pollution	heat energy discharged by power plants or factory cooling water, or by urban wastewater effluent (warm wastewater)	Yes – but localized
Oil pollution	human activities, including the flushing of ocean vessel bilges, leakage from undersea oil wells, and runoff or discharges from land-based facilities	Yes – significant for breached coastal nuclear and industrial facilities
Declining fishery resources	Marine environmental change and the fishery industry effects on environmental disruption	Yes – Death Assemblages and large numbers of eggs or fry of certain fish species

7.3 Macro Algae Dominated Habitat (MA)

Kelp Forests are temperate nearshore habitats that support diverse macroinvertebrate communities due to both the primary productivity and the structural complexity of their fronds.

Well adapted to be resilient against strong currents, they are tolerant to storm surges, but are prone to concentration radioactive material, after the tsunami of the Indian Ocean in 2010 radioactive compounds were found in the kelp off the California coast in the weeks after the tsunami event in Japan, but did not remain in the kelp for a long period of time suggesting they had been expelled into the biotope to presumably be taken up by other organisms (Cone 2012; Betina et al. 2011; Bagulayan et al. 2012; Manley and Lowe 2012).

7.4 Unconsolidated Sediment Dominated Habitat (US)

Mud Flats and other fine sediment habitats support infaunal macrobenthos that turns the sediments and process organics. These fine soils and the high degree of organics and detritus associated can be harmed by strong surges and deposited elsewhere smothering other areas with hypoxic sludge (Lenihan and Micheli 2001).

Sand habitats are teeming with diversity despite the common assumption that they do not, the macroinvertebrates present show resilience to storm events and recover quickly after a Tsunami event (Obdura 2008).

7.5 Hard Substrate Dominated (HB)

Macrofauna of hard substrates are generally in competition for space to attach, after a tsunami or hurricane boulders and cobble have been shown to be scattered, rocky shores could be denuded of life is scoured by thermal pollution, and new human created habitat might occur in the form of unintentional and artificial reef type habitat. Little is known about the specific effects on these types of macroinvertebrate assemblages, in general with the high larval settling etc. these coastal assemblages may be the first to recover after a storm event (Fox et al. 2003; Sasaki and Shepherd 2001).

7.6 Coral Reef Dominated Habitat (CS)

Coral reefs and related invertebrate reef macroinvertebrate assemblages are the most diverse and provide the most ecosystems services, yet are also some of the most delicate and threatened habitats. Coral bleaching can occur as the result of numerous stressors and tsunami events can devastate large regions from both abiotic and biotic stressors (Lee et al. 2005; Obdura 2008; Rudi and Siregar 2005; Halfpenny 2002; Hawkins et al. 1999). Of all the marine coastal biotopes, literature suggests it is the coral dependent fauna that can be devastated from a tsunami event, but more investigation is needed to determine if recovery is possible.

Macroinvertebrate assemblages recover after a tsunami or severe weather event differentially. More work is needed to verify the longer term impacts that tsunami events have from habitat perturbation to ecosystem service losses. In general, vascular plant dominated biotopes seem resilient (except for mangroves) after a tsunami event with recovery well underway in one annual cycle. Macroalgae/protista dominated biotopes may be impacted even at great distance from source of perturbations or related contamination little is known about the effects on the fauna they support. Both unconsolidated substrate dominated and hard substrate dominated biotopes are noted to have recovery times close to that identified for sea grass areas. Reef species dominated areas are subject to many environmental stressors, the physical and chemical changes that result from a tsunami event impact the corals species negatively but the fauna that rely on the physical structural components may shift in diversity but do persist. Defining recovery in terms of the macroinvertebrate assemblage would seem to suggest that recovery occurs relatively quickly, with mangroves might being the exception as it is suggested that they may never fully recover once the integrity of the habitat is destroyed.

As a recommendation for further study and documentation of impacts to coastal macroinvertebrate assemblages, methodologies which can both categorized and quantify impacts of tsunami events should be synthesized globally. Once a baseline is established for the expected changes to these communities, predictions regarding the changes to ecosystem services will be further understood and mechanism to limit loss of supplied ecosystem services could be put in place. Only with further societal and scientific endeavors can it be possible to identify post tsunami effects to ecosystem services provided by the macroinvertebrate assemblages. It will be necessary to take such action to understand how to aid the recovery of these communities and restore these important human related services.

References

Alongi D (2002) Present state and future of the world's mangrove forests. Environ Conserv 29:331–349

Bagulayan A, Bartlett-Roa JN, Carter AL, Inman BG, Keen EM, Orenstein EC, Patin NV, Sato KNS, Sibert EC, Simonis AE, Van Cise AM (2012) Journey to the Center of the Gyre: the fate of the Tohoku Tsunami Debris Field 2012. Oceanography 25(2):200–207. http://dx.doi.org/10.5670/oceanog.2012.55

Batker D et al (2010) Wetlands, hurricanes and the economy: the value of restoring the Mississippi River Delta. Earth Economics, Tacoma

Bellwood DR, Hughes TP, Folke C, Nystrom M (2004) Confronting the coral reef crisis. Nature 429:827–833

Berg H, Ohman MC, Troeng S, Linden O (1998) Environmental economics of coral reef destruction in Sri Lanka. Ambio 27:627–634

Betina J. Lomovasky, Fausto N. Firstater, Alex Gamarra Salazar, Jaime Mendo, Oscar O. Iribarne. 2011. Macro benthic community assemblage before and after the 2007 tsunami and earthquake at Paracas Bay, Peru, Journal of Sea Research, Volume 65, Issue 2, February 2011, Pages 205–212, ISSN 1385–1101, 10.1016/j.seares. 2010.10.002. (http://www.sciencedirect.com/science/article/pii/S138511011000122X).

Bolam SG, Fernandes TF, Huxham M (2002) Diversity, biomass, and ecosystem processes in the marine benthos. Ecol Monogr 72:599–615

Bryant D, Burke L, McManus J, Spalding M (1998) Reefs at risk. World Resources Institute, Washington, DC

Burke L, Maidens J (2004) Reefs at risk in the Caribbean. World Resources Institute, Washington, DC, p 80

Burke L, Selig E, Spalding M (2002) Reefs at risk in southeast Asia. World Resources Institute, Washington, DC

Castilla JC, Manríquez PH, Camaño A (2010) Effects of rocky shore coseismic uplift and the 2010 Chilean mega-earthquake on intertidal biomarker species. Mar Ecol Prog Ser 418:17–23

Cebrian J (2002) Variability and control of carbon consumption, export and accumulation in marine communities. Limnol Oceanogr 47(1):11–22

Cone M (2012) Fukushima's radioactivity found in Californian's kelp; levels spike then disappeared. Environmental Health News. Accessed 30 Mar 2012. http://www.environmentalhealthnews.org/ehs/news/2012/radioactive-iodine-from-fukushima-in-california-kelp. Last access 29, 2012

Constable AJ (1999) Ecology of benthic macroinvertebrates in soft-sediment environments: a review of progress toward quantitative models and predictions. Aust J Ecol 24(4):452–476

Dahdouh-Guebas F, Jayatisse LP, Di Nitto D, Bosire JO, Lo Seen D, Koedam N (2005) How effective were mangroves as a defence against the recent tsunami? Curr Biol 15(12): R443–R447

de Graaf GJ, Xuan TT (1998) Extensive shrimp farming, mangrove clearance and marine fisheries in the southern provinces of Vietnam. Mangroves Salt Marshes 2:159–166
Ewel KC, Twilley RR, Ong JE (1998) Different kinds of mangrove forests provide goods and services. Glob Ecol Biogeogr Lett 7:83–94
Fast AW, Menasveta P (2003) Mangrove forest recovery in Thailand. World Aquac 34(3):6–9
Fernando HJS, Mendis SG, McCulley JL, Perera K (2005) Coral poaching worsens tsunami destruction in Sri Lanka. Eos Trans AGU 86:301–304
Fox HE, Pet JS, Dahuri R, Caldwell RL (2003) Recovery in rubble fields: long-term impacts of blast fishing. Mar Pollut Bull 46:1024–1031
Gandomi Y, Shadi A, Savari A (2003) Classification of Gomishan Lagoon (Caspian Sea, Iran) by Using the Coastal and Marine Ecological Classification Standard (CMECS). Middle East J Sci Res 8(3):611–615, 2011
Halfpenny E (2002) Marine ecotourism: international guidelines and best practice case studies – a resource for tourism operators and coastal planners and managers. The International Ecotourism Society, Burlington, p 96
Hall D (2001) Trends in ocean and coastal tourism: the end of the last frontier? Ocean Coast Manag 44(9–10):601–618
Hawkins JP, Roberts CM, Van't Hof T, de Meyer K, Tratalof J, Aldam C (1999) Effects of scuba diving on Caribbean coral and fish communities. Conserv Biol 13(4):888–897
Jaramillo E, Dugan JE, Hubbard DM, Melnick D, Manzano M et al (2012) Ecological implications of extreme events: footprints of the 2010 earthquake along the Chilean coast. PLoS ONE 7(5): e35348. doi:10.1371/journal.pone.0035348
Johnston EL, Roberts DA (2009) Contaminants reduce the richness and evenness of marine communities: a review and meta-analysis. Environ Pollut 157(6):1745–1752
Kearney MS, Rogers AS, Townshend JRG, Lawrence WT, Dorn K, Eldred K, Lindsay F, Rizzo E, Stutzer D (1995) Developing a model for determining coastal marsh "health". Proceedings of the third conference on Remote Sensing for Marine and Coastal Environments, Environmental Research Institute of Michigan, Ann Arbor, pp II527–II537
Kontar Y, Peter Swarzenski, Adina Paytan, Ramesh Singh, Gwynne Rife, Bethany Henderson-Dean, Timothy Murphy, Hiro Kawamura, Koji Fujima, Yosuke Yamashiki, Shunichi Koshimura, Tomoyuki Takahashi, Swadhin Behera, Vicente Santiago-Fandiño, Yih-Chi Tan, DanLing Tang, Viacheslav Gusiakov, Yuriy Ozorovich, Tom Gleeson (2012) OS07-17-A025 Groundwater/surface water exchange in Tsunami affected areas in Japan – ecological and societal significance. In AOGS-AGU Joint Assembly, p 139
Lacerda LD, Abrao JJ (1984) Heavy metal accumulation by mangrove and saltmarsh intertidal sediments. Revista Brasiliera de Botanica 7:49–52
Lee YL, Affendi YA, Tajuddin BH, Yusuf YB, Alfian AAK, Anuar EA (2005) Resources of Langkawi Archipelago, Peninsular Malaysia NAGA. WorldFish Center Newsletter, vol 28, no. 1 & 2 Jan-Jun 2005
Lenihan HL, Micheli F (2001) Soft-sediment communities. In: Bertness MD, Gaines SD, Hay ME (eds) Marine community ecology. Sinauer Associates, Sunderland, pp 253–287
Madden CJ, Goodin KL (2007) Ecological classification of Florida bay using the coastal marine ecological classification standard (CMECS). NatureServe, Arlington
Madden C, Goodin K, Allee B, Finkbeiner M, Bamford D (2008) Coastal and marine ecological classification standard. NOAA and NatureServe, p 77
Manley SL, Lowe CG (2012) Canopy-forming kelps as California's coastal dosimeter: 131I from Damaged Japanese Reactor Measured in *Macrocystis pyrifera*. American Chemical Society 3731 dx.doi.org/10.1021/es203598r l Environ Sci Technol 2012 46:3731–3736
Millennium Ecosystem Assessment (2005) Ecosystems and human well-being: biodiversity synthesis. World Resources Institute, Washington, DC
Nakaoka M, Tanaka Y, Mukai H, Suzuki T, Aryuthaka C (2007) Tsunami impacts on biodiversity ofseagrass communities in the Andaman Sea, Thailand: (1) Seagrass abundance and diversity. In: Rigby PR and Shirayama Y (eds) Selected papers of the NaGISA world congress 2006. Publications of the Seto Marine Biological Laboratory, Special Publication Series, vol VIII, pp 49–56

Obdura DO, Tamelander J, Linden O (eds) (2008) Ten years after bleaching – facing the consequences of climate change in the Indian Ocean. CORDIO status report 2008. CIORDIO (Coastal Ocean Research and Development in the Indian Ocean)/Sida-SAREC Mombasa

Peterson CH, Kennicutt MC II, Green RH, Montagna P, Harper DE Jr, Powell EN, Roscigno PF (1996) Ecological consequences of environmental perturbations associated with offshore hydrocarbon production: a perspective on long-term exposures in the Gulf of Mexico. Can J Fish Aquat Sci 53(11):2637–2654

Quarto A, Suryadiputra N (2005) The Asian Tsunami: a protective role for coastal vegetation. Science 310:643

Rudi MSE, Siregar AM (2005) Acehnese reefs in the wake of the Asian Tsunami. Curr Biol 15:1926–1930

Sasaki R, Shepherd SA (2001) Ecology *and* post-settlement survival *of the* Ezo abalone. Haliotis discus hannai. *on* Miyagi coasts. Japan. J Shellfish Res 20:619–626

Schiel DR, Steinbeck JR, Foster MS (2004) Ten years of induced ocean warming causes comprehensive changes in marine benthic communities. Ecology 85:1833–1839. http://dx.doi.org/10.1890/03-3107

Short F, Carruthers T, Dennison W, Waycott M (2007) Global seagrass distribution and diversity: a bioregional model. J Exp Mar Biol Ecol 350:3–20

Vásquez JA (1993) Effects on the animal community of dislodgment of holdfasts of Macrocystis pyrifera. Pac Sci 47(2):180–184

Whanpetch N, Nakaoka M, Mukai H, Suzuki T, Nojima S, Kawai T, Aryuthaka C (2010) Temporal changes in benthic communited of seagrass beds impacted by the tsynami in the Andaman Sea, Tahiland. Estuar Coast Shelf Sci 87(2):246–252

Chapter 8
Discussion About Tsunami Interaction with Fringing Coral Reef

Jean Roger, Bernard Dudon, Yann Krien, and Narcisse Zahibo

Abstract The recent catastrophic tsunamis show that it is now more than ever necessary to assess tsunami hazard for all coastal communities. In fact, facing the dangerous increase of population in low-lying coastal areas during the last decades directly linked to the reduction of the natural defences against sea assaults, including tsunamis, and considering the economy of most of the concerned countries, solutions should be found quickly to protect those populations and/or mitigate the hazard. In that way, recent studies and post-event field observations have highlighted the protective role played by coral reefs and the consequences of their destructions on the tsunami amplitudes. In this study previous results about the effect of fringing coral reef geometry on the tsunami amplitude are discussed using numerical modeling of nonlinear shallow water equations (NAMI-DANCE code). For this purpose, a set of different artificial Digital Elevation Models has been prepared in agreement with real bathymetric profiles and results of simulations are compared and discussed together with the conclusions obtained by the other authors.

Keywords Tsunami • Coral fringing reef • Numerical modeling

8.1 Introduction

After the 2004 Indian Ocean event causing a record death toll of about 300,000, several tsunamis have highlighted again the waves' capability of destruction, leading also to significant loss of life especially in the American Samoa, 2009 (Okal et al. 2010), in Chile, 2010 (Fritz et al. 2011), in Japan, 2011 (Stimpson 2011), and above all in Indonesia that has been the target of several other tsunamis

J. Roger (✉) • B. Dudon • Y. Krien • N. Zahibo
Laboratoire de Recherches en Géosciences et Energie, Université des Antilles et de la Guyane, Guadeloupe, France
e-mail: jeanrog@hotmail.fr

Y.A. Kontar et al. (eds.), *Tsunami Events and Lessons Learned: Environmental and Societal Significance*, Advances in Natural and Technological Hazards Research 35, DOI 10.1007/978-94-007-7269-4_8,

since the big one (McAdoo et al. 2006; Fritz et al. 2007; Lay et al. 2011). Globally coastal communities attempt to find some solutions to face potential tsunami impacts (Rahman 2012) using different methods. According to recent works it is now clear that population protection depends mainly on education and awareness about the phenomenon and what to do in priority (Alexandra et al. 2009; Orcutt et al. 2011). Nevertheless it is also important to protect infrastructures besides human beings (Fraser et al. 2012).

Numerous research projects appeared in the early times after the 26th December 2004 tsunami, national as well as international, aiming at resolving major questions like which parts of the world coastlines are under tsunami threat, what is the potential of tsunami hazard, and how to protect the concerned coastal communities?

As in some places in Japan, the easiest way would be to build concrete sea defences (seawalls, tetrapod blocks, etc.) along the coasts but even in the case it could be feasible economically (Prasetya et al. 2008), the consequences on both environment and tourism especially in places annually frequented by hundreds of thousand people for their postcard sandy beaches could be irreversible (Schleupner 2005; Phillips and Jones 2006). In this way, Vuren et al. (2004) ask how coastal defences and societal activities in the coastal zone are compatible. In addition to this, Airoldi et al. (2005) try to propose an ecological perspective for the deployment of coastal defence structures but they would certainly be reserved for economically-rich countries, most of time less concerned by tsunami impacts. So to avoid this, people look at available natural means which could be rehabilitate and/or preserved in order to protect human beings against potential destructive tsunami waves (Tanaka 2009).

Studies dealing with the impact of several tsunamis after 2004 shows clearly that mangroves, coastal forests (like she-oak forests for example) and coral reefs represent natural barriers (Chatenoux and Peduzzi 2007; Kerr and Baird 2007; Cochard et al. 2008; Yanagisawa et al. 2009). The present problem is that the forests and mangroves tend to disappear globally, due to the dramatic increasing of coastal population (<100 km from shoreline) during the twentieth Century imposing a conversion of these flat and fertile coastal landscapes into agricultural and more generally industrial purposes (Valiela et al. 2001, 2009; Valiela 2006). Coral reefs seem to be less impacted even if they also tend to disappear for reasons like anthropogenic impacts as pollution and overfishing, but also because of storms and global warming (Wilkinson 1999; Bouchon et al. 2008a, b; Sale 2011). The main interest of coral reefs and finally, the reason of this study, is that it is generally accepted that they represent an efficient mean of protection against wave assaults (wind waves, swells, tsunamis) by most of coastal communities (Clark 1991; Frihy et al. 2004; SDMRI Report 2005; Liu and Ghidaoui 2009), being able to reduce classic wave energy until 71 % between the forereef and the reef crest (Lugo-Fernandez et al. 1998). But are they as efficient for tsunami waves as for wind-driven waves? What is the reality according to recent tsunami observations? What are the limitations of this free protection?

Nott (1997) shows that the tsunami triggered by the 1994 East Java Mw = 7.6 earthquake has been able to penetrate through the Australian eastern fringing coral reef off Cairns due to substantial gaps (funneling effect in 5–10 km wide passages) and impact the coast in front of these gaps as in the case of storm-generated waves (Young and Hardy 1993).

Several recent studies have been led to demonstrate the role played by coral reefs on tsunami waves, focusing on field observations and/or numerical modeling in order to assess which parameter of the geometry or the bottom friction would have the worst consequences on the tsunami amplitude and frequency content (Baba et al. 2008), flow speed (Fernando et al. 2008) and coastal run-up (Kunkel et al. 2006; Liu and Ghidaoui 2009; Gelfenbaum et al. 2011). In the following we discuss the impact of parameters as reef width, lagoon width, water depth, friction or the presence of gaps, using an artificial bathymetric model (a digital elevation model, D.E.M.) of a coral reef facing a sloping beach on which we model tsunami generation and propagation with NAMI-DANCE modeling code. This study follows principally the work of Kunkel et al. (2006) and Liu and Ghidaoui (2009).

Definition

Commonly, a fringing reef is a reef located close to the shore with a maximum separation of several hundred meters (i.e. the backreef channel or shallow lagoon width, also called 'boat channel') with a depth of maximum 5–10 m, to distinguish with a barrier reef, separated from the coast by a deep water-lagoon (Kennedy and Woodroffe 2002; Smithers et al. 2006). As indicated by Kennedy and Woodroffe (2002), the simplest fringing reef shows a reef crest directly attached to the shoreline, without backreef channel. Figure 8.1 shows an example of the typical scheme of a fringing coral reef located behind a barrier reef (Martinique, French Caribbean Island).

A depth profile of high resolution multibeam bathymetric data from the SHOM (Service Hydrographique et Océanographique de la Marine, France) reveals the typical coastal morphology of such coral environment. Geographic location of Martinique Island is indicated on the Google Earth view.

8.2 Tsunami Modeling

8.2.1 Modeling Code

NAMI-DANCE is a numerical modeling code used in this study. It is a modified version of the Japanese TUNAMI N2 numerical code (Imamura 1989, 1995) based on the solution of nonlinear shallow water equations (Zaytsev et al. 2009). The initial deformation calculation is based on elastic dislocation computed through Okada's formula (1985). This method assumed an instantaneous displacement of the sea surface identical to the vertical deformation of the seafloor (transmitted

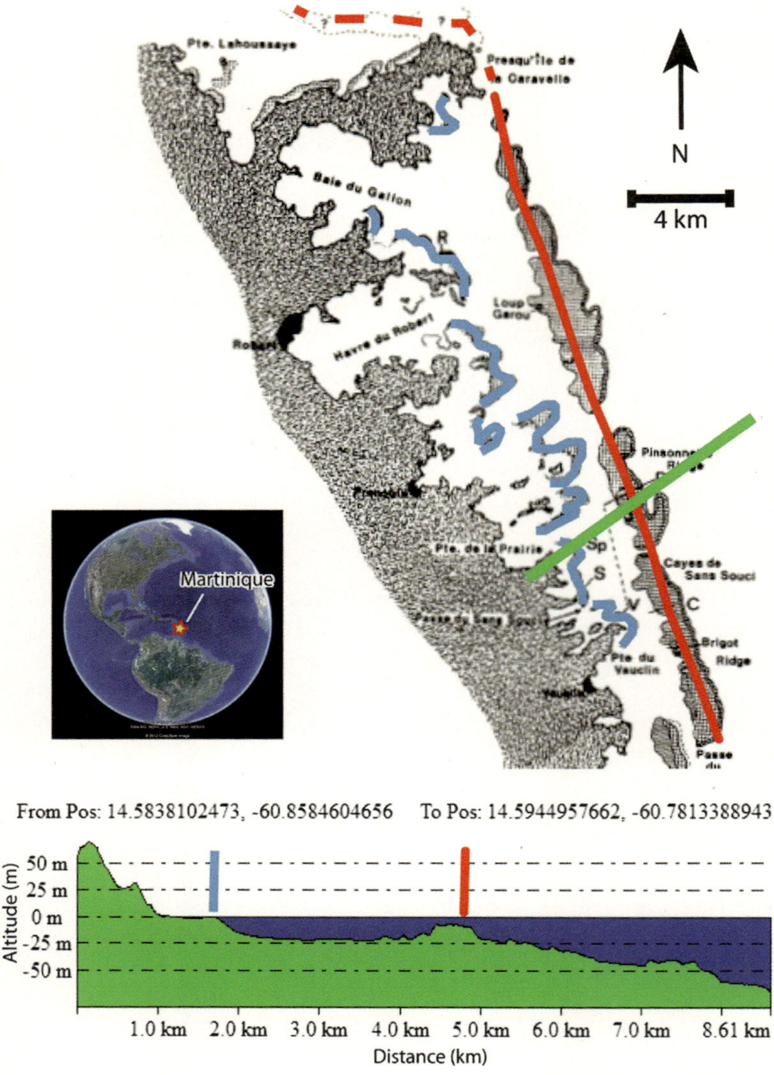

Fig. 8.1 Example of a combine coral reef along the eastern coast of Martinique Island: the *blue* and *red lines* highlight respectively the fringing and barrier reefs (Over a picture from Adey et al. 1977)

without losses to the entire water column), and solves the hydrodynamical equations 8.1, 8.2 and 8.3 of shallow water written in cartesian or spherical coordinates (Imamura et al. 2006). Non-linear terms are taken into account, and the resolution is carried out using a second-order explicit leap-frog finite different scheme. Wave dispersion is also considered.

$$\frac{\partial \eta}{\partial t} + \frac{\partial M}{\partial x} + \frac{\partial N}{\partial y} = 0 \tag{8.1}$$

$$\frac{\partial M}{\partial t} + \frac{\partial}{\partial x}\left(\frac{M^2}{D}\right) + \frac{\partial}{\partial y}\left(\frac{MN}{D}\right) + gD\frac{\partial \eta}{\partial x} + \frac{\tau_x}{\rho} = 0 \tag{8.2}$$

$$\frac{\partial N}{\partial t} + \frac{\partial}{\partial x}\left(\frac{MN}{D}\right) + \frac{\partial}{\partial y}\left(\frac{N^2}{D}\right) + gD\frac{\partial n}{\partial y} + \frac{\tau_y}{p} = 0 \tag{8.3}$$

D = **h** + **η** corresponds to the total water depth where h is the still water depth and η the sea surface elevation; v is the horizontal velocity vector; M and N are the water velocity fluxes in the x and y directions; τ_x and τ_y correspond to the bottom friction in x and y directions; g is the acceleration due to gravity (for more details see Dao and Tkalich 2007).

As most of tsunami modeling codes, this one allows the introduction of a specific initial disturbance like a single leading wave (solitary wave) as we will show in the following. It allows also the adjustment of the bottom friction coefficient f, linked to the Manning's roughness coefficient n by relation 8.4; the value of the Manning's roughness coefficient is set to 0.025 s/m$^{1/3}$ by default, a value commonly used, corresponding to a sandy bottom or bed rock cut channel (Linsley and Franzini 1979; Venturato et al. 2004), i.e. mildly rough interface. The value for coral reefs is not well-known as indicated by Kunkel et al. (2006). Imamura (2009) notices that this coefficient should be considered principally when the spatial grid size is larger than the scales of structures: in that case the bathymetric features are not correctly reproduced and thus the interaction between them and the waves is not well reproduced. In our case, the spatial resolution of the grid being 2 m, it encompasses largely the reef structure wavelength. Thus the role played by the friction coefficient will not be shown in the following as it has already been discussed by Kunkel et al. (2006): the authors conclude that the frictional effect lead to an energy dissipation of tsunami waves underlined by a run-up decrease of about 50 % for a variation of the drag coefficient of 0.03 to 0.1; the relation between Manning's and drag coefficients is explained in Rosman and Hench (2011). Fernando et al. (2008) reached the same conclusion of a considerable impact of coral friction on wave propagation and tsunami flow speed using a flume experiment with a synthetic coral reef showing a gap or not. Another part of energy dissipation is also due to wave breaking or reflection when passing through the reef, especially in the case of a reef located far offshore (typically a barrier reef).

$$n = \sqrt{\frac{fD^{\frac{1}{3}}}{2g}} \tag{8.4}$$

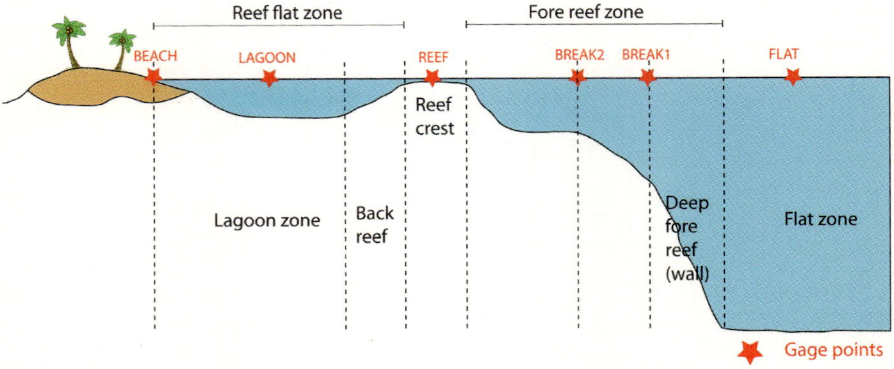

Fig. 8.2 Schematic profile of a coral reef in front of a sloping beach (Adapted from http://geology.uprm.edu/Morelock/reef.htm)

8.2.2 Artificial DEM: Scenarios

For the purpose of this study, because of the multiplicity of existing geometries of fringing reefs, a typical fringing reef profile presented in several previous studies concerning coral reefs (Fig. 8.2) has been chosen to build an artificial D.E.M. with adaptive geometry (Fig. 8.3).

Tsunami propagation is calculated over a 2 m – resolution bathymetric grid (i.e. D.E.M.) of dimension 1,500 × 1,500 m of a schematic coral reef in front of a sloping beach. Aiming to determine the role played by the main parameters as the reef width, the lagoon width, the channel width, the water depth upon the reef and the link between all of them, a set of different grids georeferenced in geographic coordinates has been prepared using a MATLAB subroutine. The resolution has been chosen with respect to real coastal feature (coral reef) wavelengths in order to reproduce as well as possible the shoaling effect, resonance phenomenon, etc.

The water thickness above the reef crest allows to test the case of a tsunami occurring at the same time of a storm surge or to consider the tide (low or high tide).

Here we only show the main results obtained with a handful of scenarios whose characteristics are presented in Table 8.1. An example of 3-dimensionnal D.E.M. is presented on Fig. 8.4: it corresponds to the case of a 100 m-wide reef with a 100 m-wide channel enclosing a 300 m-wide lagoon.

Tsunami propagation is calculated over each D.E.M. using the same initial deformation. For this preliminary study the initial deformation of the sea surface corresponds to a 3 m-high leading wave (only the positive peak) generated in the grid domain (Fig. 8.5) and showing a shape mimicking roughly a real tsunami wave. Six synthetic virtual tide gages (mareographs) have been located on this grid in order to record wave profiles as a function of time in strategic sites.

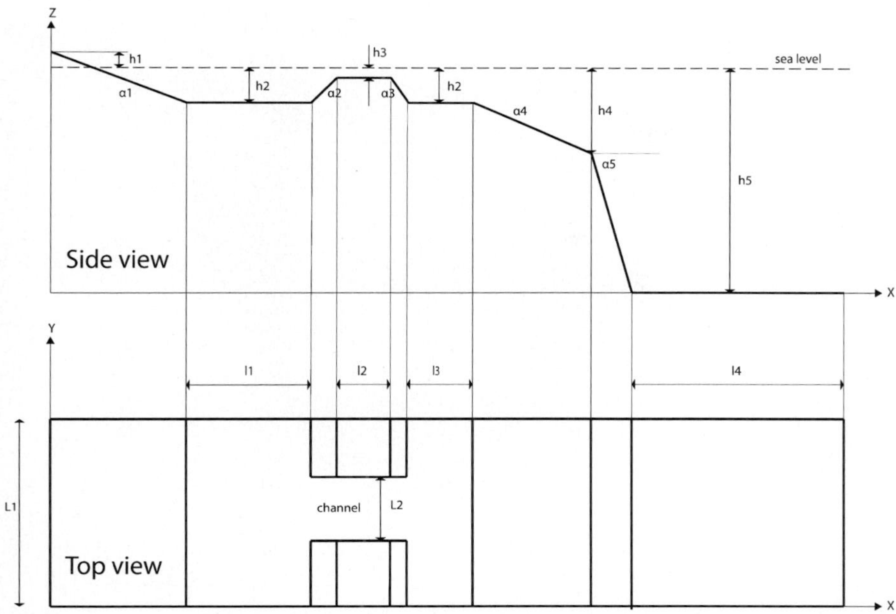

Fig. 8.3 Profile and top view of an idealized coral reef showing a gap in front of a sloping beach

Table 8.1 Interesting parameters of several tested cases: l_lagoon, h_reef, l_reef, l_channel correspond respectively to the lagoon width, the water thickness upon the reef crest, the reef width and the channel or gap width

model_number	l_lagoon (l1)	h_reef (h3)	l_reef (l2)	l_channel (L2)
1	50	0	10	10
2	50	0	10	50
3	50	0	10	100
4	50	0	50	10
7	50	0	100	10
10	50	1	10	10
12	50	1	10	100
19	100	0	10	10
37	300	0	10	10

8.2.3 Results

Propagation of a tsunami-like wave (the leading wave) has been done upon 60 different bathymetric grids. The interaction between this wave and the coral reef is shown on Fig. 8.6. It highlights the wave shoaling on the forereef because of depth decreasing, overtopping of the reef crest, refraction through the reef due to the presence of a gap and reflection on the shoreline. Run-up calculations are not shown here as they have already been discussed by Kunkel et al. (2006).

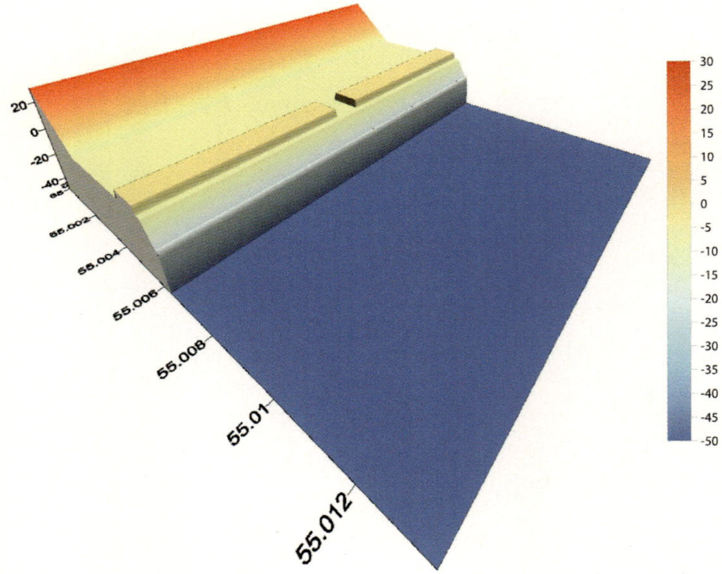

Fig. 8.4 Example of a D.E.M. prepared for tsunami propagation

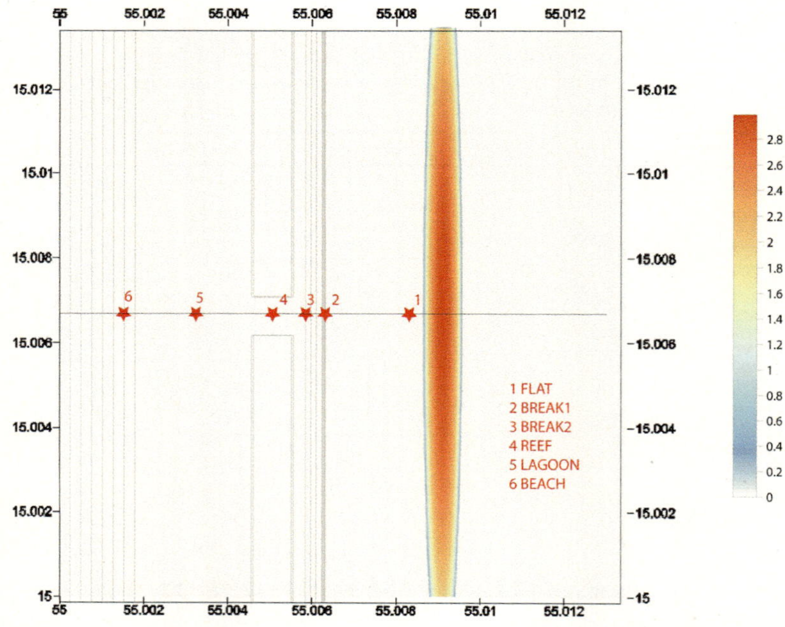

Fig. 8.5 Initial surface deformation (*top view*) obtained with NAMIDANCE in front of the reef (the *black rectangular lines* represent the bathymetric and topographic isohypses with a step of 5 m). The tides gages used to compare the signals are located with *red crosses*

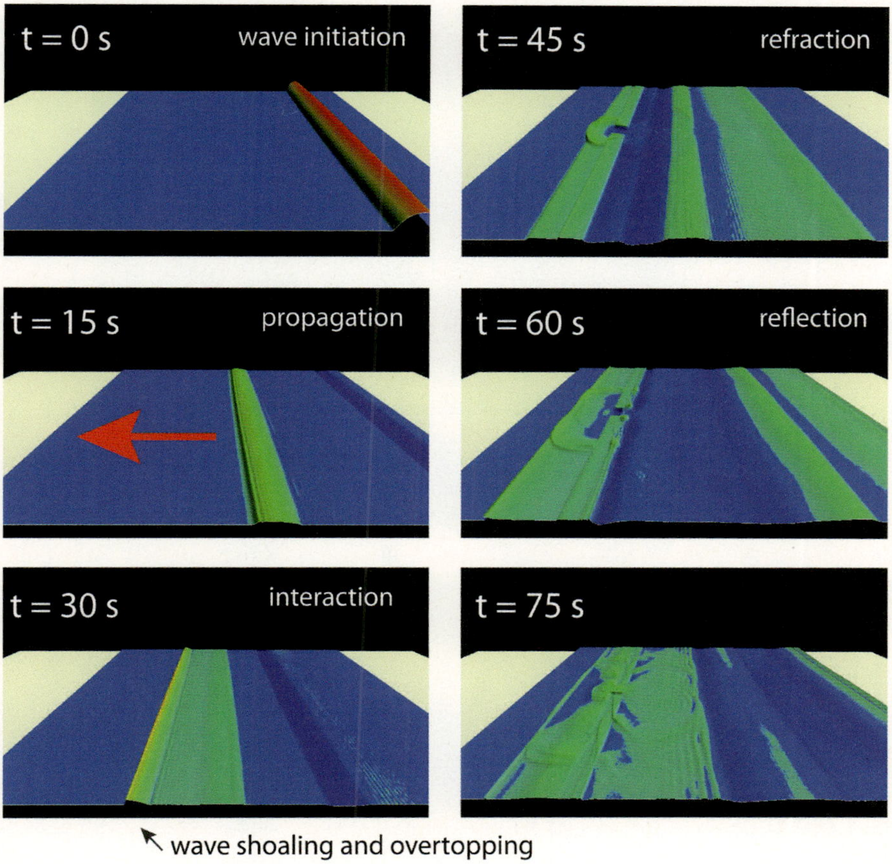

Fig. 8.6 Tsunami initiation and propagation (*red arrow* indicates direction) sketch towards a fringing coral reef showing interaction at a time step of 15 s

Figure 8.7 represents the maximum wave heights reach on each point of the grid after the propagation time. It reveals that the maximum wave heights reached in the near region behind the gap are more important than in region located behind healthy reef (without gap) as shown by Liu and Ghidaoui (2009). The gap (channel) in the reef leads to the diffraction of the incident wave-train which is followed by two main wave paths propagating in the lagoon towards the beach. They could be explained by wave interference between the refracted wave in the gap, the reflected wave on the beach and the overtopping wave. Amongst the 60 tests done during this study, we will concentrate on the most important related to post-event field observations:

(a) The sensitivity of the gap width is highlighted on Fig. 8.8a with 3 results of tsunami propagation at the same point (gage 5, Fig. 8.5) inside the lagoon for respectively 10, 50 and 100 m-wide channel. It indicates that an identical

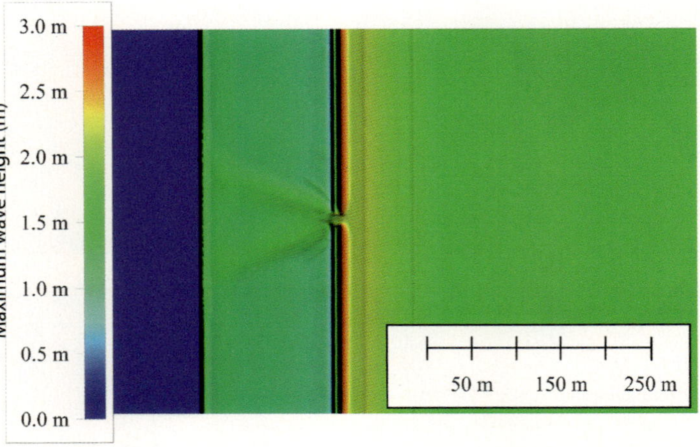

Fig. 8.7 Maximum wave height map of a scenario with a gap in the reef. It highlights two main wave paths inside the lagoon

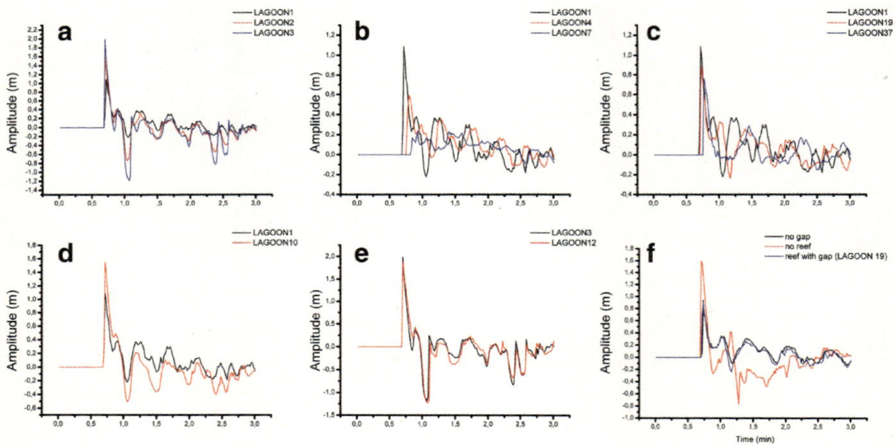

Fig. 8.8 Tide gage comparisons; the name of each gage LAGOON# refers to a number of gage in Table 8.1. In each case only one parameter changes: (**a**) reef channel width: 10, 50 and 100 m; (**b**) reef width: 10, 50 and 100 m; (**c**) lagoon width: 10, 50 and 100 m; (**d**) reef crest depth over a 10 m-wide reef: 0 and 1 m; (**e**) reef crest depth over a 100-m wide reef: 0 and 1 m; (**f**) no gap, no reef and a reef with a 10 m-wide gap

incident wave would have more energy, in terms of maximum wave amplitude, passing through the coral reef and thus able to reach the shore and inundate it

(b) Variation of the reef's width from 10 to 50 and 100 m (Fig. 8.8b) highlights that the wave amplitude near the shore will be less important with a larger reef. The incoming wave showing a wavelength of about 100 m is reduced by about 25 % and 85 % over a 10-m wide and 100-m wide reef respectively (Fig. 8.9). It is in

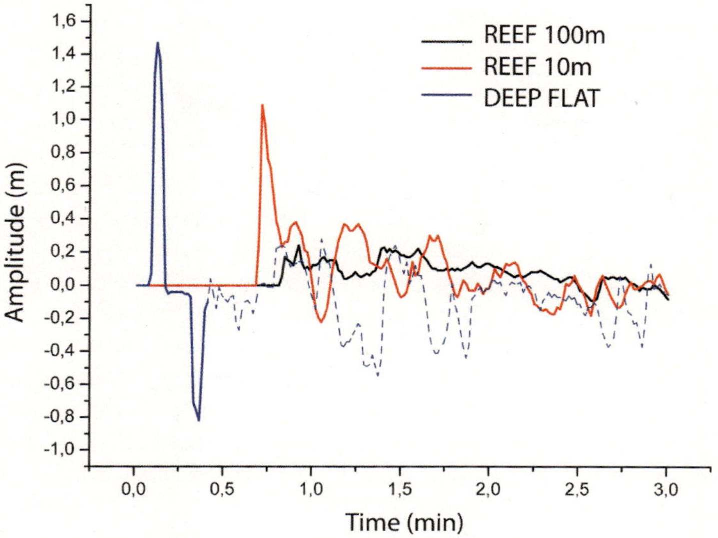

Fig. 8.9 Wave attenuation as a function of reef width: *blue curve* represents the initial signal recorded near the source; *red* and *black curves* represent respectively the recorded signal in the lagoon after passing over a over 10 m and 100 m-wide reefs

agreement with results of numerical modeling carried out by Mohandie and Teng (2009) who indicate that the reef is effective as a natural barrier for tsunamis with a width of about the same order of magnitude as the incoming wavelength

(c) In the same way, Fig. 8.8c shows that wave amplitude (of the first peak) will be less important with a larger lagoon

(d) and (e) The tests of the role played by the water depth upon the reef crest indicate that this water depth has not a so-significant impact on the wave amplitude as the precedent parameters if the reef is 100 m-wide, but upon a 10 m-wide reef, a water depth of 1 m leads to larger wave amplitude in the lagoon than with a water depth of 0 (Fig. 8.8d, e)

(f) Figure 8.8f illustrates the variation of signal with or without a gap in a 10 m-wide reef, and with and without a reef. As it has been already demonstrated, the presence of the reef has a direct impact on the amplitude of the tsunami near the coast. With unchanging roughness parameter the presence of a reef nearly divides by two the wave height in the lagoon

The tsunami height inside the lagoon after passing upon a reef without gap or with a 10 m-wide gap does not show substantial difference than in the case of a 50 or 100 m-wide gap (Fig. 8.8a) but however, the wave height without gap is less important, in agreement with Fernando et al. (2008) or Marris (2005).

8.3 Discussion and Conclusion

Our methodology of using an artificial D.E.M. that could be adapted for all the situations provides a useful tool to test the role played by each parameter of the geometry of a coral platform.

Our results agree with recent numerical experiments and several reported witnesses of a tsunami particular behavior linked to the existence of a fringing coral reef over the past two decades.

The role played by gaps is strongly underlined and converged on all the previous results. Indeed gaps allow a larger part of wave energy to pass through the reef contrarily to good-health reef (without gaps) in addition to the speed increase of the water flow due to a funneling effect inside the gaps (Liu and Ghidaoui 2009; Tanaka 2009). Besides, Nott (1997) indicates that the 1994 tsunami was probably amplified when passing through those gaps or as a result of resonance, diffraction or refraction phenomenon between the reef and the coast. He concludes that Cairns coast is finally not protected against such waves. Despite this, the role played by gaps and coral friction in general has been confirmed recently several researsch teams as Fernando et al. (2008) who discuss about the 2004 Sumatra tsunami impacting Sri Lanka coastlines. In their study they clearly demonstrate that the variation of friction underlined by the poaching and/or destruction of corals, equivalent to the creation of gaps within the reefs, leads to a substantial increase of tsunami flow velocity, because of reducing the bottom drag coefficient or roughness coefficient (Rosman and Hench 2011).

This is in agreement with the work of Lowe et al. (2005) who study the energy dissipation over a reef for classic waves; they conclude that the attenuation is due to the bottom friction often prevailing on wave breaking (on the contrary of what happens on a sandy beach). This is further demonstrated for tsunami waves by Kunkel et al. (2006) whose numerical experiment of tsunami propagation over a reef allow to propose a relation between run-up and drag coefficient (directly linked to the roughness coefficient; Wu et al.,1999).

The previous authors, which work has been partially tested with another method by Liu and Ghidaoui (2009), show also that the run-up over an idealized topography located behind the reef is directly linked to the reef width; but they are cautious with the results interpretation underlining the dependence of the run-up with the incident wavelength and amplitude as well as the geometry and health of the reef.

In the same way, Baba et al. (2008) model the propagation of the 2007 Solomon Islands tsunami through the Australian North-eastern coastline with and without the Great Barrier Reef and conclude that the reef reflects much of this low-amplitude tsunami energy, that the energy passing through is divided because of the gaps, and above all, that in addition to wave shoaling and breaking, the reef slows the waves down, delaying the tsunami impact.

They also indicate that the bottom friction of the reef should influence the tsunami as previously tested by Kunkel et al. (2006), hypothesis that has been confirmed more recently by the work of Gelfenbaum et al. (2011) for the 2009

American Samoa tsunami. On the contrary, the effectiveness of the reefs protective role is debated theoretically by Lynett (2007) who concludes that for very small obstacle lengths, i.e. typically a fringing coral reef compared with travelling tsunami wavelength, the reduction induced by the reef on the tsunami run-up and the maximum velocity will be inconsequential.

In that way, Roeber et al. (2010) demonstrate that the shallow reefs surrounding Tutuila Island were not enough to protect the coastline from the 2009 Samoa tsunami and they add the report of local resonances of short-period dispersive waves due to energy trapping within shallow lagoons, triggering more catastrophic consequences, highlighted on site by large disparities of impact along the coast. Nonetheless, the different conclusions reached by all these studies seem to agree globally with the fact that everything depends primarily on the reef width which induce dissipation through bottom friction, the presence and size of gaps, and on the incident wave height, especially if it exceeds the average depth of the top of the reef.

To summarize, in this study we show that the geometry and the location of the fringing coral reef (more or less close to the coast) including gaps or not plays an important role on the tsunami behavior in agreement with the existing studies (Kunkel et al. 2006; Fernando et al. 2008; Liu and Ghidaoui 2009; Baba et al. 2008; Gelfenbaum et al. 2011).

Tsunami waves seem to behave as classical waves as long as their characteristics stay within the same range of amplitude and respects the water depth over the reef. The tide or the weather condition (occurrence of a storm surge) should be considered accordingly to this fact. The greatest protection from destructive tsunamis will come from wide and high rough coral reefs, showing as little gaps as possible (Gelfenbaum et al. 2011).

Furthermore, the presence of gaps and the so-enclosed water body surrounded by the coral reef and the coast could induced indirect effects of tsunami arrival like flow speed increase or resonance phenomenon leading to a considerable rise of wave amplitude as shown by Roeber et al. (2010).

It follows from this work that rehabilitation and protection of coral reefs, leading to recover man-made gaps principally, should be considered as a natural mean of tsunami defence structure together with economically interesting purposes (touristic diving, fish nesting, etc.).

Prospects

A more accurate study should be realized using real bathymetric data to compare to well-known events including friction coefficient variations, tests of different incident waves and a numerical code using Boussinesq solution to reproduce as well as possible wave breaking and dispersion phenomenon (Roeber and Cheung 2012).

Acknowledgements The authors would like to thank Ahmet Yalciner (METU, Turkey) for sharing NAMI-DANCE software. They also thank anonymous referees for constructive comments.

This work has been funded by the European INTERREG IV TSUNAHOULE Project which aims at determining tsunami and waves hazard principally for the French Caribbean Islands and neighboring islands.

References

Adey WH, Adey PJ, Burke R, Kaufman L (1977) The Holocene reef systems of Eastern Martinique, French West Indies. Atoll Research Bulletin 218. http://www.sil.si.edu/DigitalCollections/atollresearchbulletin/issues/00218.pdf

Airoldi L, Abbiati M, Beck MW, Hawkins SJ, Jonsson PR, Martin D, Moschella PS, Sundelöf A, Thompson RC, Aberg P (2005) An ecological perspective on the deployment and design of low-crested and other coastal defence structures. Coast Eng 52:1073–1087

Alexandra K, Cain G, Iwasaki P (2009) Tsunami education: a blueprint for coastal communities. Report of the pacific tsunami museum and county of Hawaii planning program, Honolulu, 90 p. http://www.tsunami.org/pdf/Final_Document_Tsunami_Blueprint.pdf

Baba T, Mleczko R, Burbidge D, Cummins PR, Thio HK (2008) The effect of the Great Barrier Reef on the propagation of the 2007 Solomon Islands tsunami recorded in Northeastern Australia. Pure Appl Geophys 165:2003–2018

Bouchon C, Portillo P, Bouchon-Navaro Y, Max L, Hoetjes P, Braithwaite A, Roach R, Oxenford H, O'Farrel S, Day O (2008a) Status of the coral reefs of the Lesser Antilles after the 2005 bleaching event. In: Wilkinson C, Souter D (eds) Status of Caribbean coral reefs after bleaching and hurricanes in 2005, Global Coral Reef Monitoring Network, and Reef and Rainforest Research Centre, Townsville 152 p., 85–103

Bouchon C, Portillo P, Bouchon-Navaro Y, Louis M, Hoetjes P, de Meyer K, Macrae D, Armstrong H, Datadin V, Hardin S, Mallela J, Parkinson R, Van Bochove J-W, Wynne S, Lirman D, Herlan J, Baker A, Collado L, Nimrod S, Mitchell J, Morrall C, Isaac C (2008b) Status of coral reefs of the Lesser Antilles: The French West Indies, The Netherlands Antilles, Anguilla, Antigua, Grenada, Trinidad and Tobago. In: Wilkinson C (ed) Status of coral reefs of the world: 2008. Global Coral Reef Monitoring Network/Reef and Rainforest Research Center, Townsville, pp 265–279

Chatenoux B, Peduzzi P (2007) Impacts from the 2004 Indian Ocean Tsunami: analysing the potential protecting role of environmental features. Nat Hazards 40:289–304

Clark JR (1991) Coastal zone management. Land Use Policy 8(4):324–330

Cochard R, Ranamukhaarachchi SL, Shivakoti GP, Shipin OV, Edwards PJ, Seeland KT (2008) The 2004 tsunami in Aceh and Southern Thailand: a review on coastal ecosystems, wave hazards and vunerability. Perspect Plant Ecol Evol Syst 10:3–40

Dao MH, Tkalich P (2007) Tsunami propagation modelling – a sensitivity study. Nat Hazard Earth Syst Sci 7:741–754

Fernando HJS, Samarawickrama SP, Balasubramanian S, Hettiarachchi SSL, Voropayev S (2008) Effects of porous barriers such as coral reefs on coastal wave propagation. J Hydro Environ Res 1(3–4):187–194

Fraser S, Raby A, Pomonis A, Goda K, Chian SC, Macabuag J, Offord M, Saito K, Sammonds P (2012) Tsunami damage to coastal defences and buildings in the March 11th 2011 M_w 9.0 Great East Japan earthquake and tsunami. Bull Earthq Eng. doi:10.1007/s10518-012-9348-9

Frihy OE, El Ganaini MA, El Sayed WR, Iskander MM (2004) The role of fringing coral reef in beach protection of Hurghada, Gulf of Suez, Red Sea of Egypt. Ecol Eng 22:17–25

Fritz HM, Kongko W, Moore A, McAdoo B, Goff J, Harbitz C, Uslu B, Kalligeris N, Suteja D, Kalsum K, Titov V, Gusman A, Latief H, Santoso E, Sujoko S, Djulkarnaen D, Sunendar H, Synolakis C (2007) Extreme runup from the 17 July 2006 Java tsunami. Geophys Res Lett 34: L12602. doi:10.1029/2007GL029404

Fritz HM, Petroff CM, Catalan PA, Cienfuegos R, Winckler P, Kalligeris N, Weiss R, Barrientos SE, Meneses G, Valderas-Bermejo C, Ebeling C, Papadopoulos A, Contreras M, Almar R, Dominguez JC, Synolakis C (2011) Field survey of the 27 February 2010 Chile tsunami. Pure Appl Geophys 168(11):1989–2010

Gelfenbaum G, Apotsos A, Stevens AW, Jaffe B (2011) Effects of fringing reefs on tsunami inundation: American Samoa. Earth Sci Rev 107:12–22

Kennedy DM, Woodroffe CD (2002) Fringing reef growth and morphology: a review. Earth Sci Rev 57:255–277

Kerr AM, Baird AH (2007) Natural barriers to natural disasters. Bioscience 57(2):102–103

Kunkel CM, Hallberg RW, Oppenheimer M (2006) Coral reefs reduce tsunami impact in model simulations. Geophys Res Lett 33:L23612. doi:10.1029/2006GL027892

Imamura F (1989) Tsunami numerical simulation with the staggered leap-frog scheme (Numerical code of TUNAMI-N1). School of Civil Engineering, Asian Inst. Tech. and Disaster Control Research Center, Tohoku University

Imamura F (1995) Tsunami numerical simulation with the staggered leap-frog scheme (Numerical code of TUNAMI-N1 and N2). Disaster Control Research Center, Tohoku University, 33 p

Imamura F (2009) Tsunami modeling, calculating inundation and hazard maps. In: Bernard EN, Robinson AR (eds) The sea, tsunamis, 1st edn. Harvard University Press, Cambridge, 321–332 p

Imamura F, Yalciner AC, Ozyurt G (2006) Tsunami modelling manual (TUNAMI model). 58 p. http://www.tsunami.civil.tohoku.ac.jp/hokusai3/J/projects/manual-ver-3.1.pdf

Lay T, Ammon CJ, Kanamori H, Yamazaki Y, Cheung KF, Hutko AR (2011) The 25 October 2010 Mentawai tsunami earthquake (Mw 7.8) and the tsunami hazard presented by shallow megathrust ruptures. Geophys Res Lett 38:L06302. doi:10.1029/2007GL029404

Linsley RK, Franzini JB (1979) Water resources engineering. McGraw-Hill, New-York, 716 p

Liu H, Ghidaoui MS (2009) Investigations of the effects of coral reefs on coastal wave propagation by model simulations. In: Proceedings of the 33rd IAHR Congress – Water Engineering for a Sustainable Environment, Vancouver, 9–14 August

Lowe RJ, Falter JL, Bandet MD, Pawlak G, Atkinson MJ, Monismith SG, Koseff JR (2005) Spectral wave dissipation over a barrier reef. J Geophys Res 110:C04001. doi:10.1029/2004JC002711

Lugo-Fernandez A, Roberts HH, Suhayda JN (1998) Wave transformations across a Caribbean fringing-barrier coral reef. Cont Shelf Res 18:1099–1124

Lynett PJ (2007) Effect of a shallow water obstruction on long wave runup and overland flow velocity. J Waterw Port Coast Ocean Eng 133(6):455–462

Marris E (2005) Tsunami damage was enhanced by coral theft. Nature 436:1071

McAdoo BG, Dengler L, Prasetya G, Titov V (2006) How an oral history saved thousands on Indonesia's Simeulue Island during the December 2004 and March 2005 tsunamis. Earthq Spectra 22(3):661–669

Mohandie RK, Teng MH (2009) Numerical and experimental study on the effect of coral reef and beach vegetation on reduction of long wave run-up. American Geophysical Union, Fall Meeting Abstract NH31B-1115

Nott J (1997) Extremely high-energy wave deposits inside the Great Barrier Reef, Australia: determining the cause – tsunami or tropical cyclone. Mar Geol 141:193–207

Okal EA, Fritz HM, Synolakis CE, Borrero JC, Weiss R, Lynett PJ, Titov VV, Foteinis S, Jaffe BE, Liu PL-F, Chan I-C (2010) Field survey of the Samoa tsunami of 29 September 2009. Seismol Res Lett 81(4):577–591

Orcutt JA, Committee on the Review of the Tsunami Warning and Forecast System and Overview of the Nation's Tsunami Preparedness, National Research Council (2011) Tsunami warning and preparedness: an assessment of the U.S. tsunami program and the nation's preparedness efforts. The National Academies Press, Washington, DC, 296 p

Phillips MR, Jones AL (2006) Erosion and tourism infrastructures in the coastal zone: problems, consequences and management. Tour Manag 27(3):517–524

Prasetya GS, Healy TR, Lange (de) WP, Black KP (2008) Extreme tsunami run up and inundation flows at Banda Aceh, Indonesia: are there any solutions to this type of coastal disaster? In: Proceedings of the Solutions to Coastal Disasters Congress, Turtle Bay, Oahu, 13–16 Apr 2008. American Society of Civil Engineers, ISBN: 978-0-7844-0978-7

Rahman HA (2012) Community and tsunami disaster. Proceeding of the international conference on environment science and engineering, IPCBEE, 32. IACSIT Press, Singapore, http://www.ipcbee.com/vol32/029-ICESE2012-D30015.pdf

Roeber V, Cheung KF (2012) Boussinesq-type model for energetic breaking waves in fringing reef environments. Coast Eng 70:1–20

Roeber V, Yamazaki Y, Cheung KF (2010) Resonance and impact of the 2009 Samoa tsunami around Tutuila, American Samoa. Geophys Res Lett 37:L21604. doi:10.1029/2010GL044419

Rosman JH, Hench JL (2011) A framework for understanding drag parameterizations for coral reefs. J Geophys Res 116:C08025. doi:10.1029/2010JC006892

Sale PF (2011) Our dying planet. An ecologist's view of the crisis we face. University of California Press, Berkeley

Schleupner C (2005) Evaluation of coastal squeeze and beach reduction and its consequences for the Caribbean island Martinique. Working paper FNU-72, Research Unit Sustainability and Global Change, Hamburg University, 25 p. http://www.fnu.zmaw.de/fileadmin/fnu-files/publication/working-papers/Evaluation_of_coastal_squeeze_and_beach_reduction.pdf

SDMRI Report (2005) Report on the rapid assessment of status of corals in Gulf of Mannar after tsunami. http://www.reefcheck.org/PDFs/reports/India%202005.pdf

Smithers SG, Hopley D, Parnell KE (2006) Fringing and nearshore coral reefs of the Great Barrier Reef: episodic Holocene development and future prospects. J Coast Res 22(1):175–187

Stimpson I (2011) Japan's Tohoku earthquake and tsunami. Geol Today 27(3):96–98. doi:10.1111/j.1365-2451. 2011.00793.x

Tanaka N (2009) Vegetation bioshields for tsunami mitigation: review of effectiveness, limitations, construction, and sustainable management. Landsc Ecol Eng 5(1):71–79

Valiela I, Bowen JL, York JK (2001) Mangrove forests: one of the world's threatened major tropical environments. Bioscience 51(10):807–815

Valiela I (2006) Global coastal change. Blackwell Publishing, Oxford

Valiela I, Kinney E, Culbertson J, Peacock E, Smith S (2009) Global losses of mangroves and salt marshes. In: Duarte CM (ed) Global loss of coastal habitats: magnitudes, causes and consequences. Fundacion BBVA, Madrid

Venturato AJ, Titov VV, Mofjeld H, Gonzalez FI (2004) NOAA TIME Eastern Strait of Juan de Fuca, Washington, Mapping project: procedures, data sources, and products. NOAA Technical Memorandum OAR PMEL-127, contribution 2713, 26 p. http://www.pmel.noaa.gov/pubs/PDF/vent2713/vent2713.pdf

Vuren (van) S, Kok M, Jorissen RE (2004) Coastal defence and societal activities in the coastal zone: compatible or conflicting interests? J Coast Res 20(2):550–561

Wilkinson C, Souter D (eds) (2005) Status of Caribbean coral reefs after bleaching and hurricanes in 2005, Global Coral Reef Monitoring Network/Reef and Rainforest Research Centre, Townsville 152 p., 85–103

Wilkinson CR (1999) Global and local threats to coral reef functioning and existence: review and predictions. Mar Freshw Res 50(8):867–878

Wu F-C, Shen HW, Chou Y-J (1999) Variation of roughness coefficients for unsubmerged and submerged vegetation. J Hydraul Eng 125:934–942

Yanagisawa H, Koshimura S, Goto K, Miyagi T, Imamura F, Ruangrassamee A, Tanavud C (2009) The reduction effects of mangrove forest on a tsunami based on field surveys at Pakarang Cape, Thailand, and numerical analysis. Estuar Coast Shelf S 81:27–37

Young IR, Hardy TA (1993) Measurement and modelling of tropical cyclone waves in the Great Barrier Reef. Coral Reefs 12:85–95

Zaytsev AI, Kovalev DP, Kurkin AA, Levin BV, Pelinovsky EN, Chernov AG, Yalciner A (2009) The tsunami on Sakhalin on August 2, 2007: Mareograph evidence and numerical simulation. Russ J Pac Geol 3(5):437–442. doi:10.1134/S1819714009050054

Chapter 9
Ecological Status of Sandy Beaches After Tsunami Events: Insights from Meiofauna Investigations After the 2011 Tohoku-oki Tsunami, Sendai Bay, Japan

Katarzyna Grzelak, Witold Szczuciński, Lech Kotwicki, and Daisuke Sugawara

Abstract Tsunami may strongly impact beach ecosystems. To assess its magnitude five beaches along the Sendai Bay, Japan, were studied 2 months after the 11th March 2011 Tohoku-oki tsunami with focus on their recovery and meiofauna assemblages within few weeks after the event. The beaches recovered and new meiofauna assemblages established, which were strongly correlated to sediment grain size. The new data and review of previous works suggest that for beach ecosystems tsunami plays a role of ecosystem disturbance, not a catastrophe.

Keywords Tohoku-oki tsunami • Beach recovery • Meiofauna • Sandy beach • Japan

9.1 Introduction

Tsunami is a series of waves characterised by small amplitude and very long wavelengths. They are caused by displacement of a large volume of water, usually attributed to earthquake, landslide or volcanic eruptions. As tsunami approaches the coast and the waters become shallow the wave amplitude significantly increases and seawater may inundate the coastal zone up to several tens of meters above sea level. Large tsunami waves (e.g. 2004 Indian Ocean tsunami, 11th March 2011 Tohoku-oki tsunami) caused death of thousands of humans, generate massive

K. Grzelak (✉) • L. Kotwicki
Institute of Oceanology, Polish Academy of Sciences, Powstanców Warszawy 55, Sopot 81-712, Poland
e-mail: kgrzelak@iopan.gda.pl

W. Szczuciński
Institute of Geology, Adam Mickiewicz University in Poznań, Maków Polnych 16, 61-606 Poznań, Poland

D. Sugawara
Disaster Control Research Center, Tohoku University, Sendai, Miyagi 980-8579, Japan

Y.A. Kontar et al. (eds.), *Tsunami Events and Lessons Learned: Environmental and Societal Significance*, Advances in Natural and Technological Hazards Research 35, DOI 10.1007/978-94-007-7269-4_9,

damages of houses and infrastructure, erode, transport and deposit huge amounts of sediments, and affect the coastal zone environment. The environmental impact of tsunami includes for instance translation of large amount of saltwater on land, deposition of thick layer of salty sediment covering plants and former soils and damage of natural habitats due to erosion.

Coastal ecosystems are affected in various ways by tsunami (e.g. Szczuciński et al. 2006). Apart from widely studied impacts on coral reefs (e.g. Baird et al. 2005; Brown 2005) and mangrove forests (e.g. Dam Roy and Krishnan 2005; Cochard et al. 2008; Yanagisawa et al. 2009), the most evident effect is an alteration of distribution and abundance of living benthic organisms. Biological investigations of tsunami impact on benthic communities have focused mainly on the larger macrofauna organism, which can readily be counted and identified (Kendall et al. 2009; Whanpetch et al. 2010; Lomovasky et al. 2011). Smaller organisms (meiofauna) received less attention, despite their significant roles in ecosystem functions, such as nutrient cycling and energy flow (for details see Giere 2009). Sandy beaches provide the habitat with the most diverse meiofauna communities. At the same time, beaches are one of the most tsunami affected parts of the coastal zone.

Reports from the previous tsunamis documented almost complete beach erosion (MacInnes et al. 2009; Paris et al. 2009), but also relatively fast beach recovery, completed its reversion back to the equilibrium stage within 1 year after the tsunami disturbance (Choowong et al. 2009). Both processes obviously have an effect on beach dwellers. So far the studies of possible impacts on meiofauna of episodic events such as tsunami are limited to few studies after 2004 Indian Ocean tsunami (Altaff et al. 2005; Kotwicki and Szczuciński 2006) suggesting quick recovery of meiofauna organisms.

Here we report the ecological status of meiofaunal assemblages at five sandy beaches shortly after the 2011 Tohoku-oki tsunami, with the aim of evaluating the natural meiofauna resilience and recovery after tsunami damage. To our knowledge this is the first report investigating the impact of the Tohoku-oki tsunami on beach faunal biota. The available records on tsunami impact on beach meiofauna are discussed in terms of its ecological significance.

9.2 Study Area and 11th March 2011 Tsunami

The devastating tsunami generated by the Mw 9.0 earthquake (Simons et al. 2011) struck northeastern Japan on 11th March 2011. The survey area includes five sandy beaches located at the Sendai Plain, directly exposed to tsunami waves impact (Figs. 9.1 and 9.2). Tsunami inundation of coastal plain in the investigated area extended up to 5 km inland. Flow heights (tsunami elevation above a mean sea-level datum) were over 10 m near the coastal dune ridges, and decreased landward to less than 3 m close to the inundation limit (Mori et al. 2011; The 2011 Tohoku Earthquake Tsunami Joint Survey Group 2011).

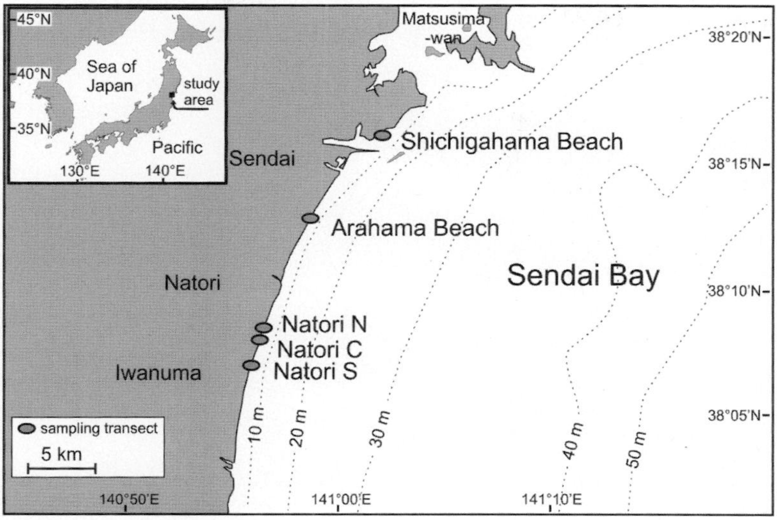

Fig. 9.1 Study area, location of the studied transects (at each transect meiofauna and sediment were sampled at three positions- at low tide, mean sea level and high tide position). Simplified bathymetry after Saito (1989). Inset map shows location of the study area in Japan

Changes in daily tidal range are app. 1 m (Goto et al. 2012). Offshore of the Sendai coast the seabed is covered by sand or silt with no bedrock exposed (Miyagi Prefecture 2000).

9.2.1 Beach Erosion and Recovery

Although more than 10 m high tsunami inundated the Sendai beaches, there was only minor beach erosion (Goto et al. 2011). The recovery of Arahama beach was studied by Goto et al. (2012) based on the analyses of the field and DEM (digital elevation model) data measured in April, June and July. Moreover, the topographic profile before (2005–2006) and after (March 19–23, 2011) the tsunami using 5 m DEM data were extracted. These data clarified that the berm was fully eroded and the beach profile was smoothed by the tsunami. However, new berm, which is nearly the same scale to that before the tsunami, was quickly recovered until the early June. In fact, the beach width was not markedly affected by tsunami erosion with difference of only 10 m in width before (app. 130 m) and after (app. 120 m) the tsunami (Goto et al. 2012).

Although the beach was one of the major sediment sources for tsunami deposits left on land (e.g. Szczuciński et al. 2012) the observations from the Sendai Bay coast by Goto et al. (2012), Richmond et al. (2012), Tappin et al. (2012) and Udo et al. (2012), point that most of the eroded beaches in the investigated area were rebuilt with sand within 1 month after the tsunami.

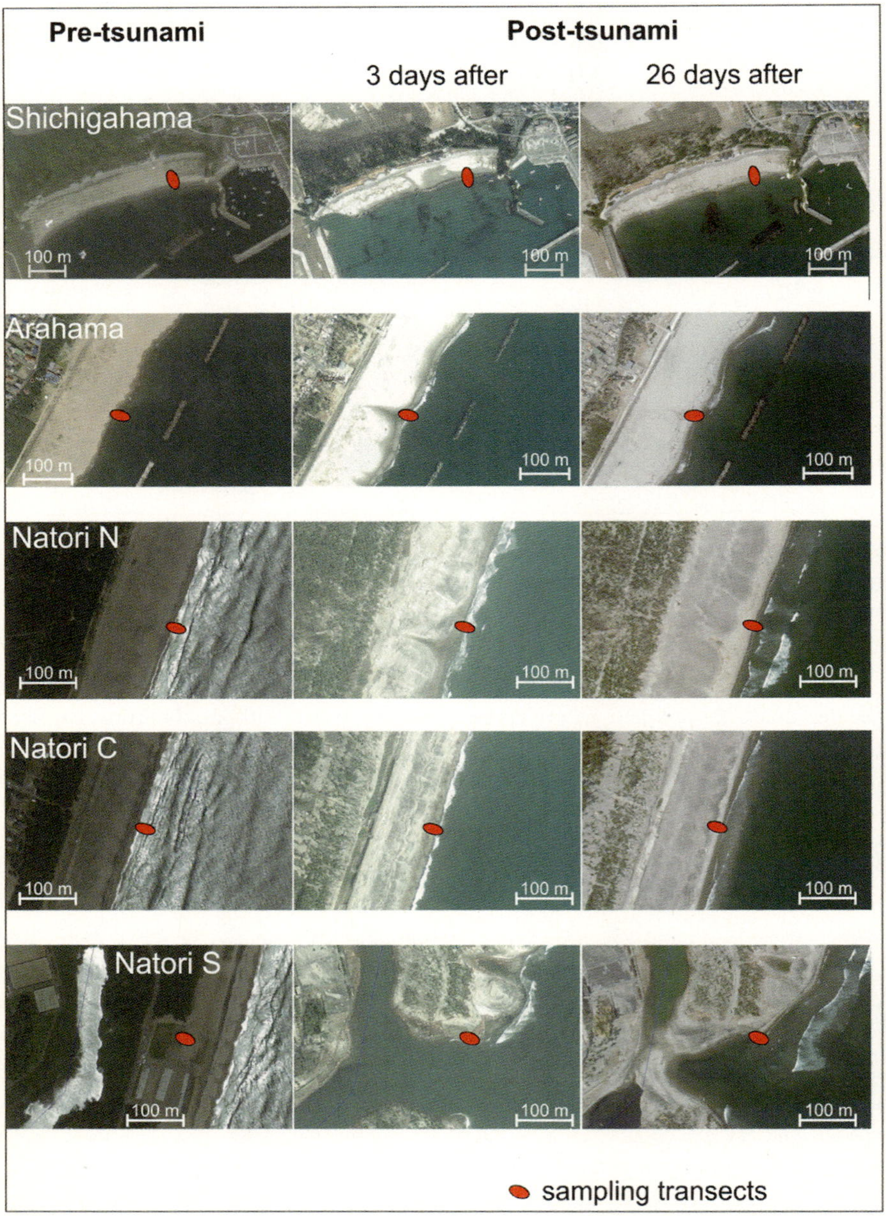

Fig. 9.2 Satellite images (source Google Earth) showing pre- and post-tsunami beach changes with indication of sampling transects

9.3 Material and Methods

9.3.1 Sampling and Samples Processing

The five studied beaches (Table 9.1) were sampled approximately 50 days after the 11th March Tohoku tsunami event (4th–10th May 2011). At each beach's transect meiofauna and sediments were sampled in three positions – at low tide, mean sea level and high tide position using a meiocore sampler (inner diameter 3.6 cm). The upper 5 cm of sediment was taken and, for meiofauna analysis, immediately fixed in 70 % alcohol. In laboratory, a standard decantation technique was used to extract the animals from the sediment. Samples were washed over a 500 μm mesh, then sieved on a 32 μm mesh to retain the meiofauna fraction and stained with Bengal Rose. All metazoan meiobenthic organisms were classified at higher taxon level following Higgins and Thiel (1988) and counted under a stereo-microscope. Meiofaunal abundance was calculated per 10 cm^2.

Sediment samples were subjected to grain size analysis with optical diffractometry method on laser diffraction based Mastersizer 2000 Particle Analyzer. All the results are presented in phi scale (larger value means finer sediments). Conversion of metric scale into phi values is based on:

$$\text{phi}\,(\Phi) = -log_2\text{D}$$

where D is the size in mm. The grain size statistics (mean, sorting, skewness and kurtosis) were calculated using logarithmic method of moments with Gradistat software (Blott and Pye 2001).

The multivariate community data on major taxa level (square-root transformed, Bray-Curtis similarity was used to calculate resemblance) was analysed by means of non-parametric permutational ANOVA (PERMANOVA) to assess differences between investigated beaches and sampling position within the beach. In the case of

Table 9.1 Basic data on sampling sites, sediment type and mean values of grain size statistics (Φ) of the analyzed samples

Beach	Geographical coordinates	Beach protection structures	Sediment type	Mean	Sorting	Skewness
Shichigahama	38° 16′ N 141° 2′ E	Pier	Medium sand	1.68	0.82	1.62
Arahama	38° 13′ N 140° 59′ E	Concrete breakwaters	Medium sand	1.56	0.43	0.38
Natori N	38° 8′ N 140° 56′ E	None	Coarse sand	0.75	0.63	−0.13
Natori C	38° 7′ N 140° 56′ E	Seawall	Medium sand	1.13	0.56	−0.43
Natori S	38° 6′ N 140° 56′ E	Wooden breakwaters	Coarse sand	0.89	0.58	0.00

a significant result in the PERMANOVA main test, pairwise tests for significant term were performed. Due to a restricted number of possible permutations in pairwise tests, p-values were obtained from Monte Carlo sampling (Anderson and Robinson 2003). Principal coordinates analysis (PCO) was used to visualise the PERMANOVA results. Total densities per 10 cm^2, richness (S), and diversity (Shannon diversity index-H'$\log_e$) were calculated for each investigated beach. For analysis of selected biological variables separate univariate PERMANOVA tests were performed. To analyse and model the relationship between meiofauna assemblages and the environmental variables (grain size statistics: mean, sorting, skewness, kurtosis) DISTLM (distance-based linear model) routine was used (log transformed, Euclidean distance was used to calculate resemblance). All the multivariate analyses and calculations of S and H′ were performed using the PRIMER v6 with PERMANOVA+ add-on software package (Anderson et al. 2008).

9.4 Results

9.4.1 Beach Sediments

The studied surface beach sediments were composed mostly of medium-grained sand (Table 9.1). The mean grain size ranged from 0.67Φ at Natori N to 1.84Φ at Shichigahama site (Fig. 9.3).

Coarse sand was encountered only in Natori S and Natori N beaches. In general the finer sediments were found on beaches protected by artificial structures (pier in case of Shichigahama, breakwaters at Arahama beach). Moderately well sorted sand constituted most of the beach sediments, apart from the Arahama site where sediments were composed of well sorted sand. The grain size distribution was mostly symmetrical, with some exception of fine and very fine skewed sediments encountered at Arahama and Shichigahama sites, and mesokurtic. Only Shichigahama beach sediments exhibited different grain distribution being characterised by very leptokurtic grained sediments (Table 9.1).

9.4.2 Meiofauna Analysis

The meiofauna in the study area was composed of 12 higher taxa, three represented by larval stages – copepods nauplii, polychaetes and insects (Table 9.2). Between one to eight higher meiofaunal taxa were recorded at single sampling point and significant differences in number of taxa were observed between the investigated beaches (Table 9.3A). Five taxa accounted for 96 % of meiofaunal communities: nematodes, copepods nauplii, gastrotrichs, turbellarians and ostracods. Nematodes were by far the most abundant group (~73 % of the total individuals), presented the

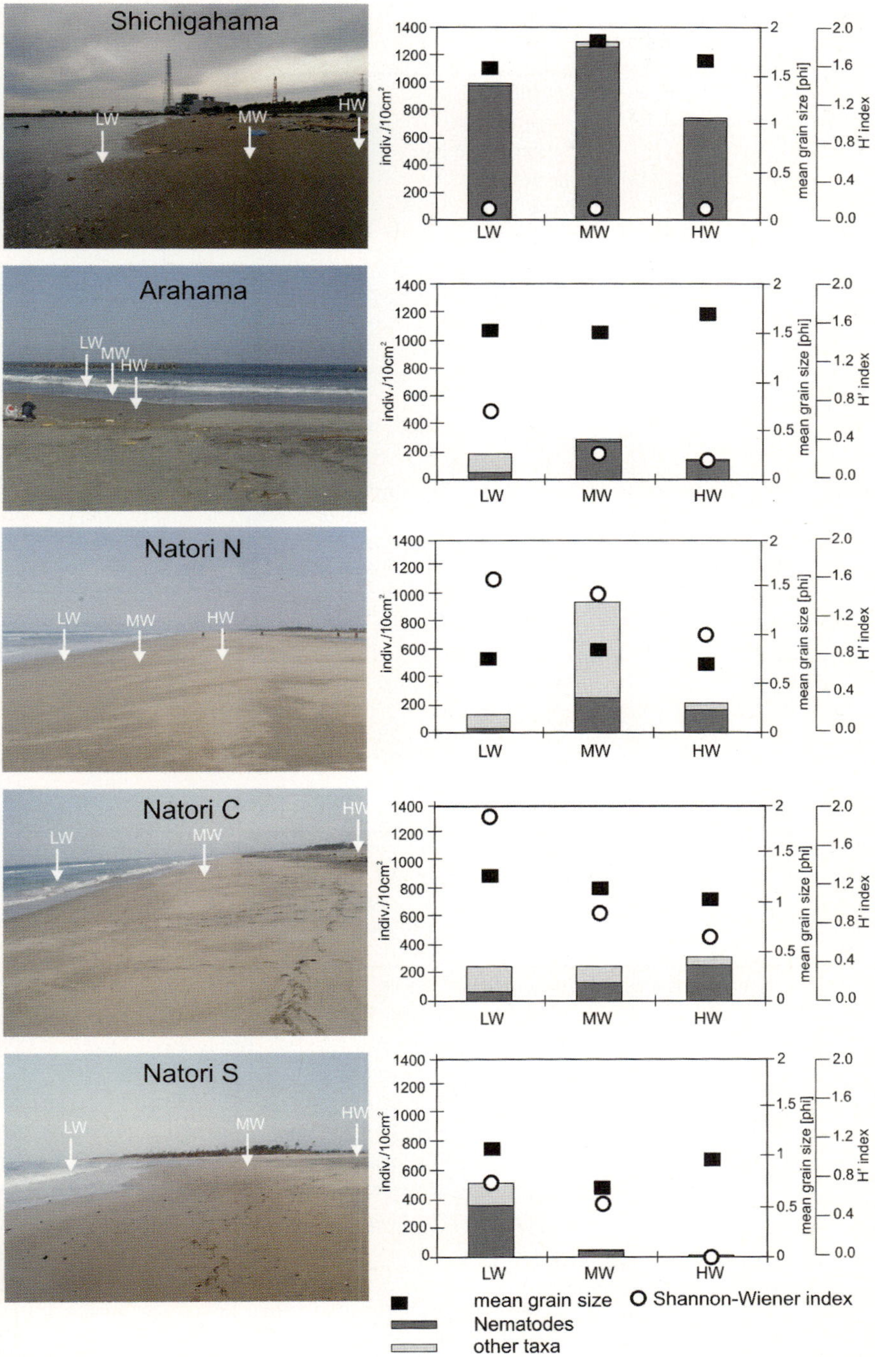

Fig. 9.3 Images of investigated beaches with indication of sampling position (*left panel*), *LW* low tide, *MW* mean water, *HW* high tide. Total meiofauna density, mean grain size and Shannon-Wiener index (*right panel*)

Table 9.2 Abundance distribution of major meiofaunal taxa from investigated beaches. Five the most abundant taxa are in *bold*

	No of individuals/10 cm^2		
	Mean	SD	%
Nematoda	307.3	379.6	**73.37**
Harpacticoida	4.0	8.9	0.95
Copepoda nauplii	39.2	111.3	**9.36**
Gastrotricha	27.9	53.3	**6.65**
Turbellaria	16.7	33.8	**3.99**
Polychaeta larvae	7.9	20.9	1.89
Oligochaeta	2.1	5.7	0.49
Ostracoda	11.6	31.3	**2.77**
Tardigrada	0.5	1.6	0.11
Bivalvia	0.1	0.4	0.03
Acari	0.1	0.5	0.03
Insecta larvae	0.6	1.8	0.14

Table 9.3 PERMANOVA results for (A) the univariate (factor BEACH) and (B) multivariate (factor BEACH and WATER LEVEL) descriptors of the meiobenthic community, with pair-wise comparisons for significant term. *Bold* values stand for significant differences. BEACH – beach location, WATER LEVEL – sampling position in relation to high, mean and low tide level

(A)

		df	MS	Pseudo-F	P(perm)
Meiofauna community descriptors	Density	4	3077.7	3.028	**<0.01**
	No of taxa	4	976.2	2.621	**<0.05**
	H′	4	4269.7	4.008	**<0.01**

(B)

	Source	df	MS	Pseudo-F	P(perm)
Meiofauna structure	Beach	4	2513.8	2.894	**<0.01**
	Water level	2	1350.8	1.555	0.146
	Res	8	868.7		

	Beach	t	P(MC)
Pair-wise comparisons	Natori S vs. Natori C	1.3192	0.219
	Natori S vs. Arahama	0.8388	0.546
	Natori S vs. Natori N	1.3943	0.185
	Natori S vs. Shichigahama	1.8696	0.064
	Natori C vs. Arahama	1.5304	0.084
	Natori C vs. Natori N	0.6228	0.754
	Natori C vs. Shichigahama	2.8571	**<0.01**
	Arahama vs. Natori N	1.9557	**<0.05**
	Natori N vs. Shichigahama	3.1164	**<0.01**
	Arahama vs. Shichigahama	2.0212	**<0.05**

highest density (mean abundance 307 indiv./10 cm^2) and as the only taxon occurred in all the samples. Turbellarians and gastrotrichs were also very common taxa, as they were present in 80 % of the analysed samples. Nematodes dominated at almost all sampling stations. Only at Natori N (LW, MW) and Arahama beach (LW) percentage

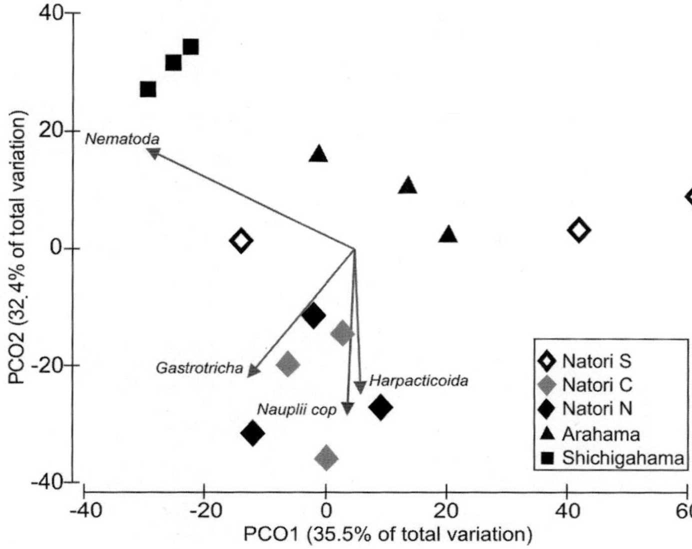

Fig. 9.4 Principal coordinates ordination plot for meiofauna higher taxa showing all taxa which were significantly correlated with the PCO coordinates (r > 0.7). Vectors length corresponds to the correlation values

of this taxon was lower (Fig. 9.3), and meiofauna community was dominated by polychaetes, copepods nauplii and turbellarians, respectively. Other taxa such as water mites (Acari) and Bivalvia occurred only occasionally and/or in low numbers per sampling station, e.g. Harpacticoida. Total meiofaunal abundances varied significantly between the investigated beaches, being the highest at the Shichiganama (Table 9.3A, Fig. 9.3). Densities at other beaches were much lower, with great differences between sampling position (Fig. 9.3). At Arahama, Natori N and Shichiganama the highest densities were recorded at mean water level, while at Natori S and Natori C this position was characterised by lower abundance.

The variability of the Shannon-Wiener diversity index (H') at the studied beaches was high and statistically significant (Table 9.3A). Consistent spatial meiofaunal diversity pattern was observed, with decreasing value of H' from low to high water level. The only exception was Shichiganama, where generally H' was very low and almost no differences between sampling points were observed (Fig. 9.3).

Results of multivariate data showed that investigated beaches harbour distinct meiofaunal assemblages (Table 9.3B). The most different from other beaches was Shichiganama. Pair-wise comparisons showed that also Arahama and Natori N were significantly different from each other.

When analysing meiofaunal structure in a PCO (Fig. 9.4), the first two components explained almost 68 % of the total variability. The graph showed a clear separation between Shichigahama, Arahama and other beaches. Observed differences are mainly connected with presence of nematodes and their gradual increase in abundance from Natori beaches to Shichigahama, while Natori beaches are characterised by increasing relationship with gastrotrichs, copepods nauplii and harpacticoids.

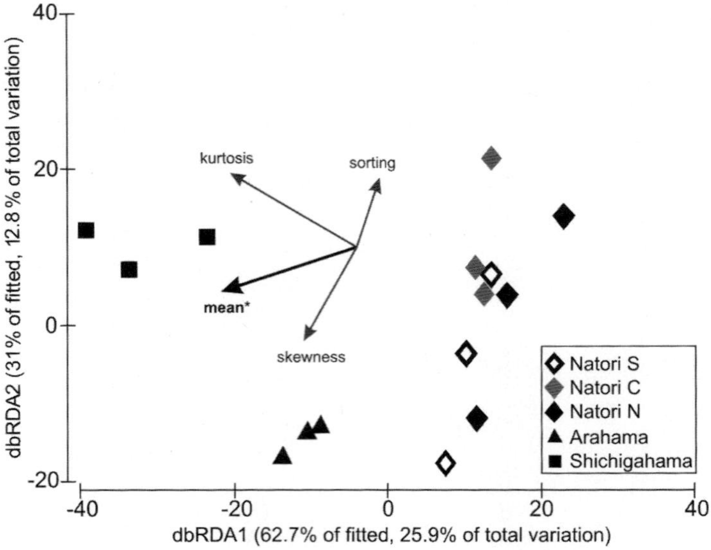

Fig. 9.5 Distance based redundancy plot illustrating the DISTLM model based on the higher meiofauna taxa assemblage data and fitted environmental variables with their vector (strength and direction of effect of the variable on the ordination plot)

Four environmental variables were included in the DISTLM procedure. The best fitted model (illustrated in Fig. 9.5) that explains the meiofaunal structure at the investigated beaches includes all of the variables, together explaining 41 % of total variation. However, only mean grain size, which is the main contributor in explaining observed variability in meiofauna assemblages (19.8 %) was verified as statistically significant.

9.5 Discussion and Conclusions

Many studies have shown meiofauna to be useful in assessing environmental disturbance, with the general conclusions that the resilience of meiofauna is relatively strong; even though, the response depends on the intrinsic resilience of the particular community, intensity of disturbance, or such stochastic factors as seasonal or trophic conditions in the disturbed area. Regardless of nature of the disturbance (anthropogenic, natural, chronic, episodic), such events generally result in a reduction in meiofauna density and/or diversity (e.g. Schratzberger and Warwick 1999; Lee et al. 2001). Nevertheless, some studies indicate that significant meiofauna population can occur even within 1 week of disturbance (deposited dredging material, Schratzberger et al. 2006). Similar observation was made by Altaff et al (2005) at an Indian beach soon after tsunami in December 2004, which flooded the coastline with a tsunami flow depth of up to 6 m. Five days after tsunami meiofauna density achieved typical values for that beach.

These reports and own observations may indicate that almost 2 months period, from tsunami event to sampling time, was long enough to colonize the newly built sandy beaches, and observed densities resulted from fast growth rates, typical for meiofaunal species life cycles. According to Heip et al. (1985) some of common nematode species, dominant taxon in most habitats, have annual number of generations even up to 17. However, since no data from pre-tsunami period are available, it is difficult to predict if established meiofauna assemblages at the investigated beaches are stable.

Probably, the composition of the post-disturbed meiofauna community will never be exactly the same as it was before, although the general structural traits may be predictable (Altaff et al. 2005; Grzelak et al. 2009). This suggestion seems even more reliable particularly when we look at the beach recovery process, which has been very rapid in comparison to previous studies (Goto et al. 2012), such as the 2004 Indian Ocean tsunami. For instance at Arahama transect the beach had almost completely reverted back to the pre-tsunami conditions within 3 months after the event, while beach recovery on the western coast of Thailand was estimated on 60 % within 6 months after tsunami 2004 event (Goto et al. 2012; Choowong et al. 2009, respectively). Moreover, the sediment characteristics of investigated beaches indicate relatively stable conditions.

In Thailand, beach sediments shortly after the 2004 Indian Ocean tsunami (app. 50 days after) were generally poorly sorted and very coarse skewed (Kotwicki and Szczuciński 2006). Comparable sediment characteristics to present result (better sorted and grain size distribution much closer to symmetrical) were achieved in Thailand within 1 year. Sediment properties and composition of sand particles, which form interstices, are undoubtedly one of the most important factors for the abundance, composition and colonization processes of meiofauna. Those variables determine spatial and structural conditions, as well as indirectly determine the physical and chemical regime of the sediment. Thus, little variability of the sediment, indicated by its grain size statistics, pointed at the existence of relatively favourable conditions for meiofauna organisms' development and re-entry to beach ecosystem.

Observed differences between investigated beaches in meiofauna densities and composition may result from natural spatial variation and differences in complex of environmental conditions, for instance wave regime and sediment characteristic. As was shown by present study significant impact on meiofauna assemblages at investigated beaches has grain size (Fig. 9.5). Although it explained only 20 % of total variation in meiofauna assemblages, what means that several varied factors influence those communities (e.g. quantities and qualities of food available, predation) and should be taken into consideration during similar future investigations, grain size remains an important foundation for the living conditions and ecological status of meiobenthos.

High variability of meiofauna abundance was comparable to observation made by Kotwicki and Szczucinski (2006) at three Thai beaches impacted by tsunami in 2004. Generally speaking, the spatial distribution of meiofauna is extremely patchy and unpredictably variable (Giere 2009). Nevertheless, densities recorded in present study are of the same range as densities reported for other marine subtropical and tropical regions (Kotwicki et al. 2005; Giere 2009 and references therein). The same is true in terms of taxonomical composition.

The results on meiofauna assemblages in newly (after the tsunami) rebuild parts of the studied beaches suggest that recolonisation process occurs very fast. It is not surprising, since experimental studies have shown that the first stage of the recolonisation proceeds rapidly, as long as the ecological niches remain wide and can be occupied by the range of opportunistic generalists (e.g. De Troch et al. 2005).

Nematodes were dominant group at all investigated beaches. This taxon is characterised by the greatest species richness in the benthic zone, what resulting in utilization of all trophic resources. Set of morphological features together with their frequent occurrence make nematodes possible to occupy wide range of ecological niches with different trophic requirements and sediment properties. Moreover, nematodes appear to be among the most abundant rafting organisms (Thiel and Gutow 2005), what makes them good colonizers. For this reason nematodes are usually observed in high densities, even after severe disturbance events and are considered to be the most valuable tools for pollution studies (e.g. Sherman and Coull 1980; Coull and Chandler 1992; Lee et al. 2001; Altaff et al. 2005; Kotwicki and Szczuciński 2006; Szymelfenig et al 2006; Hua et al. 2010).

Harpacticoids are usually second the most abundant taxa in meiobenthic samples. Representatives of this group are considered as the classical 'emergers' among the meiobenthos. It means that they have high ability to intensive dispersal, redistribution and colonization of new or disturbed habitats. As was shown by experimental studies dispersal rates of harpacticoids might be even 65-fold higher than those of nematodes (Commito and Tita 2002). That is why, their very low abundance or even absence at Shichigahama beach might suggest lack of donor communities in the proximity of investigated beaches, which is important factor in colonization potential and success (Derycke et al. 2007).

Meiofauna is usually considered as less sensitive to disturbance and less persistently affected than macrofauna (Kennedy and Jacoby 1999; Giere 2009). In particular, responds rapidly to environmental changes, both of natural and anthropogenic origin. Meiofauna can resist disturbance such as tsunami waves because their small body size may allow them to be re-suspended and short life cycles allow them to withstand high mortality. Thus, it appears that the initial respond of meiofauna assemblages to tsunami event is similar, regardless of the climatic zone, and benthic meiofaunal communities are highly developed, even shortly after disturbance.

Although the studies of environmental impact of tsunami on beach environment are still very limited it seems that although the beach is usually subjected to strong erosion, the new beach sediments are rapidly colonized by meiofauna organisms both in tropical and temperate climates. Consequently, in terms of beach ecosystems tsunami events shall be considered as ecological disturbance not a catastrophe.

Acknowledgements The fieldwork was supported by Adam Mickiewicz University in Poznań, Poland and Tohoku University in Sendai, Japan. The help of International Tsunami Survey Team members (C. Chagué-Goff, S. Fujino, J. Goff, K. Goto, B. Jaffe, Y. Nishimura, B. Richmond, D. Tappin, R. Witter and E. Yulianto) during the fieldwork and of Michał Rzeszewski during the lab measurements is greatly acknowledged.

References

Altaff K, Sugumaran J, Naveed MS (2005) Impact of tsunami on meiofauna of Marina beach, Chennai, India. Curr Sci 89(1):34–38

Anderson MJ, Robinson J (2003) Generalized discriminant analysis based on distances. Aust NZ J Stat 45:301–318

Anderson MJ, Gorley RN, Clarke KR (2008) PERMANOVA+ for PRIMER: guide to software and statistical methods. PRIMER-E, Plymouth

Baird AH, Campbell SJ, Anggoro AW, Ardiwijaya RL, Fadli N, Herdiana Y, Kartawijaya T, Mahyiddin D, Mukminin A, Pardede ST, Pratchett MS, Rudi E, Siregar AM (2005) Acehnese reefs in the wake of the Asian tsunami. Curr Biol 15(21):1926–1930

Blott SJ, Pye K (2001) Gradistat: a grain size distribution and statistics package for the analysis of unconsolidated sediments. Earth Surf Proc Land 26:1237–1248

Brown BE (2005) The fate of coral reefs in the Andaman Sea, eastern Indian Ocean following the Sumatran earthquake and tsunami, 26 December 2004. Geogr J 171:372–374

Choowong M, Phantuwongraj S, Charoentitirat T, Chutakositkanon V, Yumuang S, Charusiri P (2009) Beach recovery after 2004 Indian Ocean tsunami from Phang-Nga, Thailand. Geomorphology 104:134–142

Cochard R, Ranamukhaarachchi SL, Shivakoti GP, Shipin OV, Edwards PJ, Seeland KT (2008) The 2004 tsunami in Aceh and Southern Thailand: a review on coastal ecosystems, wave hazards and vulnerability. Perspect Plant Ecol, Evol Syst 10(1):3–40

Commito JA, Tita G (2002) Differential dispersal rates in an intertidal meiofauna assemblage. J Exp Mar Biol Ecol 268:237–256

Coull BC, Chandler GT (1992) Pollution and meiofauna: field, laboratory, and mesocosm studies. Oceanogr Mar Biol Annu Rev 30:191–271

Dam Roy S, Krishnan P (2005) Mangrove stands of Andamans vis-à-vis tsunami. Curr Sci 89 (11):1800–1804

De Troch M, Vandepitte L, Reas M, Suarez-Morales E, Vincx M (2005) A field colonization experiment with meiofauna and seagrass mimics: effect of time, distance and leaf surface area. Mar Biol 148:73–86

Derycke S, Van Vynckt R, Vanaverbeke J, Vincx M, Moens T (2007) Colonization patters of nematoda on decomposing algae in the estuarine environment: community assembly and genetic structure of the dominant species Pellioditis marina. Limnol Oceanogr 52:992–1001

Giere O (2009) Meiobenthology. The microscopic motile fauna of aquatic sediments. Springer, Berlin

Goto K, Chagué-Goff C, Fujino S, Goff J, Jaffe B, Nishimura Y, Richmond B, Sugawara D, Szczuciński W, Tappin DR, Witter R, Yulianto E (2011) New insights into tsunami hazard from the 2011 Tohoku-oki event. Mar Geol 290:46–50

Goto K, Sugawara D, Abe T, Haraguchi T, Fujino S (2012) Liquefaction as an important source of the 2011 Tohoku-oki tsunami deposits at Sendai Plain, Japan. Geology 40(10):887–890

Grzelak K, Kotwicki L, Szczuciński W (2009) Monitoring of sandy beach meiofaunal assemblages and sediments after the 2004 tsunami in Thailand. Pol J Environ Stud 18(1):43–51

Heip C, Vincx M, Vranken G (1985) The ecology of marine nematodes. Oceanogr Mar Biol Ann Rev 23:399–489

Higgins RP, Thiel H (1988) Introduction to the study of meiofauna. Smithsonian Institution Press, London

Hua E, Zhang ZN, Yu ZS, Zhang Y (2010) Preliminary study on the immediate response of the nematode community to Typhoon Soudelor. Deep-Sea Res II 57:1064–1070

Kendall MA, Aryuthaka C, Chimonides J, Daungnamon D, Hills J, Jittanoon C, Komwachirapitak P, Kongkaew V, Mittermeyr A, Monthum Y, Nimsantijaroen S, Paterson GLJ, Foster-Smith R, Foster-Smith J, Thongsin N (2009) Post-tsunami recovery of shallow water biota and habitats on Thailand's Andaman coast. Pol J Environ Stud 18(1):69–76

Kennedy AD, Jacoby CA (1999) Biological indicators of marine environmental health: meiofauna- a neglected benthic component? Environ Monit Assess 54:47–68
Kotwicki L, Szczuciński W (2006) Meiofauna assemblages and sediment characteristic of sandy beaches on the west coast of Thailand after the 2004 tsunami event. Phuket Mar Biol Centre Res Bul 67:39–47
Kotwicki L, Szymelfenig M, De Troch M, Urban-Malinga B, Weslawski JM (2005) Latitudinal biodiversity patterns of meiofauna from sandy litoral beaches. Biodivers Conserv 14:461–474
Lee H, Vanhove S, Peck LS, Vincx M (2001) Recolonisation of meiofauna after catastrophic iceberg scouring in shallow Antarctic sediments. Polar Biol 24:918–925
Lomovasky BJ, Firstater FN, Salaraz AG, Mendo J, Iribarne OO (2011) Macro benthic community assemblage before and after the 2007 tsunami and earthquake at Paracas Bay, Peru. J Sea Res 65:205–212
MacInnes BT, Bourgeois J, Pinegina TK, Kravchunovskaya EA (2009) Tsunami geomorphology: erosion and deposition from the 15 November 2006 Kuril Island tsunami. Geology 37:995–998
Miyagi Prefecture (2000) Report of the committee for the technical survey at Sendai coast. 73 pp
Mori N, Takahashi T, Yasuda T, Yanagisawa H (2011) Survey of 2011 Tohoku earthquake tsunami inundation and run-up. Geophys Res Lett 38, L00G14. doi:10.1029/2011GL049210
Paris R, Wassmer P, Sartohadi J, Lavigne F, Barthomeuf B, Desgages E, Grancher D, Baumert P, Vautier F, Brunstein D, Gomez C (2009) Tsunamis as geomorphic crises: lessons from the December 26, 2004 tsunami in Lhok Nga, West Banda Aceh (Sumatra, Indonesia). Geomorphology 104:59–72
Richmond B, Szczuciński W, Chagué-Goff C, Goto K, Sugawara D, Witter R, Tappin DR, Jaffe B, Fujino S, Nishimura Y, Goff J (2012) Erosion, deposition and landscape change on the Sendai coastal plain, Japan, resulting from the March 11, 2011 Tohoku-oki tsunami. Sed Geol 282:27–39
Saito Y (1989) Modern storm deposits in the inner shelf and their recurrence intervals, Sendai Bay, northeast Japan. In: Taira A, Masuda F (eds) Sedimentary facies in the active plate margin. Terra Scientific Publishing Company, Tokyo, pp 331–344
Schratzberger M, Warwick RM (1999) Differential effects of various types of disturbances on the structure of nematode assemblages: an experimental approach. Mar Ecol Prog Ser 181:227–236
Schratzberger M, Bolam S, Whomersley P, Warr K (2006) Differential response of nematode colonist communities to the intertidal placement of dredged material. J Exp Mar Biol Ecol 334:244–255
Sherman K, Coull BC (1980) The response of meiofauna to sediment disturbance. J Exp Mar Biol Ecol 46:59–71
Simons M, Minson SE, Sladen A, Ortega F, Jiang J, Owen SE, Meng L, Ampuero JP, Wei S, Chu R, Helmberger DV, Kanamori H, Hetland E, Moore AW, Webb FH (2011) The 2011 magnitude 9.0 Tohoku-oki earthquake: mosaicking the megathrust from seconds to centuries. Science 332:1421–1425
Szczuciński W, Chaimanee N, Niedzielski P, Rachlewicz G, Saisuttichai D, Tepsuwan T, Lorenc S, Siepak J (2006) Environmental and geological impacts of the 26 December 2004 tsunami in coastal zone of Thailand – overview of short and long-term effects. Pol J Environ Stud 15(5):793–810
Szczuciński W, Kokociński M, Rzeszewski M, Chagué-Goff C, Cachão M, Goto K, Sugawara D (2012) Sediment sources and sedimentation processes of 2011 Tohoku-oki tsunami deposits on the Sendai Plain, Japan – insights from diatoms, nannoliths and grain size distribution. Sed Geol 282:40–56
Szymelfenig M, Kotwicki L, Graca B (2006) Benthic re-colonization in post-dredging pits in the Puck Bay (Southern Baltic Sea). Estuar Coast Shelf Sci 68:489–498
Tappin DR, Evans HM, Jordan CJ, Richmond B, Sugawara D, Goto K (2012) Coastal changes in the Sendai area from the impact of the 2011 Tōhoku-oki tsunami: interpretations of time series satellite images, helicopter-borne video footage and field observations. Sed Geol 282:151–174

The 2011 Tohoku Earthquake Tsunami Joint Survey Group (2011) Nationwide field survey of the 2011 off the Pacific coast of Tohoku earthquake tsunami. J Japan Soc Civil Eng, Ser B2 (Coast Eng) 67:63–66

Thiel M, Gutow L (2005) The ecology of rafting in the marine environment. II. The rafting organisms and community. Oceanogr Mar Biol Annu Rev 43:279–418

Udo K, Sugawara D, Tanaka H, Imai K, Mano A (2012) Impact of the 2011 Tohoku earthquake and tsunami on beach morphology along the northern Sendai coast. Coast Eng J 54, art. no. 1250009

Whanpetch N, Nakaoka M, Muaki H, Suzuki T, Nojima S, Kawai T, Aryuthaka C (2010) Temporal changes in benthic communities of seagrass beds impacted by a tsunami in the Andaman Sea, Thailand. Estuar Coast Shelf Sci 87:246–252

Yanagisawa H, Koshimura S, Goto K, Miyagi T, Imamura F, Ruangrassamee A, Tanavud C (2009) The reduction effects of mangrove forest on a tsunami based on field surveys at Pakarang Cape, Thailand and numerical analysis. Estuar Coast Shelf Sci 81(1):27–37

Chapter 10
Impact of Tsunami Inundation on Soil Salinisation: Up to One Year After the 2011 Tohoku-Oki Tsunami

Catherine Chagué-Goff, Henri K.Y. Wong, Daisuke Sugawara, James Goff, Yuichi Nishimura, Jennifer Beer, Witold Szczuciński, and Kazuhisa Goto

Abstract The long-term effect of tsunami inundation on soil salinisation was assessed following the 2011 Tohoku-oki tsunami in two areas on the Sendai Plain, near Sendai airport in the Miyagi Prefecture and Matsukawa-ura near Soma in the Fukushima Prefecture. Data gathered over four sampling seasons 2, 5, 9 and 11 months after the tsunami near Sendai airport show that the salt content generally decreased with time. Concentrations were nevertheless higher in February 2012 than in October 2011, probably due to capillary action and evaporation following long periods with little precipitation in the winter, while the lower concentrations in October were attributed to dilution due to intense rainfall prior to the sampling period. In February 2012, the area with chloride concentrations over

C. Chagué-Goff (✉)
School of Biological, Earth and Environmental Sciences,
University of New South Wales, Sydney 2052, NSW, Australia

Australian Nuclear Science and Technology Organisation,
Locked Bag 2001, Kirrawee DC, NSW 2232, Australia
e-mail: c.chague-goff@unsw.edu.au

H.K.Y. Wong
Australian Nuclear Science and Technology Organisation,
Locked Bag 2001, Kirrawee DC, NSW 2232, Australia

D. Sugawara • K. Goto
International Research Institute of Disaster Science, Tohoku University,
Aoba 6-6-11-1106, Aramaki, Aoba-ku, Sendai 980-8579, Japan

J. Goff • J. Beer
School of Biological, Earth and Environmental Sciences,
University of New South Wales, Sydney 2052, NSW, Australia

Y. Nishimura
Institute of Seismology and Volcanology, Faculty of Science, Hokkaido University,
N10W8 Kita-ku, Sapporo, Hokkaido 060-0810, Japan

W. Szczuciński
Institute of Geology, A. Mickiewicz University in Poznań, Poznań 61-606, Poland

Y.A. Kontar et al. (eds.), *Tsunami Events and Lessons Learned: Environmental and Societal Significance*, Advances in Natural and Technological Hazards Research 35,
DOI 10.1007/978-94-007-7269-4_10,

the guidelines for the establishment of rice seedlings still extended for nearly 1 km between 2.45 and 3.33 km inland. Chloride concentrations also reached the guideline values at the land surface 1.71 km inland. This corresponded to the limit of the area deemed not suitable for rice production by local rice farmers. However, recent observations revealed that rice crops were not only halted in 2011 but also in 2012, probably due to high salinisation of soil and/or surface and groundwater. Our study shows that soil salinisation was still recorded to nearly 15 cm depth in areas with fine-grained organic-rich soil ~2.5 km from the shoreline 11 months after the tsunami, and that water-leachable ions were preferentially retained in organic-rich muddy sediment and soil, reflecting the long-term impact of tsunami inundation. In Matsukawa-ura, salt crusts still covered the area flooded by the tsunami in February 2012 and both the soil and muddy tsunami deposit were characterised by high chloride and sulphate concentrations. The latter might also lead to sulphide toxicity. Remediation measures have been implemented in certain areas, but further research needs to be carried out to test the effectiveness of the measures being used to allow rice production to resume.

Keywords Chloride • Rice crops • Salt • Sulphate • 2011 Tohoku-oki tsunami

10.1 Introduction

The 11 March 2011 Tohoku-oki tsunami generated by a M_W 9.0 megathrust earthquake affected ~2,000 km of the Pacific Coast of Japan, and inundated over 400 km^2 of land (Mori et al. 2012). The tsunami reached over 5 km inland on the Sendai Plain (Mori et al. 2012), depositing sand and mud, ranging between ~30 cm and a few mm thick (e.g., Goto et al. 2011; Chagué-Goff et al. 2012a; Szczuciński et al. 2012). Seawater ponded over vast areas of farmland, in natural and man-made depressions, as a result of the nearly flat topography on the low-lying Sendai Plain and subsidence associated with the earthquake (e.g., Goto et al. 2011). Research carried out within 2 months of the tsunami (May 2011) along a 5.1 km shore-perpendicular transect north of Sendai airport in the Miyagi Prefecture revealed that not only was the ponded water saline to brackish, mostly due to evaporation, but that the tsunami deposit and the underlying soil had been contaminated by saltwater (Chagué-Goff et al. 2012b). Five months after the tsunami (August 2011), contamination of sediment and soil was still notable (Chagué-Goff et al. 2012b). A marine chemical signature was also reported in May 2011 beyond the extent of the mud deposit up to the limit of tsunami inundation (Chagué-Goff et al. 2012a, b), which matched the limit mapped immediately after the tsunami (Association of Japanese Geographers 2011).

Contamination of the tsunami deposit and underlying soil by metals and metalloids has been reported in Thailand following the 2004 Indian Ocean Tsunami (2004 IOT) (e.g., Szczuciński et al. 2005) and in coastal areas of Honshu Island following the 2011 Tohoku-oki tsunami (e.g., Komai et al. 2012). However, in our study area near Sendai airport on the Sendai Plain, metals and metalloids were found to occur in concentrations within the background levels of uncontaminated Japanese soils, and this was partly attributed to the minimal contribution of offshore marine sediment to the tsunami deposit (Chagué-Goff et al. 2012b; Szczuciński et al. 2012).

Recent studies carried out after the 2004 IOT (e.g., FAO 2005; Szczuciński et al. 2005; UNEP 2005; Raja et al. 2009; McLeod et al. 2010), 2009 South Pacific Tsunami (Chagué-Goff et al. 2011), 2010 Maule Tsunami (Yoshii et al. 2013) and 2011 Tohoku-oki Tsunami (e.g., Fujikawa et al. 2011; Goto and Inagaki 2011; Chagué-Goff et al. 2012b; Komai et al. 2012; Yoshii et al. 2013) have shown that tsunami inundation results in salinisation of both tsunami sediment and underlying soil, or even soil beyond the extent of the tsunami deposit (Chagué-Goff et al. 2012b; Yoshii et al. 2013). Much has been reported on the effects of tsunami inundation on agriculture, including erosion and scouring of top soil, changed land levels and drainage patterns, deposition of sediment and debris on land, contamination by excess salts, increased soil sodicity (amount of available sodium), deterioration of soil fertility due to changes in soil texture and structure (e.g., FAO 2005; UNEP 2005). A few studies also investigated temporal changes in water leachable ions following the 2004 IOT. The high rainfall in the Andaman and Nicobar Islands (>3,000 mm/year) was found to result in leaching of salt in the soil profile to near pre-tsunami levels (Raja et al. 2009). Szczuciński et al. (2007) made a similar suggestion after investigating changes in Thailand, although they focussed their research on the surface tsunami and/or soil layer. As for McLeod et al. (2010), they used soil apparent electrical conductivity to assess soil salinity and leaching processes with time in the Aceh province, Indonesia. Most studies and observations following the 2004 IOT thus revealed that the high rainfall in these tropical climates generally led to relatively fast leaching of salt, although it also depended upon the length of saltwater ponding and other factors, such as soil type and texture (e.g., IAARD and NSW DPI 2008).

Our present study focusses on the effect of tsunami inundation on soil salinisation in a temperate climate characterised by moderate precipitation (~1250 mm/year; NINS 2012) on Honshu Island, Japan, and its impact on rice crops. Further to the May and August 2011 surveys near Sendai airport, two follow-up surveys were carried out in October 2011 and February 2012 to assess the spatial and temporal extent of tsunami inundation on soil salinisation. Samples were also collected from three sites at Matsukawa-ura, near Soma, Fukushima Prefecture in February 2012 and these results are provided as a comparison. This site is located ~42 km N of the Fukushima Daiichi Nuclear Power Plant (FDNPP) that was damaged due to the 2011 Tohoku-oki tsunami, leading to the Fukushima Daiichi nuclear disaster (e.g., Ohnishi 2012). The site was not accessible in May 2011 to the first author, a resident of Australia, as it was inside the 80 km exclusion zone set by the Australian Department of Foreign Affairs and Trade (DFAT 2011) at the time of the first survey. It was however visited in February 2012, when the first author was allowed to travel to the site, as determined by DFAT (2012), and samples could be collected to assess the amount of water-leachable salts in the tsunami deposit and underlying soil. As the aim of this study is to evaluate the effect of tsunami inundation on rice production on the Sendai Plain, data were plotted against the guidelines provided by the Crop Production Division of Miyagi Prefecture (CPDMP) (2011), based on the handbook by the Agricultural and Forestry Department, Chiba Prefectural Government (AFDCPG) (1973). The AFDCPG (1973) guidelines indicate that chloride concentrations of 500–700 mg kg^{-1} result in brine damage of rice seedlings and CPDMP (2011) also showed that a chloride concentration of 500–700 mg kg^{-1} corresponds to a soil electrical specific conductivity of 0.3 mS cm^{-1} (based on a 1:5 soil:water ratio, and assuming that the content of ions in a soil is the same as that of the standard sea water). The electrical conductivity of soil is most often determined to estimate the amount of salt (e.g., FAO 2005; IAARD and NSW DPI 2008), as it is a faster method not requiring analysis. However, in this study, we determined the concentrations of soluble ions. Thus, as we are referring to these guidelines, we focus our results on changes in chloride concentrations, although it is well known that salt contamination is reflected in other parameters, such as increases in sodium and other elements, and increases in pH (e.g., FAO 2005). For example, Goto and Inagaki (2011) reported boron concentrations of 10–20 mg kg^{-1} in available form in tsunami sediments, which were largely in excess of the optimal concentration of 0.5–2.0 mg kg^{-1}, and were thus likely to result in crop damage.

10.2 Methods

A shore-perpendicular transect extending from the shoreline to c. 5.1 km inland was established north of Sendai airport, in the central part of the Sendai Plain (Fig. 10.1). Samples of tsunami deposit and underlying soil were collected along the transect at 30 sites in May 2011. The sampling depth intervals in the tsunami deposit ranged from 2 mm (mud drapes) to 4 cm. The top 2 cm of the underlying soil was also sampled, as well as soil beyond the landward limit of the tsunami deposit (Fig. 10.1c). In addition, at four sites (WP325, WP326, WP327 and WP329), not only the tsunami deposit (sand, sand and mud, or mud) but also the underlying soil was sampled up to a maximum depth of 20 cm to assess the extent of salt contamination in the soil (Fig. 10.1c). The stratigraphy of the tsunami deposit varies at these four sites, but is typical of the variability observed along the transect. The first 4–5 cm of the soil profile were sampled at 1 cm intervals and the remaining profile every 5 cm. A limited number of sites were sampled again in August 2011, October 2011 and February 2012, in order to assess the temporal changes in the effects of salinisation on the sediment and soil in the area of study. At two of the sites in Matsukawa-ura (Ma1 and Ma2), muddy and sandy intervals of the tsunami deposit ranged between 0.5 and 5 cm, while the top 2 cm of the underlying soil was sampled. Soil samples were also collected beyond the limit of tsunami inundation (Ma3) (Fig. 10.1d).

Grain size analysis of samples collected in May 2011 ($n = 133$) was carried out by laser diffraction using a Malvern Mastersizer 2000 after pre-treatment with hydrogen peroxide. The instrument can be used for sediment with a grain size

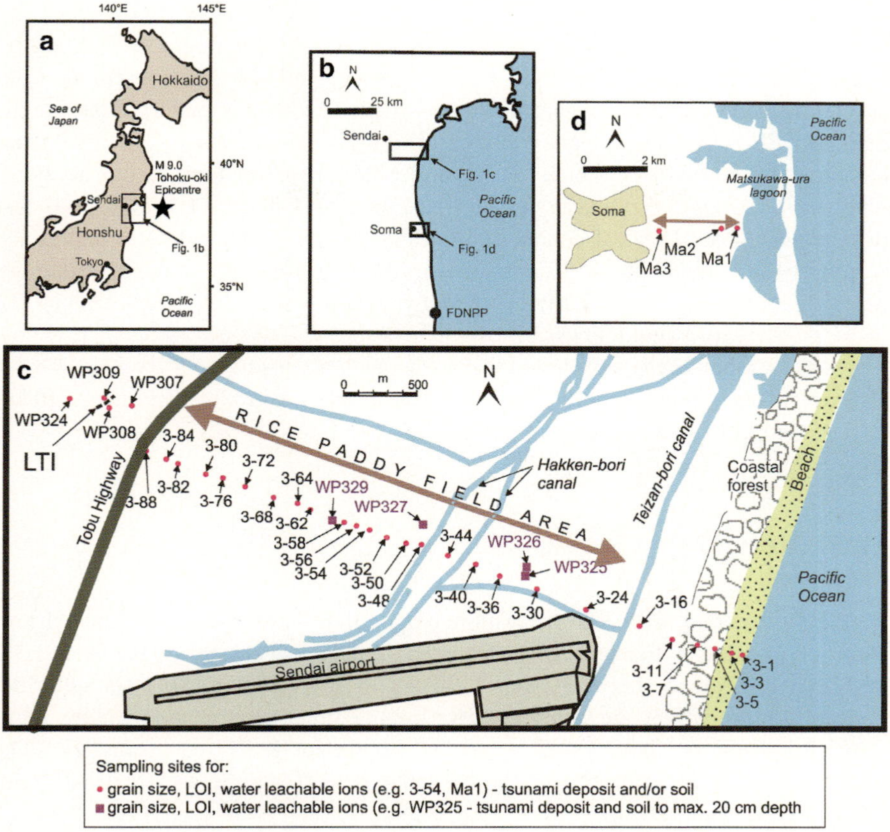

Fig. 10.1 (**a**) Map of Japan showing the epicentre of the M 9.0 Tohoku-Oki earthquake and regional map with studies areas. (**b**) Regional map showing study areas in Miyagi Prefecture near Sendai and in Fukushima Prefecture near Soma (*FDNPP* = Fukushima Daiichi Nuclear Power Plant). (**c**) Location map with sampling sites near Sendai airport, Miyagi Prefecture. The extent of rice paddy fields along the transect prior to the 2011 tsunami (between ~1.1 and 4.5 km from the shoreline, although this area also includes roads, canals and residential areas), and the limit of tsunami inundation (*LTI*) are shown. See text for explanation. (**d**) Location map of study area at Matsukawa-ura, near Soma, Fukushima Prefecture, with sampling sites. The *brown double-ended arrow* indicates the area covered by rice paddy fields prior to the tsunami

<1,400 μm, which represented >98 % of all samples, except at two sites (3-3 and 3-16), where the coarser fraction amounted to 5 % and was analysed using sieves. Results for these samples were combined using corrections due to density and weight/volume percentages. The organic content of all samples collected during the four field surveys near Sendai airport and in Matsukawa-ura (n = 341) was determined by ashing at 550 °C for 4 h after drying overnight at 105 °C. All samples were processed for water-leachable ions, by drying at 105 °C, adding high-purity water (18.2 MΩ cm) (ratio soil:water of 1:10) before being placed on a shaker at 125 rpm for 24 h. The supernatant was then filtered using a 0.2 μm disposable filter

and split for anions and cations. Water leachable anions (Cl, Br, F, SO_4, NO_3, PO_4 and NO_2) were determined by Dionex DX-600 Ion Chromatograph (IC), while water-leachable cations (Ca, K, Mg, Mn, Na, P, S and Sr) preserved with sub-boiling nitric acid were determined by Varian VISTA AX (axial) CCD Simultaneous ICP-AES (see Chagué-Goff et al. 2012b, for details on all analytical techniques used here). Although a 1:10 soil:water ratio was used in this study to obtain a large enough volume for analysis, as opposed to a 1:5 ratio used for measurements of electrical conductivity (CPDMP 2011), chloride concentrations were calculated by taking the dilution factor into effect and reported on a sediment dry weight basis. Therefore, they can directly be compared to the AFDCPG (1973) and CPDMP (2011) guidelines.

Datasets from each period sampled were analysed using SPSS v. 21 (2012). However, data are only reported for all samples analysed in May 2011 near Sendai airport, and in Matsukawa-ura in February 2012.

10.3 Results

Most ions reported in May 2011 exhibited a similar distribution along the 5.1 km shore-perpendicular transect, with distinctly higher concentrations between 1.55 and 4.03 km inland (between sites 3-30 and 3-80) (see Fig. 10.2 for chloride concentrations, which are representative of the other ions, as shown by high positive correlations between all elements, except for P, PO_4, NO_3 and F; NO_2 data were not included as they are mostly below detection limit; see Table 10.1 and Chagué-Goff et al. 2012b).

Results also indicate that concentrations were in general higher at the surface of the tsunami deposits compared to intervals below, in particular where salt crusts were visible (Fig. 10.3; May 2011). Lower concentrations were measured in the sandy tsunami deposit and soil (except ~1.5–1.85 km inland), as well as on the beach from the shoreline to 1.24 km inland (site 3-24) and from 4.27 km inland (site 3-84) to beyond the limit of tsunami inundation. Nevertheless, the soluble chloride content was 100 times higher (~2,000 mg kg^{-1}) within the limit of tsunami inundation (site WP308) than immediately beyond it (~20 mg kg^{-1}, site WP309), despite the fact that the land had been ploughed since the event and no sedimentological evidence nor debris were visible (Chagué-Goff et al. 2012b). Chloride concentrations beyond the limit of tsunami inundation (17–23 mg kg^{-1}, $n = 4$, at sites WP309 and WP324) are thus used as background concentrations. In February 2012, chloride concentrations were similar and close to background at both sites (8–26 mg kg^{-1}, $n = 4$).

The tsunami deposit and underlying soil generally fined inland, with a discontinuous sand deposit varying in thickness from ~30 to 0.5 cm recorded up to ~2-8-2.9 km inland (Chagué-Goff et al. 2012a; Szczuciński et al. 2012). It consisted mostly of medium to fine sand (Figs. 10.2 and 10.3) overlain by a discontinuous mud drape or cap from ~1.4 km inland. The 3 cm to 1–2 mm thick mud-dominated deposit covered the paddy fields up to 4.65 km inland west of Tobu Highway (site WP307 in Fig. 10.1; see also Chagué-Goff et al. 2012a; Szczuciński et al. 2012).

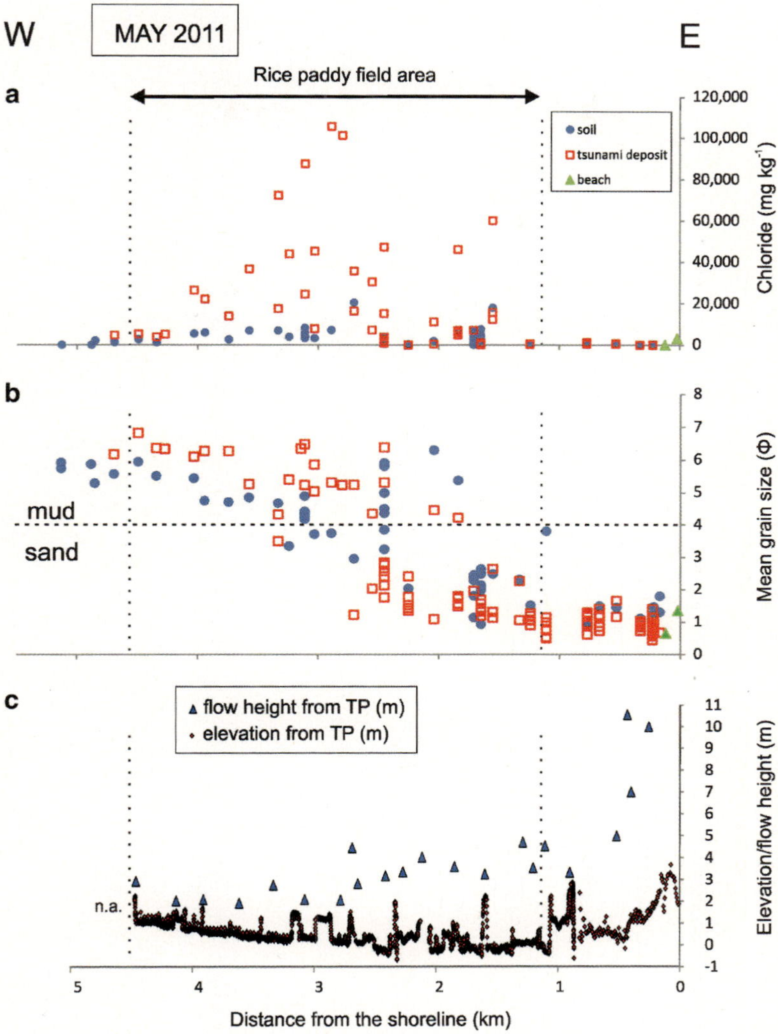

Fig. 10.2 (**a**) Chloride concentrations (May 2011) in the tsunami deposit (*red open squares*), underlying soil and soil not covered by the tsunami deposit (*blue dots*) and beach (*green triangles*) along the transect extending to 5.13 km inland near Sendai airport. The area covered by rice paddy fields (~1.1 to ~4.5 km inland) is marked. (**b**) Mean grain size of the tsunami deposit (*red open squares*), underlying soil and soil not covered by the tsunami deposit (*blue dots*) and beach (*green triangles*) along the transect. The horizontal dashed line indicates the limit between the sand (mean grain size <4 Φ) and mud (mean grain size >4 Φ). (**c**) Land elevation (m) above the Tokyo Peil (*TP*), which is the reference sea level, and flow height (m) above TP, recorded in May 2011 along the transect up to Tobu Highway (4.5 km inland) (Modified after Goto et al. 2011). Data show the nearly flat topography between the coastal forest and Tobu Highway (see Fig. 10.1c for location of transect). *n.a.* = not available

Table 10.1 Correlation coefficients for non parametric test of Spearman rank correlation for water leachable ions, organic matter content (LOI), distance from the shoreline (Dist.), and grainsize expressed as Φ (data from May 2011, n = 133). Datasets were first tested for normality using Kolmogorov – Smirnov and Shapiro-Wilks tests. Spearman rank correlation tests were applied to the soil variable datasets because the data were not normally distributed

	Ca	K	Mg	Mn	Na	P	S	Sr	F	Cl	Br	NO_3	SO_4	PO_4	LOI	Dist.	Φ
Ca	1.000	0.928**	0.954**	0.852**	0.916**	0.242**	0.942**	0.975**	0.089	0.927**	0.923**	−0.102	0.943**	0.058	0.798**	0.744**	0.780**
K		1.000	0.966**	0.802**	0.957**	0.276**	0.937**	0.944**	0.087	0.955**	0.964**	−0.084	0.940**	0.094	0.771**	0.674**	0.752**
Mg			1.000	0.831**	0.956**	0.187*	0.945**	0.984**	0.067	0.971**	0.974**	−0.129	0.952**	−0.015	0.745**	0.663**	0.737**
Mn				1.000	0.817**	0.184*	0.860**	0.855**	0.103	0.826**	0.809**	−0.155	0.850**	−0.105	0.799**	0.771**	0.790**
Na					1.000	0.231**	0.957**	0.944**	0.106	0.976**	0.975**	−0.142	0.961**	0.038	0.766**	0.696**	0.763**
P						1.000	0.254**	0.186*	0.163	0.176*	0.175*	0.320**	0.236**	0.667**	0.361**	0.354**	0.270**
S							1.000	0.955**	0.106	0.957**	0.947**	−0.084	0.998**	0.005	0.825**	0.753**	0.843**
Sr								1.000	0.071	0.961**	0.957**	−0.146	0.960**	−0.011	0.769**	0.691**	0.759**
F									1.000	0.043	0.062	−0.035	0.090	0.051	0.211*	0.254**	0.212*
Cl										1.000	0.991**	−0.152	0.965**	−0.048	0.730**	0.651**	0.734**
Br											1.000	−0.166	0.955**	−0.021	0.728**	0.638**	0.727**
NO_3												1.000	−0.094	0.232**	0.038	0.087	0.049
SO_4													1.000	−0.010	0.800**	0.734**	0.821**
PO_4														1.000	0.070	0.114	−0.002
LOI															1.000	0.848**	0.905**
Dist.																1.000	0.862**
Φ																	1.000

** indicates that the correlation is significant for $p < 0.01$; * indicates that the correlation is significant for $p < 0.05$

Fig. 10.3 Photograph of site 3-36 (~1.85 km inland, north of Sendai airport) showing the salt residues on the surface muddy sand, underlain by sand (taken in May 2011). The darker soil can be seen below the tsunami deposit. See Fig. 10.1c for site location

Readers are referred to Szczuciński et al. (2012) for further details on the sedimentological characteristics of the deposit and underlying soil in this area.

In the following, we focus on ion concentrations and temporal changes in the area inundated by the tsunami that was mostly occupied by rice paddy fields before the 2011 event (except for housing, roads and canals), between ~1.1 and ~4.5 km from the shoreline (Fig. 10.1c; from ~100 m seaward of site 3-24 to site 3-88, ~20 m from Tobu Highway). We also provide results for the sites at Matsukawa-ura that were only sampled in February 2012 (Fig. 10.1d).

As shown in Fig. 10.4, chloride concentrations decreased with time, both in the tsunami deposit and underlying soil. Nevertheless, a similar pattern is observed at all sampling times (May, August and October 2011, February 2012). The higher concentrations are recorded between about 2.5 km and 3.5 km inland, with a general decrease from that point both seaward and landward (Fig. 10.4). The dotted line in Figs. 10.4a–d marks 500–700 mg kg^{-1} chloride concentrations, the AFDCPG (1973) and CPDMP (2011) guidelines. Our data show that the extent of the tsunami deposit and soil characterised by chloride concentrations over 500–700 mg kg^{-1} is much smaller in February 2012, almost a year after the tsunami, than in May 2011 (2 months after the event). However, it still covers an area nearly 1 km along the transect, extending from 2.45 to 3.33 km inland, where the deposit and soil are mostly mud-dominated. In addition, the surface sample at site WP326 (1.71 km inland), contained 569 mg kg^{-1} chloride, thus with concentrations that could be detrimental to rice seedlings. It is also interesting to note that chloride concentrations were higher in February 2012 (9,250 mg kg^{-1} at the surface characterised by fine organic-rich material (algal mat) at site 3-54, 2.8 km inland) (Fig. 10.4d) than in October 2011 (maximum of 3,520 mg kg^{-1} in the underlying soil at site 3-56, 2.89 km inland) (Fig. 10.4c).

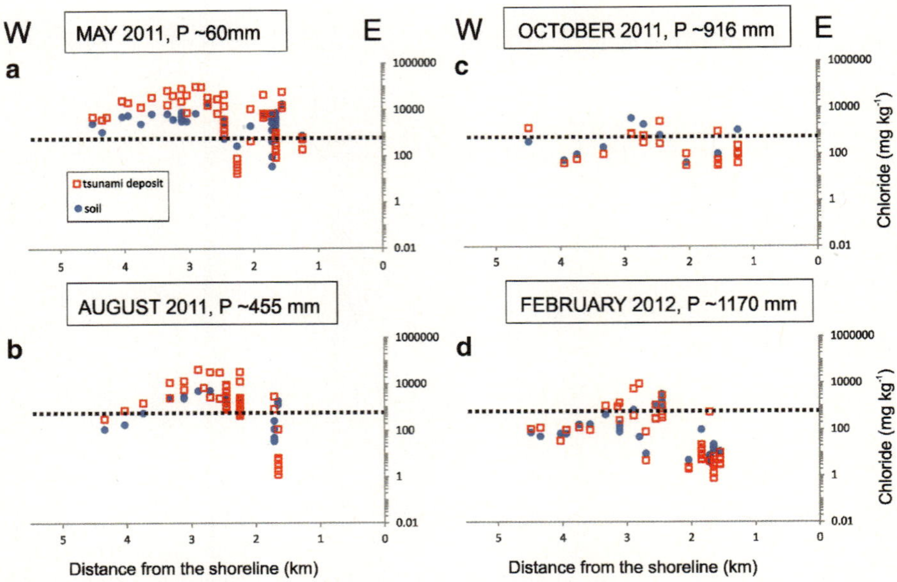

Fig. 10.4 Chloride concentration in the tsunami deposit (*red open squares*) and underlying soil (*blue dots*) in May 2011 (**a**), August 2011 (**b**), October 2011 (**c**) and February 2012 (**d**). The total precipitation (P) from the 11 March 2011 tsunami to the period of sampling at Sendai airport (After Japan Meteorological Agency 2012) is indicated after the sampling date. Note that data are expressed using a logarithmic scale. Each *square* or *dot* represents a sample (see Figs. 10.5, 10.6, 10.7 and 10.8 for details at particular sites). The *dotted line* marks the guidelines (500–700 mg kg^{-1} chloride) of the Agricultural and Forestry Department, Chiba Prefectural Government (1973) and Crop Production Division of Miyagi Prefecture (2011), above which brine damage occurs in rice seedlings. Data show that while the area unsuitable for rice production gets generally smaller with time, both the overlying muddy tsunami deposit and underlying soil are still considered contaminated almost 1 year after the tsunami over an area covering nearly 1 km (between 2.45 and 3.33 km inland), as well as at the surface 1.71 km inland

The downward leaching of salt through the sandy tsunami deposit into the underlying soil is illustrated in Fig. 10.5 (site WP325), where concentrations of water leachable ions were fairly low in the 20 cm thick sandy deposit even in May 2011 (<1,070 mg kg^{-1} chloride). Chloride concentrations however reached 7,620 mg kg^{-1} at the top of the soil layer and decreased downward to 2,080 mg kg^{-1} at 17 cm depth (in the soil profile). At site WP326, characterised by a 1 cm-thick sandy deposit, ion concentrations were higher in the sand than in the underlying soil (Fig. 10.6), probably because the seawater ponded and could not leach downwards as easily through the underlying soil. Where the sandy deposit was overlain by a mud cap, like at site WP327 (Fig. 10.7), ions exhibited highest concentrations in the surface layer (e.g., 47,080 mg kg^{-1} chloride), leached through the sand and were higher in the underlying soil than in the sand unit. At site WP329, where the tsunami deposit consisted of a thin mud layer, ion concentrations were highest at the surface (e.g., 87,500 mg kg^{-1} chloride), and then decreased downward in the underlying mud layer and underlying soil (Fig. 10.8). Most ions that

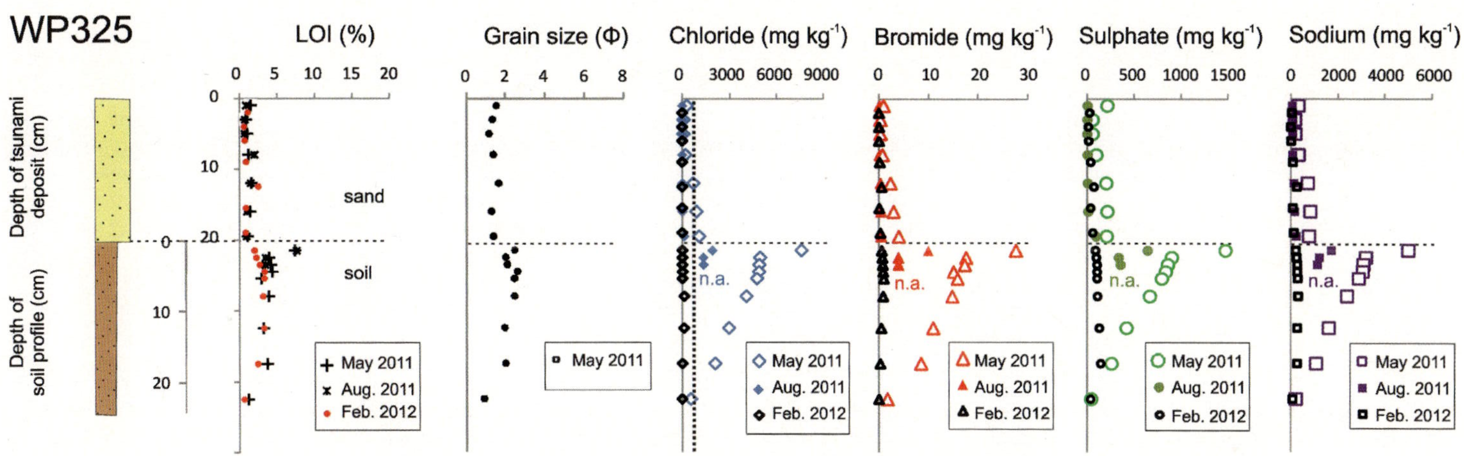

Fig. 10.5 Stratigraphy, organic content (expressed as loss on ignition LOI (%)), grainsize (Φ) (May 2011 only), water-leachable (soluble) chloride, bromide, sulphate and sodium concentrations in tsunami sediment and underlying soil at site WP325. See Fig. 10.1c for site location. The *horizontal dashed line* shows the limit between the tsunami deposit and underlying soil, and the *vertical dotted line* marks 500–700 mg kg^{-1} chloride, the AFDCPG (1973) and CPDMP (2011) guidelines. Each symbol represents a sample collected at a different depth. Data are provided at three sampling periods (May and October 2011, February 2012), except for grainsize. (*n.a.* = not available, as only top soil profile sampled in August 2011)

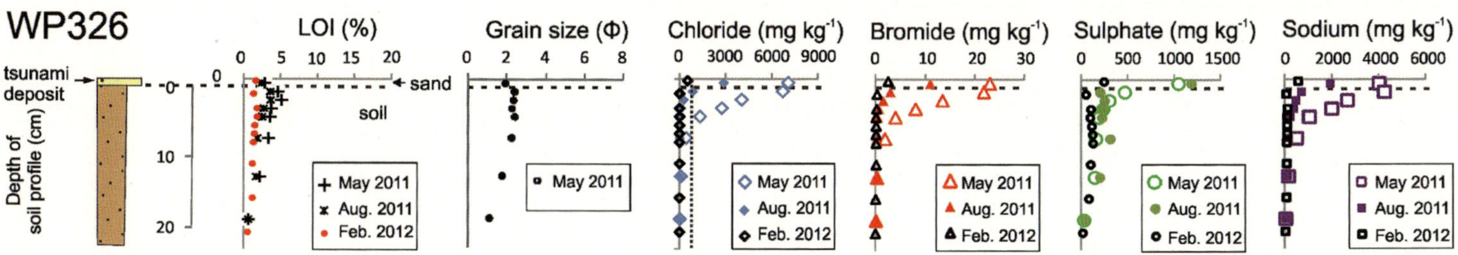

Fig. 10.6 Stratigraphy, organic content (expressed as loss on ignition LOI (%)), grainsize (Φ) (May 2011 only), water-leachable (soluble) chloride, bromide, sulphate and sodium concentrations in tsunami sediment and underlying soil at site WP326. See Fig. 10.1c for site location. The *horizontal dashed line* shows the limit between the tsunami deposit and underlying soil, and the *vertical dotted line* marks 500–700 mg kg^{-1} chloride, the AFDCPG (1973) and CPDMP (2011) guidelines. Each symbol represents a sample collected at a different depth. Data are provided at three sampling periods (May and October 2011, February 2012), except for grain size

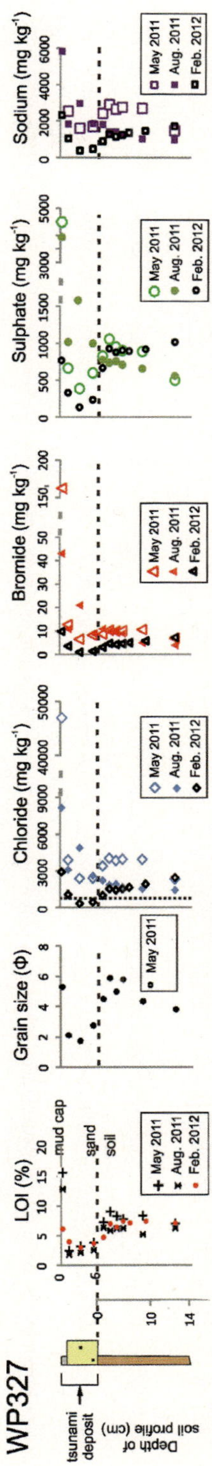

Fig. 10.7 Stratigraphy, organic content (expressed as loss on ignition LOI (%)), grainsize (Φ) (May 2011 only), water-leachable (soluble) chloride, bromide, sulphate and sodium concentrations in tsunami sediment and underlying soil at site WP327. See Fig. 10.1c for site location. The *horizontal dashed line* shows the limit between the tsunami deposit and underlying soil, and the *vertical dotted line* marks 500–700 mg kg^{-1} chloride, the AFDCPG (1973) and CPDMP (2011) guidelines. Each symbol represents a sample collected at a different depth. Data are provided at three sampling periods (May and October 2011, February 2012), except for grainsize. Note the discontinuous dashed lines and changes of scale for chloride, bromide and sulphate

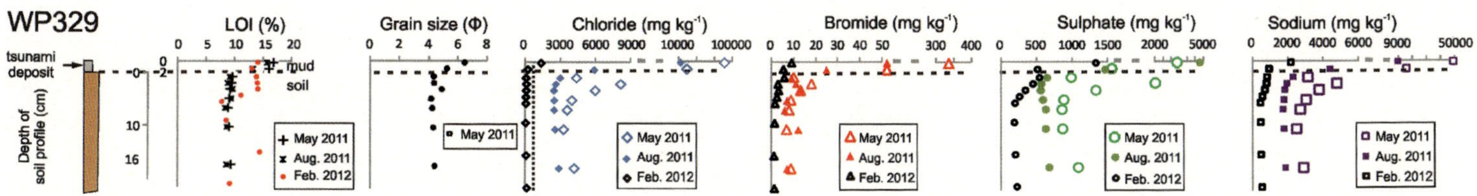

Fig. 10.8 Stratigraphy, organic content (expressed as loss on ignition LOI (%)), grainsize (Φ) (May 2011 only), water-leachable (soluble) chloride, bromide, sulphate and sodium concentrations in tsunami sediment and underlying soil at site WP329. See Fig. 10.1c for site location. The *horizontal dashed line* shows the limit between the tsunami deposit and underlying soil, and the *vertical dotted line* marks 500–700 mg kg^{-1} chloride, the AFDCPG (1973) and CPDMP (2011) guidelines. Each symbol represents a sample collected at a different depth. Data are provided at three sampling periods (May and October 2011, February 2012), except for grainsize. Note the discontinuous *dashed lines* and changes of scale for chloride, bromide, sulphate and sodium

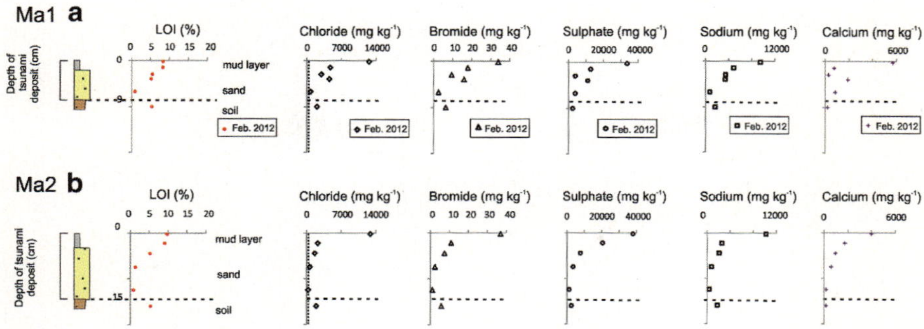

Fig. 10.9 Stratigraphy, organic content (expressed as loss on ignition LOI (%)), water-leachable (soluble) chloride, bromide, sulphate, sodium and calcium concentrations in tsunami sediment and underlying soil at sites Ma1 (**a**) and Ma2 (**b**) in Matsukawa-ura, recorded in February 2012. See Fig. 10.1d for site location. The *horizontal dashed line* shows the limit between the tsunami deposit and underlying soil, and the *vertical dotted line* marks 500–700 mg kg^{-1} chloride, the AFDCPG (1973) and CPDMP (2011) guidelines. Each symbol represents a sample collected at a different depth

occur in high concentrations in seawater (e.g., Na, SO_4, Br, Mg, Ca, K) followed a similar pattern, reflecting downward saltwater leaching following tsunami inundation (see Figs. 10.5, 10.6, 10.7 and 10.8 for distributions of Na, SO_4, Br).

Temporal changes in ion concentrations are illustrated in Figs. 10.5, 10.6, 10.7 and 10.8 at these four sites. Soluble ion contents decreased with time, in particular where the tsunami deposit and underlying soil were sandy (WP325 and WP326). There, chloride concentrations were below the AFDCPG (1973) and CPDMP (2011) guidelines in February 2012 (Figs. 10.5 and 10.6). At site WP327, where the tsunami deposit consisted of a sandy layer overlain by a mud cap, and the underlying soil was predominantly silty (Φ of 4–6), data show that except in the sandy layer, chloride concentrations were well above the AFDCPG (1973) and CPDMP (2011) guidelines in February 2012. Sulphate concentrations were even higher in February 2012 than in August 2011 in the soil. It is thus inferred that salt might have leached downward, as shown by a slight downward increase in Cl, Br, SO_4 and Na, and concentrations near the base of the soil profile higher in February 2012 than in August 2011 (Fig. 10.7). Nevertheless this might also suggest that further downward leaching is prevented. At site WP329, chloride concentrations (1,390 mg kg^{-1}) twice the AFDCPG (1973) and CPDMP (2011) guidelines were only recorded in the surface muddy layer, although measurable sulphate concentrations were noted in the soil (Fig. 10.8). The downward decrease in concentrations in the soil profile and lower concentrations with time suggest dilution and possible associated downward leaching.

At Matsukawa-ura, the tsunami deposit was characterised by a mud layer overlying a sandy deposit at both sites within the inundation limit (Figs. 10.9 and 10.10a). The area was flooded by seawater at least until June 2011, and this was reflected by the presence of extensive salt crusts on the land in February 2012 (Fig. 10.10b), while no salt crusts were visible near Sendai airport at that time.

Fig. 10.10 (**a**) Photograph of site Ma1 (labelled WP502) in Matsukawa-ura, showing the salt residues on the mud and rice plant remains, overlying the mud and sand tsunami deposit and soil (taken in February 2012); (**b**) Photograph of site Ma2 showing the extensive salt crust (taken in February 2012). See Fig. 10.1d for site locations

Chloride concentrations were above the AFDCPG (1973) and CPDMP (2011) guidelines in both the muddy and sandy tsunami deposit and the underlying soil, with the highest content recorded at the surface of both sites (12,600 and 12,900 mg kg^{-1}), except in the lowest sandy intervals at site Ma2. The high sulphate (up to 38,000 mg kg^{-1}) (Fig. 10.9) and calcium concentrations suggest that salt residues were dominated by calcium sulphate or gypsum ($CaSO_4.H_2O$) in this area, although it does not exclude other salts, such as NaCl and KCl. This is also suggested by the strong positive correlation between these elements (Table 10.2). Beyond the limit of tsunami inundation, chloride concentrations were close to background (18–35 mg kg^{-1}, $n = 3$).

10.4 Discussion

As reported in Chagué-Goff et al. (2012b), soluble ion concentrations were highest in May 2011, 2 months after the tsunami, mostly where seawater had ponded for a long time, both where the deposit was sand- (~1.5–1.85 km inland) or mud- (~2.9–3.4 km inland) dominated (Fig. 10.2).

Data gathered along the transect north of Sendai airport over four sampling periods from May 2011 to February 2012, 11 months after the tsunami, show that salt concentrations in the tsunami deposit and underlying soil generally decreased with time (Fig. 10.4). This is mostly attributed to rainfall, which probably resulted in dilution and downward and/or lateral leaching of salt, as also reported in previous studies (e.g., Szczuciński et al. 2007; Raja et al. 2009; McLeod et al. 2010; Chagué-Goff et al. 2012b). However, higher concentrations were recorded in algal mats and surface organic-rich mud in February 2012 than in October 2011, despite additional precipitation of ~250 mm between both surveys.

Table 10.2 Correlation coefficients for non parametric test of Spearman rank correlation for water leachable ions, organic matter content (LOI), and distance from the shoreline (Dist.) for samples collected at Matsakawa-ura in February 2012 (n = 17). Datasets were first tested for normality using Kolmogorov – Smirnov and Shapiro- Wilks tests. Spearman rank correlation tests were applied to the soil variable datasets because the data were not normally distributed

	Ca	K	Mg	Mn	Na	P	S	Sr	Cl	Br	NO_3	SO_4	PO_4	LOI	Dist.
Ca	1.000	0.581*	0.936**	0.887**	0.826**	−0.124	0.946**	0.985**	0.824**	0.865**	−0.739**	0.946**	0.311	0.411	−0.500*
K		1.000	0.603*	0.561*	0.723**	−0.368	0.588*	0.627**	0.711**	0.698**	−0.297	0.588*	0.210	0.070	−0.585*
Mg			1.000	0.968**	0.936**	0.005	0.993**	0.936**	0.914**	0.942**	−0.798**	0.993**	0.245	0.589*	−0.500*
Mn				1.000	0.917**	0.036	0.968**	0.895**	0.868**	0.919**	−0.770**	0.968**	0.175	0.620**	−0.389
Na					1.000	−0.026	0.929**	0.868**	0.978**	0.985**	−0.685**	0.929**	0.175	0.571*	−0.593*
P						1.000	−0.005	−0.133	−0.044	−0.025	−0.015	−0.005	−0.127	0.765**	0.502*
S							1.000	0.951**	0.904**	0.940**	−0.798**	1.000**	0.280	0.578*	−0.482
Sr								1.000	0.865**	0.899**	−0.699**	0.951**	0.276	0.415	−0.519*
Cl									1.000	0.985**	−0.644**	0.904**	0.210	0.532*	−0.667**
Br										1.000	−0.699**	0.940**	0.210	0.561*	−0.621**
NO_3											1.000	−0.798**	−0.257	−0.488*	0.353
SO_4												1.000	0.280	0.578*	−0.482
PO_4													1.000	0.053	−0.104
LOI														1.000	0.114
Dist.															1.000

**indicates that the correlation is significant for $p < 0.01$; *indicates that the correlation is significant for $p < 0.05$

The October survey occurred 2 weeks after Typhoon Roke, which resulted in 330 mm precipitation, and 60 mm of rain also fell during the October sampling (Chagué-Goff et al. 2012b). This is likely to have led to dilution and/or downward leaching of salt at the time. Elevated salt contents at the surface of fine sediment and in algal mats in February 2012 are on the other hand probably attributed to capillary action and evaporation, as the precipitation (snow fall) was only 22 and 44 mm in January and February 2012, respectively (Japan Meteorological Agency 2012).

Winds from the SE and SSE are prevalent in the region (Japan Meteorological Agency, 2012) and are likely to contribute to evaporation and resulting increasing concentration of salts at the land surface. Our data therefore suggest that not only is salt preferentially retained in organic-rich fine sediment, but that the total precipitation is not the only governing factor regarding retention or leaching of salt, and that these processes depend upon variations in precipitation and other climatic parameters. Wind, air temperature, solar radiation, vapour pressure deficit and relative humidity affect evaporation (e.g., Morton 1968), and thus indirectly the retention or leaching of salt. McLeod et al. (2010) indicated that loss of functional drainage following the tsunami could also have affected the rate of salt leaching. As in Aceh, Indonesia, the low topography on the Sendai Plain (Fig. 10.2c) is also likely to result in slow leaching.

Downward leaching of salt occurred more readily in the sandy deposit and/or sandy underlying soil (e.g., WP 325, Fig. 10.5), due to the higher porosity of the material compared to mud (e.g., Szczuciński et al. 2005; Chagué-Goff et al. 2012b). It is however interesting to note that the downward leaching reported by McLeod et al. (2010) in a number of fields in Aceh, resulting in a surficial decrease in salinity associated with an increase at depth, was not observed at any of the sites studied here (Figs. 10.5, 10.6, 10.7 and 10.8).

At site WP327, salt was preferentially retained in the fine organic-rich sediment and underlying soil, at least down to nearly 15 cm depth. The creation of a hard pan by puddling (compacting), harrowing (breaking up the soil) and levelling the soil under wet conditions (e.g., Wopereis et al. 1992) is a common practice in rice farming. It results in a zone of low hydraulic conductivity which helps reduce water loss. Here, the presence of the compact hard pan underneath the rice paddy soil appears to impede downward leaching below this impervious surface and the resulting downward increase in ion concentrations (Fig. 10.7). However, where there is no hard pan, soluble salts might leach further downward, as seen at sites WP325 (Fig. 10.5) and WP329 (Fig. 10.8), and contribute to contamination of groundwater. We measured the ion concentrations in the soil profile down to 20 cm at a few sites, and at various times to assess the temporal changes and effect on the soil. However, further study would be required to ascertain the fate of soluble salts within the deeper soil profile and possible impact on groundwater, in addition to the effect of saltwater intrusion.

With regards to the effect of tsunami inundation on soil salinisation and impact on rice production, our study shows that 11 months after the tsunami (February 2012), the area with chloride concentrations over the guidelines was nearly 1 km long, and ranged mostly between 2.45 km and 3.33 km inland. At the surface of the thin sandy deposit 1.71 km inland, the chloride concentration was also near the guideline values. Thus, this suggests that the area up to 3.33 km inland was still not suitable for rice production in February 2012, and that contamination by salt was still recorded at a depth of nearly 15 cm in the soil profile in some areas (such as at site WP327, 2.45 km inland). Indeed, we found in February 2012 that rice production not only had been halted following the tsunami in 2011, but could not resume before 2013 on rice paddy fields within 3.4 km of the shoreline. This was because the soil electrical conductivity exceeded 0.3 mS cm^{-1} and the salinity of the groundwater at 4 m depth (1–4 mS cm^{-1}) was also too high to allow rice farming (Mr Sato, pers. comm. 2012; Chagué-Goff et al. 2012b). Remediation measures using freshwater flushing were to be implemented to flush the salt out of the soil within the 3.4 km zone (Mr Sato, pers. comm. 2012). We measured an electrical conductivity of 2 mS cm^{-1} in irrigation channels at the seaward limit of the remediation zone, and just under 1.3 mS cm^{-1} landward of it. While it was within the CPDMP (2011) guidelines, which recommend an EC of 2.2 mS cm^{-1} for paddy water, we observed soy bean crops between 3.4 and 4 km inland, instead of rice crops, during an additional visit at the end of September 2012. This might suggest that although soil salinity was within the recommended AFDCPG (1973) and CPDMP (2011) guidelines as reported in this study (Fig. 10.4d), the quality of surface and/or ground-water was probably still not adequate for rice production.

At Matsukawa-ura, no attempts appeared to have been made to resume rice production at the sites investigated (Ma1 and Ma2, ~1.75 and 2.20 km from the shoreline, and ~0.2 and 0.65 km from the landward edge of Matsukawa-ura lagoon, respectively), although flushing of rice paddy fields was observed further inland in this area. Matsukawa-ura is only ~42 km N of the FDNPP. Joint US/Japan survey data gathered by the Japanese National Nuclear Security Administration and US Department of Energy suggest that Matsukawa-ura was not directly affected by the nuclear disaster, based on ^{134}Cs and ^{137}Cs deposition activities (US Department of Energy 2012). Thus, rice production might be able to resume, once salinity and sodicity have decreased to background levels. However, the high levels recorded nearly 1 year after tsunami inundation, which are probably partly due to the extended period of ponding by seawater, suggest that this area is likely to be unsuitable for rice production for a considerably longer period than near Sendai airport. The extremely high sulphate concentrations recorded there might also inhibit rice cropping, due to the possible conversion of sulphate to sulphide under highly reducing conditions (e.g., as a result of flooding). Thus, rice production might be inhibited due to sulphide toxicity, although previous studies have shown that oxidation around the rice roots might counteract this effect (e.g., De Datta 1981). Further study would be recommended to ascertain the possible adverse effect of excess sulphate on the rice paddy fields.

Co-seismic subsidence of 17–21 cm was reported in the area near Sendai airport (Goto et al. 2011), and it could impede flushing measures and contribute to increased soil salinisation from groundwater. Hulugalle et al. (2009) studied the soil physical properties in Aceh following the 2004 Indian Ocean Tsunami and reported a significant physical degradation of the soil, which can also affect crops. Although the soil physical characteristics were not investigated in this study, tsunami inundation is likely to have had a negative impact, partly due to the high sodicity (e.g., So and Aylmore 1993). High sodicity also affects rice production and the application of gypsum and/or compost in association with flushing is recommended in saline-sodic soils to improve rice yield (e.g., Nayak et al. 2008; Abdel-Fattah 2012). Whether the large amount of calcium sulphate present on the rice paddy fields in Matsukawa-ura might be beneficial is not clear as yet as concentrations might be too high resulting in an ion imbalance over and above the possible conversion to sulphide and resulting sulphide toxicity. Wang et al. (2010) also suggest altering the irrigation water depth to increase rice yield. Therefore, a number of measures might be applied to remediate salinisation of the soil, as well as removing the tsunami deposit if too thick, in order to be able to resume rice production in both areas. In the meantime, alternative farming has been suggested and started, such as the Tohoku cotton project, as cotton is less sensitive to salinisation (Tohoku Cotton Project 2012). Further research is recommended to assess the longer term effect of salinisation due to tsunami inundation on rice growing, as little data are available in temperate climates such as Honshu Island, where rainfall is significantly lower than in tropical regions. Remediation measures should also be tested and carried out.

10.5 Conclusions

Two months after the tsunami salt contamination was significant in areas where seawater had ponded for a long time, independent of the nature of the tsunami deposit (sand or mud). However, data gathered over a period of almost a year after the event show that salt is preferentially retained in organic-rich fine sediment, either at the surface of the deposit or in the soil profile, down to about 15 cm depth. Our data also revealed that although salinisation of the muddy sediment and soil generally decreased with time, it was also dependent on the variability of precipitation and other climatic parameters, which can result in evaporation and concentration near the surface of the deposit. At Matsukawa-ura, where seawater ponded for much longer, salt crusts were still visible in February 2012, and chloride concentrations were well above the guidelines for the establishment of rice seedlings. Salt residues were probably dominated by forms of calcium sulphate. While calcium sulphate is often suggested as remediation for high sodicity, the high concentrations at Matsukawa-ura might result in acute sulphide toxicity if the soil becomes flooded and reducing conditions are established.

While the area with chloride concentrations over the suggested guidelines for the establishment of rice seedlings has decreased since May 2011, it still extended nearly 1 km between 2.45 and 3.33 km near Sendai airport in February 2012, and concentrations near the guideline levels were also recorded at the surface of the tsunami deposit 1.71 km inland. The landward limit corresponds to the area still deemed unsuitable for rice production in 2012. It also appears that conditions were still not suitable for rice crops to resume in 2012 as they were instead replaced by soy beans. Remediation measures are suggested and should be implemented to allow rice production to resume on the Sendai Plain.

Acknowledgements Collection of samples in February 2012 was carried out when CCG was a visiting professor at the Institute of Seismology and Volcanology, University of Hokkaido, Japan. Brett Rowling and Matthew Dore (ANSTO) are thanked for assisting in sample preparation and analysis. We acknowledge Mr. Sato for his assistance during our field survey in February 2012, and thank the anonymous reviewer for their comments that greatly improved the manuscript.

References

Abdel-Fattah MK (2012) Role of gypsum and compost in reclaiming saline-sodic soils. J Agric Vet Sci 1:30–38

AFDCPG (Agricultural and Forestry Department, Chiba Prefectural Government) (1973) Handbook of agricultural and forestry pollution, 306 pp (in Japanese)

Association of Japanese Geographers (2011) Map of tsunami inundation reported by Tsunami Damage Team. http://danso.env.nagoya-u.ac.jp/20110311/map/574017Sendaikukou.jpg

Chagué-Goff C, Schneider J-L, Goff JR, Dominey-Howes D, Strotz L (2011) Expanding the proxy toolkit to help identify past events – lessons from the 2004 Indian Ocean tsunami and the 2009 South Pacific tsunami. Earth Sci Rev 107:107–122

Chagué-Goff C, Andrew A, Szczuciński W, Goff J, Nishimura Y (2012a) Geochemical signatures up to the maximum inundation of the 2011 Tohoku-Oki tsunami – implications for the 869AD Jogan and other palaeotsunamis. Sediment Geol 282:65–77

Chagué-Goff C, Niedzielski P, Wong HKY, Szczuciński W, Sugawara D, Goff J (2012b) Environmental impact assessment of the 2011 Tohoku-oki tsunami on the Sendai plain. Sediment Geol 282:175–187

CPDMP (Crop Production Division of Miyagi Prefecture) (2011) Technical note for crops after the Great East Japan Earthquake, 2nd advisory. http://www.pref.miyagi.jp/noenkan/nousansyokuryou/gijyutu-2.pdf (in Japanese)

De Datta SK (1981) Principles and practices of rice production. Wiley, Singapore, 618 pp

DFAT (Australian Department of Foreign Affairs and Trade) (2011) Travel advice for Japan. http://www.smartraveller.gov.au/zw-cgi/view/Advice/Japan. Accessed 19 Apr 2011

DFAT (Australian Department of Foreign Affairs and Trade) (2012) Travel advice for Japan. http://www.smartraveller.gov.au/zw-cgi/view/Advice/Japan. Accessed 30 Jan 2012

FAO (United Nations Food and Agriculture Organization) (2005) Field guide on salinity in Aceh-Draft publication RAP 05. www.fao.org/ag/tsunami/docs/saltwater-guide.pdf. Accessed 8 Feb 2012

Fujikawa T, Okazawa H, Nakamura T, Takeuchi Y, Komamura M (2011) Physical and chemical properties of tsunami deposits in the northeast area of Fukushima prefecture after the Tohoku-Kanto earthquake. Int J GEOMATE 1:44–49

Goto I, Inagaki K (2011) Countermeasures for tsunami-damaged farmlands from salt water by the Great East Japan Earthquake Disaster. Annu Rep Agric Acad Jpn 16:109–122 (in Japanese)

Goto K, Chagué-Goff C, Fujino S, Goff J, Jaffe B, Nishimura Y, Richmond B, Sugawara D, Szczuciński W, Tappin DR, Witter R, Yulianto E (2011) New insights of tsunami hazard from the 2011 Tohoku-Oki event. Mar Geol 290:46–50

Hulugalle NR, Jaya R, Luther GC, Ferizal M, Daud S, Yatiman I, Yufniati ZA, Feriyanti F, Tamrin H, Han B (2009) Physical properties of tsunami-affected soils in Aceh, Indonesia: 2½years after the tsunami. Catena 77:224–231

IAARD (Indonesian Agency for Agricultural Research and Development, Indonesia), NSW DPI (NSW Department of Primary Industries, Australia) (2008) A practical guide to restoring agriculture after a tsunami, 57 pp. http://www.dpi.nsw.gov.au/research/projects/06P302/a-practical-guide. Accessed 3 Jan 2013

Japan Meteorological Agency (2012) Weather, climate & earthquake information. http://www.data.jma.go.jp/obd/stats/data/en/smp/index.html. Accessed 6 Dec 2012
Komai T, Kawabe Y, Hara J, Sakamoto Y, Zhang M (2012) Geochemical survey of tsunami sediments and transport of toxic elements from offshore environment – urgent investigations for earthquake March 11, 2011, Proceedings of the Twenty-second (2012) International Offshore and Polar Engineering Conference: Rhodes, Greece, International Society of Offshore and Polar Engineers (ISOPE), pp 49–53
McLeod M, Slavich P, Irhas Y, Moore N, Rachman A, Ali N, Iskandar T, Hunt C, Caniago C (2010) Soil salinity in Aceh after the December 2004 Indian Ocean tsunami. Agric Water Manag 97:605–613
Mori N, Takahashi T, The 2011 Tohoku Earthquake Tsunami Joint Survey Group (2012) Nationwide post event survey and analysis of the 2011 Tohoku earthquake tsunami. Coast Eng J 54:1250001-1–1250001-27
Morton FI (1968) Evaporation and climate: a study in cause and effect, scientific series no. 4. Inland Water Branch, Department of Energy, Mines and Resources, Ottawa
Nayak AK, Sharma DK, Mishra VK, Minhas PS, Verma CL (2008) Reclamation of saline-sodic soil under a rice–wheat system by horizontal surface flushing. Soil Use Manag 24:337–343
NINS (National Institute of Natural Sciences), National Astronomical Observatory of Japan (2012) Chronological scientific tables. Maruzen CO. LTD, Tokyo, http://www.rikanenpyo.jp/ (in Japanese)
Ohnishi T (2012) The disaster at Japan's Fukushima-Daiichi nuclear power plant after the March 11, 2011 earthquake and tsunami, and the resulting spread of radioisotope contamination. Radiat Res 177:1–14
Raja R, Chaudhuri S, Ravisankar N, Swarnam T, Jayakumar V, Srivastava R (2009) Salinity status of tsunami-affected soil and water resources of South Andaman, India. Curr Sci 96:152–156
So H, Aylmore L (1993) How do sodic soils behave – the effects of sodicity on soil physical behavior. Aust J Soil Res 31:761–777
SPSS v. 21 (2012) http://www-01.ibm.com/support/docview.wss?uid=swg21608060
Szczuciński W, Niedzielski P, Rachlewicz G, Sobczyński T, Zioła A, Kowalski A, Lorenc S, Siepak J (2005) Contamination of tsunami sediments in a coastal zone inundated by the 26 December 2004 tsunami in Thailand. Environ Geol 49:321–331
Szczuciński W, Niedzielski P, Kozak L, Frankowski M, Ziola A, Lorenc S (2007) Effects of rainy season on mobilization of contaminants from tsunami deposits left in a coastal zone of Thailand by the 26 December 2004 tsunami. Environ Geol 53:253–264
Szczuciński W, Kokociński M, Rzeszewski M, Chagué-Goff C, Cachão M, Goto K, Sugawara D (2012) Sediment sources and sedimentation processes of 2011 Tohoku-Oki tsunami deposits on the Sendai plain, Japan – insights from diatoms, nannoliths and grain size distribution. Sediment Geol 282:40–56
Tohoku Cotton Project (2012) http://www.tohokucotton.com/en/. Accessed 26 Jan 2013
UNEP (United Nations Environment Programme) (2005) After the Tsunami. Rapid environmental assessment, 140 pp. http://www.unep.org/tsunami/reports/TSUNAMI_report_complete.pdf
US Department of Energy (2012) Radiation monitoring data from Fukushima area – 5/13. http://energy.gov/downloads/radiation-monitoring-data-fukushima-area-51311. Accessed 18 Dec 2012
Wang M-M, Wang Z-W, Huang L-H, Ma H-Y, Liu M, Gu X-y (2010) Effect of irrigation water depth on rice growth and yield in a saline-sodic soil in Songnen Plain, China. J Food Agric Environ 8:530–534
Wopereis MCS, Wösten JHM, Bouma J, Woodhead T (1992) Hydraulic resistance in puddled rice soils: measurement and effects on water movement. Soil Tillage Res 24:199–209
Yoshii T, Imamura M, Matsuyama M, Koshimura S, Matsuoka M, Mas E, Jimenez C (2013) Salinity in soils and tsunami deposits in areas affected by the 2010 Chile and the 2011 Japan tsunamis. Pure Appl Geophys 170:1047–1066

Chapter 11
Estimating the 2004 Indian Ocean Tsunami Wave Height and Period from Boulders' Distribution at Pakarang Cape, Thailand

Kazuhisa Goto, Kiyohiro Okada, and Fumihiko Imamura

Abstract We propose a numerical method to estimate the local wave height and period of a tsunami from the distributions of boulders. The method was applied to boulders (<23 t) at Pakarang Cape, Thailand that were displaced from the reef slope onto the tidal bench by the 2004 Indian Ocean tsunami. They were deposited characteristically below the high-tide line, irrespective of size and no boulders were deposited on land. These features were used as the constraint of the calculation. We conducted over 10,000 cross-sectional calculations to satisfy above mentioned constraints, showing that the wave height and period at the shoreline were calculated to be 4–6 m and 18–37 min, respectively, which well concur with observed values. The input parameters for this calculation are the sizes and initial positions of boulders with seaward/landward limits, which are obtainable through the geological survey for historical and pre-historical tsunamis. This method is useful to estimate the local wave height and period of historical and pre-historical tsunamis from boulders reported throughout the world.

Keywords Tsunami • Boulder • Pakarang Cape • 2004 Indian Ocean tsunami

K. Goto (✉) • F. Imamura
International Research Institute of Disaster Science, Tohoku University,
6-6-11 Aoba, Aramaki, Aoba-ku, Sendai 980-8579, Japan
e-mail: goto@irides.tohoku.ac.jp

K. Okada
School of Engineering , Tohoku University, 6-6-11 Aoba, Sendai 980-8579, Japan

Present organization: Pacific Consultants Co., Ltd.,
Present address: 1-9-1 Ichiban-cho, Aoba-ku, Sendai 980-8579, Japan

Y.A. Kontar et al. (eds.), *Tsunami Events and Lessons Learned: Environmental and Societal Significance*, Advances in Natural and Technological Hazards Research 35,
DOI 10.1007/978-94-007-7269-4_11,

11.1 Introduction

Estimating the local wave height and period of the historical tsunamis is critically important to understand their behavior and power. Such estimation will greatly aid future disaster mitigation efforts. However, such values for historical tsunamis that occurred prior to the start of modern tidal measurement systems are usually difficult to estimate solely from historical documents (e.g., Goto et al. 2010a). Even for recent tsunami events, the local wave period is generally difficult to estimate if tidal record is not available, although the wave heights can be measured through the field survey.

Boulders of lengths of a few meters deposited on the reef or rocky platform by the tsunami, so-called tsunami boulders, are useful markers to characterize historical and pre-historical tsunamis. Such boulders, reported around the world (e.g., Goff et al. 2006; Goto et al. 2007; Kelletat et al. 2007; Higman and Bourgeois 2008; Paris et al. 2009, 2010; Bourgeois and MacInnes 2010; Lamarche et al. 2010; Etienne et al. 2011; Morton et al. 2011; Richmond et al. 2011a, b; Spiske and Bahlburg 2011), are believed to be useful for estimating hydrodynamic features of tsunamis (e.g., Imamura et al. 2008; Nandasena et al. 2011a). However, presently proposed theoretical or empirical models can estimate the minimum wave height or minimum current velocity necessary to move the boulder (Nott 2003; Noormets et al. 2004; Nandasena et al. 2011b; Weiss 2012) or possible inundation area (Pignatelli et al. 2009). These models were proposed to differentiate the boulders deposited by the tsunami and storm waves and are not suit to estimate the tsunami wave characteristics. In fact, no proper method has been proposed to date to estimate the wave height and period of paleotsunamis.

In this paper, we propose a numerical method for estimation of the local tsunami wave height and period from clast size and spatial distributions of boulders. We test its validity for application to the well-studied tsunami boulders at Pakarang Cape, Thailand (Fig. 11.1), which were deposited by the 2004 Indian Ocean tsunami (IOT).

11.2 Tsunami Boulders at Pakarang Cape, Thailand

The observed tsunami wave heights (inundation heights) around Pakarang cape were 4–7 m (Matsutomi et al. 2006; Yanagisawa et al. 2009). No information about the wave period exists at the cape, but tidal gauge records at Phuket, approximately 100 km south of Pakarang Cape (Matsutomi et al. 2006), indicate that the period was 35 and 25 min for first and second waves. This wave period is consistent with the numerical results, which showed approximately 30 min for the first wave (Goto et al. 2009).

Tsunami boulders at the cape were investigated immediately after the tsunami (Goto et al. 2007). At least 1,000 boulders, some as heavy as 23 t, were deposited around the cape. These boulders were fragments of reef and coral rocks and were

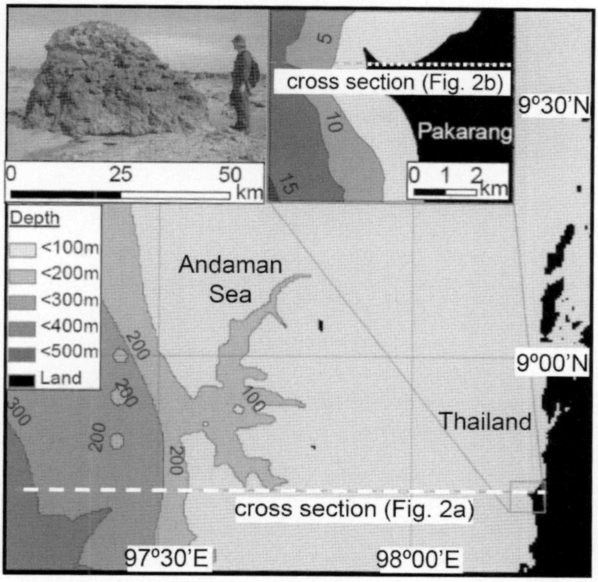

Fig. 11.1 Location map of the studied area at Pakarang Cape, Thailand and a photograph of the tsunami boulder

scattered on the tidal bench: they had originated from the reef slope shallower than 10 m water depth. Goto et al. (2007) reported that no boulders were found on land and that all boulders were stopped below the high-tide line, irrespective of size. In fact, the largest boulder was deposited very close to the shoreline together with numerous small boulders. Numerical modeling for boulder transport revealed that this feature is explained by the drastic reduction of the hydraulic force of the tsunami attributable to the propagation of the seaward reflected wave generated at the beach slope (Goto et al. 2010b).

11.3 Numerical Methods

11.3.1 Tsunami Inundation

We conducted a cross-sectional calculation for simplification because previous modeling results revealed that boulders were displaced toward the east by the first run up wave and that they were not re-transported a long distance by the backwash flow because the backland topography is very flat (Goto et al. 2009, 2010b). Moreover, considering the future application of the method to the historical and pre-historical tsunamis, the method should be simple for wider application.

Shallow-water theory was used for numerical calculation of tsunami propagation in the shallow region and the run-up after Goto et al. 2009.

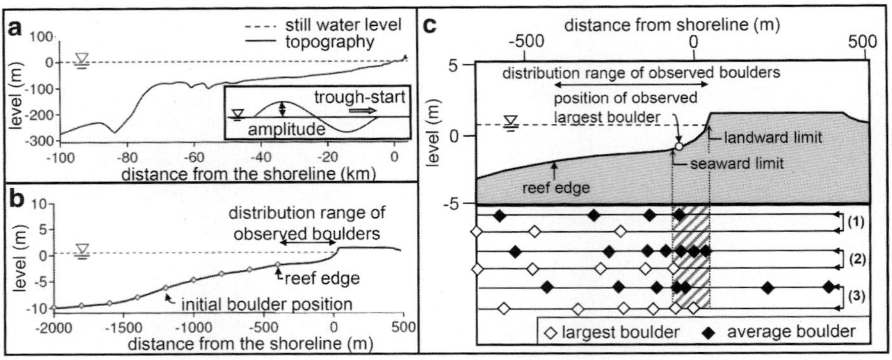

Fig. 11.2 (**a**) Schematic diagram of the coastal profile and incident (*trough-start*) wave used for this study. (**b**) Schematic diagram of the coastal profile near Pakarang Cape. Initial positions of boulders are also shown. (**c**) Schematic diagrams of the coastal profile near the shoreline at Pakarang Cape (*upper panel*). Present position of the observed largest boulder, distribution range of boulders, and assumed landward/seaward limits are shown. Final stop positions of average and largest boulders for three examples (*lower panel*): (1) 2.5 m and 15 min for initial wave amplitude and period. The landward limit is satisfied but no largest boulder reaches the seaward limit. (2) 3.5 m and 18 min. Landward/seaward limits are both satisfied. (3) 2.5 m and 25 min. The seaward limit is satisfied but two average boulders reached beyond the landward limit

$$\frac{\partial \eta}{\partial t} + \frac{\partial M}{\partial x} = 0 \tag{11.1}$$

$$\frac{\partial M}{\partial t} + \frac{\partial}{\partial x}\left(\frac{M^2}{D}\right) + gD\frac{\partial \eta}{\partial x} + \frac{gn^2}{D^{7/3}}M|M| = 0 \tag{11.2}$$

In those equations, η denotes the vertical displacement of the water surface above the still water surface, M is the discharge flux in the x direction, D represents the total water depth ($= h + \eta$), and n is Manning's roughness coefficient. Manning's roughness coefficient was estimated as 0.025 for the sea bottom and on land. The staggered leap-frog method, which is a finite-difference method, was used to solve these equations numerically.

We generated a single incident wave (sine wave) from 100 km distance from the shoreline (Fig. 11.2a). The water depth at the generation point of initial sine wave is 300 m. The typical coastal profile is adopted for this analysis; the length of the tidal bench is 400 m (Fig. 11.2b). The grid-cell size is 10 m and the calculation time is 200 min. We input a trough-start wave because the wave trough arrived first at Thailand (Matsutomi et al. 2006). Initial wave period and amplitude are varied for 15–90 min with a 1 min interval, and 1.5–15 m with a 0.1 m interval. We calculated 10,336 cases.

11.3.2 Boulder Transport

For numerical analyses of boulder transport, we used a one-dimensional model proposed by Imamura et al. (2008) and applied to boulders at Pakarang Cape by Goto et al. (2009, 2010b).

The largest (4.1 × 2.5 × 2.2 m) and average size (1.8 × 1.3 × 1.1 m) boulders in ellipsoidal shape with densities 1.62 g/cm^3 were used for these analyses. We set these boulders at nine positions with 200 m intervals offshore from the reef edge, which corresponds to the possible source area (Fig. 11.2b). Following Goto et al. (2009), we ignored the detachment process of boulders and set the boulders on the sea floor. No interaction (e.g., collision or shielding effect) took place among these boulders. The simulation was made for largest and average size boulders separately.

11.3.3 Constraint Conditions

Based on the field observations, the following two constraint conditions can be inferred: (1) the hydraulic force of the tsunami need not have been strong and/or long lasting to cast ashore the small boulders above the high-tide line (landward limit), but (2) the force must have been strong and/or long-lasting to displace the largest boulder close to the shoreline across the hundred-meter-wide tidal bench (seaward limit). Regarding (2), the largest boulder is deposited approximately 60 m from the mean shoreline. Petroff et al. (2001) and Imamura et al. (2008) reported based on their hydraulic experiments that the advection distance can vary even when the same bore strikes the same particle. Considering this uncertainty, here we assumed the seaward limit as 90 m from the mean shoreline (Fig. 11.2c): at least one largest boulder should be reached within this limit and all the boulders should be below the landward limit. We also conducted sensitivity analysis by assuming the seaward limit as 130 m from the mean shoreline.

11.4 Wave Characteristics

11.4.1 Wave Characteristics

Immediately before the arrival of the first wave crest, the tidal level decreases: the tidal bench was exposed above the tidal level (Goto et al. 2009). Then, the first crest of the wave reached the shore and inundated to the land. The maximum current velocity at the reef edge and the shoreline during the calculation time was highly variable depending on the initial wave amplitude and period. For example, if the initial wave amplitude is fixed as 2 m, then the velocity is the highest when the initial wave period was approximately 30 min. However, if the initial wave amplitude is fixed as 5 m, then the velocity is the highest when the initial wave period was approximately 40 min. On the other hand, if the initial wave period is constant, then the velocity is generally increases with increasing the initial wave amplitude.

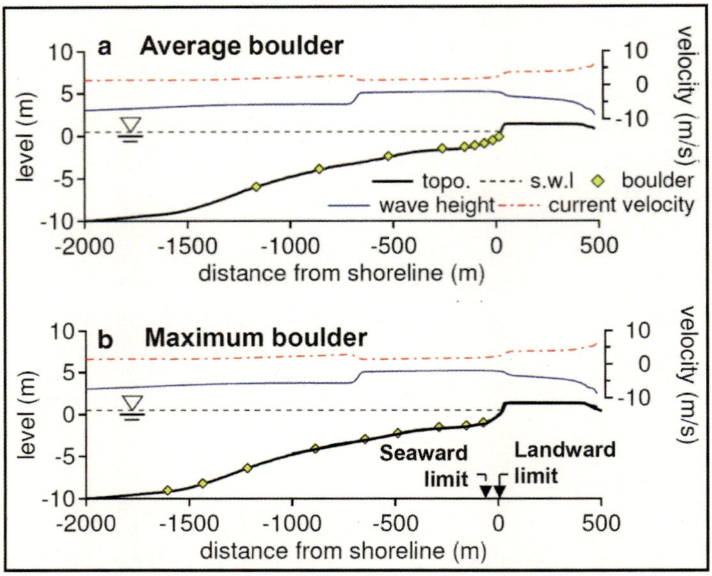

Fig. 11.3 Distributions of (**a**) average and (**b**) maximum boulders 81 min after the tsunami generation when we input 3.5 m and 18 min initial wave amplitude and period. Current velocity (m/s), maximum water level (m), and final stop positions of average and maximum boulders are also shown. Landward/seaward limits are both satisfied in this case

11.4.2 Estimation of the Wave Height and Period

The boulder movements are highly controlled by the time series variation of the wave force acting on the boulders, which is mainly characterized by the current velocity. Therefore, only few combinations of the initial wave amplitude and period can satisfy the constraint condition in Sect. 11.3.3.

Figures 11.2c-(2) and 11.3 show an example of numerical result that satisfies the landward and seaward limit. Alternatively, Fig. 11.2c-(1) shows an example that the tsunami wave force is too weak to transport maximum boulder at the seaward limit. Figure 11.2c-(3) shows an example that the tsunami wave force is too strong and few boulders were deposited on land beyond the landward limit. In this way, the conditions of the wave height and period to satisfy both the seaward and landward limits are very limited.

Black dots in Fig. 11.4 show combinations of the initial wave amplitude and period that satisfy both landward/seaward limits (only 45 cases among 10,336 cases), which respectively correspond to the ranges of 1.7–4.0 m and 18–37 min. This range of the amplitude corresponds to the wave height at the shoreline of approximately 4–6 m. Consequently, our numerical results are well consistent with those observed wave heights and periods (4–7 m and 25–35 min).

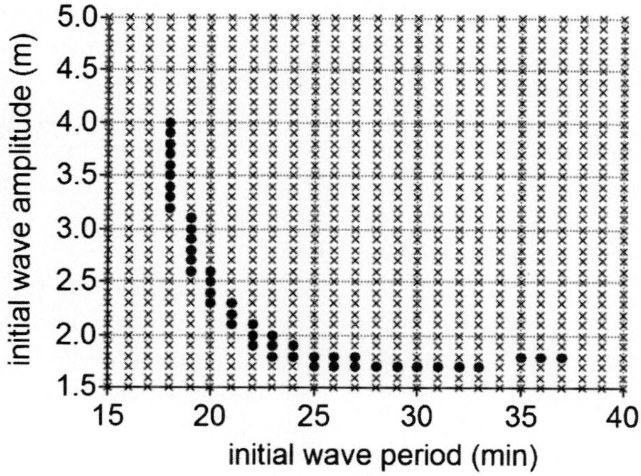

Fig. 11.4 Diagram showing the relation between initial wave amplitude (m) and period (min) that satisfy the landward/seaward limits (*black dots*)

11.4.3 Sensitivity of the Method

The combination of the wave amplitude and period that satisfies the landward/seaward limits is determined based on the sensitive balance of the local topography, hydrodynamic features of the tsunami, and the initial distribution of boulders at the source. The distribution is characterized by the inverse proportion (Fig. 11.4). This is because the boulders are generally more mobile for both higher wave height (stronger hydrodynamic force) and longer period (longer time that the force acting on the boulder) within the range of period between 15 and 40 min. Above the line in Fig. 11.4, the hydraulic force is too strong; several average boulders were reached above the high-tide line beyond the landward limit. Below this line, the force is too weak to reach the largest boulder within the seaward limit.

Wave period is sensitive to the assumption of the seaward/landward limits; it is 17–46 min if we assumed the seaward limit as 130 m from the mean shoreline, although the range of wave amplitude does not change.

11.5 Discussion and Conclusions

The input parameters for this calculation are the sizes and initial positions of boulders and topography, which are obtainable through the geological survey for historical tsunamis. Moreover, historical tsunami boulders, which were once deposited far beyond the transport limit by the storm waves, are difficult to move again. Consequently, the clast size and spatial distributions of such boulders have probably preserved the original distributions that prevailed immediately after deposition by the tsunamis (e.g., Goto et al. 2010a). Therefore, under the proper assumptions of the seaward/landward limits, our method is useful to estimate local wave height and the period of the historical tsunamis from the boulders reported throughout the world.

For instance, the 1771 Meiwa Tsunami deposited boulders of <220 t below the high-tide line, irrespective of size, along the southern and eastern coasts of Ishigaki Island, Japan (e.g., Goto et al. 2010a). In this and similar cases, similar assumptions for limits with this study can be made to estimate the local wave height and period. On the other hand, several historical tsunami boulders have been shown to be displaced on land (e.g., Goff et al. 2006). In such a case, other assumptions such as a landward fining feature or the distribution range of boulders should be made as a constraint condition.

Our method is also applicable to historical and pre-historical tsunamis, whose source mechanism are uncertain. However, it is important to note that the estimation accuracy of the wave height and period might have greater error than that of this study if the tsunami source model, which is related with the fault model (e.g., megathrust earthquake vs. interplate earthquake), is entirely uncertain. This is because the transport distance of boulders varies depending on the initial waveform of the tsunami (trough-start or crest-start) (Goto et al. 2009), which is determined by the source model. Therefore, the required wave height and period might vary.

Acknowledgments This research was supported by a Grant-in-Aid from MEXT (K. Goto: no. 23684041) and JSPS (F. Imamura: no. 22241042).

References

Bourgeois J, MacInnes B (2010) Tsunami boulder transport and other dramatic effects of the 15 November 2006 central Kuril Islands tsunami on the island of Matua. Z Geomorph NF 54 (Suppl):175–195

Etienne S, Buckley M, Paris R, Nandasena AK, Clark K, Chagué-Goff C, Goff J, Richmond B (2011) The use of boulders for characterizing past tsunamis: lessons from the 2004 Indian Ocean and 2009 South Pacific tsunamis. Earth Sci Rev 107:75–89

Goff J, Dudley WC, deMaintenon MJ, Cain G, Coney JP (2006) The largest local tsunami in 20th century Hawaii. Mar Geol 226:65–79

Goto K, Chavanich SA, Imamura F, Kunthasap P, Matsui T, Minoura K, Sugawara D, Yanagisawa H (2007) Distribution, origin and transport process of boulders deposited by the 2004 Indian Ocean tsunami at Pakarang Cape, Thailand. Sediment Geol 202:821–837

Goto K, Okada K, Imamura F (2009) Importance of the initial waveform and coastal profile for the tsunami transport of boulders. Pol J Environ Stud 18:53–61

Goto K, Kawana T, Imamura F (2010a) Historical and geological evidence of boulders deposited by tsunamis, Southern Ryukyu Islands, Japan. Earth Sci Rev 102:77–99

Goto K, Okada K, Imamura F (2010b) Numerical analysis of boulder transport by the 2004 Indian Ocean tsunami at Pakarang Cape, Thailand. Mar Geol 268:97–105

Higman B, Bourgeois J (2008) Deposits of the 1992 Nicaragua tsunami. In: Shiki T, Tsuji Y, Yamazaki T, Minoura K (eds) Tsunamiites features and implications. Elsevier, Amsterdam

Imamura F, Goto K, Ohkubo S (2008) A numerical model for the transport of a boulder by tsunami. J Geophys Res 113:C01008. doi:10.1029/2007JC004170

Kelletat D, Scheffers SR, Scheffers A (2007) Field signatures of the SE-Asian megatsunami along the west coast of Thailand compared to Holocene paleo-tsunami from the Atlantic region. Pure Appl Geophys 164:413–431

Lamarche G, Pelletier B, Goff J (2010) Impact of the 29 September 2009 South Pacific tsunami on Wallis and Futuna. Mar Geol 271:297–302

Matsutomi H, Sakakiyama T, Nugroho S, Matsuyama M (2006) Aspects of inundated flow due to the 2004 Indian Ocean tsunami. Coast Eng J 48:167–195

Morton RA, Gelfenbaum G, Buckley ML, Richmond BM (2011) Geological effects and implications of the 2010 tsunami along the central coast of Chile. Sediment Geol 242:34–51

Nandasena NAK, Paris R, Tanaka N (2011a) Reassessment of hydrodynamic equation: minimum flow velocity to initiate boulder transport by high energy events (storms, tsunamis). Mar Geol 281:70–84

Nandasena NAK, Paris R, Tanaka N (2011b) Numerical assessment of boulder transport by the 2004 Indian Ocean tsunami in Lhok Nga, West Banda Aceh (Sumatra, Indonesia). Comput Geosci 37:1391–1399

Noormets R, Crook KAW, Felton EA (2004) Sedimentology of rocky shorelines: 3: hydrodynamics of megaclast emplacement and transport on a shore platform, Oahu, Hawaii. Sediment Geol 172:41–65

Nott J (2003) Waves, coastal boulder deposits and the importance of the pre-transport setting. Earth Planet Sci Lett 210:269–276

Paris R, Wassmer P, Sartohadi J, Lavigne F, Barthomeuf B, Desgages E, Grancher D, Baumert P, Vautier F, Brunstein D, Gomez C (2009) Tsunamis as geomorphic crises – lessons from the December 26, 2004 tsunami in Lhok Nga, West Banda Aceh (Sumatra, Indonesia). Geomorphological 104:59–72

Paris R, Fournier J, Poizot E, Etienne S, Morin J, Lavigne F, Wassmer P (2010) Boulder and fine sediment transport and deposition by the 2004 tsunami in Lhok Nga (western Banda Aceh, Sumatra, Indonesia) – a coupled offshore-onshore model. Mar Geol 268:43–54

Petroff CM, Moore AL, Arnason H (2001) Particle advection by turbulent bores – orientation effects. In: Proceedings of the International Tsunami Symposium 2001, pp 897–904

Pignatelli C, Sanso P, Mastronuzzi G (2009) Evaluation of tsunami flooding using geomorphologic evidence. Mar Geol 260:6–18

Richmond BM, Buckley M, Etienne S, Strotz L, Chagué-Goff C, Clark K, Goff J, McAdoo B (2011a) Geologic signatures of the September 2009 South Pacific tsunami in the Samoan Islands. Earth Sci Rev 107:37–50

Richmond BM, Watt S, Buckley M, Jaffe BE, Gelfenbaum G, Morton RA (2011b) Recent storm and tsunami coarse-clast deposit characteristics, Southeast Hawaii. Mar Geol 283:79–89

Spiske M, Bahlburg H (2011) A quasi-experimental setting of coarse clast transport by the 2010 Chile tsunami (Bucalemu, Central Chile). Mar Geol 289:72–85

Weiss R (2012) The mystery of boulders moved by tsunamis and storms. Mar Geol 295–298:28–33

Yanagisawa H, Koshimura S, Goto K, Miyagi T, Imamura F, Ruangrassamee A, Tanavud C (2009) The reduction effects of mangrove forest on a tsunami based on field surveys at Pakarang Cape, Thailand and numerical analysis. Estuar Coast Shelf S 81:27–37

Chapter 12
A Note on Imbricated Granite Boulders on NW Penang Island, Malaysia: Tsunami or Storm Origin?

Sharad Master

Abstract The Batu Feringgi area (N coast, Penang Island), with a high concentration of beach hotels, is critical to the tourist economy of Malaysia. Three large imbricated granite boulders were discovered on the NE end of the beach, at 5°28′51.77″N, 100°15′72″E. These boulders, dipping 45°–70° seaward, are shaped as tabular parallelepipeds with rounded corners, with maximum masses of 1.1–2.4 t, based on a density of 2.71 g/cm^3. The boulder shapes were dictated by the presence of joints in the coastal outcrops, which represent an uplifted and exhumed tropically-weathered granite-tor landscape.

In order to produce imbrication of several boulders, the mode of transport has to be rolling/overturning, rather than by sliding or saltation. The hydrodynamic equations for the initiation of boulder transport used in this study are the modified Nott equations, from Nandasena et al. (Marine Geology 281:70–84, 2011). Calculations were made using slopes of 2°and 5°.

The results of the calculations indicate that the minimum velocities required to transport the boulders under free-rolling transport modes were 6.07 and 6.12 m/s for 2° and 5° slopes respectively. For joint-bounded boulders, the minimum velocities are 9.39 and 9.53 m/s for 2° and 5° slopes respectively. These velocities are higher than the maximum velocities experienced at this particular locality during the great 26 December 2004 Indian Ocean tsunami, the largest known tsunami in recorded history. Because this tsunami flooded the area but did not result in appreciable damage to infrastructure, it is concluded that the imbricated boulders on Batu Feringgi beach are the result of tropical storm activity in the past, rather than from recent or past tsunamis. The N coast of Penang is thus regarded as safe from the hazard of damaging tsunamis resulting from mega-earthquakes in the Sumatra-Andaman subduction zone, but the area is prone to tropical storm damage (with a return frequency of about 1 in 400 years), with wave velocities exceeding 6–9.5 m/s.

S. Master (✉)
Economic Geology Research Institute, School of Geosciences, University of the Witwatersrand, P. Bag 3, WITS 2050, Johannesburg, South Africa
e-mail: sharad.master@wits.ac.za

Y.A. Kontar et al. (eds.), *Tsunami Events and Lessons Learned: Environmental and Societal Significance*, Advances in Natural and Technological Hazards Research 35,
DOI 10.1007/978-94-007-7269-4_12,

Keywords Imbricated boulders • Penang island • Malaysia • Hydrodynamic calculations

12.1 Introduction

Penang Island (Figs. 12.1 and 12.2) is one of the fastest developing regions of Malaysia, with growth especially in tourism and hi-tech manufacturing sectors (Lateh et al. 2011). With increasing development and concomitant population growth has come the increased risk of natural hazards, resulting in a renewed focus on the recognition and mitigation of such hazards on Penang. For example, there have been recent studies on the effects of seismicity on built structures (Fadzli 2007), on tsunami hazard on the west (Colbourne 2005) and north (Jahromi 2009) coasts, and on landslide hazards in Penang (Ahmad et al. 2006; Lee and Pradhan 2006; Pradhan and Saro 2010; Lateh et al. 2011). The Batu Feringgi area on the NW coast of Penang has one of the highest concentrations of tourist beach hotels and related infrastructure on the island, and is thus critical to the tourist economy of Penang and Malaysia.

Batu Feringgi beach was affected, like most of the coastline of the entire region, by the Great Indian Ocean tsunami generated by the great Sumatra-Andaman earthquake of 26th December 2004 (Lay et al. 2005, Fig. 12.3; Roy and Ismail 2006;

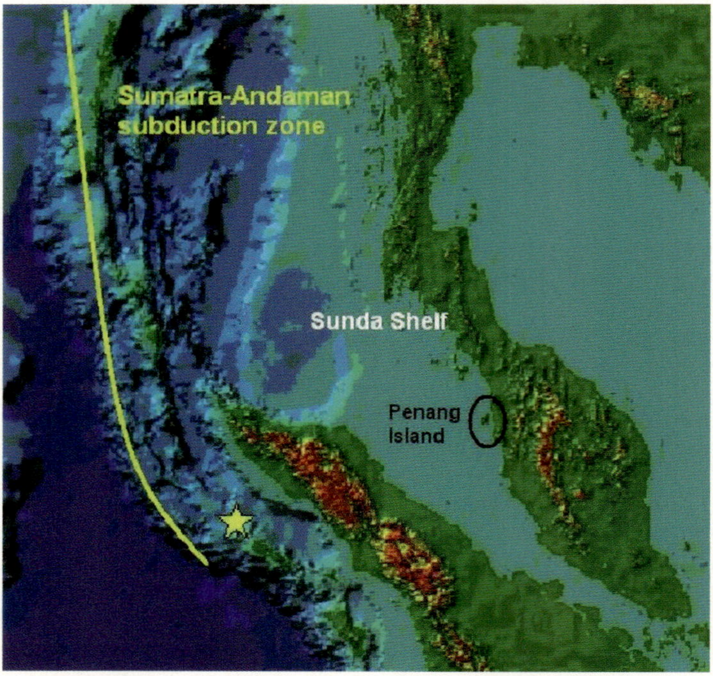

Fig. 12.1 The location of Penang Island, Malaysia (*enclosed in oval*), in relation to the stable Sunda Shelf, and the tectonically active Sumatra-Andaman subduction zone. The *yellow star* represents the epicentre of the great 26 December 2004 Sumatran earthquake, which then propagated northwards for 900 km along the Sumatra-Andaman fault (Lay et al. 2005; Neetu et al. 2005), and generated the great Indian Ocean tsunami of 26 December 2004

Fig. 12.2 Google Earth satellite view of Penang Island, with Batu Feringgi located at its northernmost end

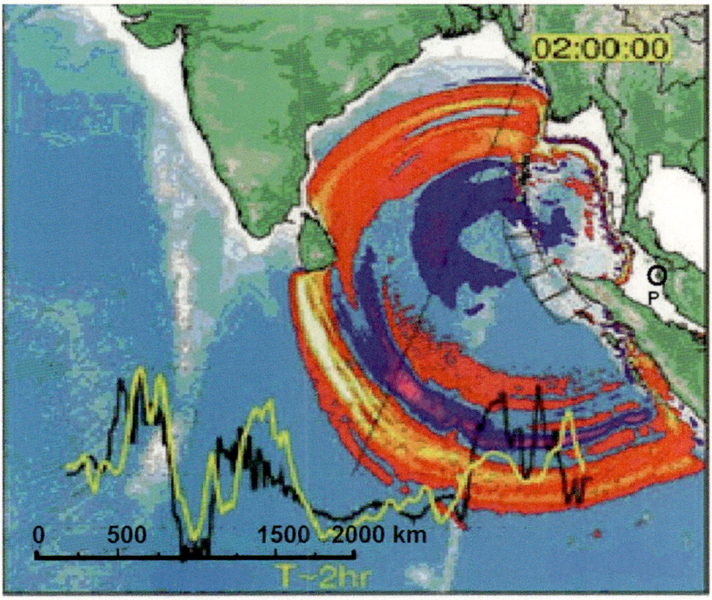

Fig. 12.3 Tsunami propagation model at time of 2.0 h after the great Sumatra-Andaman earthquake of 26 December 2004 (From Lay et al. 2005). The *red* colours are peaks and *blue* colours are troughs in the tsunami waves. Note the location of Penang Island (*circle with P*), in the Sunda Shelf, where much smaller tsunami waves were experienced, compared to the Bay of Bengal and the open Indian Ocean

Fig. 12.4 (**a**) Imbricated rock slabs, thought to have been produced by ancient tsunamis. (**a**) Imbricated slabs, situated 32 m above sea level, at Mermaids Inlet, Jervis Bay, New South Wales, Australia. Photo courtesy of Ted Bryant, University of Woollongong. (**b**) Imbricated granite slabs, Clifton Beach, Cape Town, South Africa (After Master 2011). Measuring tape is 1 m long

Roy et al. 2007). However, unlike areas further north in Thailand, which were severely affected by the tsunami (Goto et al. 2010a), the Batu Feringgi area suffered little infrastructural damage, even though the area was inundated by large waves up to 3.3 m high (Roy and Ismail 2006; Roy et al. 2007) which flooded beach hotel grounds, and about a dozen or more people were reported as having been killed. The author discovered large imbricated granite boulders on the NE end of the tourist beach in the Batu Feringgi area. Since seaward-dipping imbricated boulders have been used as evidence for tsunami activity in other areas (Bryant et al. 1992; Young et al. 1996; Master 2011) (Fig. 12.4), it was decided to investigate the Batu Feringgi imbricated boulders further, and to study their relevance to tsunami hazards in NW Penang. In particular, it was important to know if such imbricate boulders were the result of the 26th December 2004 tsunami, or earlier tsunamis, or some other process, such as storm activity, which has recently been proposed as a source of much boulder movement along coastlines (Williams and Hall 2004; Switzer and Burston 2010; Etienne and Paris 2010; Cox et al. 2012), since both tsunamis and storms can deposit imbricated boulders.

12.2 Geological Setting

Penang Island is situated off the west coast of Peninsular Malaysia, and is composed of rocks similar to those which constitute the mainland (Kwan et al. 1992; Ong 1993). The rocks of the Malay peninsula have been divided into three tectonic domains, the granitic West Coast and East Coast provinces (Bignell and Snelling 1977), separated by the thin medial Central Province. These are now called the

Western, Central and Eastern Belts of Malaysia (Tate et al. 2008). The Western and Central Belts are separated by the Bentong-Raub suture zone, representing a collision zone between the two provinces (Hutchison 1977). The Western Belt is part of the North-Thailand-West Malaya Main Range province, which has mainly S-type biotite granites and abundant tin mineralization of Mid to Late Triassic age (Bignell and Snelling 1977; Liew and McCulloch 1985; Azman 2000; Searle et al. 2012). The Central and Eastern Belts (or East Malaya province) consist predominantly of Permian-Triassic I-type hornblende-biotite granites (Searle et al. 2012).

Three episodes of granite emplacement at 307 ± 8 Ma, 251 ± 7 Ma and 211 ± 2 Ma have been suggested for Penang and the NW Main Range province, based on Rb-Sr dating (Bignell and Snelling 1977; Kwan et al. 1992). However, all the granites have been affected by a hydrothermal conductive convection system related to the late Triassic intrusions (Kwan et al. 1992). Recent U-Pb zircon dating of granites from near Kuala Lumpur has yielded ages of 215 ± 7 to 210 ± 7 Ma (Searle et al. 2012).

Penang Island is underlain by two main granite plutons, the North Penang Pluton, composed of biotite-orthoclase granite, and the South Penang Pluton, composed of biotite-muscovite-microcline granite (Ong 1993). The North Penang Pluton has been divided into three major units known as the Tanjung Bungah granite, the Feringgi granite, and the Muka Head microgranite (Ong 1993). The granites on Penang have been dated at 215 ± 6 Ma and 209 ± 2 Ma (Liew and McCulloch 1985). The biotite and muscovite granites of Penang Island have been extensively weathered (Geological Survey of Malaysia 1992; Ahmad et al. 2006). Fission track work on Penang granites has shown that the island suffered a major episode of tilting and uplift in the Oligocene/Miocene (Kwan et al. 1992), since when there has been uplift and erosion of a formerly buried granite body. The western and southern parts of the island have a thin veneer of late Cenozoic sedimentary cover, and there are thick soils of Pleistocene to Holocene age (Geological Survey of Malaysia 1992; Ahmad et al. 2006). At the present day, the Malaysian Peninsula and Penang Island are part of the stable and aseismic Sunda Shelf. The closest tsunamigenic fault zone is the Sumatran-Andaman subduction zone reverse fault along the plate boundary between the Indo-Australian Plate and the SE Asian Plate. The island has rugged topography in its interior, with a highest elevation of 735 m, and a narrow coastal zone where the beaches are, with the elevation rising rapidly to over 100 m within a few hundred metres from the coastline. The beaches on Penang are situated in those few places where the elevation is less than 15 m (Fig. 12.5).

The Batu Feringgi area of Penang Island is underlain by the Feringgi biotite granite (Ong 1993). Petrographically, this granite is composed of quartz, orthoclase, microcline, plagioclase, biotite, and minor muscovite (Ahmad et al. 2006). Some varieties are porphyritic, with large alkali feldspar phenocrysts; other varieties contain large pockets of tourmaline.

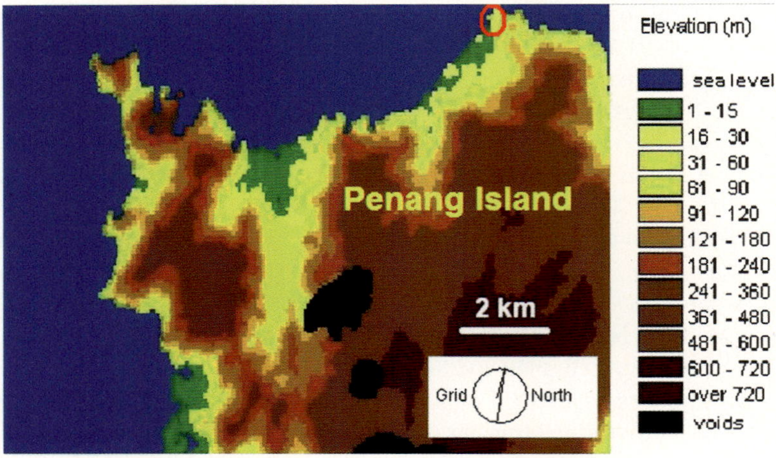

Fig. 12.5 Digital elevation model for NW Penang Island, after Anon (2011), modelled from ASTER satellite data obtained from GDS (2011). The beaches are found in areas with average elevation less than 15 m (shown *green*), along the west and north coasts. The locality of the imbricated boulders on Batu Feringgi beach is in the extreme north of the island, indicated with a *red oval*

12.3 Imbricate Boulders on Batu Feringgi Beach

The locality of the imbricate boulders is at 5°28′51.77″N, 100°15′72″E, in the Batu Feringgi district, northwest Penang, at the northeast end of the long sandy beach of Batu Feringgi, at the start of the rocky headland (having an elevation of up to 30 m) that forms the northernmost point of Penang Island (Figs. 12.5 and 12.6). Here loose granite boulders are arranged landward of a granitic outcrop which contains steep northward-dipping joints (Fig. 12.7).

There are three boulders (labeled 1, 2 and 3, Fig. 12.8) which are arranged in an imbricate fashion, with steep northward (seaward) dips of 45° 65° and 70° respectively. The boulders are tabular in shape, and are roughly rectangular or in the shape of parallelepipeds with rounded corners. Although their exact shapes and hence masses are not known, a rectangular box derived from their axial dimensions a, b and c (with $a > b > c$) would indicate the maximum masses of the boulders, which are calculated as 2.3, 1.1 and 2.4 tonnes for boulders 1, 2 and 3 respectively (Fig. 12.8, Table 12.1). The shapes of the boulders were dictated by the presence of joints in the coastal outcrops, which represent an uplifted and exhumed tropically-weathered granite-tor landscape with residual soils (Ahmad et al. 2006). Boulders that have moved recently can be recognized from the occurrence of striae, crush zones, fresh scars or percussive marks (Etienne and Paris 2010). Such features generally occur when boulders move over a rock platform or limestone terrace, which is absent at Batu Feringgi beach. No such features were found on the imbricated boulders in the present study, indicating that they were probably not of recent origin, but had been subjected to marine erosion for long enough to have

Fig. 12.6 Google Earth satellite view of northern Penang Island, showing Batu Feringgi Beach and the locality of the imbricated granite boulders discussed in this study

Fig. 12.7 Loose boulders (numbered) at the northeast end of Batu Feringgi Beach, which are banked up against a rocky jointed granite headland. Note sitting person (on *left* hand side) for scale

Fig. 12.8 Imbricate arrangement of three boulders (1, 2 and 3), which are analysed in this study. The large granite mass behind boulder 1 is an outcrop, rather than a loose boulder

Table 12.1 Physical parameters for imbricated boulders

Boulder	a (m)	b (m)	c (m)	c/b	c^2/b^2	Density (g/cm^3)	Max. mass (tons)
1	1.50	1.32	0.44	0.33	0.11	2.71	1.2
2	1.25	0.75	0.44	0.59	0.34	2.71	1.1
3	2.20	0.85	0.47	0.55	0.31	2.71	1.3

obliterated any trace of such damage resulting from transport. The absence of adhering barnacles, limpets or seaweed attachment scars rules out the recent transport of the boulders from the intertidal zone, and they may have been moved only a few metres.

12.4 Hydrodynamic Calculations

Boulders transported by moving water can be moved by three different locomotion modes- sliding, rolling, and saltation, and different hydrodynamic factors are involved in each mode. The hydrodynamic equations for the initiation of boulder transport were originally obtained by Nott (1997, 2003), and are called the Nott equations. These equations have been used widely to model boulder transport by storms and tsunamis (Mastronuzzi and Sansò 2000; Kennedy et al. 2007; Scicchitano et al. 2007; Imamura et al. 2008; Etienne and Paris 2010; Switzer and Burston 2010; Goto et al. 2010b; Costa et al. 2011). Nandasena et al. (2011)

modified the Nott equations, based partly on the work of Voropayev et al. (2003) and Nanayama and Shigeno (2006), and produced a new set of hydrodynamic equations governing boulder transport initiation.

The conditions for the initiation of transport of a submerged or subaerial boulder under sliding mode are given by the modified Nott equation (Nandasena et al. 2011), as follows:

$$\mathbf{u^2 \geq 2([\rho_s/\rho_w] - 1)gc(\mu_s \cos\theta + \sin\theta)/[C_d(c/b) + \mu_s C_l]} \quad (12.1)$$

where

u = fluid velocity required to initiate boulder movement
ρ_w = density of fluid
c = length of shortest axis of boulder
b = length of intermediate axis of boulder
μ_s = coefficient of static friction
ρ_s = density of boulder
g = acceleration due to gravity
θ = angle of slope
C_d = coefficient of drag
C_l = coefficient of lift

In order to produce imbrication of several boulders, the mode of transport has to be rolling/overturning. Initiation of boulder transport under rolling mode requires higher water velocities than the initiation of boulder movement under sliding conditions. For the case of rolling or overturning boulder movement initiation, the modified Nott equation (Nandasena et al. 2011) is as follows:

$$\mathbf{u^2 \geq 2([\rho_s/\rho_w] - 1)gc(\cos\theta + [c/b]\sin\theta)/\left[C_d(c^2/b^2) + C_l\right]} \quad (12.2)$$

According to Nandasena et al. (2011), initial transport of a submerged or subaerial boulder will be in saltation mode when:

$$\mathbf{u^2 \geq 2([\rho_s/\rho_w] - 1)gc\ \cos\theta/C_l} \quad (12.3)$$

For joint-bounded boulders, movement is initiated when

$$\mathbf{u^2 \geq 2([\rho_s/\rho_w] - 1)gc(\cos\theta + \mu_s \sin\theta)/C_l} \quad (12.4)$$

Following Noormets et al. (2004) and Nandasena et al. (2011), the fluid density is that of seawater, taken as 1.02 g/cm^3; the acceleration due to gravity is 9.81 m/s^2; and the coefficients of drag (C_d), static friction (μ_s) and lift (C_l) are taken as 1.95, 0.7 and 0.178, respectively. The mean density of the Batu Feringgi granite is 2.71 g/cm^3 (Ahmad et al. 2006). The slope on the Teluk Bahang beach is very flat, and is taken as 2°. Because the offshore topography is unknown, additional calculations were made for a slope of 5°. The physical parameters for the imbricated boulders are given in Table 12.1, and the results of the calculations are given in Table 12.2.

Table 12.2 Calculated flow velocities of seawater needed to initiate boulder movement (u_{min}) under different transport modes (sliding, rolling/overturning and saltating), for loose boulders in submerged and subaerial settings only, at slopes of $\theta = 2°$ and $\theta = 5°$

	Sliding mode		Rolling mode		Saltating mode		Joint-bounded	
	u_{min} (m/s)	u_{min} (m/s)	u_{min} (m/s)	u_{min} (m/s)	u_{min} (m/s)	u_{min} (m/s)	u_{min} (m/s)	u_{min} (m/s)
Bldr	$\theta = 2°$	$\theta = 5°$	$\theta = 2°$	$\theta = 5°$	$\theta = 2°$	$\theta = 5°$	$\theta = 2°$	$\theta = 5°$
1	3.72	3.83	6.07	6.12	8.97	8.96	9.08	9.23
2	2.88	2.97	4.17	4.23	8.97	8.96	9.08	9.23
3	3.07	3.17	4.46	4.52	9.27	9.26	9.39	9.53

12.5 Discussion

Because the boulders are imbricated, their transport mode must have been rolling/overturning. The imbrication is produced by bedload transport, and boulder to boulder interaction, and the imbricated boulders dip upstream (in this case, offshore), in the direction from which the currents were coming. For each slope angle ($\theta = 2°$ and $\theta = 5°$), the highest velocity needed to initiate movement of any of the three boulders is taken as the minimum velocity of flow that would have moved all three boulders together (in order to produce the imbricate stacking observed). Since the imbricated boulders are stacked among jointed outcrops of granites, calculations were also made for the minimum velocities required to initiate boulder movement in joint-bounded mode, in addition to the calculations made for the case of free-standing boulders (Table 12.2).

The calculated minimum velocities required to initiate rolling/overturning of the imbricated boulders range from 6.07 ($\theta = 2°$) to 6.12 ($\theta = 5°$) m/s for rolling mode, to 9.39 ($\theta = 2°$) and 9.53 ($\theta = 5°$) m/s for joint-bounded mode. It could be argued that the particular values of the coefficients of lift (C_l), and static friction (μ_s) used in the equations of Nandasena et al. (2011) may not be valid for this case. So a sensitivity analysis was made, in which the coefficients of static friction (μ_s) and of lift (C_l) were varied by a factor of two (separate calculations were done using half and twice the values used by Nandasena et al. 2011, which originate from Noormets et al. 2004). As an example of these calculations, the u_{min} value for boulder 1 at $\theta = 5°$, under joint-bounded conditions, which was calculated as 9.23 m/s using the μ_s value of 0.7 from Noormets et al. (2004) (Table 12.2), yielded u_{min} values of 9.08 and 9.48 for μ_s values of 0.35 and 1.4 respectively. So varying the coefficient of static friction (μ_s) by a factor of two results only in a ±3 % change in the values of minimum velocities required to initiate boulder movement. The effect of changing the coefficient of lift (C_l) is much larger, yielding u_{min} values of 6.52 m/s for a C_l values of 0.356, and 16.04 m/s for a C_l value of 0.059, compared to the Noormets et al. (2004) value of 0.178 used in Table 12.2. If the density of seawater is changed from the 1.02 g/cm^3 used in the calculations (Table 12.2) to a value of 1.5 g/cm^3, to allow for a turbulently mixed muddy ocean, then the u_{min} value for boulder 1 falls from 9.23 to 6.48 m/s. The sensitivity analysis shows that even if many of the coefficients used in the calculations presented in Table 12.2 were varied by a factor of 2, the results would still yield minimum velocities in the range 6–9.5 m/s in order to initiate the movement of the imbricated boulders under free standing and joint-bounded conditions. These velocities are higher than the maximum velocities experienced at this specific locality during the great 26 December 2004 Indian Ocean tsunami, the largest known tsunami in recorded history.

A number of video recordings of the incoming tsunami waves on 26th December 2004 were captured by tourists on the north coast of Penang Island, and are available for viewing on YouTube (Videos 12.1, 12.2, and 12.3). Analysis of Video 12.1, taken at Tanjung Bungah some 3 km east of Batu Feringgi, and which includes footage of tsunami waves passing by the mosque called Masjid Terapung, which is built on the coast on a promontory at right angles to the main trend of the coastline, has allowed for relatively accurate measurement of wave velocities. The wave passes by the Masjid Terapung, which has dimensions of 77 m (obtained from Google Earth), in 15 s, which gives an open-water incoming wave velocity (before run-up), of about 5.1 m/s. The video also shows that as the waves reached the shoreline, the velocities decreased rapidly in shallow water (although the actual velocities could not be quantified from the video because of a lack of measurable landmarks). The measured wave heights recorded at Tanjung Bungah were 3.0 m (Roy and Ismail 2006; Roy et al. 2007). Since the velocity of tsunami waves is equal to the square root of the depth multiplied by the acceleration due to gravity (9.81 m/s^2), it means that as the waves approach the shore (i.e., during run-up) their velocities decrease exponentially, and their amplitudes increase (e.g., Okal 1988). Another video (Video 12.2), taken on Batu Feringgi beach, shows a narrow shoreline of about 10 m, which was traversed by the first incoming tsunami wave in 3 s, giving a wave velocity of about 3.3 m/s. The water brought in by the first wave was about 0.5 m deep, which would have been just sufficient to cover the imbricated boulders. A third video, taken from a hotel window, shows tsunami waves coming in parallel to a sea-wall, and moving at a rate of about one car-length per second, or about 3 m/s. Videos 12.2 and 12.3 do not record the largest waves to have hit Batu Feringgi, since the highest measured wave heights there were recorded as 3.3 m above mean sea level (Roy and Ismail 2006; Roy et al. 2007).

The run-up wave heights and wavelengths of tsunami waves hitting a shoreline are strongly dependent on local conditions. The best studied tsunami is the 11 March 2011 Japanese Tohoku tsunami, for which the instrumentally-recorded data set contains the most comprehensive collection of coastal and deep-ocean observations ever made during a single tsunami event. From this event, more than 350 coastal water level records show regions of signal attenuation and amplification due to refraction, reflection, and bathymetric focusing (Eble et al. 2012).

Similarly, during storm events, wave behavior is strongly influenced by local conditions. According to Regnauld et al. (2004), bathymetric complexities exercise significant controls on wave refraction patterns during storms, and have resulted in significant variability in coastal response during storms in western France. Another factor of great importance is the storm surge, resulting from the pile up

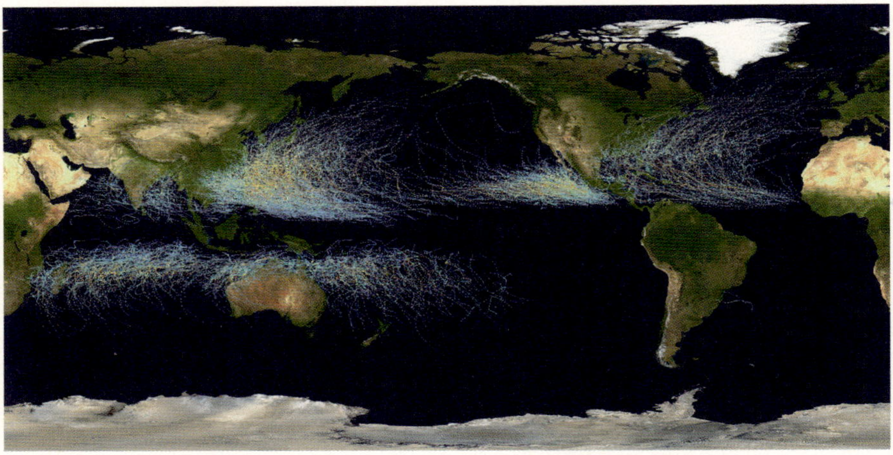

Fig. 12.9 Map of global tropical cyclone tracks recorded from 1985 to 2005. Compiled by Nilfanion (2006), with background image from NASA

of wind-driven waters from the continental shelves onto the coastal areas (Coch 1994; Flather 2003). Bathymetric factors such as water depth and slope are thus expected to control the velocities, wavelengths and propagation directions of waves on Batu Feringgi beach during storms, as well as during tsunami events.

Penang Island is less than 5.5° from the Equator, and hence is generally thought to be relatively free from the effects of major tropical storms or hurricanes, which, for reasons to do with global atmospheric vorticity, are mainly confined in subequatorial belts between 10° and 30° north and south of the Equator (McBride 1995) (Fig. 12.9). However, on 26th December 2001, typhoon Vamei was formed at 1.4°N in the southern South China Sea in what was previously deemed a cyclone-free equatorial zone, strengthened quickly, and made landfall along the southeast coast of Malaysia 60 km northeast of Singapore, and dissipated over Sumatra on 28th December 2001 (Fig. 12.10). The wind speeds were calculated to be as high as 36 m/s in the southern South China Sea, and would have generated wind waves with wave heights as high as 7.5 m on the southeast coast of Malaysia (Mohammed et al. 2012). An analysis by Chang et al. (2003) suggested that the return frequency for such a typhoon is approximately 1 in 400 years, which could also be taken as the return frequency for similar typhoons or tropical storms affecting Penang Island.

The question of whether boulders are moved by tsunami waves or storm waves is still a hotly debated issue, because the processes are quite similar, and it is difficult to unequivocally prove one mechanism of formation to the exclusion of the other (Williams and Hall 2004; Switzer and Burston 2010; Etienne and Paris 2010; Cox et al. 2012; Lau et al. 2012; Seshachalam et al. 2012). The Batu Feringgi beach appears to be fortuitously situated far enough from the seismogenic and tsunamigenic Sumatra-Andaman fault zone to be on the fringe of the area most prone to devastating effects of megatsunamis generated by the largest earthquakes generated in that region. Yet it contains boulders that were moved by waves with a

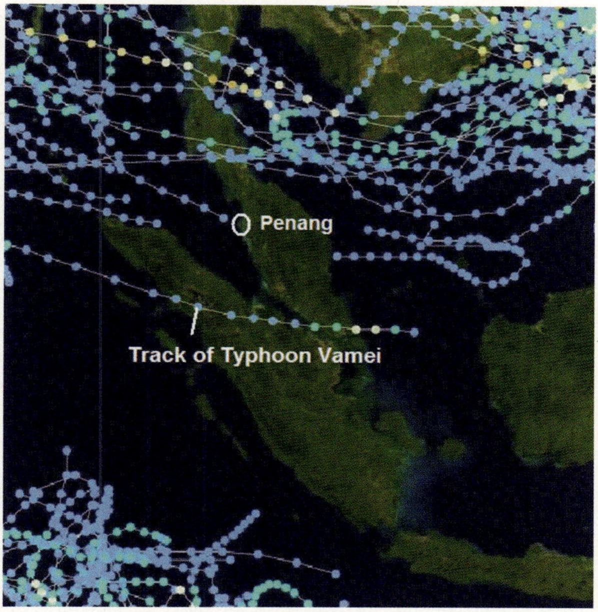

Fig. 12.10 Close-up view of SE Asia region in global map of tropical cyclone tracks from 1985 to 2005 (From Nilfanion 2006). The track in the centre of the image is from typhoon Vamei, 26–28 December 2001 (Chang et al. 2003)

minimum calculated velocity which was much higher than experienced in that particular location during the largest tsunami event ever recorded. It is possible that some previous pre-historic tsunami could have moved these boulders, but the chances are rather small, since the return frequency for tsunamigenic mega earthquakes, similar to or larger than the magnitude 9.2 earthquake of 26 December 2004 along the Sumatra-Andaman subduction zone, is of the order of millennia (McCaffrey 2008). The imbricated boulders, located at sea level, would be unlikely to have escaped erosion by attrition in this high-energy marine environment over such a length of time. The tectonic situation of Penang Island in the stable Sunda Straits also means that it is not prone to tsunamis generated by landslides on the continental shelf, most of which are triggered by seismic activity (Ward 2001; Nadim et al. 2005). In this case, tsunamis can effectively be ruled out as the causative agents, and it leaves the agency of rare (but not unknown) tropical storms, such as Vamei of 2001, as the most likely cause of the boulder movements on Batu Feringgi.

12.6 Conclusions

The Malaysian Peninsula and Penang Island are part of the stable and aseismic Sunda Shelf. The closest tsunamigenic fault zone is the Sumatran-Andaman subduction-zone reverse fault along the plate boundary between the Indo-Australian Plate and the SE Asian Plate. The 26 December 2004 Sumatra-Andaman earthquake produced a tsunami that flooded the Batu Feringgi area of north Penang Island, but did not result in appreciable damage to infrastructure. The measured wave run-ups were up to 3.3 m above mean sea-level on Batu Feringgi, and the incoming wave velocities were in the range of 3–5 m/s, based on video analysis. However, the minimum velocities required to initiate movement of the imbricate boulders from Batu Feringgi were calculated to be in the range of 6–9.5 m/s. Hence it is concluded that the imbricated boulders on Batu Feringgi beach are the result of tropical storm activity in the past, rather than from the recent tsunami. If the magnitude 9.2 Sumatra-Andaman earthquake of 26 December 2004 (one of the strongest ever recorded) could not produce a sizeable tsunami on the north coast of Penang, then it is unlikely that any recent tsunami could have produced the imbricated boulders at Batu Feringgi, leaving tropical storms as the most likely culprit. The north coast of Penang (including Batu Feringgi beach) is thus regarded as safe from the hazard of damaging tsunamis resulting from mega-earthquakes in the Sumatra-Andaman subduction zone, but the area is prone to rare tropical storm damage, with wave velocities exceeding 6–9.5 m/s.

Acknowledgements I am grateful to Vicente Santiago-Fandiño and Yevgeniy Kontar for their comments. I also thank Grahame Oliver and An Yi Lau (National University of Singapore) and Ray Durrheim (CSIR/University of the Witwatersrand) for discussions. I am indebted to two anonymous reviewers who helped to improve the paper. Financial support from the South African National Research Foundation to attend the AOGS-AGU Joint Assembly in Singapore (August, 2012) is gratefully acknowledged.

References

Ahmad F, Yahaya AS, Farooqi MA (2006) Characterization and geotechnical properties of Penang residual soils with emphasis on landslides. Am J Environ Sci 2(4):121–128

Anon (2011) Penang- landforms. http://www.geoscience-environment.com/es771/penang.htm

Azman AG (2000) The western belt granite of Peninsular Malaysia: some emergent problems on granite classification and its implication. Geosci J 4(4):283–293

Bignell JD, Snelling NJ (1977) Geochronology of Malayan granites. Overseas geology and mineral resources, vol 47. IGS, London

Bryant EA, Young RW, Price DM (1992) Evidence of tsunami sedimentation on the southeastern coast of Australia. J Geol 100:753–765

Chang CP, Liu C-H, Kuo H-C (2003) Typhoon Vamei: an equatorial tropical cyclone formation. Geophys Res Lett 30:1150. doi:10.1029/2002GL016365, 4 pp

Coch NK (1994) Geologic effects of hurricanes. Geomorphology 10:37–63

Colbourne FW (2005).Tsunami impact on the West Coast of Penang Island, Malaysia. Research Project Report, M.S. in Physical Sciences, Emporia State University, Emporia, Kansas, USA

Costa PJM, Andrade C, Freitas MC, Oliveira MA, da Silva CM, Omira R, Taborda R, Baptista MA, Dawson AG (2011) Boulder deposition during major tsunami events. Earth Surf Proc Land 36:2054–2068

Cox R, Zentner DB, Kirchner BJ, Cook MS (2012) Boulder ridges on the Aran Islands (Ireland): recent movements caused by storm waves, not tsunamis. J Geol 120:249–272

Eble M, Mungov G, Rabinovich A, Harris E ,Titov V (2012) Spatial and temporal characterization of the 11 March 2011 tsunami. Abstracts, AOGS-AGU (WPGM) Joint Assembly, 13–17 Aug 2012, Singapore, Abstract No. OS07-17-A029 (CD-ROM)

Etienne S, Paris R (2010) Boulder accumulations related to storms on the south coast of the Reykjanes Peninsula, Iceland. Geomorphology 114:55–70

Fadzli MN (2007) Development of design response spectra for Penang Island. M.Sc. thesis (unpublished), Universiti Sains Malaysia, Penang, Malaysia

Flather RA (2003) Storm surges. In: Holton JR et al (eds) Encyclopedia of atmospheric sciences. Academic, New York, pp 109–117

GDS (2011) ASTER GDS Web Site, Earth Remote Sensing Data Analysis Center. http://gds.aster.ersdac.jspacesystems.or.jp/gds_www2002/ index_e.html

Geological Survey of Malaysia (1992) Quaternary geological map of Penang, Seberang Prai and Kuala Kurau

Goto K, Okada K, Imamura F (2010a) Numerical analysis of boulder transport by the 2004 Indian Ocean tsunami at Pakarang Cape, Thailand. Mar Geol 268:97–105

Goto K, Kawana T, Imamura F (2010b) Historical and geological evidence of boulders deposited by tsunamis, southern Ryukyu Islands, Japan. Earth-Sci Rev 102:77–99

Hutchison CS (1977) Granite emplacement and tectonic subdivisions of Peninsular Malaysia. Geol Soc Malaysia Bull 9:187–207

Imamura F, Goto K, Ohkubo S (2008) A numerical model for the transport of a boulder by tsunami. J Geophys Res 113, C01008. doi:10.1029/2007JC004170

Jahromi BI (2009) Design of a tsunami barrier to the north of Penang Island. M. Eng. (Civil-Hydraulics and Hydrology) dissertation, Faculty of Civil Engineering, Universiti Teknologi Malaysia, 86 pp

Kennedy DM, Tannock KL, Crozier MJ, Rieser U (2007) Boulders of MIS 5 age deposited by a tsunami on the coast of Otago, New Zealand. Sediment Geol 200:222–231

Kwan TS, Krähenbühl R, Jager E (1992) Rb-Sr, K-Ar and fission track ages for granites from Penang Island, West Malaysia: an interpretation model for Rb-Sr whole-rock and for actual and experimental mica data. Contrib Mineral Petrol 111(4):527–542

Lateh H, Khan MMA, Jefriza (2011) Monitoring of shallow landslide in Tun Sardon 3.9 km, Pinang Island, Malaysia. Int J Phys Sci 6(12):2989–2999

Lau AYA, Etienne S, Terry J, Lee YS, Switzer S (2012) Preliminary findings on recent and ancient large-wave signatures in coastal reef-platform boulder fields from Makemo Atoll, French Polynesia. Abstracts, AOGS-AGU (WPGM) Joint Assembly, 13–17 Aug 2012, Singapore, Abstract No. OS14-A001 (CD-ROM)

Lay T, Kanamori H, Ammon CJ, Nettles M, Ward SN, Aster RC, Beck SL, Bilek SL, Brudzinski MR, Butler R, DeShon HR, Ekström G, Satake K, Sipkin S (2005) The Great Sumatra-Andaman earthquake of 26 December 2004. Science 308:1127–1133

Lee S, Pradhan B (2006) Probabilistic landslide hazards and risk mapping on Penang Island, Malaysia. J Earth Syst Sci 115(6):661–672

Liew TC, McCulloch MT (1985) Genesis of granitoid batholiths of Peninsular Malaysia and implications for models of crustal evolution: evidence from a Nd-Sr isotopic and U-Pb zircon study. Geochim Cosmochim Acta 49:587–600

Master S (2011) Transported megaboulders and the recognition of palaeotsunamites at Clifton Beach and surrounding areas, Cape Town, South Africa. Abstract Book, GeoSynthesis 2011 Conference and Exhibition, Integrating the Earth Sciences, 30 August–1 September 2011, Cape Town, pp 188–189

Mastronuzzi G, Sansò P (2000) Boulders transport by catastrophic waves along the Ionian coast of Apulia (Southern Italy). Mar Geol 170:93–103

McBride JL (1995) Tropical cyclone formation. In: Elsberry RL (ed) Global perspectives on tropical cyclones, World Meteorological Organization, Geneva, Report No. TCP-38. http://derecho. math.uwm.edu/classes/TropMet/GPTC/tcclimo.pdf

McCaffrey M (2008) Global frequency of magnitude 9 earthquakes. Geology 36(3):253–266

Mohammed A, Tkalich P, Vinod Kumar K, Vethamony P (2012) Hindcasting of wind waves generated by typhoon Vamei. Abstracts, AOGS-AGU (WPGM) Joint Assembly, 13–17 Aug 2012, Singapore, Abstract No. OS011-16-A006 (CD-ROM)

Nadim F, Kvalstad T, Guttormsen T (2005) Quantification of risks associated with seabed instability at Ormen Lange. Mar Pet Geol 22:311–318

Nanayama F, Shigeno K (2006) Inflow and outflow facies from the 1993 tsunami in southwest Hokkaido. Sed Geol 187:139–158

Nandasena NAK, Paris R, Tanaka N (2011) Reassessment of hydrodynamic equations: minimum flow velocities to initiate boulder transport by high energy events (storms, tsunamis). Mar Geol 281:70–84

Neetu S, Suresh I, Shankar R, Shankar D, Shenoi SSC, Shetye SR, Sundar D, Nagarajan B (2005) Comment on "The Great Sumatra-Andaman earthquake of 26 December 2004". Science, 310, 1431, doi:10.1126/science.1118950

Nilfanion (2006) Global tropical cyclone tracks. http://en.wikipedia.org/wiki/File:Global_tropical_cyclone_tracks-edit2.jpg

Noormets R, Crook KAW, Felton EA (2004) Sedimentology of rocky shorelines: 3. Hydrodynamics of megaclast emplacement and transport on a shore platform, Oahu, Hawaii. Sedim Geol 172:41–65

Nott J (1997) Extremely high wave deposits inside the Great Barrier Reef, Australia: determining the cause- tsunami or tropical cyclone. Mar Geol 141:193–207

Nott J (2003) Waves, coastal boulders, and the importance of the pre-transport setting. Earth Planet Sci Lett 210:269–276

Okal EA (1988) Seismic parameters controlling far-field tsunami amplitudes: a review. Nat Hazards 1:67–96

Ong WS (1993) The geology and engineering geology of Pulau Pinang. Geological survey of Malaysia, Map report 7

Pradhan B, Saro L (2010) Delineation of landslide hazard areas on Penang Island, Malaysia, by using frequency ratio, logistic regression and artificial neural network models. Env Earth Sci 60(5):1037–1054

Regnauld H, Pirazzoli PA, Morvan G, Ruz M (2004) Impacts of storms and evolution of the coastline in western France. Mar Geol 210:325–337

Roy GD, Ismail AIM (2006) Numerical modelling of tsunami along the coastal belt of Penang using a polar coordinate shallow water model. Far East J Appl Math 23(3):241–261

Roy GD, Karim MF, Ismail AIM (2007) A nonlinear polar coordinate shallow water model for tsunami computation along North Sumatra and Penang Island. Cont Shelf Res 27:245–257

Scicchitano G, Monaco C, Torcorici L (2007) Large boulder deposits by tsunami waves along the Ionian coast of south-eastern Sicily (Italy). Mar Geol 238:75–91

Searle MP, Whitehouse MJ, Robb LJ, Ghani AA, Hutchison CS, Sone M, Ng SW-P, Roselee MH, Chung S-L, Oliver GJH (2012) Tectonic evolution of the Sibumasu-Indochina terrane collision zone in Thailand and Malaysia: constraints from new U-Pb zircon chronology of SE Asian tin granitoids. J Geol Soc, London 169(4):489–500

Seshachalam S, Karthikeyan A, Switzer A, Gouramanis C (2012) Sedimentological characteristics of tsunami and storm deposits: a modern analog from Southeast Indian coast. Abstracts, AOGS-AGU (WPGM) Joint Assembly, 13–17 Aug 2012, Singapore, Abstract No. OS14-A013 (CD-ROM)

Switzer AD, Burston JM (2010) Competing mechanisms for boulder deposition on the southeast Australian coast. Geomorphology 114:42–54

Tate RB, Tan DK, Ng TF (2008) Geological map of Peninsular Malaysia. In: Hutchison CS, Tan NK (eds) Geology of Peninsular Malaysia. Geological Society of Malaysia, Kuala Lumpur, 479 p

Voropayev SI, Testik FY, Fernando HJS, Boyer DL (2003) Morphodynamics and cobbles behaviour in and near the surf zone. Ocean Eng 30:1741–1764

Video 1. Tsunami 2004 at Tanjung Bungah Penang. http://www.youtube.com/watch?feature=player_detailpage&v=lkPw0WSgWqc

Video 2. Raw tsunami video Penang Beach Malaysia 2004. Uploaded by PubDom on 14 February 2007. http://www.youtube.com/watch?feature=player_detailpage&v=wJlDvzdB-zg

Video 3. Penang Tsunami. Uploaded by Adda58 (Jeff Addinsall) on 27 May 2008, shot at 2 PM, 26 December 2004. http://www.youtube.com/watch?feature=player_detailpage&v=NCWd21ortdk
Ward S (2001) Landslide tsunami. J Geophys Res 106(6):11201–11215
Williams DM, Hall AM (2004) Cliff-top mega-clast deposits of Ireland, a record of extreme waves in the North Atlantic- storms or tsunamis? Mar Geol 206:101–117
Young RW, Bryant EA, Price DM (1996) Catastrophic wave (tsunami?) transport of boulders in southern New South Wales, Australia. Zeitschrift für Geomorphologie 40(2):191–207

Chapter 13
Effects of Tsunami Wave Erosion on Natural Landscapes: Examples from the 2011 Tohoku-oki Tsunami

Goro Komatsu, Kazuhisa Goto, Victor R. Baker, Takashi Oguchi, Yuichi S. Hayakawa, Hitoshi Saito, Jon D. Pelletier, Luke McGuire, and Yasutaka Iijima

Abstract The 2011 Tohoku-oki Tsunami affected approximately 600 km of the northeastern coast of the Japanese Honshu Island, leaving traces of destruction on man-made buildings and depositing mud- to boulder-sized sediment. Our field observations at Aneyoshi along the Sanriku "ria" coast, where a maximum run-up height of 39.2 m was recorded, add to the limited number of studies of tsunami wave effects on natural landscapes. We found evidence for (1) tsunami wave erosion that exposed bare rock by stripping basal hillslopes of regolith and vegetation, including trees, (2) transport and deposition of coarse gravel, and (3) scour-hole generation around a large boulder and a large sea wall fragment. Computer simulations indicate that the highest first wave reaching the Aneyoshi coast may have been about 20 m high, that the combined duration of the first three waves was tens of minutes to 1 h, and that the maximum wave velocity on land reached over 10 m/s and probably exceeded 20 m/s in the lower, wide reach of the Aneyoshi valley. We hypothesize that hillsides along the Sanriku Coast have been stripped by

G. Komatsu (✉)
International Research School of Planetary Sciences, Università d'Annunzio, Pescara, Italy
e-mail: goro@irsps.unich.it

K. Goto
Planetary Exploration Research Center, Chiba Institute of Technology, Narashino, Japan

V.R. Baker
Department of Hydrology and Water Resources, University of Arizona, Tucson, USA

T. Oguchi • Y.S. Hayakawa • H. Saito
Center for Spatial Information Science, University of Tokyo, Kashiwa, Japan

J.D. Pelletier
Department of Geosciences, University of Arizona, Tucson, USA

L. McGuire
Program in Applied Mathematics, University of Arizona, Tucson, USA

Y. Iijima
Department of Earth Science, Tohoku University, Sendai, Japan

Y.A. Kontar et al. (eds.), *Tsunami Events and Lessons Learned: Environmental and Societal Significance*, Advances in Natural and Technological Hazards Research 35, DOI 10.1007/978-94-007-7269-4_13,

erosion of numerous ancient tsunami events recurring at century or even decadal scales, since at least the mid-Holocene. The cumulative effects of tsunami erosion on the hillslopes and their long-term evolution are important potential topics for future studies.

Keywords 2011 Tohoku Tsunami • Aneyoshi • Erosion • Hillslopes • Ria coast • Sanriku • Scour holes • Sediment

13.1 Introduction

The devastating tsunami waves of March 11, 2011 (hereinafter the 2011 Tohoku-oki Tsunami), along the northeastern coast of Japan caused severe damage to coastal communities, taking the lives of tens of thousands of people. The tsunami inundated up to 4.5 km inland on the Sendai Plain (Goto et al. 2011), reaching a maximum run-up height of about 40 m for local sites along the Sanriku Coast (Mori et al. 2012).

Tsunami waves leave important geomorphological evidence of their impacts on the land surface through their erosional and depositional processes. However, tsunami erosion of natural coastal landscapes has received relatively little formal study (e.g., Dawson 1994; Kench et al. 2008; Paris et al. 2009; Goto et al. 2012). Controversial coastal features attributed to paleo-tsunami processes include a variety of bedrock-sculpted forms (Bryant and Young 1996; Nott and Bryant 2003), but a lack of direct linkage to observed and well-documented tsunamis hampers any effective assessment of associated cause-consequence relationships.

The extraordinary scale of the 2011 Tohoku-oki Tsunami provides a rare opportunity for documenting effects of tsunami waves on coastal landscapes, but previous investigations have focused mainly on impacts of the tsunami on high population centers or on depositional processes. In this paper, we present results of our preliminary study of tsunami erosional effects observed at a Sanriku "ria" coast locality, where there were only a few buildings prior to the 2011 Tohoku-oki Tsunami. We also hypothesize about the consequences of repeated tsunamis along the Sanriku "ria" coast as the basis for future investigations and discussions.

13.2 Methodology

Our field study consists of observations conducted in July 2011 and the construction of a digital elevation model (DEM) later in the same year using a terrestrial laser scanner. The laser scanner used was GLS-1500 by Topcon Inc., which is capable of measuring up to 300 m with an accuracy of 4 mm at 150 m distance. The area was measured from nine different scanning positions within the study site, and sets of point cloud data for each position were merged using a tie-point methodology and distance resections based on targets for registration. Geographic coordinates in UTM Zone 54N of scanning positions were obtained using both a GNSS receiver

(Nikon-Trimble GeoXH Explorer 6000) and base station data from the GEONET (GPS Earth Observation Network System) operated by Geospatial Information Authority of Japan. The three-dimensional point cloud data were then converted into a 0.2-m resolution digital elevation model (DEM) using linear triangulated irregular network interpolation.

The results of the field study were compared with computer simulations of incoming waves and inundation on land in order to gain insights on the relationships between tsunami wave behaviors and erosional effects on natural landscapes. To reproduce tsunami propagation in the open sea, TUNAMI-CODE (Goto et al. 1997), which is based on shallow-water theory, was used for the modeling. The modeling reproduced the tsunami propagation and inundation for 180 min after the generation. The modeling was applied to a spatial grid size of 150 m. For the wave source of the Tohoku-oki tsunami we used composite fault model (Tohoku University model version 1.1.), originally proposed by Imamura et al. (2011). The wave heights near the Aneyoshi shore at 1 m depth, were extrapolated from the waves computed at 70 m depth using Green's law.

Run-up of the tsunami wave on land was simulated using the FLO-2D model (FLO-2D Software Inc. 2009). Numerical simulations were conducted on a square grid with 5 m/pixel resolution using a uniform Manning's roughness coefficient of 0.05 (appropriate for high-energy turbulent flow over a rough bed). Ground topography input to the model was derived from airborne LiDAR data. The input wave was obtained using a one-dimensional numerical solution of the shallow water equations solved for the off-shore run-up of a wave with height and frequency similar to that recorded by the GPS-mounted wave buoy off the Sanriku Coast. The final input wave profile at the coastline was obtained by amplifying the predicted height of the approaching wave at 70 m depth to account for the near-shore effects that were likely to have increased the wave height. The input tsunami wave is qualitatively similar in height and frequency to the most significant incoming wave predicted by two dimensional shallow water simulations by the TUNAMI-CODE.

13.3 Observations and Interpretations in Aneyoshi

The southern part of the Sanriku coastline of northeastern Honshu, Japan, consists of multiple embayments, or inlets, into the Kitakami Mountains (Fig. 13.1a). This produces a "ria" (drowned river valley) configuration that is particularly susceptible to the enhancement of the erosion effects from tsunamis generated at the active Pacific-Asiatic plate-boundary subduction zone, lying approximately 200 km to the east. If the tsunami is large enough, it could lead to severe erosion of the valley sides because these type of bays can focus waves causing them to achieve great heights within the embayments. Our post-Tohoku-oki Tsunami survey in July 2011 of areas with limited coastal defense infrastructure, such as sea walls, revealed that the highest intensity erosion of valleys and hillslopes occurred in the narrowest coastal embayments, particularly those with funnel-shaped inlets that open most directly to the incident tsunami wave crests.

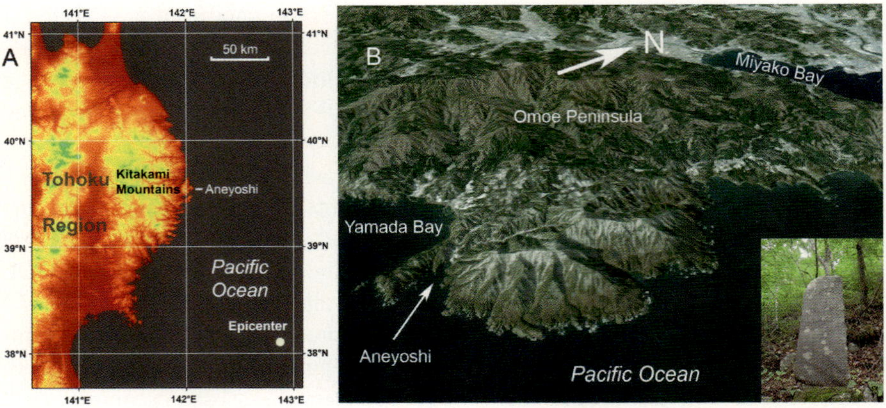

Fig. 13.1 Location map of study site in the Tohoku Region, Japan. (**a**) Aneyoshi is located along the Sanriku Coast characterized by a "ria" (drowned river valley) configuration. (**b**) Google Earth image (Image NASA, Image © 2012 DigitalGlobe) of Omoe Peninsula and Aneyoshi. (Inset) Learning from the 1896 Meiji-Sanriku and 1933 Showa-Sanriku tsunami events, this Aneyoshi tsunami stone monument was placed at about 60 m a.s.l. It recommends having dwellings at high elevations for the peace and happiness of descendants, reminding people about the catastrophic outcomes of tsunamis and warning against building houses below the elevation of the monument

Aneyoshi is a small village in the Iwate Prefecture (Figs. 13.1 and 13.2). Prior to the 2011 Tohoku-oki Tsunami, there were only a few buildings associated with the Aneyoshi campground and a small fishing port near the shore. The port was protected by a sea wall.

Along the valleys around Aneyoshi, where wave run-up reached a maximum of 39.2 m (Mori et al. 2012), there was considerable erosion of the valley sides, transport and deposition of coarse gravel, and scour-hole generation (Fig. 13.3). The tsunami waves that destroyed the sea wall protecting the port ran up along the main valley and into multiple side valleys. In the lower main valley, the lower portions of hillsides were stripped of regolith and vegetation, exposing the light-toned volcanic bedrock. Hillslope erosion extends up almost to the maximum run-up height that was defined by small debris in each valley, although with decreasing extent and intensity with elevation. This indicates that the tsunami waves that traveled along the Aneyoshi valleys maintained their erosional capacity, even as they slowed in achieving their maximum run-up heights.

The northern tributary valley (Figs. 13.2 and 13.3) floor is filled with gravel-rich sediment, much of which, at least on the surface, was probably emplaced by the 2011 Tohoku-oki Tsunami, as indicated by the presence of modern anthropogenic debris. Gravel-rich sedimentary infilling of side valleys is common for tsunamis (e.g., Nanayama et al. 2000; Nanayama and Shigeno 2006), and the northern tributary was probably the site of tsunami-transported sediment deposited as a result of decreasing wave velocity. The gravel was probably at least partially derived from the eroded beach.

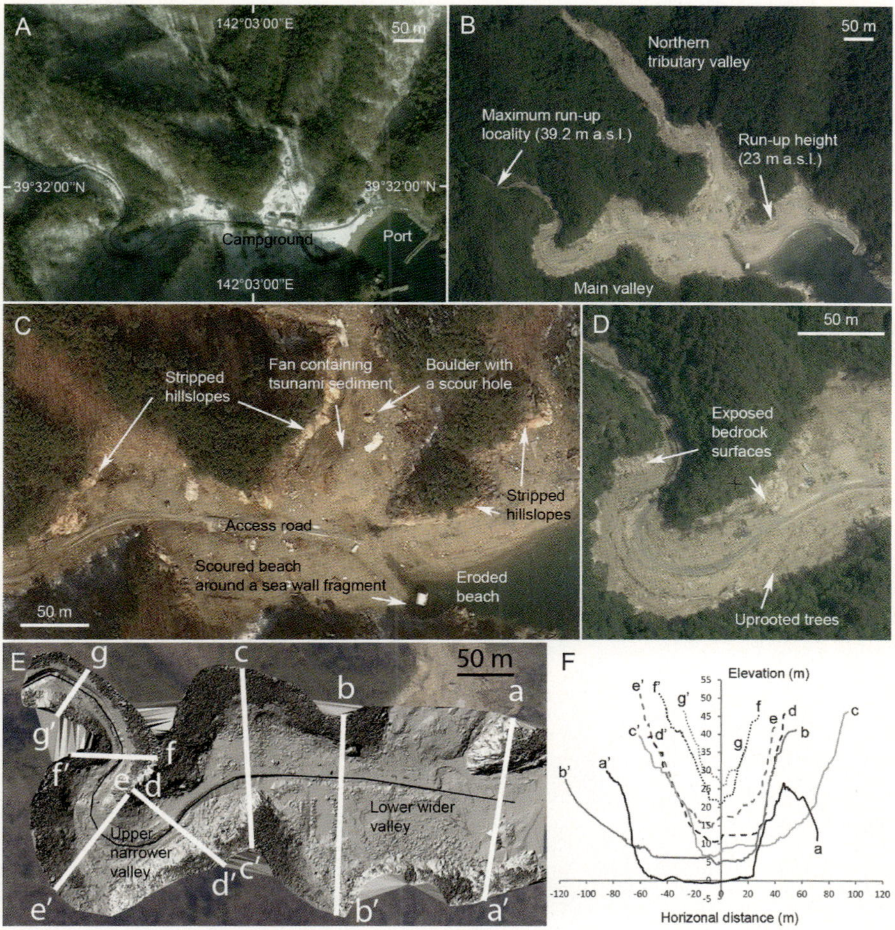

Fig. 13.2 Satellite images and aerial photographs of the Aneyoshi study site: Pre-2011 Tohoku-oki Tsunami (**a**: January 27th, 2005, Google Earth image, Image © 2012 DigitalGlobe); and Post-2011 Tohoku-oki Tsunami (**b**: May–June, 2011, Geospatial Information Authority of Japan; **c**: March 24th, 2011, Google Earth image, Image © 2012 GeoEye; and **d**: May–June, 2011, Geospatial Information Authority of Japan). Cross-section profiles of the main Aneyoshi valley (**e**: positions of the cross-section profiles, shown over a shaded-relief map produced from the terrestrial LiDAR DEM. The black line indicates the longitudinal reference line; and **f**: cross-section profiles across the main Aneyoshi valley)

A large boulder in the northern tributary valley sedimentary fill is partly surrounded by a prominent erosional scour hole (Fig. 13.3f). The better-defined and deeper portion (~30 cm) of the hole is more pronounced on the down-valley side of the boulder, indicating that the scour was generated mainly by the tsunami waves running up the valley. Scour hole formation by return flows also occurred, as indicated by prominent erosion on the up-valley side of a large section of sea wall (8 m long axis) deposited on the pre-tsunami beach face (Fig. 13.2c).

Fig. 13.3 Erosional and depositional features observed at Aneyoshi. (**a**) The Aneyoshi port and a remnant of the destroyed sea wall. The gravel-rich beach was eroded and probably provided some of the sediment that was deposited in the northern tributary valley. The electric lines were installed after the tsunami. (**b**) A large concrete section of the sea wall protecting the Aneyoshi port. The section was broken off, transported and deposited by the tsunami onto the former beach. Note the stripped hillslopes. (**c**) Stripped hillslopes along the main valley. (**d**) A tree knocked over by the tsunami high above the valley floor. (**e**) Gravel-rich sediment, visible in the fan deposit. This was at least partially deposited by the tsunami waves as velocities decreased up the northern tributary valley. (**f**) Scour hole generated around a large boulder over 5 m in size, located in the northern tributary valley

Seven transverse profiles across the main Aneyoshi valley were extracted from the 0.2-m resolution DEM to document the bare rock morphology (Fig. 13.2e, f). A major change in the morphology is noted, with a lower, wider valley segment that is relatively straight and flat floored, and an upper, narrower, V-shaped segment that displays a tight meandering morphology.

13.4 Wave Conditions

According to the Japan Meteorological Agency (JMA), almost all the tide gauges along the Sanriku Coast were destroyed during the run-up of the first wave. On the other hand, the tsunami waveform was recorded by a GPS-mounted wave buoy, located about 10–20 km offshore (in approx. 200 m water depth) from the Sanriku Coast (PARI 2011). The buoy recorded seven large waves (PARI 2011), of which the first was up to 6.7 m in amplitude (PARI 2011).

Our TUNAMI-CODE model, computing properties of the tsunami waves approaching Aneyoshi, indicates that the first wave reached about a 20 m height relative to the original water level near the shore (Fig. 13.4a, b), and this is reasonably consistent with the run-up (23 m) measured on land close to the shore (Fig. 13.2b: Mori et al. 2012). Such a wave height would most likely lead to the maximum run-up conditions of 40 m in elevation recorded further inland. The subsequent second and third waves only reached heights of about 5 m. The combined duration of the first

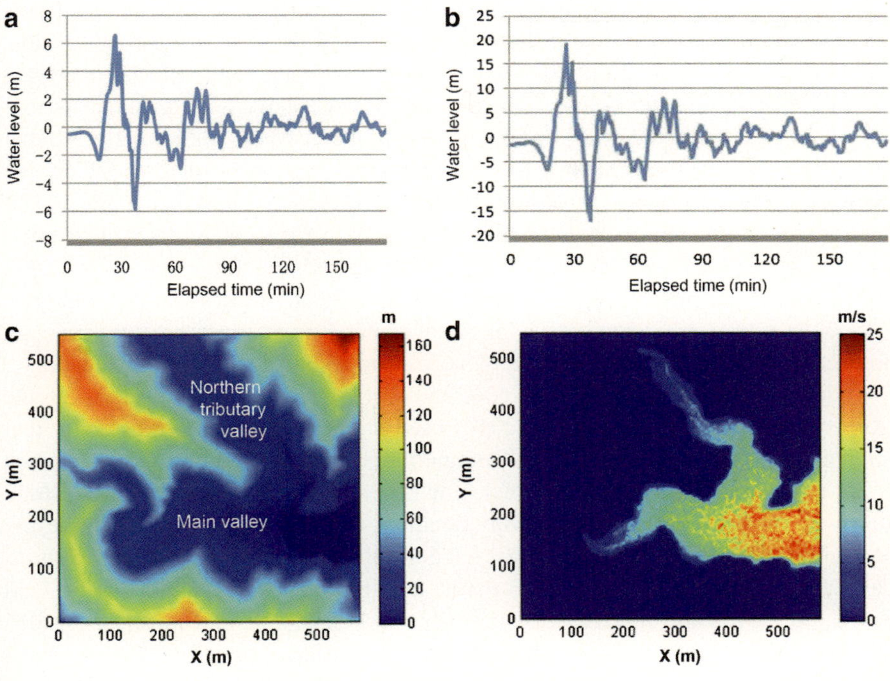

Fig. 13.4 2011 Tohoku-oki Tsunami simulations at Aneyoshi. (**a**) Amplitude of the waves near the shoreline at 70 m water depth, computed by TUNAMI-CODE. (**b**) Amplitude of the waves near the shoreline at 1 m water depth. Note that these waves were extrapolated from the waves computed using Green's law at the depth of 70 m, corresponding to the position in the sea closest to the shore, where the simulation is possible with the available grid data. (**c**) Topography within the computational domain at the study site. (**d**) Predicted maximum flow velocities computed by the FLO-2D model over the topography shown in (**c**)

three waves was tens of minutes to 1 h. Numerical simulations of the tsunami wave run-up on land based on the FLO-2D model, estimated the maximum overland velocity at over 20 m/s in the lower, wider segment of the main Aneyoshi valley (Fig. 13.4c, d). The modeled wave velocity still remained over 10 m/s along at low elevations of the narrow, meandering segment of the main valley.

13.5 Discussion and Conclusions

13.5.1 Conditions for Tsunami Wave Erosion

Although stripping of regolith and vegetation and related bedrock erosion have been observed after other recent tsunamis (e.g., Fritz et al. 2009; Paris et al. 2009), it remains uncertain as to what conditions are required to generate these effects. The effectiveness of erosion by water flow is a complex function of numerous factors, including velocity and duration, the mode of flow (wave front vs. continuous flow), and the type of sediment load (Shuto et al. 2007). In the case of bedrock erosion, the lithology and presence of fractures within the bedrock could also be important factors. Catastrophic flood flow dynamics and related impacts on natural environments, including bedrock erosion have been extensively studied (e.g., Baker et al. 1993; Komatsu et al. 2009). Depending on rock type and structural defects, the indicated flow velocities of 10–20 m/s for flow depths of no more than a few 10 s of meters, could prove to be very effective for erosion, perhaps even generating cavitation erosion of bedrock (Baker and Costa 1987). In addition, the impact of the 2011 Tohoku-oki Tsunami on the Aneyoshi site may indicate a threshold phenomenon, whereby the energy expenditures by incident tsunami waves have to be high enough to exceed resistance factors imposed by vegetation-stabilized regolith on hillsides.

While there is no direct record of the 2011 tsunami wave velocity at the Aneyoshi study site, Fritz et al. (2012) used a survivor's video to estimate that the velocity of backwash flow in Kesennuma Bay on the Sanriku Coast reached 11 m/s. The only estimate from a real-time observation in the vicinity of the study site was obtained from Yamada Bay south of Aneyoshi (Fig. 13.1), where the incoming wave front velocity may have reached as much as 32 m/s (Ohishi 2011). Our simulation (Fig. 13.4c) of the maximum run-up velocity on the land surface of Aneyoshi indicates that it easily reached 10 m/s and locally exceeded 20 m/s, values that are consistent with the extraordinary speed of the incoming wave front observed in the Yamada Bay. Although tsunami stripping of regolith and vegetation are evident at Aneyoshi, direct evidence for erosion of intact bedrock is lacking because there are no appropriate pre- and post-tsunami photograph sets for comparison. However, we do observe that the high-energy conditions of the waves, achieving velocities over 10 and 20 m/s locally during a period of tens of minutes to 1 h were strong enough to cause significant stripping of regolith and vegetation, thereby exposing the underlying bedrock interface.

13.5.2 Implications for Coastal Landscape Evolution

Although erosional features have been reported for recent tsunamis (e.g., Kench et al. 2008; MacInnes et al. 2009), considerable uncertainty remains as to which geomorphic features can be expected to persist over long time periods (e.g. Goff et al. 2009; Goto et al. 2012). This poses difficulties for the identification of paleo-tsunamis from geomorphic features (except for tsunami deposits). This also applies to mega-tsunamis caused by mechanisms such as landslides (e.g., Fritz et al. 2009), submarine volcanic eruptions (e.g., Simkin and Fiske 1984) and asteroid impacts (e.g., Wünnemann et al. 2007). Our observations at Aneyoshi can thus provide some insights as to what to assess when studying the erosional effects of large tsunamis.

Older tsunamis affecting the Sanriku Coast include well known events such as the 1933 Showa-Sanriku, the 1896 Meiji-Sanriku, the 1611 Keicho-Sanriku, and possibly the 869 Jogan (e.g., Mori et al. 2012) tsunamis, and less known probably minor ones (1960, 1915, 1894, 1867, 1856, ca. 1793, 1751, 1693, 1689, 1687, 1677, 1616, 1585; Shuto et al. 2007) recorded in historical documents during the last 400–500 years (this corresponds roughly to the Edo Period–present day). Little is known about the exact magnitude of each tsunami, but some tsunamis appear to have been significant, even comparable to the 2011 Tohoku-oki Tsunami for their impacts on the Sanriku coast (e.g., the1896 Meiji-Sanriku Tsunami, Shuto et al. 2007).

Combining our observations at Aneyoshi with these past tsunami records leads us to hypothesize that century- and possibly even decade-scale tsunami events constitute a recurrent hillslope erosion process in the Sanriku Coast. Moreover, the erosional impacts of tsunamis may extend further back in time at least until the mid-Holocene (c.a. 5,000 year B.P.) when the sea level peaked rapidly following deglaciation (e.g., Chida et al. 1984).

The time-scales of soil formation and its associated vegetation recovery are estimated to be in the order of 100–10,000 years (e.g., Friend 1992). The stripped and unprotected hillsides may become subject to enhanced long-term hillslope erosion, including mass wasting processes. Thus, the stripping-induced hillslope instability due to frequent large tsunamis, combined with possible bedrock erosion caused by high-energy tsunami waves, may be an important factor in geomorphological evolution of the Sanriku "ria" coastal landscapes. However, we currently lack sufficient data with which to evaluate this inference.

In conclusion, knowledge gained from the 2011 Tohoku-oki Tsunami suggests that tsunamis can locally be an important erosional geomorphic process along certain types of coastlines. The cumulative effects of tsunami erosion on the Sanriku Coast hillslopes and their long-term evolution are important potential topics for future studies.

Acknowledgements We thank an anonymous reviewer for the useful comments that improved the manuscript. This research was supported by the National Science Foundation-funded Rapid Response Research (RAPID) project (EAR-1138061), and by a research grant from the Tohoku University for an emergency field survey following the 2011 Tohoku-oki Tsunami.

References

Baker VR, Costa JE (1987) Catastrophic flooding. Allen and Unwin, Boston, pp 1–21

Baker VR, Benito G, Rudoy AN (1993) Paleohydrology of late Pleistocene superflooding Altai Mountains, Siberia. Science 259:348–350

Bryant EA, Young RW (1996) Bedrock-sculpturing by tsunami South Coast, New South Wales, Australia. J Geol 104:565–582

Chida N, Matsumoto H, Obara S (1984) Recent alluvial deposit and Holocene sea-level change on Rikuzentakata coastal plain, northeast Japan. Ann Tohoku Geogr Assoc 36(4):232–239 (in Japanese)

Dawson AG (1994) Geomorphological effects of tsunami run-up and backwash. Geomorphology 10:83–94

FLO-2D Software Inc (2009) FLO-2D reference manual, version 2009. FLO-2D Software Inc, Nutrioso

Friend JA (1992) Achieving soil sustainability. J Soil Water Conserv 47:156–157

Fritz H, Mohammed F, Yoo J (2009) Lituya Bay landslide impact generated mega-tsunami 50th anniversary. Pure Appl Geophys 166:153–175

Fritz HM, Phillips DA, Okayasu A, Shimozono T, Liu H, Mohammed F, Skanavis V, Synolakis CE, Takahashi T (2012) The 2011 Japan tsunami current velocity measurements from survivor videos at Kesennuma Bay using LiDAR. Geophys Res Lett 39:L00G23. doi:10.1029/2011GL050686

Goff JR, Lane E, Arnold J (2009) The tsunami geomorphology of coastal dunes. Nat Hazards Earth Syst Sci 9:847–854

Goto C, Ogawa Y, Shuto N, Imamura F (1997) IUGG/IOC time project, numerical method of tsunami simulation with the leap-frog scheme: IOC manuals and guides. UNESCO, Paris, 130 p

Goto K, Chague-Goff C, Fujino S, Goff J, Jaffe B, Nishimura Y, Richmond B, Suguwara D, Szczucinski W, Tappin DR, Witter R, Yulianto E (2011) New insights of tsunami hazard from the 2011 Tohoku-Oki event. Mar Geol 290:46–50

Goto K, Sugawara D, Abe T, Haraguchi T, Fujino S (2012) Liquefaction as an important local source of the 2011 Tohoku-Oki tsunami deposits at Sendai Plain, Japan. Geology 40:887–890

Imamura F, Koshimura S, Oie T, Mabuchi Y, Murashima Y (2011) Tsunami simulation for the 2011 off the Pacific coast of Tohoku Earthquake (Tohoku University model version 1.0). 12 pp

Kench PS, Nichol SL, Smithers SG, McLean RF, Brander RW (2008) Tsunami as agents of geomorphic change in mid-ocean reef islands. Geomorphology 95:361–383

Komatsu G, Arzhannikov SG, Gillespie AR, Burke RM, Miyamoto H, Baker VR (2009) Quaternary paleolake formation and cataclysmic flooding along the upper Yenisei River. Geomorphology 104:143–164. doi:10.1016/j.geomorph.2008.08.009

MacInnes BT, Bourgeois J, Pinegina TK, Kravchunovskaya EA (2009) Tsunami geomorphology: erosion and deposition from the 15 November 2006 Kuril Island tsunami. Geology 37:995–998

Mori N, Takahashi T, The 2011 Tohoku earthquake tsunami joint survey group (2012) Nationwide post event survey and analysis of the 2011 Tohoku earthquake tsunami. Coast Eng J 54:27. doi:10.1142/S0578563412500015

Nanayama F, Shigeno K (2006) Inflow and outflow facies from the 1993 tsunami in southwest Hokkaido. Sediment Geol 187:139–158

Nanayama F, Shigeno K, Satake K, Shimokawa K, Koitabashi S, Miyasaka S, Ishii M (2000) Sedimentary differences between the 1993 Hokkaido-nansei-oki tsunami and the 1959 Miyakojima typhoon at Taisei, southwestern Hokkaido, northern Japan. Sediment Geol 135:255–264

Nott J, Bryant EA (2003) Extreme marine inundation (tsunamis?) of coastal Western Australia. J Geol 111:691–706

Ohishi M (2011) Time-lapse photography of tsunami at Kawashiro, Omoe Peninsula, Miyako City and the maximum run-up height at Aneyoshi: The geological Society of Japan web site http://www.geosociety.jp/hazard/content0054.html

Paris R, Wassmer P, Sartohadi J, Lavigne F, Barthomeuf B, Desgages É, Grancher D, Baumert P, Vautier F, Brunstein D, Gomez C (2009) Tsunamis as geomorphic crisis: lessons from the December 26, 2004 tsunami in Lhok Nga, west Banda Aceh (Sumatra, Indonesia). Geomorphology 104:59–72

PARI (Port and Airport Research Institute) (2011) Results of the GPS-mounted wave buoys at Kuji, Miyako, and Onahama. http://www.pari.go.jp/files/3609/130613169.pdf

Shuto N, Satake K, Matsutomi H, Imamura F, Koshimura S (eds) (2007) Encyclopedia of Tsunami. Asakura publishing Co, 368 p (in Japanese)

Simkin T, Fiske RS (1984) Krakatau, 1883: the volcanic eruption and its effects. Smithsonian Institution Press, Washington, D.C., 400 p

Wünnemann K, Weiss R, Hofmann K (2007) Characteristic of oceanic impact induced large water waves – reevaluation of the tsunami hazard. Meteorit Planet Sci 42:1893–1903

Chapter 14
Concatenated Hazards: Tsunamis, Climate Change, Tropical Cyclones and Floods

Tom Beer, Debbie Abbs, and Oscar Alves

Abstract Concatenated Hazards refers to situations where one extreme event precipitates one or more other extreme events. The exemplar is the tsunami and the resulting nuclear accident that occurred in Japan following the 11 March 2011 magnitude 9 earthquake.

Australia's major natural hazards are hydro-meteorological in nature and have resulted in concatenated hazard events. An example is the 2011 Cyclone Yasi. The rainfall in the Australian tropics is due to the effects of the monsoonal wet season, augmented by the extra rainfall from the occasional tropical cyclone. Though tropical cyclones themselves can produce strong winds, storm surges, and floods – the combination of a particularly wet wet-season and a tropical cyclone can intensify the disaster and amplify the consequences. This was the situation on 3 February 2011 when Cyclone Yasi made landfall in North Queensland, following on a December-January period that had seen extensive flooding in Queensland as a result of a strong La Nina. The extra concatenation from Tropical Cyclone Yasi was the increase in Australian banana prices.

The predictive ocean–atmosphere model (POAMA) of the Centre for Australian Weather and Climate Research indicated in May 2010 that the wet season in Queensland would be extensive, with large amounts of rainfall. Tropical cyclone Yasi, though intense, had a well-behaved track and from 30 January was forecast to make landfall in Northern Queensland.

Model results indicate that the effects of climate change will be to decrease the numbers of tropical cyclones affecting Northern Australia and to increase the proportion of severe tropical cyclones affecting the region.

T. Beer (✉) • D. Abbs • O. Alves
Centre for Australian Weather and Climate Research, Partnership Between the Bureau of Meteorology and CSIRO, Melbourne, Australia
e-mail: Tom.Beer@csiro.au

Y.A. Kontar et al. (eds.), *Tsunami Events and Lessons Learned: Environmental and Societal Significance*, Advances in Natural and Technological Hazards Research 35, DOI 10.1007/978-94-007-7269-4_14, © CSIRO 2014

Keywords Concatenated hazards • Natural hazards • Climate change • Floods • Tropical cyclones • Tsunami

14.1 Concatenated Hazards

Concatenated Hazards refers to situations where one extreme event precipitates one or more other extreme events. The exemplar is the tsunami and the resulting nuclear accident that occurred in Japan following the 11 March 2011 magnitude 9 earthquake.

This particular example was triggered by the magnitude $M_w = 9.03$ earthquake, known as the Great East Japan Earthquake, the Tohoku earthquake, or the Tohoku-Oki earthquake that occurred at 14:46 JST (UTC+9). This was the most powerful known earthquake to have occurred in the region of Japan and is one of the most powerful earthquakes to have occurred since modern seismic record-keeping.

Ide et al. (2011) estimate the fault plane of the earthquake to be 440 × 220 km and Simons et al. (2011) note that the distribution of co-seismic fault slip exceeded 50 m in several places. Such an earthquake, which leads to a vertical displacement of the ocean floor, will create a tsunami. This is well known, and the announcements of the Japan Meteorological Agency that issued the most serious warning on its scale, namely a "Major Tsunami Warning", immediately after the 11 March earthquake probably saved many thousands of lives.

Thus at the simplest, and most well-known level, earthquakes and tsunamis are concatenated hazards. However, in the case of the 11 March earthquake, the size of the tsunami was of such unexpected magnitude that the tsunami proceeded to trigger other disasters. The Japan Meteorological Agency rates a major tsunami as one that is at least 3 m high. However, because local undersea topography and the shape of inlets and harbours can amplify incoming waves, the size of a tsunami wave that affects a community can vary greatly. It was estimated that the tsunami reached heights of up to 40.5 m in Miyako in Tohoku's Iwate Prefecture.

The damage by surging water proved to be more deadly and destructive than the earthquake itself. The confirmed death toll in September 2012 was 15,878 of which at least 12,143 died by drowning. Although Japan has heavily invested in anti-tsunami sea walls that line at least 40 % of the coastline and stand up to 12 m high, the tsunami simply washed over the top of some seawalls, collapsing some in the process. This over-topping was captured on numerous video images and news documentaries that were shown in the immediate aftermath of the tsunami. It would appear (Heki 2011) that neither seismologists nor oceanographers had expected an earthquake of such magnitude or a tsunami of such magnitude.

Though it is well-known that fire is a consequence of earthquakes, and indeed the 11 March earthquake triggered fires at two oil refineries, the nuclear power stations were prepared for such an eventuality and automatically shut down following the earthquake. When a reactor has been shut down, cooling is needed to remove residual heat and to maintain spent fuel. The backup cooling process is normally powered by emergency diesel generators but at Fukushima tsunami waves overtopped seawalls and destroyed backup power systems. This led to severe overheating at Fukushima

Dai-ichi resulting in three large explosions, the leakage of radioactive materials and the evacuation of over 200,000 people when the Japanese Government declared a state of emergency. Altogether six nuclear reactors developed problems with Fukushima Dai-ichi being the worst.[1]

14.2 Australia

Though Australia is subject to intra-plate earthquakes, its major natural hazards are hydro-meteorological in nature and are capable of producing concatenated hazard events. An Australian poet characterised Australia as a land "of droughts and flooding rains". The droughts make the countryside prone to wildfires, which are known in Australia as bushfires and the rains, as the poet emphasises, lead to floods.

The Centre for Australian Weather and Climate Research (CAWCR) is a partnership between the Australian Bureau of Meteorology and the Commonwealth Scientific and Industrial Research Organisation (CSIRO) Division of Marine and Atmospheric Research. CSIRO is the Australian Federal Government's major research agency which means that the partnership offers the possibility of bringing both research and operational expertise to bear on issues related to weather and climate. One of the areas in which this partnership has borne fruit is in the treatment of floods and tropical cyclones in which the research base of CSIRO has combined with the forecasting arm of the Bureau of Meteorology to improve the service offered to the Australian public.

The Australian public is concerned both with the occurrence of tropical cyclones in the immediate future, and in the longer term future when it is possible that climatic change may affect the distribution, landfall location, and intensity of tropical cyclones. CAWCR, through its Climate Variability and Change Research Program has the largest group of climate scientists in Australia working together to look at seasonal prediction, climate change and climate variability.

The rainfall in the Australian tropics is due to the effects of the monsoonal wet season, augmented by the extra rainfall from the occasional tropical cyclone. Though tropical cyclones themselves can produce strong winds, storm surges, and floods – the combination of a particularly wet wet-season and a tropical cyclone can intensify the disaster and amplify the consequences.

This was the situation on 3 February 2011 when Cyclone Yasi made landfall in North Queensland, following on a December-January period that had seen extensive flooding in Queensland as a result of a strong La Nina. The predictive ocean–atmosphere model (POAMA) of CAWCR indicated in May 2010 that the wet season in Queensland would be extensive, with large amounts of rainfall. Tropical cyclone Yasi, though intense, had a well-behaved track and from 30 January was forecast to make landfall in Northern Queensland.

[1] http://en.wikipedia.org/wiki/2011_T%C5%8Dhoku_earthquake_and_tsunami

Model results, discussed in detail in the Climate Change and Downscaling section, indicate that the effects of climate change will be to decrease the numbers of tropical cyclones affecting Northern Australia and to increase the proportion of severe tropical cyclones affecting the region.

14.3 Tropical Cyclones

On 25 December 1974, Tropical Cyclone Tracy[2] destroyed virtually all of the Northern Australian city of Darwin causing the deaths of 71 people (49 on land and 22 at sea) and the evacuation of 75 % of the city's residents. This event shocked the Australian public and encouraged the serious scientific study of Australian tropical cyclones.

This was not the first time Darwin had been severely damaged by a tropical cyclone: In both January 1897 and March 1937 the city was badly damaged, but only after Tracy was more attention given to building codes and other social aspects of disaster planning. Darwin was rebuilt and is now a thriving city of 128,100 people as at June 2011.

The Bureau of Meteorology[3] provides a database of past tropical cyclones, histories of tropical cyclones, and a library of individual cyclone reports. Australian Tropical Cyclones[4] are classified according to the modified Saffir-Simpson scale shown in Table 14.1, with Category 1 tropical cyclones having winds below 42 m/s but above 33 m/s, which is the minimum wind speed needed for a tropical storm to be classified as a tropical cyclone. Category 5 tropical cyclones are the most intense and will cause catastrophic damage to structures.

There are, on average, approximately 12 tropical cyclones per year that are identified as occurring within the Australian region. Of these about 40 % (~5) make landfall over the Australian continent. Tropical cyclones and tropical storms provide a large proportion of rainfall in tropical Australia that ranges from 40 % in tropical Queensland to 60 % in tropical Western Australia (Lavender and Abbs 2013). Tropical cyclone climatologies for Australia have been used to determine the tropical cyclone hazard.

Numerical weather prediction models have not, as yet, reached sufficiently fine resolution that they can predict the formation and subsequent strengthening and motion of a tropical cyclone. They are, however, able to identify tropical lows so that a sufficiently skilled forecaster is able to use such numerical weather prediction models, along with satellite photographs of tropical cyclone clouds that position the tropical cyclone, and thus use the two items of information to assist with forecasts of tropical cyclone tracks.

---

[2] http://www.bom.gov.au/cyclone/history/pdf/tracy.pdf

[3] http://www.bom.gov.au/cyclone/history/index.shtml

[4] http://www.bom.gov.au/cyclone/about/intensity.shtml

Table 14.1 Modified Saffir-Simpson tropical cyclone scale used by the Australian Bureau of Meteorology

Category	Wind description	Wind speed km/h	Typical effects
1	Gales	118–125	Minimal house damage. Damage to some crops, trees and caravans. Boats may drag moorings
2	Destructive winds	125–164	Minor house damage. Significant damage to signs, trees and caravans. Heavy damage to some crops. Risk of power failure. Small boats may break moorings
3	Very destructive winds	164–224	Some roof and structural damage. Some caravans destroyed. Power failure likely
4	Very destructive winds	225–279	Significant roofing and structural damage. Many caravans destroyed and blown away. Dangerous airborne debris. Widespread power failures
5	Extremely destructive winds	>280	Extremely dangerous with widespread destruction

Once the tropical cyclone has made landfall, there are four particular impacts that need to be considered: strong winds; extreme rainfall; the flooding associated with the cumulative rainfall; and the short term rise of sea level (known as storm surge).

During the severe wet season of January-February 2011 the eastern coast of Australia was affected by three tropical cyclones. Severe Tropical Cyclone Zelia[5] from 14 to 18 January 2011, Tropical Cyclone Anthony[6] from 22 to 31 January 2011, and Severe Tropical Cyclone Yasi[7] from 30 January to 3 February 2011. Figure 14.1 depicts the track of Severe Tropical Cyclone Yasi. The scale shows very destructive winds in red, destructive winds in pink, and gale force winds are shaded.

14.4 Concatenated Hazards and Cyclone Yasi

Concatenated Hazards refers to situations where one extreme event precipitates one or more other extreme events. In the case of TC Yasi (Fig. 14.1), a major portion of the Australian banana crop was wiped out causing extreme spikes in the banana price (Fig. 14.2) – repeating the situation of 2006 when Tropical Cyclone Larry made landfall on 20 March 2006 and also destroyed 80–90 % of Australia's banana crop. Australia is relatively free of banana pests and diseases, and therefore does not allow bananas to be imported. Bananas were in short supply throughout Australia for the remainder of both 2011 and 2006, which increased prices across the country by 400–500 %.

[5] http://www.bom.gov.au/cyclone/history/zelia11.shtml

[6] http://www.bom.gov.au/cyclone/history/anthony.shtml

[7] http://www.bom.gov.au/cyclone/history/yasi.shtml

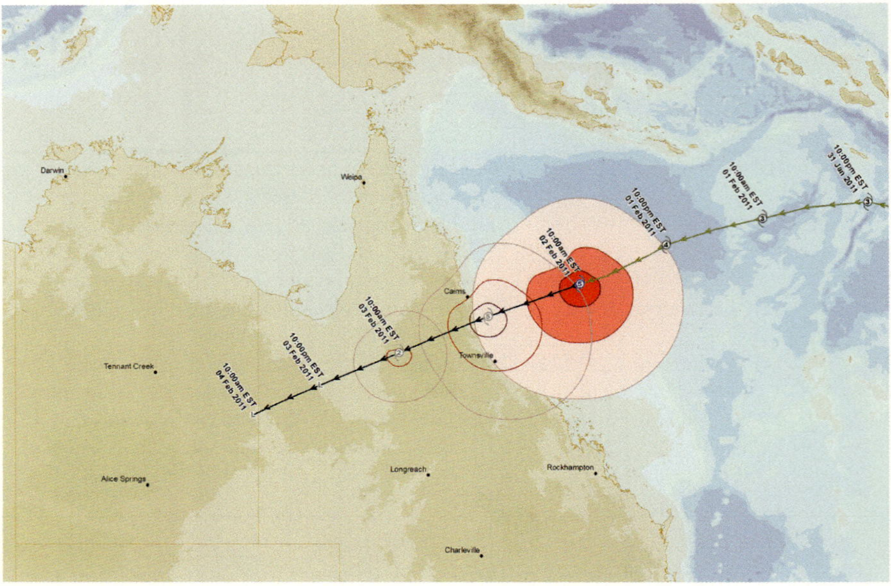

Fig. 14.1 Forecast track of TC Yasi as at 4 am on 2 February 2011. An animated version of this is available at: http://www.bom.gov.au/cyclone/history/yasi.shtml#loops

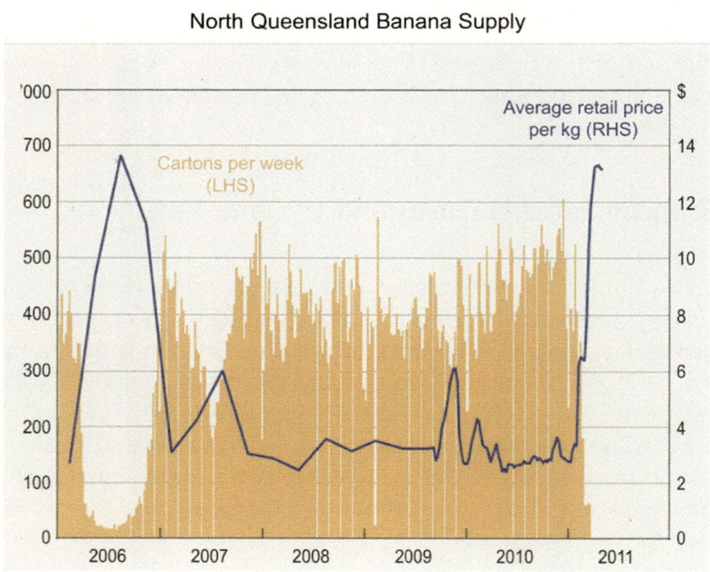

Fig. 14.2 Graph of North Queensland Banana Price and Banana supply indicating the effects of Cyclone Larry in 2006 and Cyclone Yasi in 2011 (Reserve Bank of Australia 2011)

14.5 Floods

The term "flood" covers a large variety of different hydrological events that can range from local flooding due to a blocked drain or culvert; short-term flash flooding of creeks as a result of a heavy, but localised storm; to the long-term flooding that occurs as rivers over-top their banks and inundate the floodplain that surrounds them. The dramatic effects of localised severe storms and flash flooding were graphically illustrated in January 2011 when the continuation of heavy rainfall that had started over much of Queensland in December 2010 caused severe flash flooding in the centre of the City of Toowoomba.

The Flood Commission of Inquiry in its interim report[8] states that:

> *On 10 January, the Gowrie Creek catchment experienced intense rainfall between 1.00 pm and 2.30 pm. In the city of Toowoomba itself, heavy rain began falling at about 12.45 pm, and peaked between 1.45 pm and 2.15 pm.*
>
> *The most severe rain fell in a northeast - southwest band that covered the middle and lower parts of East and Westcreeks, where they crossed Toowoomba's central business district 21. This concentration of rain in the East Creek and West Creek catchments continued for approximately 60 to 90 minutes. It had largely ceased between 2.15 pm and 2.45 pm.*
>
> *The intense rainfall over the catchment of the three creeks caused a severe flash flood in the city between 1.30 pm and 2.45 pm. Closed circuit television footage provided by Toowoomba Regional Council shows water rising at extraordinary speed and flowing over the roadways. It also demonstrates the speed with which the water rose. It is clear that this was not a situation in which any agency could have effectively warned residents of what was to come.*
>
> *Water covered all the roadway crossings of East, West and Gowrie Creeks, making them impassable to pedestrians and vehicles. The rapidity of the flooding caught people by surprise: in city streets they found themselves surrounded by water, or were trapped in their vehicles. A woman and her teenaged son lost their lives when their car was caught in the flooding in a city intersection. A number of buildings in and around the city were extensively damaged, and numerous parked cars were swept away or inundated by the flooding.*

On the same day the Lockyer creek, in the Lockyer Valley to the east of Toowoomba flooded and residents suggest that flooding in the town of Grantham occurred between approximately 3.20 pm and 4.00 pm. The flood appeared as a wave, sweeping from the Lockyer Creek across the paddocks and through the town. Thirty-six people died in the 2010/2011 floods with an estimated reconstruction cost of $5 billion.

14.6 Seasonal Forecasting

The Bureau of Meteorology has developed a seasonal prediction model known as POAMA, which is an acronym for the Predictive Ocean Atmosphere Model for Australia. It is a dynamic computer model of the climate system run on the operational super-computer that is used to generate weather forecasts. Because the

[8] http://www.floodcommission.qld.gov.au/publications/interim-report

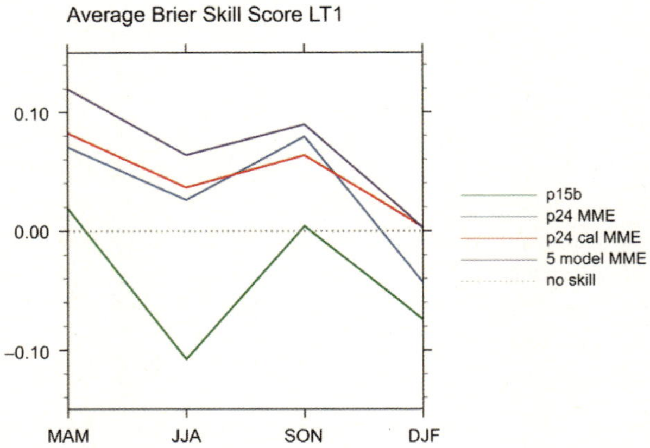

Fig. 14.3 Australian rainfall Brier skill score for POAMA1.5 (p15b), POAMA 2.4 (P24 MME), a re-calibrated version of POAMA2.4 (P24 cal MME), and a five member multi-model ensemble that incorporates POAMA 2.4a, POAMA 2.4b, POAMA 2.4c plus ECSys3 and the UKMO HadGEM2 seasonal forecast (5 model MME) initialised on 1 May 2010 for the October-December 2011 forecast

weather and climate of the eastern part of Australia is strongly influenced by El Nino (McBride and Nicholls 1983), a significant product generated by POAMA consists of model forecasts of the El Niño - Southern Oscillation. They are provided as an operational product on the web site of the Bureau of Meteorology and are included in the monthly model summary[9] of predictions from POAMA and other models operated by international organisations. Because the south-eastern Australian climate is also influenced by the Indian Ocean Dipole (IOD), forecasts of the IOD are also provided.[10]

The latest operational version of POAMA is POAMA2 (version 2.4). A 30 member ensemble of POAMA-2 forecasts is run twice per month and gives forecasts out to 9 months ahead. Probabilities are based on a 30-member ensemble from POAMA-2 starting on the 1st and the 15th of the month. The results, which are forecast out to 9 months lead, show all the runs so that the probability distributions that result provide a range of possible developments in sea surface temperature (SST) in the equatorial Pacific Ocean[11] and for the Indian Ocean. More detailed information about the model is available on the experimental POAMA page: http://poama.bom.gov.au/. Further details on POAMA 2 SST skill can be found in Wang et al. (2011). A separate system (POAMA-2 Multiweek) is now run every week in experimental mode to produce an ensemble of 33 members also out to 9 months. This system is tailored to forecasting on multi-week/monthly timescales and includes a new coupled breeding method.

The performance of the POAMA model is illustrated in Fig. 14.3 which depicts the Brier skill score[12] for POAMA1.5, POAMA 2.4, a re-calibrated version of

[9] http://www.bom.gov.au/climate/ahead/ENSO-summary.shtml

[10] http://www.bom.gov.au/climate/IOD/about_IOD.shtml

[11] NINO regions http://www.bom.gov.au/climate/enso/nino_regions_map.jpg

[12] http://en.wikipedia.org/wiki/Brier_score

POAMA2.4, and a five member multi-model ensemble that incorporates POAMA 2.4a, POAMA 2.4b, POAMA 2.4c plus ECSys3 (Stockdale et al 2011) and the UKMO HadGEM2 seasonal forecast (Arribas et al 2011) initialised on 1 May 2010 for the October-December 2011 forecast. Details of skill over Australia, including a comparison with other international models can be found in Langford and Hendon (2011).

14.7 POAMA and the Beijing Floods of 21 July 2012

On Saturday 21 July 2012 Beijing suffered the heaviest rainfall for over 60 years with the average precipitation reaching 170 mm, while a town in the suburban district of Fangshan, received 460 mm of rain. The storm was widespread. In Hebei province as at 26 July 2012, 32 people were confirmed dead and another 20 missing after the storm over the weekend. More than 2.66 million people had been directly affected by the storm that flooded 59 counties in Hebei province. The official death toll until 27 July was reported as 37 people after which it was raised to 77 people. Direct economic losses totalled more than 12.28 billion yuan ($1.92 billion).

Using archived POAMA output, we have examined (Fig. 14.4) the performance of POAMA in being able to forecast high rainfall in the Beijing area. POAMA

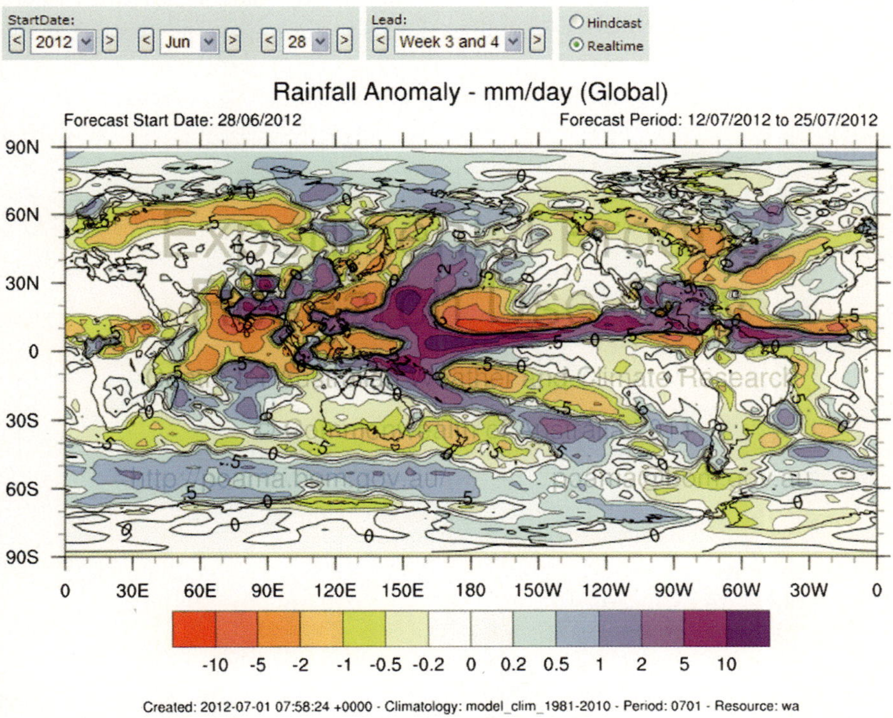

Fig. 14.4 POAMA forecast made on 28 June 2012 predicted heavy rainfall in the Beijing area for the 12 July–25 July 2012 period

consistently provided forecasts of high rainfall (exceeding 10 mm/day) for the Beijing region as from 28 June 2012 indicating that there was approximately 3 week predictability for this event from a seasonal prediction model. The Bureau of Meteorology numerical weather prediction model, called ACCESS, was able to provide high resolution analysis 48 h before the storm that indicated that localised precipitation in excess of 100 mm/day was to be expected.

14.8 May 2010 POAMA Forecast

Figure 14.5 depicts the observed (green curve) and the ensemble of POAMA2.4 forecasts indicated by the dark green curve and the shading around it for the Nino34 SST index during 2010 and early 2011. The forecasts were initialized on 1 May 2010. The strong La Niña during 2010, which contributed to extreme rainfall in eastern Australia during spring and early summer, was not predictable prior to May 2010, but by May 2010 the POAMA2.4 forecasts were clearly indicating the development of a strong cold event that would extend for at least the next 9 months.

POAMA also provides information on the spatial distribution of expected rainfall. Figure 14.6 shows the observed (Fig. 14.6a) and forecast (Fig. 14.6b) rainfall anomaly for October-December 2011 for Australia. The forecast is the

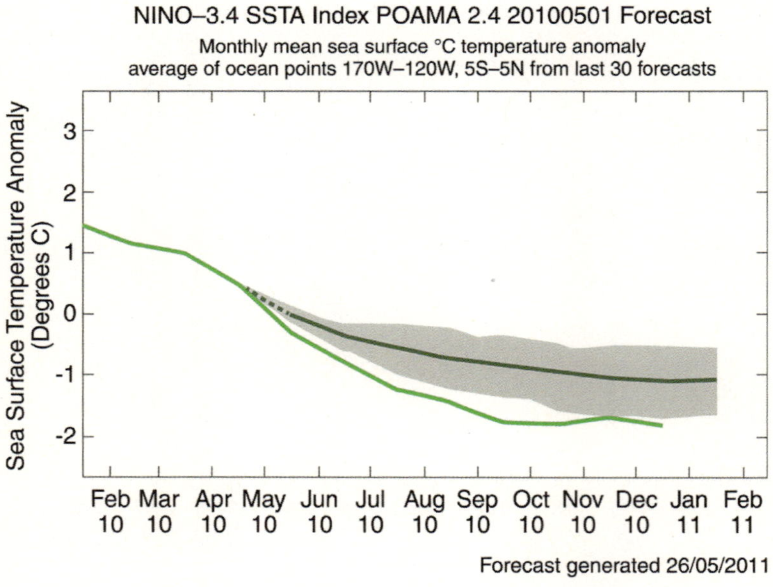

Fig. 14.5 May 2010 forecasts of the El Nino index from POAMA indicating that strong La Nina conditions were to be expected at least the next 9 months. La Nina conditions are indicative of above normal rainfall in Eastern Australia

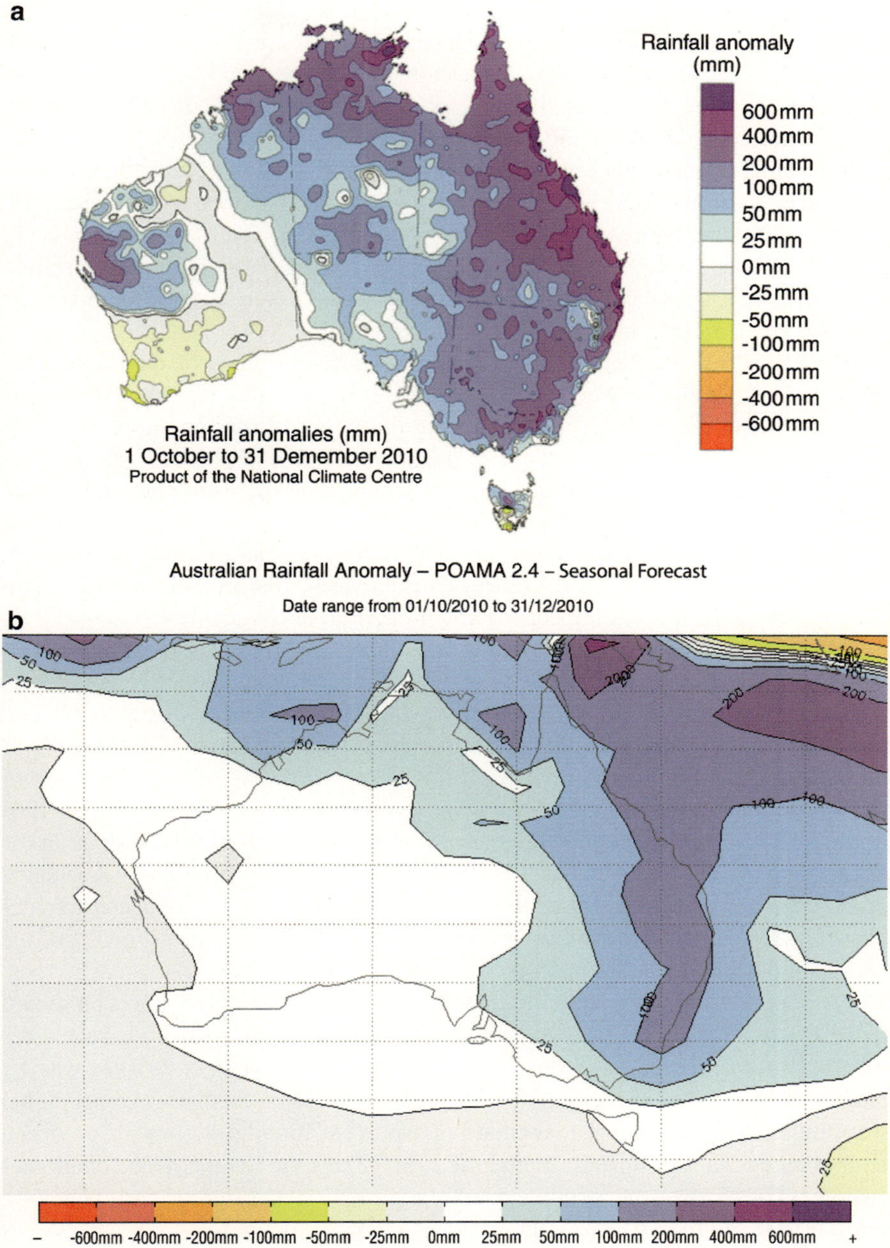

Fig. 14.6 (**a**) Observed rainfall anomalies for October to December 2011. (**b**) POAMA Forecast of rainfall anomalies for October to December 2011

30 member ensemble mean from the POAMA2.4 seasonal forecast that was initialized on 1 May 2010. The color bars are the same for Fig. 14.6a, b. Extreme wet conditions in eastern Australia for October-December 2011 were well forecast from at least 1 May 2010 due to the development of a strong La Niña event that was well predicted by the POAMA2.4 model, as shown in Fig. 14.6b.

In fact this forecast was very accurate with extreme rainfall occurring over much of Eastern Australia. The dramatic effects of localised severe storms and flash flooding on topography that had been previously moistened was graphically illustrated in January 2011 when the continuation of heavy rainfall that had started over much of Queensland in December 2010 caused severe flash flooding in the centre of the City of Toowoomba.[13]

The record rainfalls and flooding experienced across much of the region, and throughout Australia, in the 2010/11 and 2011/12 summers highlighted that the tropical Pacific ocean temperatures are not the only sea surface temperatures of importance but one needs to consider the status of all three oceans surrounding Australia – Pacific, Indian and Southern – in influencing seasonal and inter-annual rainfall variability. The spring/summer of 2010/11 saw one of the strongest La Niña events on record combined with a negative IOD event and a positive Southern Annular Mode (SAM) – i.e. all three key influences were in their wet phases in terms of expected rainfall impacts on south-eastern Australia.

SAM describes the north–south movement of the westerly wind belt that circles Antarctica, dominating the middle to higher latitudes of the Southern Hemisphere. The changing position of this westerly wind belt influences the strength and position of cold fronts and mid-latitude storm systems, and is an important driver of rainfall variability in southern Australia. In a positive SAM event, the belt of strong westerly winds contracts towards Antarctica. This results in weaker than normal westerly winds and higher pressures over southern Australia, restricting the penetration of cold fronts inland. The positive phase of SAM is typically associated with wetter and cooler conditions over much of Australia during spring and summer but with drier and cooler conditions over the southwest and southeast coasts of the continent during winter (Hendon et al. 2007).

These conditions, coupled with the warmest sea-surface temperatures on record to the north of the Australian continent, contributed to making 2010–11 Australia's wettest two-year period on record. While the extensive flooding of 2010/11 was, by and large, consistent with natural variability, it is possible that ongoing global warming contributed to the magnitude of the event through its impact on ocean temperatures. The spring and summer of 2011/12 saw the re-emergence of another (weaker) La Niña event, combined with a positive SAM during early summer, but this time the Indian Ocean played a lesser role.

[13] http://www.floodcommission.qld.gov.au/publications/interim-report

14.9 Climate Change and Downscaling

A question of continuing relevance relates to the influence of climate change on tropical cyclone numbers, intensity and landfall location. Callaghan and Power (2011) examined the statistics of past Queensland tropical cyclones and found that the number of tropical cyclones making landfall over eastern Australia declined from about 0.45 TCs/year in the early 1870s to about 0.17 TCs/year in recent times. They noted that this decline can be partially explained by a weakening of the Walker Circulation, and a natural shift towards a more El Nino-dominated era. The extent to which global warming might be also be partially responsible for the decline in land-falls – if it is at all – is unknown.

This question has been examined by using dynamical downscaling of computer models (Daloz et al. 2012). In particular the output from the 200 km resolution CSIRO Mark 3.6 general circulation model has been nested into the 65 km output of CCAM in which tropical cyclone-like vortices are detected. The details of these vortices are then elucidated using mesoscale models such as RAMS and WRF that have 5 km resolution. Using this method, simulations were undertaken for the present climate and for the end of the twenty-first century using the SRES A2 scenario. Based on 11 simulations there is a strong signal confirming a decrease in tropical cyclone numbers in the Australasian region – both north-eastern Australia and north-western Australia.

The maximum wind speed of tropical cyclones increases from about 32 m/s to about 37 m/s and associated with this increase in maximum wind speed there is a marked increase in the maximum integrated kinetic energy of tropical cyclones. At the moment the probability distribution of maximum integrated kinetic energy has a mode (consisting of 16 % of tropical cyclones) at 55 TJ. This mode lies at 75 TJ in the simulations for 2070. The radius of maximum winds is also increased from 90 km at present to 130 km in 2070.

Table 14.2 depicts the changes in precipitation between the situation in 2010 and the situation in 2070 at various distances from the centre of the tropical cyclone. In all cases except one both the average rainfall intensity and the maximum rainfall intensity increases. The exception is the average rainfall at 100 km from the centre of the tropical cyclone which decreases by 9 %. These results are all consistent with the concept that under climate change tropical cyclones will increase in size and be stronger.

Table 14.2 Projected changes in precipitation over the most intense 12 h of the tropical cyclone

Distance from tropical cyclone centre (km)	Change in average rainfall intensity (%)	Change in maximum rainfall intensity (%)
100	−9	28
200	23	30
300	33	35
400	30	44

The projected changes between 2010 and 2070 in the number, duration and location of Australian tropical cyclones obtained by downscaling seven general circulation models is shown in Table 14.3.

The models are all consistent in indicating that the number of tropical cyclones will decrease. Though the occasional model may produce different results there is also an overwhelming consensus that the duration of tropical cyclones will decrease – though all models appear to indicate overly short duration times. There is also a consensus that tropical cyclones move equatorward – both in terms of the location of their genesis, and the location of their decay (which, presumably, approximates to their landfall). These results apply both to tropical cyclones in the south western Pacific Ocean, which are the ones that affect Queensland on the east coast of Australia, and to tropical cyclones in the southern Indian Ocean, which are the ones that affect Western Australia.

14.10 Summary and Conclusion

Advances in numerical weather prediction, combined with information on the drivers of Australian climate have led to the production of a seasonal forecasting model, known as POAMA, that has displayed considerable skill in the production of multi-week and seasonal forecasts. POAMA could have been used to provide about 3 weeks warning of the 21 July 2012 heavy precipitation in Beijing.

POAMA was also successful in its May 2010 predictions of the La Nina that led to extreme rainfall in eastern Australia during the last quarter of 2010 and the first quarter of 2011. Both the magnitude and spatial extent of the rainfall were accurately reproduced. This time period corresponded to extreme flash flooding in parts of Queensland, and flooding in Brisbane, the capital of Queensland. During a La Nina event, there is a tendency for tropical cyclone numbers to increase – and there were three tropical cyclones that made landfall in Queensland over this period. Cyclone Yasi, the most memorable, destroyed the Queensland banana crop.

By using dynamical downscaling techniques it is possible to use general circulation models to investigate the impact of climate change on the likely changes in tropical cyclone numbers, intensity, duration and location.

Climate change models are all consistent in indicating that the number of tropical cyclones will decrease (Abbs 2012) such that on average for the period 2051–2090 there will be an approximately 50 % decrease compared to the 1971–2000 period. There is also an overwhelming consensus that the duration of Australian tropical cyclones will decrease – though all models appear to indicate overly short duration times. There is also a consensus that Australian tropical cyclones move equatorward – both in terms of the location of their genesis, and the location of their decay (which, presumably, approximates to their landfall).

Table 14.3 Projected changes between 2010 and 2070 in the number, duration and location of Australian tropical cyclones obtained by downscaling seven general circulation models

	South West Pacific region					Southern Indian Ocean region				
Model	Number (%)	Days (%)	Duration (Days)	Genesis (°Lat)	Decay (°Lat)	Number (%)	Days (%)	Duration (Days)	Genesis (°Lat)	Decay (°Lat)
ECHAM5	−58	−53	0.0	−2.5	−1.5	−57	−62	−0.6	−0.1	1.0
GFDL CM2.0	−51	−48	0.2	−1.3	−1.9	−54	−62	−1.0	−0.2	−0.2
GFDL CM2.1	−50	−62	−1.2	−1.7	−0.7	−37	−39	−0.9	−1.4	−1.8
MIROC3.2	−87	−90	−1.3	−0.9	−0.8	−58	−64	−0.8	−1.6	−2.5
CSIRO MK3.5	−40	−31	0.5	−0.2	0.0	−39	−49	−0.7	−0.6	−0.2
UK HADCM3	−54	−55	−0.1	−1.7	−1.4	−47	−43	0.4	−0.1	−1.9
CSIRO MK3.0	−9	11	0.7	−0.5	−2.9	−44	−49	−0.3	−0.5	−0.2
Ensemble Average	−50	−40	−0.2	−1.3	−1.3	−48	−53	−0.6	−0.6	−0.8

The POAMA and other seasonal forecast models, as well as climate change models – especially downscaled general circulation models - could provide predictions of the likelihood of concatenated hazards such as tropical cyclones and agricultural losses or severe storms and flash floods. This knowledge could be used to prepare for and reduce the effects of these types of hazards.

References

Abbs D (2012) The impact of climate change on the climatology of tropical cyclones in the Australian region. CSIRO Climate Adaptation Flagship Working Paper No. 11. http://www.csiro.au/en/Organisation-Structure/Flagships/Climate-Adaptation-Flagship/CAF-working-papers.aspx

Arribas A et al (2011) The GloSea4 ensemble prediction system for seasonal forecasting. Mon Weather Rev 139:1891–1910

Callaghan J, Power SB (2011) Variability and decline in the number of severe tropical cyclones making land-fall over eastern Australia since the late nineteenth century. Clim Dyn 37:647–662. doi:10.1007/s00382-010-0883-2

Daloz AS, Chauvin F, Walsh K, Lavender S, Abbs D, Roux F (2012) The ability of general circulation models to simulate tropical cyclones and their precursors over the north Atlantic main development region. Climate Dynam 39:1559–1576. doi:10.1007/s00382-012-1290-7

Heki K (2011) A tale of two earthquakes. Science 332(6036):1390–1391

Hendon H, Thompson DWJ, Wheeler MC (2007) Australian rainfall and surface temperature variations associated with the Southern Hemisphere annular mode, J. Clim., 20:2452–2467

Ide S, Baltay A, Beroza GC (2011) Shallow dynamic overshoot and energetic deep rupture in the 2011 Mw 9.0 Tohoku-Oki earthquake. Science 332(6036):1426–1429

Langford S, Hendon H (2011) Assessment of international seasonal rainfall forecasts for Australia and the benefit of multi-model ensembles for improving reliability. CAWCR Technical Report No. 039

Lavender SL, Abbs DJ (2013) Trends in Australian rainfall: Contribution of tropical cyclones and closed lows, Clime Dynam 40:317–326. doi:10.1007/s00382-012-1566-y

McBride JL, Nicholls N (1983) Seasonal relationships between Australian rainfall and the Southern oscillation. Mon Weather Rev 111:1998–2004

Reserve Bank of Australia (2011) Statement on Monetary Policy – Box B: An update on the impact of the natural disasters in Queensland, Reserve Bank of Australia, Sydney, May 2011

Simons M, Minson SE, Sladen A, Ortega F, Jiang J, Owen SE, Meng L, Ampuero JP, Wei S, Chu R (2011) The 2011 magnitude 9.0 Tohoku-Oki earthquake: mosaicking the megathrust from seconds to centuries. Science 332(6036):1421–1425

Stockdale TN, Anderson DLT, Balmaseda MA, Doblas-Reyes F, Ferranti L, Mogensen K, Palmer TN, Molteni F, Vitart F (2011) ECMWF seasonal forecast system 3 and its prediction of sea surface temperature. Clim Dynam 37:455–471. doi:10.1007/s00382-010-0947-3

Wang G, Hudson D, Ying Y, Alves O, Hendon H, Langford S, Liu G, Tseitkin F (2011) POAMA-2 SST skill assessment and beyond. CAWCR Res Lett 6:40–46

Chapter 15
Domino Effects and Industrial Risks: Integrated Probabilistic Framework – Case of Tsunamis Effects

Ahmed Mebarki, Sandra Jerez, Igor Matasic, Gaëtan Prodhomme, Mathieu Reimeringer, Vincent Pensee, Quang Anh Vu, and Adrien Willot

Abstract This paper presents an integrated probabilistic framework that deals with the industrial accidents and domino effects that may occur in an industrial plant. The particular case of tsunamis is detailed in the present paper: simplified models for the inundations depths and run-ups as well as their mechanical effects on industrial tanks.

The initial accident may be caused by severe service conditions in any of the tanks either under or at atmospheric pressure, or triggered by a natural hazard such as earthquake, tsunami or extreme floods for instance. This initial event generates, in general, a set of structural fragments, a fire ball, a blast wave as well as critical losses of containment (liquid and gas release and loss). The surrounding facilities may suffer serious damages and may also be a new source of accident and explosion generating afterwards a new sequence of structural fragments, fire ball, blast wave and confinement loss. The structural fragments, the blast wave form and the features of the fire ball can be described following database and feedback collected from past accidents.

The surrounding tanks might be under or at atmospheric pressure, and might be buried or not, or protected by physical barriers such as walls. The vulnerability of the potential targets should therefore be investigated in order to assess the risk of propagation of the accidents since cascading sequences of accidents, explosions and fires may take place within the industrial plant, giving rise to the domino effect that threatens any industrial plant.

A. Mebarki (✉) • I. Matasic • V. Pensee • Q.A. Vu
Université Paris-Est, Lab. Modélisation et Simulation Multi Echelle UMR 8208 CNRS, 5 Bd Descartes, 77454, Marne-La-Vallée, France
e-mail: Ahmed.Mebarki@univ-paris-est.fr

S. Jerez
Escuela Colombiana de Ingeniería, Av. 13 No.205-59, Bogotá, Colombia

G. Prodhomme • M. Reimeringer • A. Willot
INERIS, Institut National de l'Environnement Industriel et des Risques, Parc Technologique ALATA, BP 2 – 60550, Verneuil-en-Halatte, France

Y.A. Kontar et al. (eds.), *Tsunami Events and Lessons Learned: Environmental and Societal Significance*, Advances in Natural and Technological Hazards Research 35,
DOI 10.1007/978-94-007-7269-4_15,

The present research describes the risk of domino effect occurrence. The methodology is developed so that it can be operational and valid for any industrial site. It is supposed to be valid for a set of sizes, forms and kinds of tanks as well as a given geometric disposal on the industrial site. The interaction and the behavior of the targets affected or impacted by the first explosion effects should be described thanks to adequate simplified or sophisticated mechanical models: perforation and penetration of metal fragments when they impact surrounding tanks, as well as global failure such as overturning, buckling, excessive bending or shear effects, etc. The vulnerability analysis is detailed for the case of tanks under the mechanical effects generated by tsunamis.

Keywords Tsunamis • Industrial accidents • Explosions • Domino effect • Atmospheric tank • Tank under pressure • Risk of failure

15.1 Introduction

The important quantity of hazardous substances which are produced, confined or treated in industrial plants may generate, under given severe conditions, explosions, fires and fragmentations of the tanks where they are stored or of the pipelines in which they are transported. Due to internal or external causes, an initial sequence of a severe accident may be triggered and may propagate affecting the tanks, pipelines, power lines and facilities erected in their vicinity. This propagation may result in catastrophic consequences: structural damages as well as human losses or severe injuries.

Actually, the existing literature and reports on past accidents show that industrial accidents may generate at the same time blast waves, projection of structural fragments (in general, parts of exploded tanks), fireballs causing thermal radiation or thermal effects as well as loss of confinement with ejection of gases and liquids, that may be flammable or toxic. (Abbasi and Abbasi 2007; ARIA (website); Holden 1988; Lees 2005).

The analysis and modeling of the whole events and sub-sequent events and effects (mechanical, thermal, chemical, etc.) is a complex scientific and multidisciplinary challenge, (Abbasi and Abbasi 2007; Ali and Li 2008; Antonioni et al. 2009; Børvik et al. 2003; Corbet et al. 1995; Cozzani and Salzano 2004; Marhavilas et al. 2011; Mebarki et al. 2009a, b; Mingguang and Juncheng 2008; Neilson 1985; Ohte et al. 1982; Ruiz et al. 1989; Seveso Inspection Tool 2009; Talaslidis et al. 2004; TNO 2005a, b; Tsamopoulos 2004; Università degli Studi di Torino; van den Berg 1985; Xie 2007):

- *Triggering event*: description of the triggering event (the first accident in any of the existing tanks within an industrial plant). The probability of occurrence as well as the thermodynamic and mechanical conditions for the first accident has to be well known and described. Past accidents and existing databases may also be helpful. The case of tsunamis generated by earthquakes is detailed in this paper.

- *Subsequent effects and propagation*: detailed description of the sub-events such as fragments generation and ejection, blast wave and the pressure front, fireballs and the thermal front and flows, and ejection of gases and liquids (loss of confinement/containment) requires mechanical modeling and also probabilistic distribution of the involved parameters (input and output).
- *Interaction and effects on the surrounding tanks and facilities*: simplified or sophisticated behavior of the affected targets and their response to the sub-events are required either by analytical, by testing or numerical approaches. For instance, partial penetration or perforation of the impacted tanks should be adequately evaluated as well as the local or global damages suffered by the target facilities (tanks, pipelines, etc.). The same requirements hold for the other sub-events such as blast waves and fireballs. The confinement losses may cause indirect effect such as fire ignitions, for instance.
- *Successive sequences of accidents*: the series of events and sub-events that take place within the industrial plant once a triggering event hits any of the existing tanks that are potential sources for domino effect initiation require time and dynamic analysis. Numerical analyses are helpful.

Once the domino effect is studied in deep details, its socio-economic consequences combined to the probability of its occurrence result in an expected cost of the domino effect. Investments and protective measures can then objectively be evaluated through an optimization of the global generalized cost. This concept is very helpful for the stakeholders in order to mitigate the potential disaster, and protect adequately the most sensitive and highly strategic installations.

15.2 Integrated Probabilistic Framework for Industrial Explosions and Domino Effects

15.2.1 General Plant: set of Tanks as Sources of Industrial Accidents and Sequence of Accidents

Let us consider any industrial site that may contain several tanks, either under pressure or at atmospheric pressure, see Fig. 15.1. Each of these tanks can be the source of accident or explosion due to either internal or external causes:

- *Internal causes*: critical corrosion, weakened welding, excessive cracking, over pressure or a critical temperature of the stored gas or liquid, handling accidental damage to critical, etc.
- *External causes*: malicious or malevolent acts, natural events such as strong earthquakes or tsunamis as well as extreme floods, explosions or impacts, fires and thermal effects, lightning, etc.

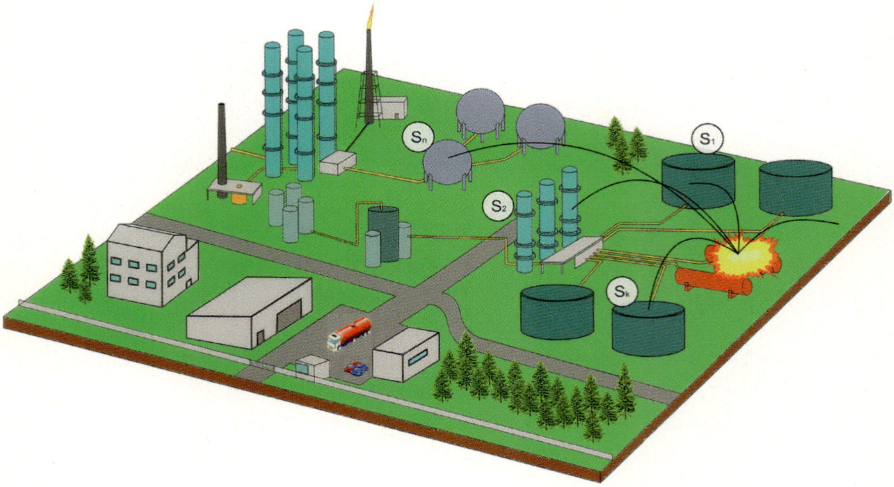

Fig. 15.1 First accident sequence within an industrial plant and its propagation

Therefore, each tank can be denoted as a Source "S" of industrial hazard as it may be damaged, may also explode and may generate threats (mechanical, chemical, thermal, etc.) to the surrounding facilities, and other tanks. Let us then denote these potential sources of industrial hazard as sources S(i) with $i = 1$ up to N_s, N_s being the number of tanks erected within the industrial site under study.

When an accident affects a given source S(i), such as an explosion or fire for instance, it can generate one or various events, i.e.:

- A set of structural fragments (plates, end-cups...): each generated fragment can be ejected from the tank source and become therefore a projectile,
- A fire ball and thermal effects,
- A blast wave, and
- A loss of confinement (containment: gas and liquid losses).

Let us denote $E_1^{\ g}$ the event that corresponds to occurrence of any first accident (explosion for instance, impact, fire, etc.) that occurs within the entire industrial plant at a given starting instant t_0. Its probability of occurrence should be evaluated during any given reference period, T_{ref}, that may correspond for instance to the expected plant lifetime. This probability is defined during the entire reference period as:

$$P\left(E_1^g\right) = P\left(E_1^g\big|_{T_{REF}}\right), \text{during the reference period } T_{REF} \tag{15.1}$$

If this first accident gives rise to any of the subsequent events among fragments, fire ball, blast wave and confinement loss, one should calculate the risk of such propagation event $E_1^{\ propa}$ since they may generate threats against the surrounding facilities (tanks, etc.). The probability of the propagation event is defined as:

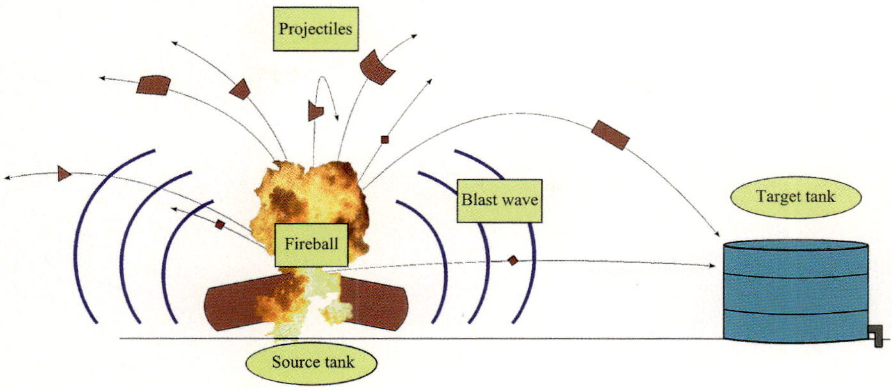

Fig. 15.2 BLEVE explosion generating projectiles, fire ball as well as blast wave

$$P\left(E_1^{propa}\right) = P\left(E_1^{propa}\big|_{E_1^g}\right).P\left(E_1^g\right) \tag{15.2}$$

In fact, each explosion may also generate simultaneously these phenomena, i.e. fragments, blast wave, fire ball and confinement loss, see Fig. 15.2.

Each individual event effect or their combined effects may therefore damage the surrounding tanks (or facilities, buildings, etc.). The event $E_1{}^{damage}$ that corresponds to damage caused to the surrounding tanks has its occurrence probability defined as:

$$P\left(E_1^{damage}\right) = P\left(E_1^{damage}\big|_{E_1^{propa}}\right)P\left(E_1^{propa}\big|_{E_1^g}\right).P\left(E_1^g\right) \tag{15.3}$$

For the case where the tanks erected in the vicinity of the initial accidents are affected by the propagation event and are therefore suffering mechanical damages, they may then give rise to a new sequence and cascading accidents for instance. One should say that a new sequence $E_2{}^g$ is taking place within the industrial plant, triggering the secondary sequence of explosions and accidents, leading then to the rise of the so-called "domino effect". Its probability of occurrence is therefore estimated as, see Fig. 15.3:

$$P\left(E_2^g\right) = P\left(E_2^g\big|_{E_1^{DAMAGE}}\right).P\left(E_1^{damage}\big|_{E_1^{PROPA}}\right).P\left(E_1^{propa}\big|_{E_1^G}\right).P\left(E_1^g\right) \tag{15.4}$$

For the general purpose, one could say that any sequence $E_n{}^g$, where n is the range of the accidents sequence ($n \geq 1$), can give rise to an additional sequence ($n + 1$) with a probability defined as:

$$P\left(E_{n+1}^g\right) = P\left(E_{n+1}^g\big|_{E_n^{damage}}\right).P\left(E_n^{damage}\big|_{E_n^{propa}}\right).P\left(E_n^{propa}\big|_{E_n^g}\right).P\left(E_n^g\right) \tag{15.5}$$

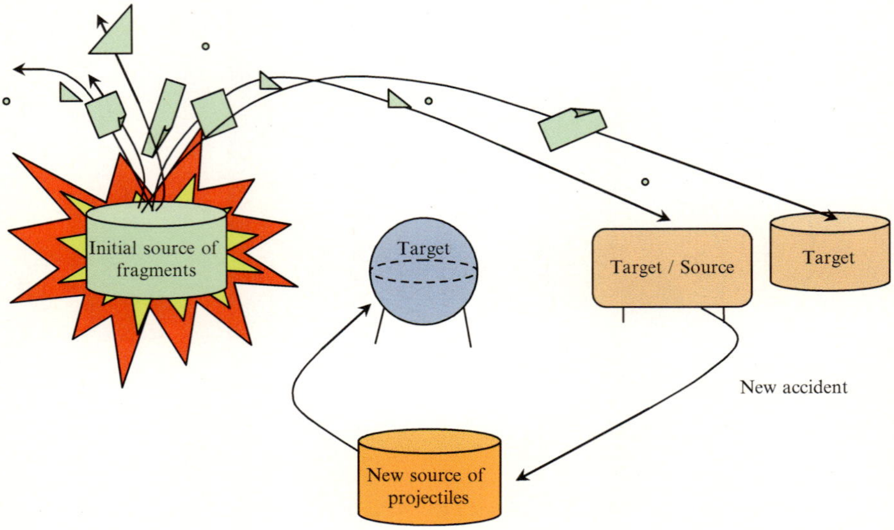

Fig. 15.3 Successive sequences of accidents/explosions within an industrial plant and domino effect: case of fragments impact, for instance

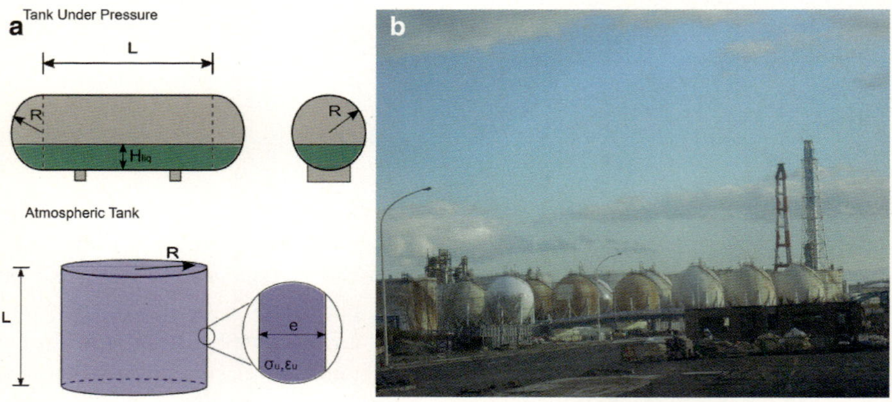

Fig. 15.4 Industrial plants: (**a**) Two main kinds of tanks; (**b**) Case of a Japanese refinery damaged by fires that occurred during the sequence earthquake-tsunami Tohuku, March 2011 (Author's pictures)

15.2.2 First Sequence of Accidents

In order to develop a general framework and use general notations, let us therefore denote the accident at the source S(i) as the event $E_s(i)$, *see* Figs. 15.1 and 15.4:

$$\begin{aligned} E_s(i) &= E_a(i) \cup E_p(i) \\ &= \begin{cases} E_a(i)\text{: if S(i)=Tank at Atmospheric Pressure} \\ E_p(i)\text{: if S(i)=Tank Under Pressure} \end{cases} \\ \text{with } i &= 1..N_s \end{aligned} \tag{15.6}$$

having a probability of occurrence defined as the scalar product:

$$P(E_s(i)) = \underline{1_S}(i).\underline{P_S}(i) \tag{15.7}$$

Where: $E_s(i)$ becomes either $E_a(i)$ for an atmospheric tank or $E_p(i)$ for an under pressure tank, if these two kinds are the only potential sources of accidents and explosions, with respective probabilities of occurrence denoted $P(E_a(i))$ and $P(E_p(i))$;

$\cup$ = events union symbol

$$1_{\underline{S}}(i) = \begin{pmatrix} 1_{Sa} \\ 1_{Sp} \end{pmatrix} \text{ : Identity vector of the tank} \tag{15.8}$$

$$1_{Sa} = \begin{cases} 1: & \text{if } S(i) = \text{Tank Atmospheric Pressure} \\ 0: & \textit{else} \end{cases} \tag{15.9}$$

$$\text{so that : } 1_{\underline{S}}(i) = 1_{\underline{a}} \text{with } 1_{\underline{a}}\binom{1}{0} \tag{15.10}$$

and,

$$1_{Sa} = \begin{cases} 1: & \text{if } S(i) = \text{Tank under Pressure} \\ 0: & \textit{else} \end{cases} \tag{15.11}$$

$$\text{So that : } 1_{\underline{S}}(i) = 1_{\underline{p}} \text{with } 1_{\underline{p}} = \binom{0}{1} \tag{15.12}$$

$$P_{\underline{S}}(i) = \begin{pmatrix} P(E_a(i)) \\ P(E_p(i)) \end{pmatrix}$$
$$\text{: Probability vector for the source tank S(i) according to its nature} \tag{15.13}$$

$$\text{So that : } P_{\underline{S}}(i) = \begin{pmatrix} P(E_a(i)) \\ 0 \end{pmatrix} \text{for a tank at atmospheric pressure} \tag{15.14}$$

$$\text{and } P_{\underline{S}}(i) = \begin{pmatrix} 0 \\ P(E_p(i)) \end{pmatrix} \text{for a tank under pressure.} \tag{15.15}$$

Within the industrial plant, under the simplified hypothesis that only two different kinds of tanks are erected within the plant under study, the number of tanks is so that:

$$\begin{aligned} N_S &= N_a + N_p \\ &= \text{total number of tanks, i.e.atmospheric as well as under pressure tanks} \end{aligned} \tag{15.16}$$

$$N_a = \sum_{i=1}^{N_S} \underline{1_S}(i).\underline{1_a} \text{ : Total number of tanks at atmospheric pressure} \quad (15.17)$$

$$N_p = \sum_{i=1}^{N_S} \underline{1_S}(i).\underline{1_p} \text{ : Total number of under pressure tanks} \quad (15.18)$$

15.2.3 Occurrence Probability of the First Accident Within the Plant During a Reference Period, T_{REF}

The first sequence of accidents that might occur in a plant can be triggered by various causes. Each triggering cause might be described by theoretical considerations or on the basis of past accidents: probabilistic distributions, fuzzy sets, expert judgement, etc.

15.2.3.1 Hypothesis 1: Case of Homogeneous Sub-populations Within the Whole Population of Tanks

Let us consider the simplified case of an industrial plant where each category of tanks (atmospheric or under pressure in the present case) is considered as "homogeneous", i.e. having same design, same construction period, same conditions of use and service, storing the same category of gas or liquids, getting in failure or explosive situation under the same conditions, etc. During a given period of existence, each category presents given ratios of severe accidents. From database of past accidents or through theoretical modelling and simulations, one may therefore consider an average annual rate of accident denoted:

$$\lambda(i) = \underline{1_S}(i).\underline{\lambda_S}(i) \text{ with } \underline{\lambda_S}(i) = \begin{pmatrix} \lambda_a(i) \\ \lambda_p(i) \end{pmatrix} \quad (15.19)$$

with: $\underline{\lambda_S}(i)$ = vector of average annual accident ratios for the source tank S(i) according to its category, i = 1 up to N_s.

NOTE: *The case of malevolent acts requires specific analysis in order to evaluate or predict the annual ratios and the risk of occurrence. In the following parts of the paper, we do not consider this particular case although its consequences could be catastrophic with devastating effects.*

Furthermore, the probability of failure of an elementary event becomes in this context:

$$\underline{P_S}(i) = \begin{pmatrix} P_a \\ 0 \end{pmatrix} \text{assuming } P(E_a(i)) = P_a, \forall i = 1..k_a, \text{ for each of the atmospheric tanks} \quad (15.20)$$

$$\underline{P_S}(i) = \begin{pmatrix} 0 \\ P_p \end{pmatrix} \text{assuming } P(E_p(i)) = P_p, \forall i = 1..k_p, \text{ for each of the under pressure tanks} \quad (15.21)$$

15.2.3.2 Hypothesis 2: Particular Case of First Sequence of Accidents

Let us consider the general case of a first accident concerning simultaneously a total number *k* of tanks:

$$k = k_a + k_p;\quad k_a \in [1; N_a];\quad k_p \in [1; N_p] \tag{15.22}$$

Where: k_a, k_p = total number of atmospheric and under pressure tanks, respectively.

The probability of this event can then be derived from the Binomial distribution. Under the hypothesis of "homogeneous" sub-categories, i.e. atmospheric or under pressure tanks, and probabilistic independency between two distinct individual events, this probability of occurrence is then:

$$\begin{aligned} P\big(E_a(i = 1..k_a) \cap E_p(j = 1..k_p)\big) &= P\left(\left(\bigcap_{i=1}^{k_a} E_a(i)\right) \cap \left(\bigcap_{j=1}^{k_p} E_p(j)\right)\right) \\ &= \left[\binom{N_a}{k_a}.(P_a)^{k_a}.(1 - P_a)^{N_a - k_a}\right].\left[\binom{N_p}{k_p}.(P_p)^{k_p}.(1 - P_p))^{N_p - k_p}\right] \end{aligned} \tag{15.23}$$

Where the binomial coefficients are:

$$\binom{N_a}{k_a} = C_{N_a}^{k_a} = \frac{N_a!}{(k_a!).(N_a - k_a)!} \tag{15.24}$$

and

$$\binom{N_p}{k_p} = C_{N_p}^{k_p} = \frac{N_p!}{(k_p!).(N_p - k_p)!} \tag{15.25}$$

Particular Case: Only One Accident Occurs Once at a Given Time t_1

The particular case is the event, denoted E_1, which corresponds to one accident taking place once in either an under pressure or an atmospheric tank. As the simultaneous accident of the two different tank categories is supposed to present a null probability of occurrence, this event $E_1{}^g$ global to the entire industrial plant, has a probability of occurrence:

$$\begin{aligned} P(E_1^g) &= \left[\binom{N_a}{1}.(P_a)^1.(1 - P_a)^{N_a - 1}\right] + \left[\binom{N_p}{1}.(P_p)^1.(1 - P_p))^{N_p - 1}\right] \\ &= \left[N_a.(P_a)^1.(1 - P_a)^{N_a - 1}\right] + \left[N_p.(P_p)^1.(1 - P_p))^{N_p - 1}\right] \end{aligned} \tag{15.26}$$

Number of Accidents that Affect One Tank: During a Reference Period T_{ref}

As the explosions are rare events, we can assume that the number $N_e(i)$ of explosions of a given tank S(i), i = 1 up to N_s, follows adequately a Poisson distribution with an average annual rate λ(i) evaluated from collected database on industrial accidents. Therefore, the probability of having $N_e(i)$ of explosions affecting the tank under study during the reference period T_{REF} (expected lifetime of the industrial plant, for instance) becomes, (Mebarki et al. 2008a):

$$P(E_s(i)) = \frac{(\lambda(i).T_{REF})^{N_e}.e^{-(\lambda(i).T_{REF})}}{N_e!} \tag{15.27}$$

leading then, respectively to $P(E_a)$ and $P(E_p)$ for the two kinds of tanks erected within the concerned plant:

$$P_a = P(E_a) = \frac{(\lambda_a.T_{REF})^{N_e}.e^{-(\lambda_a.T_{REF})}}{N_e!} \tag{15.28}$$

and

$$P_p = P(E_p) = \frac{(\lambda_p.T_{REF})^{N_e}.e^{-(\lambda_p.T_{REF})}}{N_e!} \tag{15.29}$$

Where: T_{ref} = Reference period time [*unit*: *in years*, as λ is the average annual ratio].

Particular Case: k = 1

Let us consider the case of an initial accident $E_1(i)$, occurring at the source S(i) only once during the reference period. The corresponding probabilities of occurrence becomes for each category of tanks:

$$P\left(E_1(i)|_{T_{REF}}\right) = P\left(E_s(i)|_{T_{REF}}\right) = (\lambda(i).T_{REF}).e^{-(\lambda(i).T_{REF})} \tag{15.30}$$

as $N_e = 1$, leading then to:

$$P_a = (\lambda_a.T_{REF}).e^{-(\lambda_a.T_{REF})} \text{and } P_p = (\lambda_p.T_{REF}) \tag{15.31}$$

since: $k_a = 1$ or $k_p = 1$.

Synthesis

The event corresponding to any first accident within the entire plant during the whole reference period $E_1{}^g$ is also defined as:

$$E_1^g\big|_{T_{ref}} = \bigcup_{i=1}^{N_s}\left(E_1(i)\big|_{T_{ref}}\right) \tag{15.32}$$

From the above developments, the probability of occurrence of only one accident, once within the industrial plant, during the entire reference period is therefore explicitly written as:

$$\begin{aligned} P\left(E_1^g\big|_{T_{ref}}\right) = &\left[N_a.\left((\lambda_a.T_{REF}).e^{-(\lambda_a.T_{REF})}\right).\left(1-(\lambda_a.T_{REF}).e^{-(\lambda_a.T_{REF})}\right)^{N_a-1}\right] \\ &+\left[N_p.\left((\lambda_p.T_{REF}).e^{-(\lambda_p.T_{REF})}\right).\left(1-(\lambda_p.T_{REF}).e^{-(\lambda_p.T_{REF})}\right)^{N_p-1}\right] \end{aligned} \tag{15.33}$$

15.2.4 Consequences of the First Accident Triggered Within the Plant During a Reference Period, T_{REF}

15.2.4.1 Subsequent Events and Threats Generated by a Given Accident

When an accident is triggered (such as BLEVE effect, for instance), a sequence of events might take place as explained above, see Figs. 15.1, 15.2, 15.3, 15.4 and 15.5:

- A set of structural fragments (plates, end-cups...): event $E_F(i)$ since each generated fragment can be ejected from the tank source and become therefore a projectile.
- A fire ball and thermal effects: event $E_T(i)$
- A blast wave: event $E_W(i)$
- A release or loss of liquid and gas (loss of confinement): event $E_C(i)$.

These events may generate mechanical, thermo-mechanical or chemical effects that can affect the surrounding facilities or installations, and even persons on charge of the plant control and security. Our present purpose is to analyse the unfavourable effect on the tanks erected in the vicinity and that might be severely damaged so that they might give rise to a new sequence of accidents and explosions. These successive sequences of damages and explosions, and their subsequent events and consequences, are important steps when dealing with the so-called domino effect.

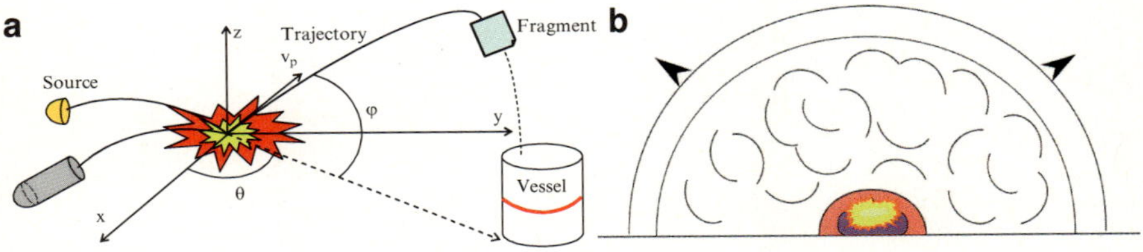

Fig. 15.5 BLEVE explosion generating projectiles, fire ball as well as blast wave: (**a**) Fragments generation and projectile impacts; (**b**) Blast wave and pressure/depression front

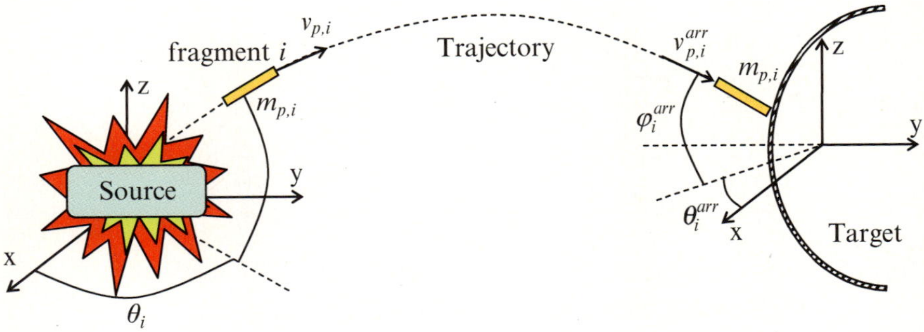

Fig. 15.6 Fragments ejected as projectiles that may impact surrounding target tanks

Therefore, let us consider the case of the first accident triggering at instant t_1 within the industrial plant under study:

$$E_1^g\big|_{T_{ref}} = \bigcup_{i=1}^{N_s}\left(E_1(i)\big|_{T_{ref}}\right),\ \text{i.e.} E_s(i) = E_1(i) \text{ triggered at an initial time } t_1(i) \tag{15.34}$$

Let us consider a global axial system as a reference system for the whole plant. The various sources are erected in their respective locations defined by the position vector $\underline{x_s}(.)$, see Figs. 15.5 and 15.6:

$$\underline{x_s}(i) = \begin{pmatrix} x_s(i) \\ y_s(i) \\ z_s(i) \end{pmatrix}, i = 1 \text{ up to } N_s \tag{15.35}$$

The elementary threat events propagating from the source accident S(i) may affect and impact the surrounding tanks that becomes, once the initial sequence triggered at time $t_1(i)$, as potential targets T(j), $j = 1$ up to N_s. The various tanks and targets erected within the considered plant, are placed in the location described by $\underline{x_t}(.)$, see Figs. 15.5 and 15.6:

$$\underline{x_t}(j) = \begin{pmatrix} x_t(j) \\ y_t(j) \\ z_t(j) \end{pmatrix}, j = 1 \text{ up to } N_s \tag{15.36}$$

In fact, all the tanks can be considered as potential source of accident S(i) as well as potential targets T(i) under the effects of the threats generated at the source, i.e.:

$$\underline{x_s}(i) = \underline{x_t}(i), j = 1 \text{ up to } N_s \tag{15.37}$$

15.2.4.2 Propagation of the First Accident and Threats to the Surrounding Tanks and Facilities

This generation of events that may threaten the neighbourhood corresponds to the propagation event, denoted $E_{propa}(i)$:

$$E_1^{propa}(i)\big|_{E_1(i)} = E_F(i) \cup E_T(i) \cup E_W(i) \cup E_C(i) \tag{15.38}$$

Therefore, the apparition of first threats to the surrounding tanks and facilities results from the first accident and propagation events:

$$E_1(i) \cap \left(E_1^{propa}(i)\big|_{E_1(i)}\right) = (E_1(i)) \cap (E_F(i) \cup E_T(i) \cup E_W(i) \cup E_C(i)) \tag{15.39}$$

The corresponding probability of propagation event is then defined as:

$$P\left(\left(E_1^{propa}(i)\right)\right) = \sum_{i=1}^{N_s} P\left(E_1^{propa}(i)\big|_{E_1(i)}\right).P(E_1(i)) \tag{15.40}$$

Propagation of the Accident: Ejected Structural Fragments as Projectiles

The first accident at the source tank S(i), i = 1 up to N_s, occurring at the starting time $t_1(i)$ may give rise to a set of $N_f(i)$ structural fragments (denoted F(i,k), k = 1 up to $N_f(i)$) that are ejected from this source and may impact other tanks or facilities that cross their trajectory, i.e.:

$$E_F(i) = \bigcup_{k=1}^{N_f(i)} E_{frag}(i, k) \tag{15.41}$$

Its occurrence probability can be estimated according to collected database inputs or from theoretical and simulation approaches that provide therefore:

$$P_F(i) = P\left(E_F(i)\big|_{E_1(i)}\right).P(E_1(i)) \tag{15.42}$$

From database analysis of past accidents and existing bibliography reports as well as theoretical approaches, one can establish the probabilistic distributions concerning the main governing parameters of the fragments motion and energy, i.e. (Mebarki et al. 2007; 2008a, b, 2009a, b):

- The number of fragments: $N_f(i)$
- The ratio $\alpha(i)$ of total internal energy $E_{internal}(i)$ within the source tank that is transformed into kinetic energy ejecting the fragments and transferred to them during their motion around the source: $E_{kin}(i)$

$$E_{kin}(i) = \alpha.E_{internal}(i) \tag{15.43}$$

- The form (among plate, end-cup, oblong end-cup, etc.) and the dimensions of each fragment
- The mass of each fragment F(i,k): $m_p(i,j)$ with $i = 1$ up to Ns and $k = 1$ up to $N_f(i)$
- The initial velocity: $v_p(i,k)$ at instant $t_1(i)$

$$\underline{v_p}(i,k) = \underline{\dot{x}}_p(i,k) = \begin{pmatrix} v_x(i,k) \\ v_y(i,k) \\ v_z(i,k) \end{pmatrix}\Bigg|_{t=t_1(i)} \tag{15.44}$$

- The initial angles of ejection: event $\theta(i,k)$

$$\underline{\theta_p}(i,k) = \begin{pmatrix} \theta_x(i,k) \\ \theta_y(i,k) \\ \theta_z(i,k) \end{pmatrix}\Bigg|_{t=t_1(i)} \tag{15.45}$$

At any instant $t \geq t_1(i)$, the motion of each fragment can be studied according to adopted values for the drag and lift coefficients, i.e. the position $\underline{x_p}(.)$, velocity $\underline{v_p}(.)$ and acceleration $\underline{g_p}(.)$:

$$\underline{x}_p(i,k) = \begin{pmatrix} x_p(i,k) \\ y_p(i,k) \\ z_p(i,k) \end{pmatrix}\Bigg|_{t\geq t_1(i);} \quad \underline{v_p}(i,k) = \begin{pmatrix} \dot{x}_p(i,k) \\ \dot{y}_p(i,k) \\ \dot{z}_p(i,k) \end{pmatrix}\Bigg|_{t\geq t_1(i);}$$

$$\underline{\gamma_p}(i,k) = \begin{pmatrix} \ddot{x}_p(i,k) \\ \ddot{y}_p(i,k) \\ \ddot{z}_p(i,k) \end{pmatrix}\Bigg|_{t\geq t_1(i)} \tag{15.46}$$

On the basis of its motion and the relative location of the whole other potential targets, i.e. the tanks T(j), $j = 1$ up to N_s, one can define the possible impacts on the tank T(j) by the impact indicator:

$$\underline{1_p}(i,k,j) = \begin{cases} 1 : \text{if fragment } F(i,k) \text{ impacts target } T(j) \text{ at instant } t_{impact}(i,k,j) \\ 0 : \text{else} \end{cases} \tag{15.47}$$

The tank T(j) is impacted by at least one or a rain of the fragments ejected from source S(i) if the impact indicator is equal to 1, i.e.:

$$\underline{1}_{\text{impact}}(\mathrm{i,k,j}) = \mathrm{Min}\left\{ \frac{1}{\sum_{k=1}^{N_f(i)} \underline{1}_p(\mathrm{i,k,j})} \right\} \tag{15.48}$$

The impact event on the tank T(j) under the set of projectiles due to source S(i) is then:

$$E_F^{\text{impact}}(j) = \bigcup_{k=1}^{N_f(i)} \left(E_F^{\text{impact}}(i,k,j) \cap E_{\text{frag}}(i,k) \right) \tag{15.49}$$

Where:

Event corresponding to tank T_j impacted by the fragment F(i,k)= $E_F{}^{\text{impact}}$(i,k,j)
Event corresponding to existence of a fragment F(i,k)= E_{frag}(i,k)

When there is an impact, the mechanical models that are adopted should provide the damage generated to the impacted tank. A set of mechanical limit state functions tells whether the impacted tank T(j) reaches its defined :

- Limit state of local damages: perforation, partial penetration, cracks, excessive stress or strains, etc.
- Limit state of global damages: over-lapping, anchors rupture, sliding, buckling, etc.
- The corresponding probability of damage can be obtained by analytical or numeric approaches. Due to the complexity of the analysis, Monte Carlo simulations are the common tool used in order to estimate this value:

$$P\left(E_F^{\text{damage}}(j)\right) = \sum_{k=1}^{N_f(i)} P\left(E_F^{\text{damage}}(j)) \Big|_{E_F^{\text{impact}}(j)}\right).P\left(E_F^{\text{impact}}(j)\Big|_{E_{\text{frag}}(i,k)}\right).P(E_{\text{frag}}(i,k)) \tag{15.50}$$

Propagation of the Accident: Fire Ball Generates Thermal Effects

The first accident at the source tank S(i), i = 1 up to N_s, may generate thermal field that may affect significantly other tanks or facilities erected in the source vicinity, see Fig. 15.7. This event and its probability of occurrence are respectively defined as:

$$E_T(i) \text{ with } P_T(i) = P\left(E_T(i)\big|_{E_1(i)}\right).P(E_1(i)) \tag{15.51}$$

This occurrence probability can be estimated according to collected database inputs or from theoretical and simulation approaches. The event generates thermal fields and flows than can be described by:

$$T\left(i, \underline{x_s}(i), \underline{x_t}(i), t - t_1(i)\right) = \text{thermal field} \tag{15.52}$$

$$\Phi_T\left(i, \underline{x_s}(i), \underline{x_t}(i), t - t_1(i)\right) = \text{thermal flow} \tag{15.53}$$

Fig. 15.7 Fires and threats to surrounding facilities: Case of a Japanese refinery damaged by fires that occurred during the sequence earthquake-tsunami Tohuku, March 2011 (Author's pictures)

The effect of the thermal event on the tank T(j) due to source S(i) can produce direct as well as indirect (such as ignition of surrounding gases or liquids and fire generation) global or local damages. The corresponding probability of damage can be obtained by analytical or numeric approaches. Due to the complexity of the analysis, Monte Carlo simulations are the common tool used in order to estimate this value:

$$P\left(E_T^{damage}(j)\right) = P\left(E_T^{damage}(j)\big|_{E_T(j)}\right).P(E_T(i)) \tag{15.54}$$

Propagation of the Accident: Blast Wave

The first accident at the source tank S(i), i = 1 up to N_s, may give rise to a blast wave that generates high positive and negative pressures around the source and may affect mechanically the surrounding tanks and facilities. This event and its probability of occurrence are respectively defined as:

$$E_W(i) \text{ with } P_W(i) = P\left(E_W(i)\big|_{E_1(i)}\right).P(E_1(i)) \tag{15.55}$$

This occurrence probability can be estimated according to collected database inputs or from theoretical and simulation approaches. The event generates a pressure field and front wave than can be described by:

$$p\left(i, \underline{x_s}(i), \underline{x_t}(i), t - t_1(i)\right) = \text{pressure field} \tag{15.56}$$

The effect of this blast wave on the tank T(j) due to source S(i) can produce global or local damages. The corresponding probability of damage can be obtained by analytical or numeric approaches. Due to the complexity of the analysis, Monte Carlo simulations are the common tool used in order to estimate this value:

$$P\left(E_W^{damage}(j)\right) = P\left(E_W^{damage}(j)\Big|_{E_W(j)}\right).P(E_W(i)) \tag{15.57}$$

Propagation of the Accident: Loss of Confinement

The first accident at the source tank S(i), i = 1 up to N_s, may cause loss of confinements with ejection of gases or liquids that may be indirect sources of fires or explosions. This event and its probability of occurrence are respectively defined as:

$$E_C(i) \text{ with } P_C(i) = P\left(E_C(i)|_{E_1(i)}\right).P(E_1(i)) \tag{15.58}$$

This occurrence probability can be estimated according to collected database inputs or from theoretical and simulation approaches. This event generates loss of confinement that releases liquids and gases and it can, therefore, give rise to a fire ignition or explosions, for instance. This confinement loss can be expressed by a critical volume of losses within a critical time (global volume or rate of volume release, in a half hour or an hour or a day, for instance) than can be described by:

$$v\left(i, \underline{x_s}(i), \underline{x_t}(i), t - t_1(i)\right) = \text{volume of confinement loss and products release} \tag{15.59}$$

The effect of this confinement loss on the tank T(j) due to source S(i) can produce global or local damages. The corresponding probability of damage can be obtained by analytical or numeric approaches. Due to the complexity of the analysis, Monte Carlo simulations are the common tool used in order to estimate this value:

$$P\left(E_C^{damage}(j)\right) = P\left(E_C^{damage}(j)|_{E_C(j)}\right).P(E_C(i)) \tag{15.60}$$

15.2.4.3 Total Energy and Momentum: Conservation Requirements

When an accident is triggered, one should verify that the conservation requirements are satisfied, i.e.:

Momentum conservation of the ejected fragments and products (liquids or gases) for the tank S(i), i =, i.e. up to N_s,

$$\vec{p}_{\text{initial}}(i) = \vec{p}_{\text{internal}}(i) + \vec{p}_{\text{external}}(i) \text{ with } \vec{p}_{\text{internal}} = \vec{0} \tag{15.61}$$

$$\vec{p}_{\text{external}}(i) \neq \vec{0} : \text{in case of a prior impact for instance} \tag{15.62}$$

$$\vec{p}_{\text{initial}}(i) = \vec{p}_{\text{fragments}}(i) + \vec{p}_{\text{fluids}}(i) \tag{15.63}$$

$$\vec{p}_{\text{fragments}} = \sum_{k=1}^{N_f} m_p(k).\vec{\boldsymbol{v}}_p(k) \text{ and } \vec{p}_{\text{fluids}} = \int_{V_g} \rho_G.\boldsymbol{v}_G.dV_g + \int_{V_l} \rho_L.\boldsymbol{v}_L.dV_l \tag{15.64}$$

Where: $\vec{p}_{\text{initial}}(i)$ = initial momentum in the considered tank S(i); $\vec{p}_{\text{external}}(i)$ = momentum provided by the external cause such as a prior external impact on the tank, for instance; $\vec{p}_{\text{fragments}}(i)$ = total momentum of the fragments; $\vec{p}_{\text{fluids}}(i)$ = total momentum of the ejected fluids (liquids and gases); m_p, v_p, N_f = mass, velocity of each fragment among the total set of N_f ejected fragments, r_G, v_G, V_g = density, velocity and volume of the gas part, respectively; r_L, v_L, V_l = density or specific weight (if g included), velocity and volume of the liquid part, respectively.

Energy conservation since the total initial energy (internal energy and external energy in case of impacts, for instance) is partly transformed into kinetic energy (fragments, gases and liquids), thermal energy, blast energy, dissipated energy and residual energy, i.e.

$$E_{\text{initial}}(i) = E_{\text{internal}}(i) + E_{\text{external}}(i) \text{ with } E_{\text{internal}} \neq 0 \tag{15.65}$$

$$E_{\text{external}}(i) \neq 0 : \text{in case of a prior impact or thermal flow, for instance} \tag{15.66}$$

$$E_{\text{initial}}(i) = E_{\text{solids}}(i) + E_{\text{fluids}}(i) + E_{\text{blast-wave}}(i) + E_{\text{thermal-effect}}(i) + E_{\text{dissipated}}(i) + E_{\text{residual}}(i) \tag{15.67}$$

$$E_{\text{fragments}} = \sum_{k=1}^{N_f} \left(\frac{1}{2} m_p(k).v_p^2(k) \right) \tag{15.68}$$

$$E_{\text{fluids}}(i) = \int_{V_g} \frac{1}{2} \rho_G.v_G^2.dV_g + \int_{V_l} \rho_L.v_L^2.dV \tag{15.69}$$

Where: E_{internal} (i) = internal energy in the considered tank S(i); E_{external} (i) = external energy provided by the external cause such as a prior external impact on the tank, for instance; E_{initial} (i) = initial total energy in the considered tank S(i); $E_{\text{fragments}}$ (i) = total kinetic energy of the fragments; E_{fluids} (i) = total kinetic energy of the ejected fluids (liquids and gases); $E_{\text{blast-wave}}$ (i) = total energy transported by the blast wave; $E_{\text{thermal-effect}}$ (i) = total energy transported by the fire ball;

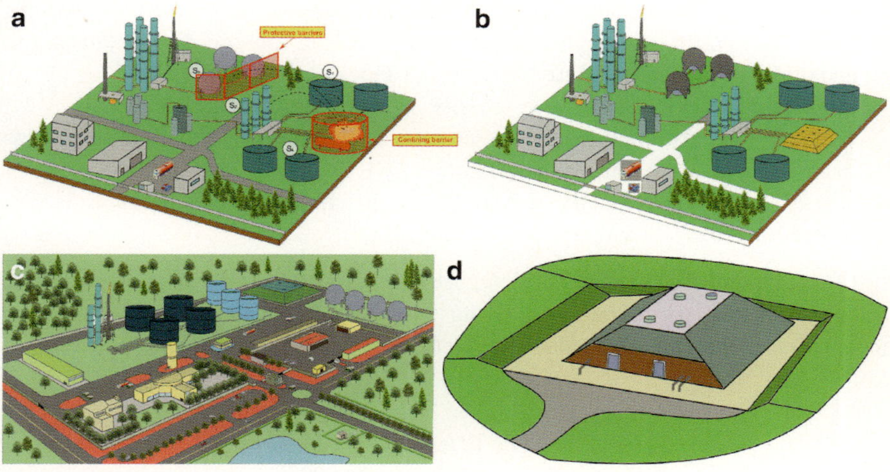

Fig. 15.8 Protective measures regarding potential sources or targets

$E_{dissipated}$ (i) = total energy dissipated in order to trigger the explosion and fragmentation of the considered tank (such as plastic deformation, cracking and propagation, fragmentation, chemical reactions and transformation, etc.); and $E_{residual}$ (i) = total residual energy remaining in the damaged tank (residual products and solids, etc.).

15.2.5 General Purpose of the Domino Effect Study and Use of the Occurrence Probability for Decision Making

The industrial accidents have in general several consequences of great importance, such as:

- Socio-economic and employees jobs losses due to production interruption, reconstruction, repair or strengthening of the industrial plant
- Environmental consequences in case of containment losses such as liquids or gases
- Fires that threaten the whole plant as well as the surrounding buildings, facilities, installations, etc.
- Threats to the health and physical integrity of the employees and inhabitants in case of toxic products release, etc.

In order to reduce or mitigate the disaster, one may consider either, see Fig. 15.8:

- Reduction of the hazards and threats by isolating the sources of possible accidents

- Reduction of the vulnerability of the potential targets by protective measures better design, barriers and protections, erection of buildings, tanks and installations with large relative security distances
- Regular inspections and severe security measures, etc.
- Protection against confinement losses such as containment and retention basins in case of petrol oil, for instance.
- Use of automatic protective systems such as shutdown systems, alarms, fire protection, etc.

From a theoretical point of view, let us consider that the site under study (industrial plant, surrounding buildings, facilities and other plants, strategic installations, operational headquarters, etc. is so that the total expected cost of losses is:

$$\mathrm{C}^{\mathrm{g}}_{\mathrm{losses}} = \sum_{k=1}^{n} \mathrm{P}\left(\mathrm{E}^{\mathrm{g}}_{k}\right).\mathrm{C}^{\mathrm{g}}_{\mathrm{k}} \quad \text{where } \mathrm{P}\left(\mathrm{E}^{\mathrm{g}}_{k}\right)$$
$$= \mathrm{P}\left(\mathrm{E}^{\mathrm{g}}_{k}\Big|_{\mathrm{E}^{\mathrm{damage}}_{\mathrm{k-1}}}\right).\mathrm{P}\left(\mathrm{E}^{\mathrm{damage}}_{\mathrm{k-1}}\Big|_{\mathrm{E}^{\mathrm{propa}}_{\mathrm{k-1}}}\right).\mathrm{P}\left(\mathrm{E}^{\mathrm{propa}}_{\mathrm{k-1}}\Big|_{\mathrm{E}^{\mathrm{g}}_{\mathrm{k-1}}}\right).\mathrm{P}\left(\mathrm{E}^{\mathrm{g}}_{\mathrm{k-1}}\right) \tag{15.70}$$

where : $\mathrm{C_k}^{\mathrm{g}}$ = socio-economic losses as consequences of the sequence k of the domino effect, k = 1 up to the range n of sequences under study (until the total plant is destroyed or reaches a given threshold of destruction, $\mathrm{C_{losses}}^{\mathrm{g}}$ = mathematical expected value of the socio-economic consequences on the whole industrial plant and its concerned vicinity.

In order to mitigate the industrial disaster, the stockholders may decide to adopt several protective solutions in order to optimise the economic investments so that to reach the optimal global cost:

$$\mathrm{C}^{\mathrm{g}}_{\mathrm{opt}} = \mathrm{Min}\left\{\mathrm{C}^{\mathrm{g}}_{0} + \Delta C^{\mathrm{g}}_{0} + \sum_{k=1}^{n}\left(\mathrm{P}\left(\mathrm{E}^{\mathrm{g}}_{k}\right) + \Delta P\left(\mathrm{E}^{\mathrm{g}}_{k}\right)\right).\mathrm{C}_{\mathrm{g}}\right\} \tag{15.71}$$

Where: $\mathrm{C_{opt}}^{\mathrm{g}}$ = optimal global cost of the entire zone (industrial plant and its affected surroundings), $\mathrm{C_0}^{\mathrm{g}}$ = initial global cost before adopting any additional protective measures so that the initial risk of domino effect $\mathrm{P}(\mathrm{E}_k{}^{\mathrm{g}})$ is reduced by $\Delta\mathrm{P}$ $(\mathrm{E}_k{}^{\mathrm{g}})$ as consequence of the protective measures.

In fact, this optimal global cost seems easy to be theoretically calculated. However, several aspects such as respect of human life, pollutions and aggressive products release, reactions of the public opinion and political decisions make this optimization not so easy to be reached in practice. However, this theoretical formulation may also be helpful in prospecting objective investments and accompanying measures (survey and early warning systems, automatic control and shutdowns, protective barriers, vicinity planning and organization) that result in risk reduction, disaster mitigation and satisfy resilience and quick recovery requirements.

15.3 Applications and Sensitivity Analysis

15.3.1 Risk of Failure

The present application considers the case of tanks with various filling levels, see Table 15.1. It is restricted to the analysis of the fragments impacts and the blast wave. For sake of simplicity, the thermal flows effects and the containment losses effects are not included in the present results. They are done separately.

Monte Carlo simulations are used in order to evaluate the probability of impacts as well as the probability of failure.

15.3.2 Structural Fragments: Probability of Impacts and Risk of Failure

The probability of failure is obtained by Monte Carlo simulations as, (Mebarki et al. 2007, 2008a, b, 2009a, b; Mingguang and Juncheng 2008):

Table 15.1 Probabilistic description of the structural fragments (Mebarki et al. 2009a, b)

Random variable	Probability density function	Probability density function formulas	Comments and details
Number of fragments (n)	Discrete exponential distribution	$P_N(n) = e$ $(-\lambda_0 - \lambda_{1n} - \lambda_2 n^2)$	$\lambda_{0,}$ λ_1 and λ_2 are multipliers of Lagrange obtained from the accidental data
Relative frequency of any projectile form (f_p)	Uniform distribution by intervals	f_p	Forms and ratios collected from past accidents and database
Projectile mass (m_p)	Uniform distribution	$m_p = V_p \times \rho$	Volume of fragments is a product of random variables and follows a uniform distribution
Projectile departure velocity (v_p)	Uniform distribution	$v_p = \sqrt{\frac{2E_c}{m_p}}$ $E_c = \alpha E_{exp}$	Velocity of projectiles depends from the kinetic energy transmitted to projectiles. Kinetic energy is a fraction of the explosion energy which can be represented by the ratio α. This factor follows a uniform distribution
Horizontal departure angle (θ)	Uniform distribution by intervals	θ	Preferential directions collected from past accidents and database
Vertical departure angle (φ)	Uniform distribution	φ with Arcsin (φ) following uniform distribution.	Experimental information unavailable: theoretical distribution is assumed

$$P_{rup} = \sum_{k=1}^{N_{sim}} \frac{1_{(\text{State of a target} \neq 0)}(k)}{N_{sim}} \tag{15.72}$$

With:

$$1_{(\text{State of a target} \neq 0)}(k) = \left[\begin{array}{c} 1 \text{ if a target is completely damaged} \\ 0 \text{ else} \end{array} \right. \tag{15.73}$$

And: N_{sim} = total number of simulations.

Furthermore, a uniform random variable, H, is considered in order to express the filling level of a source tank. Its values range within the interval] 0; H_{max}], where H_{max} is the maximal filling level of a tank.

The trajectory of the fragments and their impact on target tanks are evaluated according to existing developments (Mebarki et al. 2009a, b).

15.3.3 *Blast Waves: Effects on Target Tanks and Risk of Failure*

After an explosion, a blast wave propagates through the air and in contact with surrounding tanks it can produce mechanical effects on affected tanks, resulting in:

- Excessive bending,
- Tank overturning,
- Global buckling,
- Tank sliding on the ground surface, and
- Excessive shear and bending of the target anchors.

15.3.4 *Results and Comments*

15.3.4.1 Comparison Between the Proposed Theoretical Models and the Experimental Results

Obviously, the results depend intimately on the accuracy of the models concerning the governing parameters: fragments description, their trajectory, their impacts and perforation of the impacted targets. The number of structural fragments is described by a theoretical probabilistic model which results are reported in Fig. 15.9a, in the case of BLEVE phenomena. The kinetic energy at ejection of the fragments, as shown in Fig. 15.9b, and their initial horizontal angles are also described by adequate models, derived from experimental observations in the case of spherical as well as cylindrical metal tanks, (Mebarki et al. 2009a).

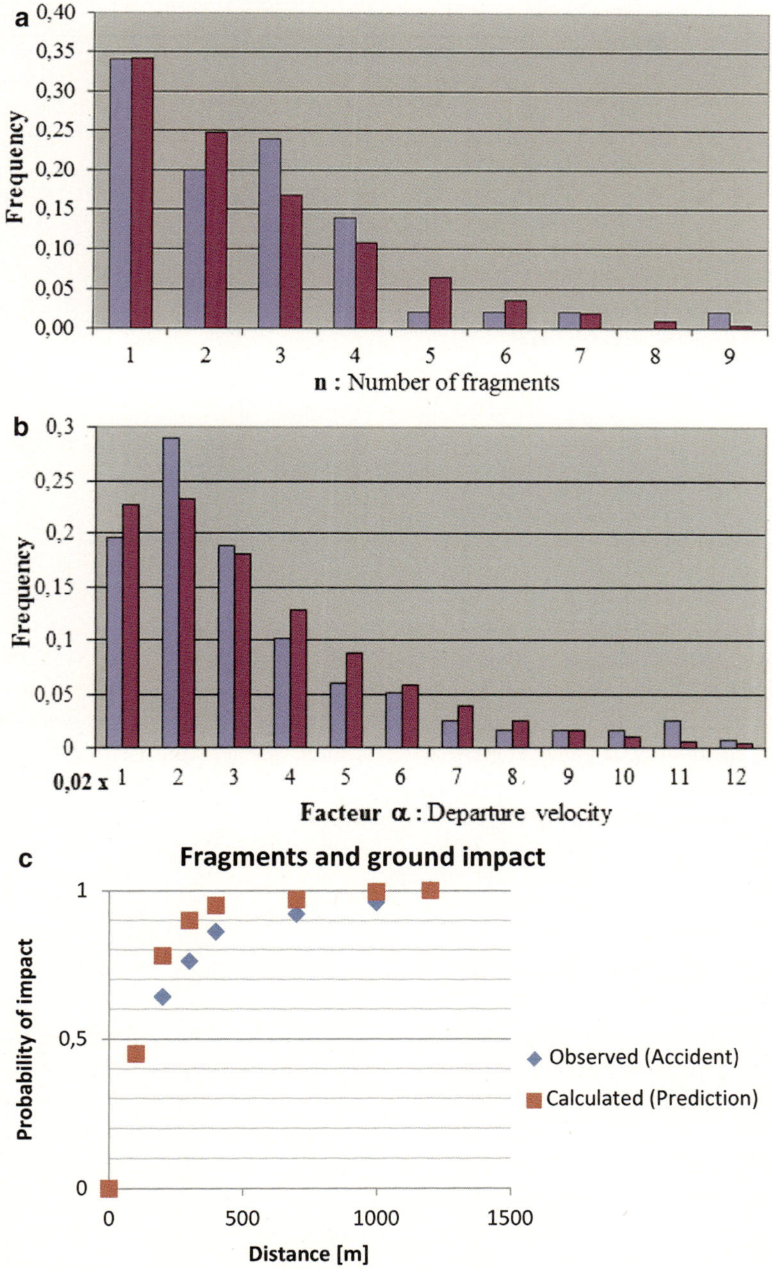

Fig. 15.9 Explosion and structural fragments. (**a**) BLEVE and Number of fragments (Mebarki et al. 2009a); (**b**) Kinetic energy of the fragments, (Mebarki et al. 2009a); (**c**) Distances of fragments impacts in the real case of Mexico accident (Mebarki et al. 2009b)

Table 15.2 Case study: source and target tanks definition

Parameter	Source tank	Target tank
Radius (R)	3.5 m	6 m
Length (L)	15 m	12 m
Capacity (V)	757 m^3	1,350 m^3
Shell thickness (e)	0.007 m	0.005 m
Burst pressure (P_e)	800,000 Pa	–
Ultimate stress (σ_u)	360 Gpa	360 Gpa
Ultimate strain (ε_u)	0.23	0.23

Table 15.3 Properties of the structural fragments ejected as projectiles impacting the target

Parameter	Value
Average speed at the impact	82.98 m/s
Average kinetic energy at the impact	14.463 MJ
Average mass of projectiles	8,954 kg
Ratio of end cups	9.09 %
Ratio of elongated end cups	45.45 %
Ratio of plates	45.45 %
P_{imp}: Probability of the impact	**5.5×10^{-3}**

Furthermore, the simulation of explosion and fragments trajectory is run for the case of Mexico accident: a good accordance is obtained between the observed distances of fragments impacts and the theoretical prediction as shown in Fig. 15.9c, (Mebarki et al. 2009b).

The interaction of the impacted tanks and the metals fragments is described by simplified models of penetration and perforation, whereas the fragments are considered as rigid rods. A set of experimental results is collected for various incidence angles and velocities at impact. The proposed simplified models for perforation and penetration provide good predictions of the penetration depth and velocities after perforation, (Mebarki et al. 2007, 2008b).

15.3.4.2 Sensitivity Analysis and Numerical Simulations

For illustrative purposes, the source tank is cylindrical and pressurized, whereas the target tank is at atmospheric pressure, see Table 15.2. The source tank content is liquefied propane. Target tank is supposed without any liquefied content, full of evaporated gas-oxygen mixture. Under this assumption, secondary effects due to liquid dynamic movement are intentionally neglected and, on the other side, a target tank is more prone to mechanical damage.

The target tank is located at a distance of 100 m from the source tank. Considering the origin of coordinate system at the centre of source tank, the centre of the target tank is then located at (100 m: horizontal distance, 0 m, 6 m: height). Table 15.3 provides the results obtained from the simulations, i.e.: projectiles impacting the target, distribution of projectiles and distribution of projectiles kinetic energy at the impact on the ground for each angular sector, see Table 15.3 and Fig. 15.10a.

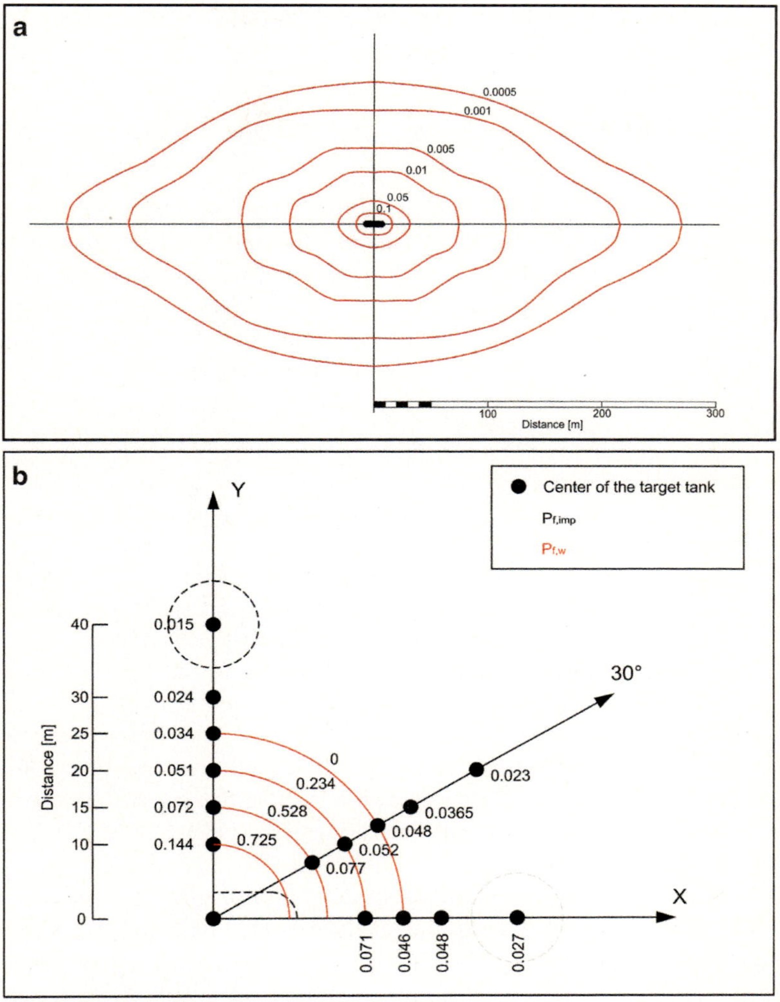

Fig. 15.10 Distribution of projectiles and failure risk. (**a**) Iso-values of impact probability around the cylindrical source tank; (**b**) Iso-values of tanks failure probability under the projectiles impact ($P_{f,imp}$) and the overpressure wave ($P_{f,w}$)

From the results of simulations for projectiles impact and overpressure wave effect on the atmospheric tank it is possible to compare levels of risk from each phenomena and for different levels of damage. It is found that overpressure waves can produce significant damages at the near field. On the other hand, risk of projectiles impact is much lower at the near field but projectiles can trigger the domino effect at much higher distance.

Figure 15.10b shows that the probability of failure from the blast wave is 3–7 times greater than the probability of failure produced by projectiles. Also, in this model only massive projectiles are considered as the potential projectiles and small, light projectiles are intentionally neglected.

15.4 Natural Hazards as Triggering Events: Tsunamis Caused by Quakes

15.4.1 Theoretical Frameworks and Required Steps

Natural hazards such as quakes, tsunamis, storms, floods, lightening, extreme winds, tornados and typhoons for instance may trigger domino effects sequences in industrial plants. Actually, the tanks and facilities may be damaged and the damages may propagate until disastrous situations take rise as it happened during the 2011 Eastern Japan Great Earthquake Disaster (Norio et al. 2011; Goto et al. 2011): quakes and tsunamis effects have caused serious and disastrous damages.

Therefore, mitigation and risk reduction for the case of coastal industrial plants, for instance, require four steps:

- Hazard modeling: description of the maximum inputs generated by the natural hazard (PGA: Peak Ground Acceleration for the quakes, PWH: Peak Water Heights and Run-ups for the tsunamis, for instance).
- System vulnerability, fragility and limit state functions: definition of the limit states that should not be reached by the systems under study in order to avoid occurrence of system failure (rupture of the industrial tanks or critical release of products after mechanical damages and impacts for instance) (Askan and Yucemen 2010; Reese et al. 2011).
- Reliability and risk analysis: assessment of the limit states occurrence probability, i.e. probability of failure. Actually, the hazard and the vulnerability are described within a probabilistic framework in order to express the error models and cover the uncertainties and heterogeneities (Eckert et al. 2012; Leone et al. 2011).
- And decision making by optimization: definition of protective measures and required investments in order to reduce the expected risks and socio-economic losses (protective barriers, coastal walls, early warning systems, real time survey, etc.), (Beltrami and Risio 2011; Grasso and Singh 2008; Jin and Lin 2011; van Zijll de Jong et al. 2011; Wilson et al. 2011).

15.4.2 Hazard Modeling: Simplified and Probabilistic Models – Case of Tsunamis

The present work focuses on the particular case of tsunamis. Several authors have investigated the description and modeling of the tsunamis in order to predict the Peak Water Heights and Run-ups (Burwell et al. 2007; Cheung et al. 2011; Constantin 2009; Demetracopoulos et al. 1994; Flouri et al. 2011; Haugen et al. 2005; Heidarzadeh et al. 2009; Helal and Mehanna 2008; Kharif and Pelinovsky 2005; Liu et al. 2009; Lovholt et al. 2011; Madsen 2010; Nandasena et al. 2011; Pophet

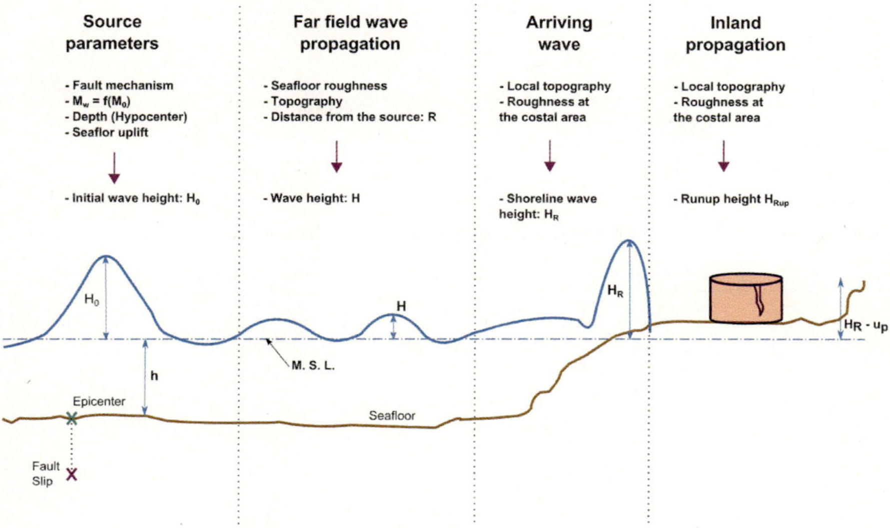

Fig. 15.11 Tsunami path from the epicentral zone towards the shore and inlands

et al. 2011; Sladen et al. 2007; Todorovska et al. 2002; Ward 2011; Wijetunge 2006; Zhang et al. 2009; Zhao et al. 2011).

Simplified models have been developed for quick evaluation of the Peak Water Heights and Run-ups. The model denoted cMD (coupled Magnitude-Distance) has proved to be efficient in predicting the Peak Ground Acceleration for various soil conditions (Mebarki 2009c):

$$\frac{A}{g} = \frac{e^{-\beta.D_h}}{1 + D_h.e^{-(M_w - M_0)}} \tag{15.74}$$

Where: A [in m/s^2] = Peak Ground Acceleration, g [in m/s^2] = gravity acceleration, D_h [in km] = hypocentral central, M_w = moment magnitude of the earthquake, and two fitting constants (β and a threshold magnitude M_0). The error model associated to the PGA is assumed to follow a Gamma distribution; a Log-Normal distribution is also acceptable.

This form of the model and the distribution of the error model are adopted to derive the Peak Water Height (PWH) of the tsunamis in the sea zone far from the shoreline, see Fig. 15.11:

$$\frac{H}{H_0} = \frac{e^{-\beta.D_h}}{1 + D_h.e^{-(M_w - M_o)}} \tag{15.75}$$

Where: H [in m] = Peak Water Height, D_h [in km] = hypocentral central, M_w = moment magnitude of the earthquake, and three fitting constants β, M_0 = threshold magnitude and H_0 [in m] = equivalent uplift height at the epicentral zone. The error model associated to the PWH is also assumed to follow a Gamma distribution.

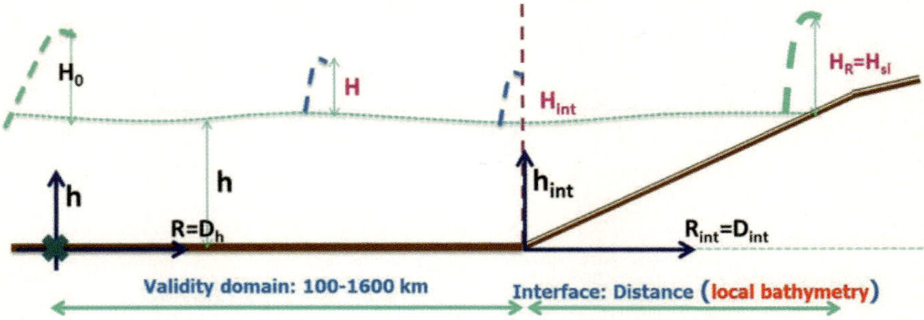

Fig. 15.12 Run-up and slopes between interface zones and shorelines

The constant H_0 is adopted according to Abe's proposal (Abe 1993):

$$\log(H_0) = 0.5M_w - 3.3 + C \tag{15.76}$$

Where: C = constant as fitting parameter depending on the kind of subduction zone.

Furthermore, the velocity of the tsunami waves is also an important parameter that governs the impact loads on the structures and facilities. It is derived from the Peak Water Height as follows:

$$V = \sqrt{g(h + H)} \tag{15.77}$$

Near the shoreline, the seabed may have a regular or disturbed slope from an interface zone (the seabed is not considered as horizontal) up to the shoreline, see Fig. 15.12. Of course, these interface zone and average slope until the shoreline depend on the local topography and bathymetry. For sake of simplification, sometimes one could assume a straight line from the interface until the shoreline. The Peak Water Height at the shoreline is obtained by energy conservation, when no attenuation is considered, i.e.:

$$H_{sl}^2 . \sqrt{H_{sl}} = H_{\text{int}}^2 . \sqrt{(h_{\text{int}} + H_{\text{int}})} \tag{15.78}$$

Due to the attenuation of energy, the final Peak Water Height at the shoreline becomes:

$$H_{sl} = \left(H_{\text{int}}^2 . \sqrt{(h_{\text{int}} + H_{\text{int}})}\right)^{2/3} . \frac{e^{-\beta . D_{\text{int}}}}{1 + D_{\text{int}} . e^{-(M_w - M_o)}} \tag{15.79}$$

Where: H_{sl} [in m] = Peak Water Height at the shoreline, H_{int} [in m] = Peak Water Height at the interface zone, h_{int} [in m] = depth of the sea at the interface zone, D_{int} [in km] = Horizontal projection of the distance from the interface zone towards the shoreline.

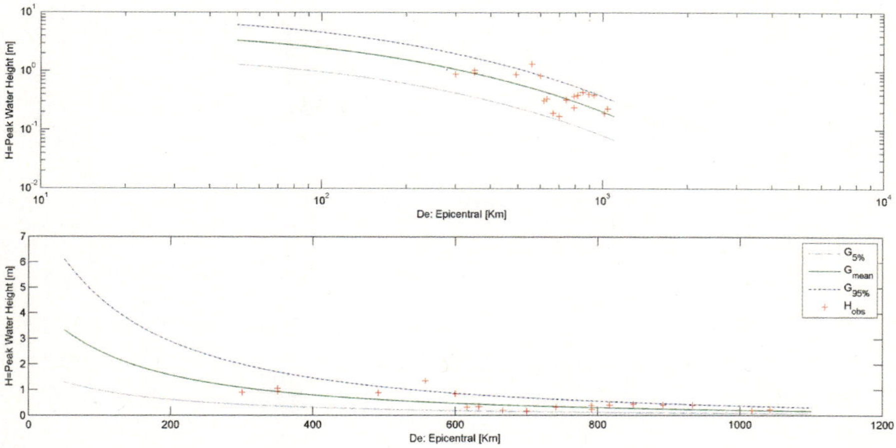

Fig. 15.13 Tsunamis height: H_{obs} = observed value; G_{mean}, $G_{5\%}$, $G_{95\%}$ = Predicted Mean value, and fractiles 5 % and 95 % respectively.

For illustrative purposes, the model is run for the case of the tsunami and Peak Water Heights observed during the earthquake Akita Oki, Japan on May25, 1983 with moment magnitude $M_w = 7.9$ (Abe 1995). According to the bathymetry collected for the zones under study (GEBCO 2012), the proposed model provides theoretical values that are in good accordance with the observed heights (H_{obs}), see Fig. 15.13. Furthermore, the theoretical confidence interval [$H_{5\%}$ up to $H_{95\%}$] contains 95 % of the experimental values, i.e. more than the acceptable ratio of 90 %. A gamma distribution is considered for the error model with a coefficient of variation $C_v = 45$ %.

15.4.3 Industrial Tanks and Vulnerability Under Tsunamis Effects

Under the quake and tsunami effects, the industrial tanks and facilities may suffer serious damages. These latters may reach serious intensities and their effect may propagate and might give rise to disastrous situation.

The tsunami may generate various effects and cause different damages to the structures and facilities at the coastal zones, under the mechanical loads, (ASCE 2010; ATC 2008, 2011; Batdorf 1974; FEMA; INERIS 2011; Nishi 2012; Goto 2008; Lukkunaprasit et al. 2009; Naito et al. 2012; Palermo and Nistor 2008; Saatçioğlu 2009; Sakakiyama et al. 2009; USGS 2011; Yeh 2008), see Fig. 15.14:

- Hydrodynamic forces
- Hydrostatic pressures
- Buoyancy forces, and

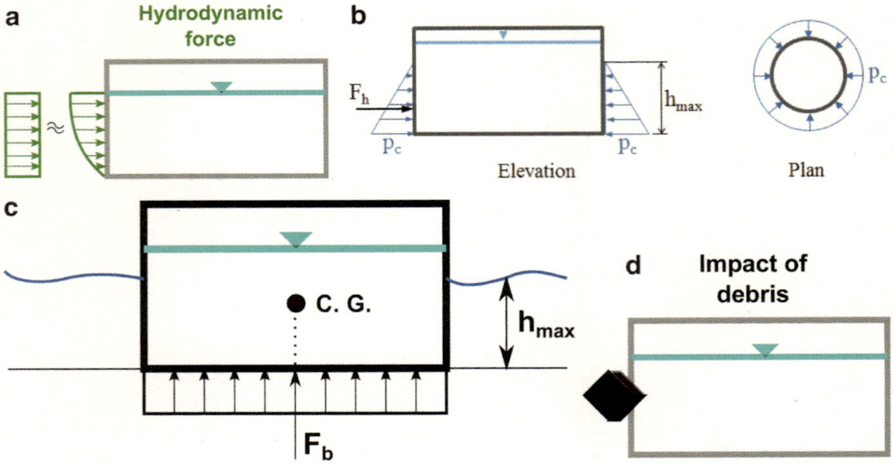

Fig. 15.14 Mechanical loadings, pressures and impacts on the industrial plants and facilities

- Debris impacts (boulders, cars, ships, etc.).

Various resulting damages may affect the industrial tanks and facilities:

- Excessive stresses under bending, shear and axial effects
- Stability, Sliding and overturning
- Lateral and longitudinal buckling
- Perforation, loss and release of flammable products (liquid, gas) that may produce fireballs and explosions.

15.4.4 Fragility Curves and Risk of Failure

Regarding any of the possible damages described above, the safety or failure of the affected structures depends on the loads, denoted S and its capacity to stand these loads, denoted R. The loads depend on the quake and tsunami inputs, i.e. PGA and PWH in the present case. The resistance depends on the geometry, external supports, the filling ratio of the tank, the thermodynamic conditions of the products contained by the tank, and the constitutive materials. Therefore, the loads as the resistances can be described by random variables. The probability of failure of the tank or facility under study is then defined as:

$$P_f = (R - S \leq 0) \tag{15.80}$$

This risk of failure is in general calculated by Monte Carlo simulations or level 2 methods, (Mebarki et al. 2008a). For industrial plants, it is worth to establish fragility curves that express the probability of exceeding given damage levels according to the governing input parameters such as PGA or PWH, for instance.

Various sophisticated or simplified methods can be used to investigate the response of cylindrical tanks to external pressures (Batford 1974; CCH 2000; CEN 2007; Chen and Rotter 2012; Godoy 2007; INERIS 2011; Koshimura and Namegaya 2009; Nistor et al. 2010a, 2010b; Suguino and Iwabuchi 2008; USGS 2011).

Figure 15.15 illustrates the fragility curve that corresponds to the case of a cylindrical tank with a slenderness (Tank height/Tank radius = L/r = 2.7), the filling ratio being considered as a uniform random variable and the limit state under study is the lateral buckling of the tank. For the present case, the risk of buckling is 0.27 for a Peak water height H = 4 m.

Once the fragility curves have been calibrated for given typologies, it becomes easy to predict the risk of disaster within an entire industrial plant against potential upcoming hazards. The protective measures can therefore be adopted according to the level of expected risks and their socio-economic consequences.

15.5 General Conclusions

The domino effect is a cascading sequence of accidents and explosions, propagating from an initial source to the surrounding tanks in an industrial plant. Under these successive sequences, the industrial plant and the facilities, constructions and also human beings are severely threatened and may suffer important and irreversible losses.

The study of this effect from the initial possible accident until the dissemination within the plant requires multi-disciplinary approaches. The purpose of the present study if to provide a theoretical formulation that describes and evaluates the risk of occurrence of the triggering event, due to external or internal causes, the propagation effect as it may produce four sub-events, mainly: fragments ejected as projectiles, blast waves, fire balls as well as loss of confinement.

Probabilistic formulation is developed in order to describe the occurrence of the first event, as well as the probability of occurrence of the four sub-events. The interaction between the surrounding vessels and facilities and the mechanical or thermal effects of these sub-events may cause global or local damage as well as possible fire ignition and explosion. They may take rise within the affected targets or their immediate vicinity such as pipelines and power lines, for instance. Mechanical, thermo-mechanical and also chemo-thermo-mechanical analyses are therefore required in order to quantify the resulting effect on the considered targets (tanks or power lines impacted, heated, blasted, etc.). In the case of impact by structural fragments for instance, penetration and perforation as well as interaction projectile-impacted tank need the use of adequate material and structural behaviours, performed in general by numerical simulations.

The present study provides the theoretical aspects that should be considered for a detailed analysis of the domino effect. Relying on these developments, numerical simulations can be performed and sensitivity analysis as well as critical scenarios can be studied.

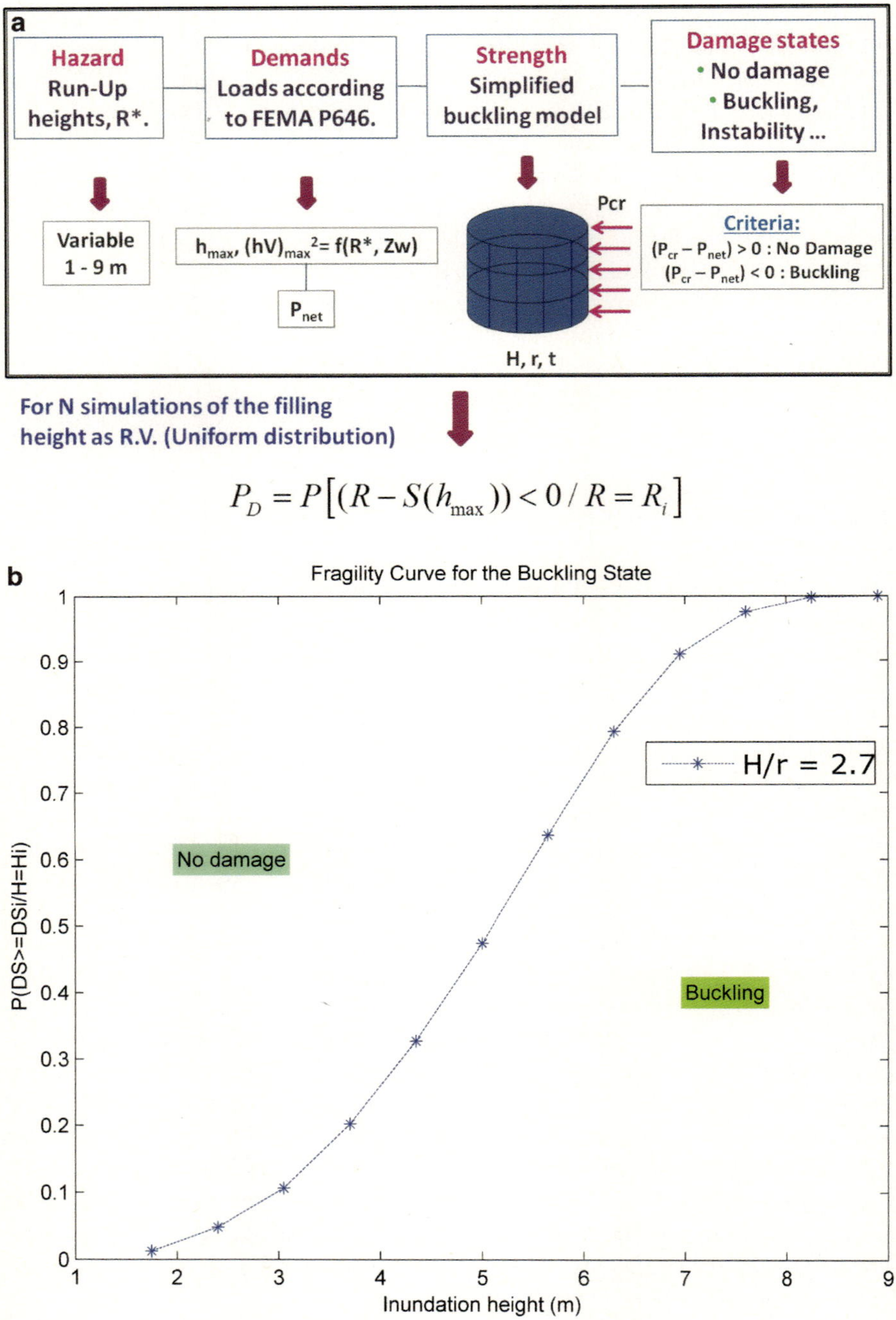

Fig. 15.15 Flowchart for fragility curve elaboration

For instance, for the case considered in this study, it is found that overpressure waves can produce significant damage at the near field. On the other hand, risk of projectiles impact is much lower at the near field but projectiles can trigger the domino effect at much higher distance. Actually, the probability of failure from the blast wave is 3 –7 times greater than the probability of failure produced by projectiles. Also, in this model only massive projectiles are considered as the potential projectiles and small, light projectiles are intentionally neglected.

For a whole industrial plant that may suffer quakes or tsunamis effects, it is required to describe the input parameters (Peak ground Acceleration for quakes, or Peak Water Height for tsunamis) by physical models affected by probabilistic distributions. As many effects and damages may be caused to the concerned components of the plant (tanks, pipelines, and other facilities), it is necessary to consider the whole possible limit states in order to describe the state of the components under study (excessive stresses, stability, sliding, perforation, etc.). Probabilistic descriptions of the inputs and components responses are helpful in order to calibrate the fragility curves of each generic category of component. The risk analysis under potential natural hazards becomes therefore easy to perform for an entire industrial plant. The protective measures derive from an optimisation process: theoretically, by balance between possible investments and socio-economic consequences of disaster occurrence.

Relying on the expected numeric results, one may perform an optimisation of the generalised utility function (or costs) resulting from initial costs and expected socio-economic consequences. Adequate investments and protective options can then be objectively decided by the stakeholders in order to mitigate the potential disasters and aim a quick recovery as well as resilience.

Acknowledgments The present study has been developed within the framework of the research projects VULCAIN and INTERNATECH, with the partial financial support by Agence Nationale de la Recherche (ANR: *PGCU 2007*, and *Flash Japon 2011*). The Chinese-French bilateral cooperation program PHC XU GUANGQI 2012 (Code Project: 27939XK) has also been helpful for the preparation and final redaction of the present paper.

References

Abbasi T, Abbasi SA (2007) The boiling liquid expanding vapour explosion (BLEVE): mechanism, consequence assessment, management. J Hazard Mater 141:489–519

Abe K (1993) Estimate of tsunami heights from earthquake magnitudes. In: Proceedings of the IUGG/IOC international tsunami symposium TSUNAMI'93, Wakayama

Abe K (1995) Modeling of the runup heights of the hokkaido-nansei-Oki tsunami of 12 July 1993. Pure Appl Geophys 144(3/4):113–124

Ali SY, Li QM (2008) Critical impact energy for the perforation of metallic plates. Nucl Eng Des 238:2521–2528

Antonioni G, Spadoni G, Cozzani V (2009) Application of domino effect quantitative risk assessment to an extended industrial area. J Loss Prev Process Ind 22:614–624

ARIA base of BARPI, France. www.aria.environnement.gouv.fr

ASCE (2010) Minimum design loads for buildings and other structures, ASCE/SEI standard. American Society of Civil Engineers, Reston, pp 7–10

Askan A, Yucemen MS (2010) Probabilistic methods for the estimation of potential seismic damage: application to reinforced concrete buildings in Turkey. Struct Saf 32:262–271, Elsevier

ATC (2008) Guidelines for Design of Structures for Vertical Evacuation from Tsunamis, FEMA P646. Applied Technology Council. Redwood City, California, For the Federal Emergency Management Agency, FEMA and the National Oceanic and Atmospheric Administration, NOAA.: 158 p

ATC (2011) Coastal Construction Manual, FEMA P-55. Applied Technology Council. Redwood City, California, For the Federal Emergency Management Agency FEMA. II: 400 p

Batdorf SB (1974) A simplified method of elastic-stability analysis for thin cylindrical shells. NACA report – 874: 25 p

Beltrami GM, Di Risio M (2011) Algorithms for automatic, real-time tsunami detection in wind-wave measurements. Part I: implementation strategies and basic tests. Coast Eng 58:1062–1071, Elsevier

Børvik T, Hooperstad OS, Langseth M, Malo KA (2003) Effect of target thickness in blunt projectile penetration of Weldox 460 E steel plates. Int J Impact Eng 28:413–464

Burwell D, Tolkova E, Chawla A (2007) Diffusion and dispersion characterization of a numerical tsunami model. Ocean Model 19:10–30, Elsevier

CCH (2000) City and county of Honolulu building code. Department of Planning and Permitting of Honolulu Hawaii, Honolulu

CEN (2007) EN 1993-1-6 eurocode 3: design of steel structures, part 1.6: strength and stability of shell structures. CEN, Brussels

Chen L, Rotter M (2012) Buckling of anchored cylindrical shells of uniform thickness under wind load. Eng Struct 41:199–208

Cheung KF, Wei Y, Yamazaki Y, Yim SCS (2011) Modeling of 500-year tsunamis for probabilistic design of coastal infrastructures in the pacific northwest. Coast Eng 58:970–985, Elsevier

Constantin A (2009) On the relevance of soliton theory to tsunami modelling. Wave Motion 46:420–426, Elsevier

Corbet GG, Reid SR, Johnson W (1995) Impact loading of plates and shells by free flying projectiles: a review. J Impact Eng 18:141–230, 0734-743X(95)00023-2

Cozzani V, Salzano E (2004) The quantitative assessment of domino effects caused by overpressure-Part I: probit models. J Hazard Mater A107:67–80

Demetracopoulos AC, Hadjitheodorou C, Antonopoulos JA (1994) Statistical and numerical analysis of tsunami wave heights in confined waters. Ocean Eng 21(7):629–643, Pergamon

Eckert S, Jelinek R, Zeug G, Krausmann E (2012) Remote sensing-based assessment of tsunami vulnerability and risk in Alexandria, Egypt. Appl Geogr 32:714–723, Elsevier

Federal Emergency Management Agency, FEMA, USA. http://www.fema.gov/photolibrary/photo_details.do?id=42405

Flouri ET, Kalligeris N, Alexandrakis G, Kampanis NA, Synolakis CE (2011) Application of a finite difference computational model to the simulation of earthquake generated tsunamis. Appl Numer Math 67:111–125. doi:10.1016/j.apnum.2011.06.003, Elsevier

GEBCO (2012) General Bathymetric Chart of the Oceans. Retrieved 15 June 2012, from www.gebco.net

Godoy LA (2007) Performance of storage tanks in oil facilities damaged by Hurricanes Katrina and Rita. J Perform Constructed Facil 21(6):441–449

Goto Y (2008) Tsunami damage to oil storage tanks. In: The 14 World Conference on Earthquake Engineering, Beijing

Goto K, Chagué-Goff C, Fujino S, Goff J, Jaffe B, Nishimura Y, Richmond B, Sugawara D, Szczucinski W, Tappin DR, Witter RC, Yulianto E (2011) New insights of tsunami hazard from the 2011 Tohoku-oki event. Mar Geol 290:46–50, Elsevier

Grasso VF, Singh A (2008) Global environmental alert service (GEAS). Adv Space Res 41:1836–1852, Elsevier

Haugen KB, Lovholt F, Harbitz CB (2005) Fundamental mechanisms for tsunami generation by submarine mass flows in idealised geometries. Mar Metroleum Geol 22:209–217, Elsevier

Heidarzadeh M, Pirooz MD, Zaker NH (2009) Modeling of the near-field effects of the worst-case tsunami in the Makran subduction zone. Ocean Eng 36:368–376, Elsevier

Helal MA, Mehanna MS (2008) Tsunamis from nature to physics. Chaos Solitons Fractals 36:787–796, Elsevier

Holden PL (1988) Assessment of missile hazards: review of incident experience relevant to major hazard plant. Safety and reliability directorate, Health & Safety Directorate

INERIS (2011) (*in French*) Note de caractérisation du comportement des équipements industriels à l'inondation. Rapport d'étude DRA-. Adrien Willot et Agnès Vallée, Institut National de l'Environnement Industriel et des Risques

Jin D, Lin J (2011) Managing tsunamis through early warning systems: a multidisciplinary approach. Ocean Coast Manag 54:189–199, Elsevier

Kharif C, Pelinovsky E (2005) Asteroids impact tsunamis. Physique 6:361–366

Koshimura S, Namegaya Y et al (2009) Tsunami fragility – a New measure to identify tsunami damage. J Disaster Res 4(6):479–490

Lees FP (2005) Loss prevention in the process industries, 3rd edn. Butterwort Heinemann, Oxford

Leone F, Lavigne F, Paris R, Denain JC, Vinet F (2011) A spatial analysis of the December 26th, 2004 tsunami-induced damages: lessons learned for a better risk assessment integrating buildings vulnerability. Appl Geogr 31:363–375, Elsevier

Liu PLF, Wang X, Salisbury AJ (2009) Tsunami hazard and early warning system in South China Sea. J Asian Earth Sci 36:2–12, Elsevier

Lovholt F, Glimsdal S, Harbitz CB, Zamora N, Nadim F, Peduzzi P, Dao H, Smebye H (2011) Tsunami hazard and exposure on the global scale. Earth-Sci Rev, Elsevier. doi:10.1016/j.earscirev.2011.10.002

Lukkunaprasit P, Thanasisathit N et al (2009) Experimental verification of FEMA P646 tsunami loading. J Disaster Res 4(6):410–418

Madsen PA (2010) On the evolution and run-up of tsunamis. J Hydrodyn 22:1–6. doi:10.1016/S1001-6058(09)60160-8, Elsevier

Marhavilas PK, Koulouriotis D, Gemeni V (2011) Risk analysis and assessment methodologies in the work sites : on a review, classification and comparative study of the scientific literature of the period 2000–2009. J Loss Prev Process Industries 24(5):477–523

Mebarki A, Mercier F, Nguyen QB, Ami Saada R, Meftah F, Reimeringer M (2007) A probabilistic model for the vulnerability of metal plates under the impact of cylindrical projectiles. J Loss Prev Process Industries 20:128–134

Mebarki A, Genatios C, Lafuente M (2008a) Risques Naturels et Technologiques : Aléas, Vulnérabilité et Fiabilité des Constructions – vers une formulation probabiliste intégrée. Presses Ponts et Chaussées, Paris, ISBN 978-2-85978-436-2

Mebarki A, Mercier F, Nguyen QB, Ami Saada R, Meftah F, Reimeringer M (2008b) Reliability analysis of metallic targets under metallic rods impact: towards a simplified probabilistic approach. J Loss Prev Process Industries 21:518–527

Mebarki A, Mercier F, Nguyen QB, Ami Saada R (2009a) Structural fragments and explosions in industrial facilities. Part I: probabilistic description of the source terms. J Loss Prev Process Industries 22(4):408–416. doi:10.1016/j.jlp.2009.02.006

Mebarki A, Mercier F, Nguyen QB, Ami Saada R (2009b) Structural fragments and explosions in industrial facilities. Part II: projectile trajectory and probability of impact. J Loss Prev Process Industries 22(4):417–425, 10.1016/j.jlp.2009.02.005

Mebarki A (2009) A comparative study of different PGA attenuation and error models: case of 1999 Chi-Chi earthquake. Tectonophysics 466:300–306

Mingguang Z, Juncheng J (2008) An improved probit method for assessment of domino effect to chemical process equipment caused by overpressure. J Hazard Mater 158:280–286

Naito C, Cox D et al (2012) Fuel storage container performance during the 2011 Tohoku japan tsunami. J Perform Constr Fac, 10.1061/(ASCE)CF.1943-5509.0000339

Nandasena NAK, Paris R, Tanaka N (2011) Reassessment of hydrodynamic equations: minimum flow velocity to initiate boulder transport by high energy events (storms, tsunamis). Mar Geol 281:70–84, Elsevier

Neilson AJ (1985) Empirical equations for the perforation of mild steel plates. J Impact Eng 3:137–142

Nishi H (2012) Damage on Hazardous Materials Facilities. In: international symposium on engineering lessons learned from the 2011 Great East Japan Earthquake, Tokyo

Nistor I, Palermo D et al (2010) Experimental and numerical modeling of tsunami loading on structures. In: International conference on coastal engineering, ASCE

Nistor I, Palermo D et al (2010b) In: Kim YC (ed) Tsunami-induced forces on structures. Handbook of coastal and ocean engineering. World Scientific Publishing Co. Pte. Ltd, Singapore, pp 261–286

Norio O, Ye T, Kajitani Y, Shi P, Tatano H (2011) The 2011 Eastern Japan great earthquake disaster: overview and comments. Int J Disaster Risk Sci 2(1):34–42

Ohte S, Yoshizawa H, Chiba N, Shida S (1982) Impact strength of steel plates struck by projectiles. Bull Japan Soc Mech Eng 25:1226–1231

Palermo D, Nistor I (2008) Tsunami-induced loading on structures. Structure Magazine 3:10–13

Pophet N, Kaewbanjak N, Asavanant J, Ioualalen M (2011) High grid resolution and parallelized tsunami simulation with fully nonlinear Boussinesq equations. Comput Fluids 40:258–268, Elsevier

Reese S, Bradley BA, Bind J, Smart G, Power W, Sturman J (2011) Empirical building fragilities from observed damage in the 2009 South Pacific tsunami. Earth Sci Rev 107:156–173, Elsevier

Ruiz C, Salvatorelli-D'Angelo F, Thompson VK (1989) Elastic response of thin-wall cylindrical vessels to blast loading. Comput Fluids 32(5):1061–1072

Saatçioğlu M (2009) Performance of structures during the 2004 Indian Ocean tsunami and tsunami induced forces for structural design. Earthquake Tsunamis 11:153–178, A. T. Tankut, Springer Netherlands

Sakakiyama T, Matsuura S et al (2009) Tsunami force acting on oil tanks and buckling analysis for tsunami pressure. J Disaster Res 4(6):427–435

Seveso Inspection Tool (2009) Réservoirs de stockage aériens atmosphériques, Deuxième version test, CRC/SIT/012-F

Sladen A, Hébert H, Schindelé F, Reymond D (2007) L'aléa tsunami en polynésie française : apports de la simulation numérique. C R Géosci 339:303–316, Elsevier

Suguino H, Iwabuchi Y et al (2008) Development of probabilistic methodology for evaluating tsunami risk on nuclear power plants. In: The 14th World Conference on Earthquake Engineering, Beijing

Talaslidis DG, Manolis GD, Paraskevopoulos E, Panagiotopoulos C, Pelekasis N (2004) The Sun website, UK: http://www.thesun.co.uk/sol/homepage/news/3615721/Four-die-in-oil-refinery-explosion.html

TNO (2005a) Methods for the calculation of possible damage to people and objects resulting from releases from hazardous materials. The Green Book CPR16E

TNO (2005b) Methods for the calculations of physical effects – due to release of hazardous materials (liquids and gases). The Yellow Book CPR14E 2005

Todorovska MII, Hayir A, Trifunac MD (2002) A note on tsunami amplitudes above submarine slides and slumps. Soil Dyn Earthq Eng 22:129–141, Elsevier

Tsamopoulos JA (2004) Risk analysis of industrial structures under extreme transient loads. Soil Dyn Earthq Eng 24:435–448

Università degli Studi di Torino. Laboratory of Molecular Electrochemistry, Italy. http://lem.ch.unito.it/didattica/infochimica/2008_Esplosivi/Explosion.html

USGS (2011) United States Geological Survey. Retrieved 13/03/2012, 2012, from www.usgs.gov

van den Berg AC (1985) The multi-energy method, a framework for vapor cloud explosion blast prediction. J Hazard Mater 12:1–10

van Zijll de Jong SL, Dominey-Howes D, Roman CE, Calgaro E, Gero A, Veland S, Bird DK, Muliaina T, Tuiloma-Sua D, Afioga TL (2011) Process, practice and priorities – key lessons learnt undertaking sensitive social reconnaissance research as part of an (UNESCO-IOC) International Tsunami Survey Team. Earth-Sci Rev 107:174–192, Elsevier

Ward SN (2011) In: Gupta HK (ed) Tsunamis. Encyclopedia of solid earth geophysics. Springer, Dordrecht, pp 1473–1492

Wijetunge JJ (2006) Tsunami on 26 December 2004: spatial distribution of tsunami height and the extent of inundation in Sri Lanka. Sci Tsunami Haz 24(3):225–240

Wilson RI, Dengler LA, Goltz JD, Legg MR, Miller KM, Ritchie A, Whitmore PM (2011) Emergency response and field observation activities of geoscientists in California (USA) during the September 29, 2009, Samoa Tsunami. Earth-Sci Rev 107:193–200, Elsevier

Xie M (2007) Thermodynamic and gas dynamic aspects of a BLEVE, Delft University of Technology, No.: 04–200708

Yeh H (2008) Maximum fluid forces in the tsunami runup zone. J Waterw Port Coast Ocean Eng 132(6):496–501

Zhang DH, Yip TL, Ng CO (2009) Predicting tsunami arrivals: estimates and policy implications. Mar Policy 33:643–650, Elsevier

Zhao BB, Duan WY, Webster WC (2011) Tsunami simulation with Green-Naghdi theory. Ocean Eng 3:389–396, Elsevier

Chapter 16
Destruction Patterns and Mechanisms of Coastal Levees on the Sendai Bay Coast Hit by the 2011 Tsunami

Tatsuki Iida, Akira Mano, Keiko Udo, and Hitoshi Tanaka

Abstract A mega tsunami hit the Sendai Bay Coast on March 11, 2011, overtopped coastal levees and intruded into far inland while sweeping houses, people and others away. Eighty percent of the levees, which rimmed the coast to protect the land from storm surges together with the wind waves were broken in various degrees of damage by the tsunami. The national and local governments decided to rebuild the levees to be durable even for mega tsunamis. This requirement motivates us to find the destruction mechanism of the coastal levees. We conducted field investigations and collected the tsunami records, aerial photos and tsunami videos. Especially, the video taken from the helicopter "Michinokugo" which flew along the Sendai Coast to the south during the attack of the tsunami's leading wave enables us to see the breaking process. Integrated analysis leads to two step mechanisms of the destruction: the first step of breaking the upper structure of the levees by the surging bore of the leading wave and the second step of expanding erosion by the return flow concentration.

Keywords Flow concentration • Local erosion • Levee breach

16.1 Introduction

The Sendai Bay Coast has 60 km long sandy beaches protected by the coastal levees of 6.0–7.2 m in height mainly against the storm surges and wind waves having the return period of about 50 years (Miyagi Prefectural Government 2004). On March 11, 2011,

T. Iida • H. Tanaka
Department of Civil and Environmental Engineering, Tohoku University, Sendai, Japan

A. Mano (✉) • K. Udo
International Research Institute of Disaster Science, Tohoku University, Sendai, Japan
e-mail: mano@civil.tohoku.ac.jp

Y.A. Kontar et al. (eds.), *Tsunami Events and Lessons Learned: Environmental and Societal Significance*, Advances in Natural and Technological Hazards Research 35, DOI 10.1007/978-94-007-7269-4_16,

Fig. 16.1 Levees broken by the tsunami on the Yamamoto Coast of the Southern Sendai Bay Coast

the Tsunamis hit the Sendai Bay Coast, and destroyed 80 % of the coastal levees, the damages ranging from slit-scattering of the armor blocks to the complete breach of the levees (Fig. 16.1). The national and local governments decided on a policy to rebuild the levees, and the levee height was determined to prevent overtopping for the higher hazards between the Level 1 tsunamis having the return period of 150 years and the storm surges together with the wind waves. The latter are higher on the most coasts of the bay area. The structure was, however, decided to be more robust to withstand Level 2 tsunamis having the return period more than 150 years, such as the 2011 Tsunami. This policy motivates us to grasp the status of destruction of the levees on the Sendai Bay Coast, and to find the destruction mechanisms of the levees.

16.2 Methodology

We collected various data such as the blue charts of the levees (Miyagi Prefectural Government 2004), observed and simulated tsunami height data, aerial photographs, and tsunami videos. Especially, the video taken from the helicopter, "Michinokugo" operated by the Ministry of Land, Infrastructure, Transport and Tourism (MLIT) was useful." Michinokugo" flew along the Sendai Bay Coast from Sendai City to Yamamoto Town during the attack of the leading wave of the tsunami and it recorded the tsunami behavior and levee state. We also conducted field investigations by measuring the tsunami height and erosion depth, and taking photos of the broken levees. We integrated these data by adjusting the spatial positions through GIS and temporal timings among the tsunami records, videos, and residents' witnesses.

16.3 Sendai Bay Coast

16.3.1 Study Area

The Sendai Bay area is characterized by the alluvial plain, sandy beaches, and shallow sea. Large cities and towns such as Higashi-Matsushima City, Shiogama City, Tagajyo City, Sendai City, Natori City, Iwanuma City, Watari Town, and Yamamoto Town have been developed in the plain. Ishinomaki City, the second largest city in Miyagi Prefecture, is located at the Kitakami River mouth as a port city and densely populated near the water front. The casualties who lost their lives and are still missing are about 3,800, accounting for 2.5 % of the city population. The Oshika Peninsula being the border of the Sendai Bay Coast and Sanriku Coast functions as a breakwater for the tsunamis generated off the Sanriku Coast. Thus, the Sendai bay area is also characterized by the very low incidence of big tsunamis. The return period of great tsunamis such as the 2011 Tsunami in the bay area is estimated to be 500–1,000 years. Low frequency of tsunamis and dense population near the coast are considered to be the major reasons for the high casualties in spite of the relatively lower tsunami height and shorter coast length compared with those of the Sanriku Coast.

16.3.2 Levees in the Bay Area

The major coastal hazards in the bay area are wind waves superimposed on storm surges. To protect the coast and the hinterland from the hazards, the coastal levees with the height of 6.0–7.2 m were constructed along the coasts. The major type of the coastal levees is a mild slope type using soil mounds as a base and covered by the concrete slabs or blocks embedded in the concrete lattices. Some coasts with poor beaches like the southern Yamamoto Coast suffer from the overtopping high wind waves. Therefore, the levees are covered by the three-sided concrete slabs and recurved parapets accompanied. The major type of the river levees is also a mild slope type which armors the river side only. At river mouths, the two types of mild slope levees are joined. The other type of the coastal levees is an upright type which is made from solid concrete (Fig. 16.2). Figure 16.3 shows the distribution of the two types of levees on the Sendai Bay Coast. The mild slope levees are dominant from the scenic and ecological viewpoint.

16.4 Features of the 2011 Tsunami

Inundation depth along the Sendai Bay Coast ranged from several to some 10 m as shown in Fig. 16.4 and almost all the coastal levees were overtopped. The inundation depth was much lower than that of Sanriku Coast which reached up to

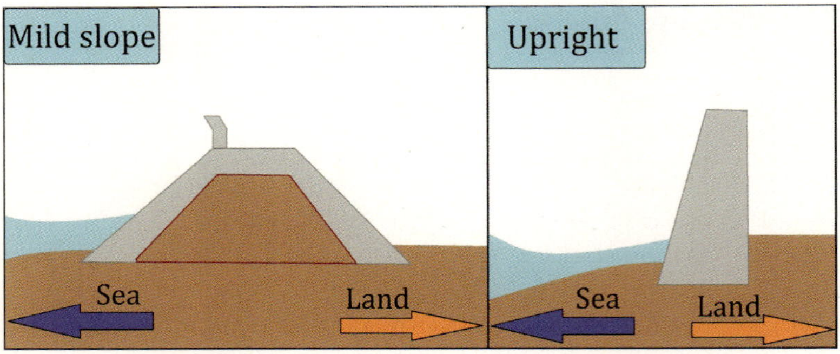

Fig. 16.2 Two structural types of levees

Fig. 16.3 Distribution of two structural types of levees on the Sendai Coast

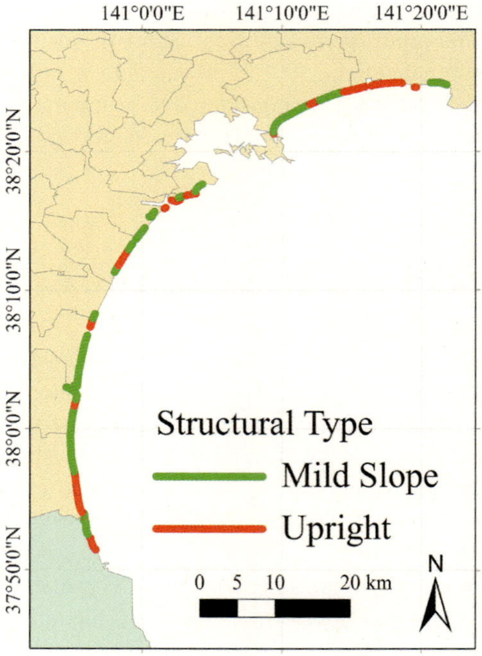

40 m. However, the damage to the levees was much greater in the bay area. The tsunami shapes in some rivers were recorded: the Kitakami River, the Naruse River, the Abukuma River, etc. The tidal gauges at the coasts in the bay area were all broken by the tsunami. Figure 16.5 shows the tsunami shape at the Abukuma Big Weir located 10 km upstream from the Abukuma River mouth. The leading wave has the highest amplitude of 5.3 m and steep front rise. This corresponds to

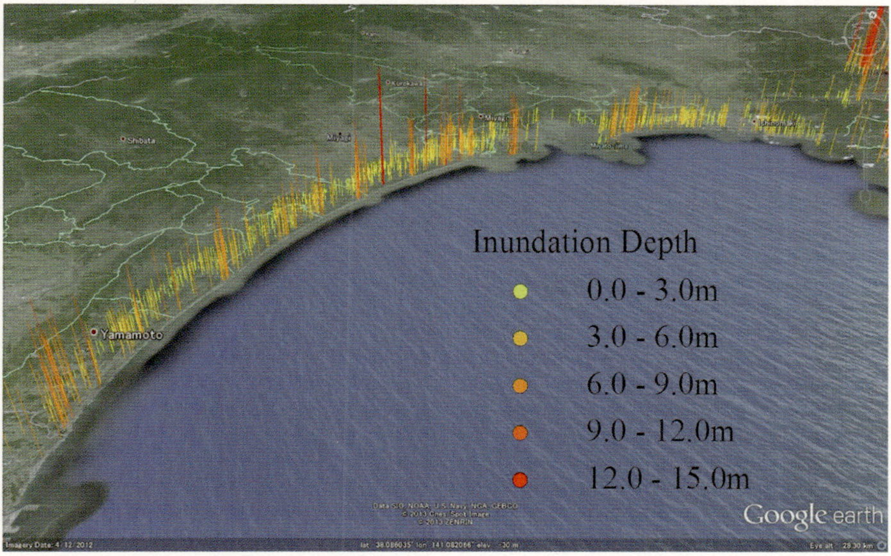

Fig. 16.4 Inundation depth on the Sendai Bay Coast by the 2011 tsunami of the Pacific Coast of Tohoku earthquake tsunami information (The Tohoku Earthquake Tsunami Joint Survey Group 2012)

Fig. 16.5 Aerial photo at Fujitsuka taken by Google Earth

a bore and agrees with the accounts by eyewitnesses who stated that the tsunami rushed as a wall of black water. The tsunami travel time from the mouth to the weir is calculated as 25 min. The tsunami shape at the Abukuma River mouth can be estimated by shifting the shape of Fig. 16.5 by 25 min and is synchronized with the tsunami video taken from "Michinokugo." The shadowed box in the figure shows the tsunami phase in which tsunami video was recorded. The phase corresponds to the back face of the leading wave.

16.5 Destruction Process

This chapter analyzes the destruction process of the levees focusing on the breach points by utilizing the tsunami video from "Michinokugo." Here breach is defined as complete loss of the levee foundation.

16.5.1 Breach at the Junction

Figure 16.6a, captured from the tsunami video, shows the return flow on the sea issued from the junction of the mild slope and upright levees (Fig. 16.6b) at Fujitsuka, 1.9 km north from the Natori River mouth. The armor blocks and wave dissipation blocks which had been placed in front of the mild slope levee were scattered behind the levee. The junction and the vicinity were widely eroded. A new channel was formed behind the upright levee with the depth deeper towards the junction. We call it "tsunami channel." The end segment of the upright levee was fallen landward due to the loss of the footing. These facts indicate the destruction process where the blocks were blown off much earlier before the video was recorded, probably by the front face bore of the leading tsunami. Then the return

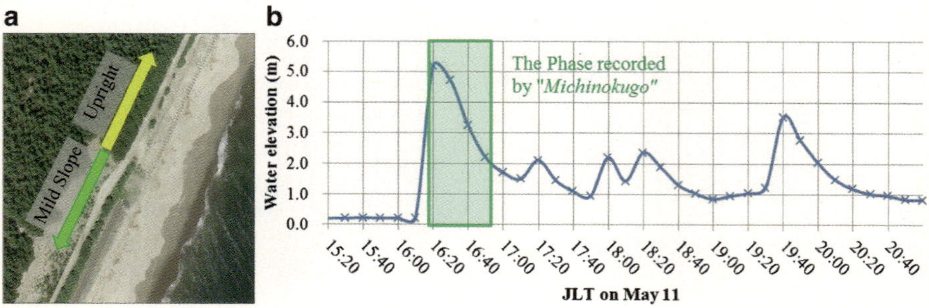

Fig. 16.6 (**a**) JLT 16:15; broken junction at the Fujitsuka taken by "Michinokugo". (**b**) Phase of the tsunamis at the Abukuma Big Weir by MLIT

Fig. 16.7 Aerial photos of the Yamamoto Coast before (*left*) and after the tsunami (*right*) taken by MLIT (March 27, 2010) and Kyodo News (March 9, 2011)

flow began and concentrated to the junction point with the major flow along the back side of the upright levee, while the flow was accelerated by the confluence of flow from the hinterland, and the levees were eroded deeply and widely. Finally the junction was breached.

16.5.2 Breaking of the Parapets

Figure 16.7 shows the aerial photos of the Yamamoto Coast before and after the tsunami. This coast had suffered from severe erosion, and thus a series of T-shape jetties were constructed. The beaches were narrow and the coastal levees with recurved parapets were protected by wave dissipation blocks in front of the levees. The tsunami breached the levees at many points and a long tsunami channel was formed along the back side of the levees. Figure 16.5 captured on the tsunami video shows that several parts of the levees were submerged and turbid water issued from the submerged parts formed large eddies on the sea. Figure 16.5 shows that the recurved parapets were broken, the top and landward slabs were destroyed, and the soil mound was eroded.

Fig. 16.8 (**a**) JLT 16:39 Breaking of the upper structures on the Yamamoto Coast taken by "Michinokugo". (**b**) Breaking of the upper structure on the Yamamoto Coast

16.5.3 Concentration of the Return Flow

Figure 16.8a show the aerial photo of the Yamamoto Coast taken on March 12, 2011, just after the tsunami, and b the damage in situ. In Fig. 16.9a, b (next page) the topographic maps of the same area before the tsunami by Geospatial Information Authority of Japan (GSI). The areas with large-scale erosion are encircled on the photos. The areas correspond to the current or old channels, and/or swamps. The number 1' in Fig. 16.9a shows an old channel with a dead

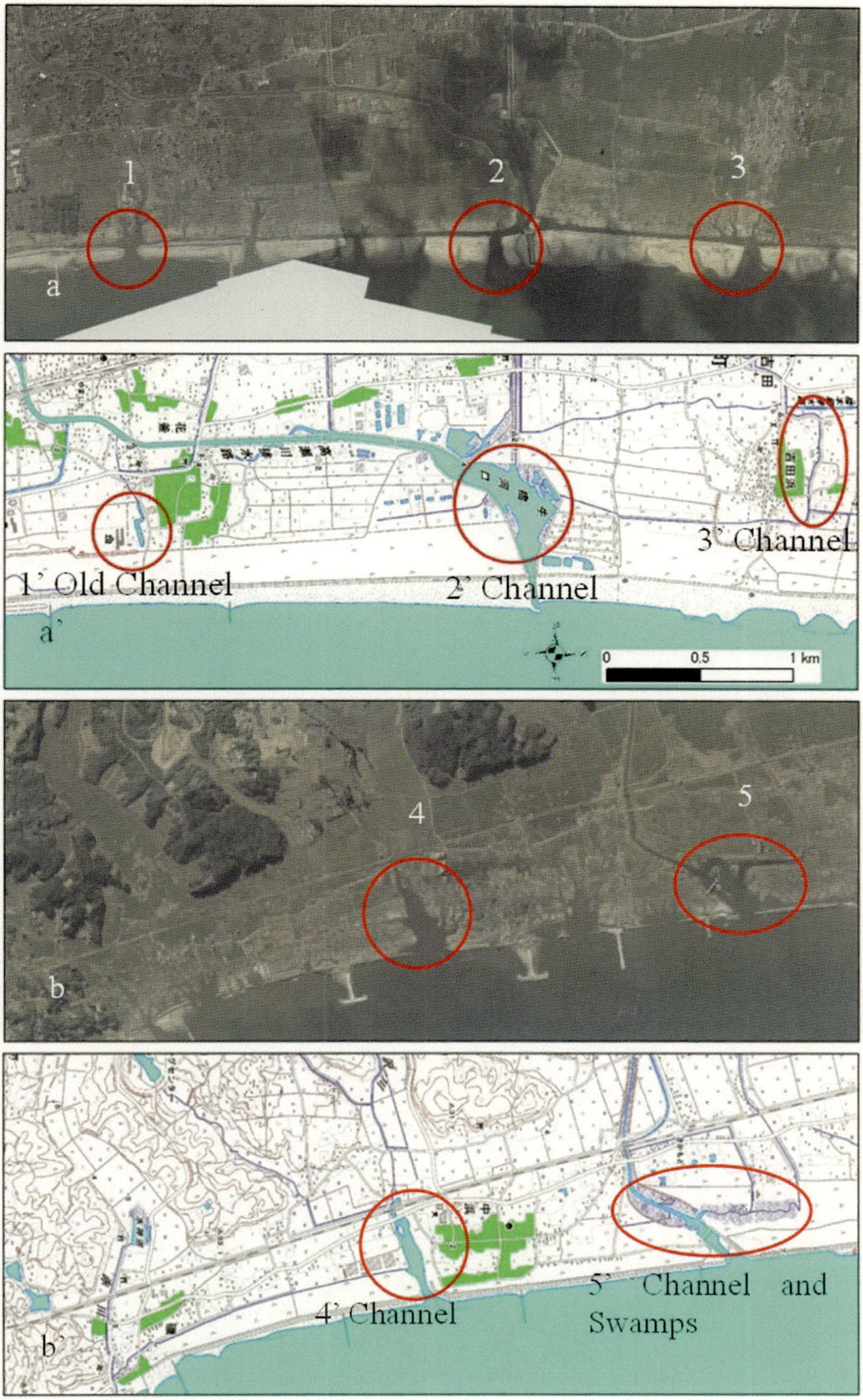

Fig. 16.9 (**a**) Large scale erosion corresponding to the channels or swamps by GSI (a: photo, a' map). (**b**) (b: photo, b' map). (**c**) (c: photo, c' map)

Fig. 16.9 (continued)

end before the tsunami. This channel was extended and connected to the sea after the tsunami. The coastal levee in the channel's extended area was widely breached. The numbers 6′ and 7′ in Fig. 16.9c show the similar status. This large scale erosion is formed by the concentration of the return flow to the channels and lowland swamps. Some of them were filled with the littoral drift. However, large ones still remained as submerged water body. This is called "tsunami bay."

16.6 Correlation of Breach

Figure 16.10 (next page) shows the distribution of the breach points along the Sendai Bay Coast, indicating dense distribution of the breach points in the southern part, especially on the Yamamoto Coast. As stated in the previous chapter, the direct cause of the breach is the erosion by the return flow concentration due to (1) breaking of the upper-structure of the levees by the leading tsunami and (2) existence of the channels or swamps in the hinterlands. The mechanism of the second cause is clear. Regarding the first cause, however, it is not certain why dense breaking occurred in the southern part. The possible factors of the breaking by

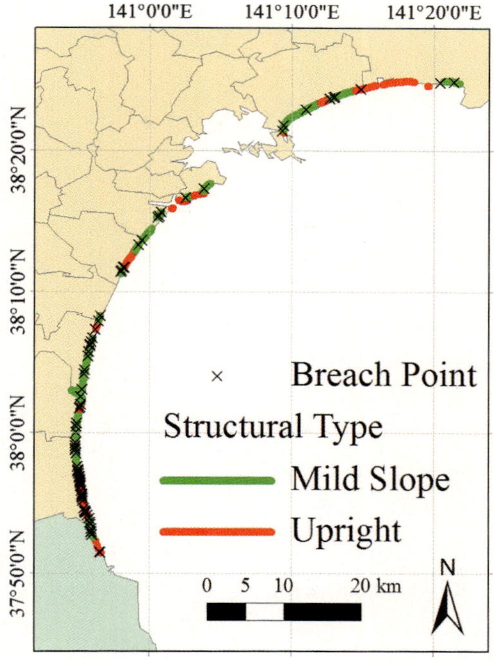

Fig. 16.10 Breach point distribution

the leading tsunami are (a) magnitude of the hazard, (b) approach to the levees, and (c) structure of the levees. To analyze the correlation with these factors, we define the breach density ρ by the reciprocal of the average interval as follows:

$$\rho = \frac{1}{(d_1 + d_2)/2} \tag{16.1}$$

in which d_1 and d_2 are the intervals of the adjacent breach points.

We choose the beach width as the approach factor with which the correlation of the breach density is shown in Fig. 16.11a. Narrow beaches strongly correlate to high breach density. This indicates that the leading tsunami gave high impact to the coastal levees on the narrow beaches. This is because the depth near the shoreline was deep due to erosion and the levees installed recurved parapets, which received the high impact. The types of the levees are shown by different marks in this figure. Upright levees have low breach density.

We choose the overflow depth as the hazard factor. Here the depth is defined as the difference of the tsunami height, which is obtained from the tsunami simulation and the levee height. Figure 16.11b shows the correlation of the overflow depth and breach density. It is obvious that high depth correlates to high breach density.

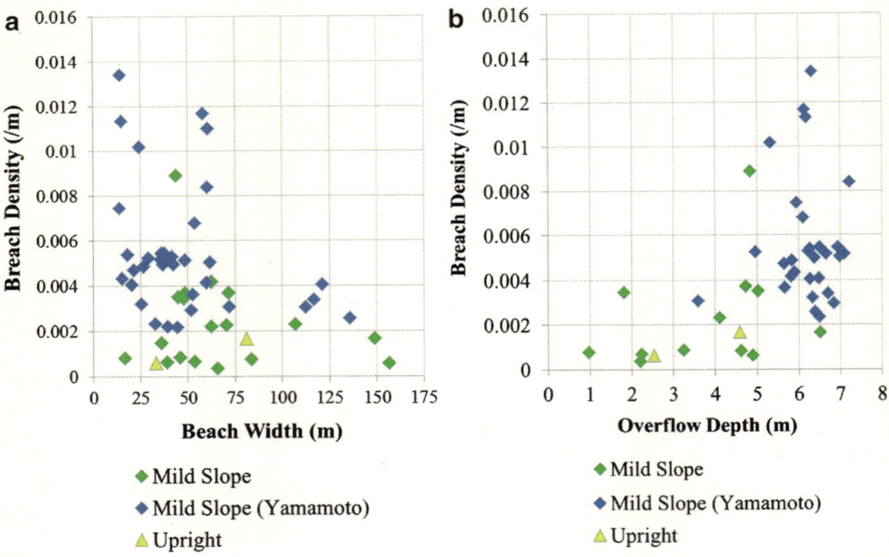

Fig. 16.11 (**a**) Beach width versus breach density (**b**) Overflow depth versus breach density

16.7 Conclusion

Our findings are summarized as follows:

1. Breach of the coastal levees densely occurred in the southern part of the Sendai Bay Coast, especially on the Yamamoto Coast.
2. The breach progressed in two steps: Firstly, the upper structure was destroyed by being hit by the leading tsunami which gave strong impact on the parapets and/or on the weak structure of junction. Secondly, the concentration of the return flow eroded and breached levees with the broken points from the first step, or with the channels or swamps in the hinterland.
3. Dense breach correlates with narrow beach and high overflow.

Acknowledgement This study was supported by the Specific Project Grant 2012 by IRIDeS and by Grant-in-Aid for Scientific Research (B), 23360193. These supports are gratefully acknowledged. The authors also wish to express their gratitude to Ms. Yuriko Takeda for the excellent GIS analysis.

References

Miyagi Prefectural Government (2004) Sendai bay coast basic protection project, chapter 5

The 2011 Tohoku Earthquake Tsunami Joint Survey Group (2012) Tohoku earthquake tsunami information. http://www.coastal.jp/ttjt/

Chapter 17
Damage and Reconstruction After the 2004 Indian Ocean Tsunami and the 2011 Tohoku Tsunami

Anawat Suppasri, Abdul Muhari, Prasanthi Ranasinghe, Erick Mas, Fumihiko Imamura, and Shunichi Koshimura

Abstract The 2004 Indian Ocean tsunami was one of the world's worst tsunamis and caused devastating damage in many Asian countries. Then, in 2011, Japan was hit by a tsunami that was generated by the greatest earthquake in the country's history. This paper discusses the damage caused by these tsunamis and subsequent reconstruction. Introduced first are the experience gained and lessons learned for future tsunami mitigation, such as tsunami awareness, proper evacuation building and the memorial parks created in the countries affected by the 2004 tsunami (Indonesia, Sri Lanka and Thailand). Second, human casualties and building damage are discussed using fatality ratios and fragility curves, respectively. These analyses show that experience and awareness help reduce human casualties in the Sanriku area, and wooden houses damaged by the 2011 tsunami fared better than in previous historical events. The damage by the 2011 tsunami to structures designed to protect against tsunamis is summarized. Most of these structures could not withstand and protect from the tsunami because they were not designed for such a large tsunami as expecting of such great event. Finally, examples of ongoing reconstruction in Japan are introduced. Most reconstruction efforts were planned after considering the lessons learned from the tsunami's impact, and the towns in question are now strengthening their disaster prevention-related plans to be better prepared for future tsunamis.

Keywords The 2004 Indian Ocean tsunami • The 2011 Great East Japan tsunami • Reconstruction • Disaster reduction • Disaster mitigation

A. Suppasri (✉) • A. Muhari • P. Ranasinghe • E. Mas • F. Imamura • S. Koshimura
International Research Institute of Disaster Science, Tohoku University, Sendai, Japan
e-mail: suppasri@irides.tohoku.ac.jp

Y.A. Kontar et al. (eds.), *Tsunami Events and Lessons Learned: Environmental and Societal Significance*, Advances in Natural and Technological Hazards Research 35,
DOI 10.1007/978-94-007-7269-4_17, © Springer Science+Business Media Dordrecht 2014

17.1 Introduction

This chapter discusses damage, reconstruction and adaptation from the two great tsunamis in this decade, the 2004 Indian Ocean tsunami and the 2011 Tohoku tsunami. The 2004 Indian Ocean earthquake and tsunami occurred on December 26th, 2004. It was the first instrumentally and historically recorded major earthquake and tsunami in the region. The earthquake's magnitude was estimated at 9.3, which shocked most scientists at the time. The estimated rupture area reached approximately 1,000 km, the longest rupture ever recorded. A tsunami was generated, and the maximum tsunami height of approximately 50 m was measured in the Banda Aceh region of Indonesia. The tsunami caused wide-ranging damage in 15 countries and approximately 230,000 deaths. Seven years later, on March 11th, 2011, a massive earthquake of magnitude 9.0 occurred off the Pacific coast of Tohoku (in the northeast region of Japan) and was followed by a devastating tsunami that again shocked many scientists with its magnitude. This time, the rupture length was estimated at 500 km, with a large slip of more than 30 m. A maximum tsunami height of approximately 40 m was measured, and approximately 20,000 deaths in Japan and one death each in the United States and Indonesia were reported. A comparison of the two tsunamis is shown in Table 17.1 (NOAA 2011).

In general, both tsunamis were generated from a massive M9-class earthquake with a large rupture area and resulted in a large tsunami of 40–50 m in height. One difference between the two events is the tenfold difference in the number of casualties due to the earthquake and tsunami source locations, which can often lead to a difference in the number of affected countries.

This chapter starts by introducing historical background of tsunami experience in Tohoku region. Comparison of human loss from the two great tsunamis was then compared using a relationship between recorded maximum tsunami height and fatality ratio reported at each location. The last section will introduce successful examples such as evacuation building, museum and memorial park from countries affected by the 2004 event and give information of present reconstruction process and their adaptation for future tsunami in tsunami affected areas by the 2011 event.

Table 17.1 Comparison between the 2004 Indian Ocean tsunami and the 2011 Great East Japan tsunami

Item	2004 tsunami	2011 tsunami
Earthquake magnitude	9.3	9.0
Size of rupture (km^2)	1,000*150	500*200
Max. tsunami height (m)	50.9	40.5
No. of death	230,000	19,000
No. of affected country	15	Mostly in Japan

17.2 Japanese Tsunami Experience in Tohoku Region

It is said that disasters always occur when people start forgetting about them. It is very important to use experience with correct understanding and remain high awareness to reduce human loss from tsunamis. This section introduces how experience affected on awareness and human loss giving Japan as an example as they had long experience on tsunamis. Location and maximum tsunami height distribution of historical tsunamis are shown in Fig. 17.1 (Suppasri et al. 2012c).

It was known as people in the past believed that tsunami comes after great shake from an earthquake. For example a great one in 1611 that caused tsunami damage in wide range of the Tohoku region. However, the earthquake in 1896 occurred at night with small shake which later classified as a tsunami-earthquake that can generate comparatively higher tsunami than a typical tsunamigenic earthquake (Satake and Tanioka 1999). As results, about 22,000 lives were killed with the maximum tsunami runup height 38.2 m reported. This was the greatest tsunami disaster in Japan before the 2011 event. In 1933, a great tsunami was generated again with the maximum runup height measured as 28.7 m but the number of death reduced to about 3,000. This was probably because this tsunami occurred just 37 years after the greatest one and people still had high tsunami awareness to perform evacuation quickly. However, there were some people that ignored an evacuation because their misunderstanding from the earthquake in 1896 that large tsunami will be generated by earthquake with small shake (Yamashita 2008a).

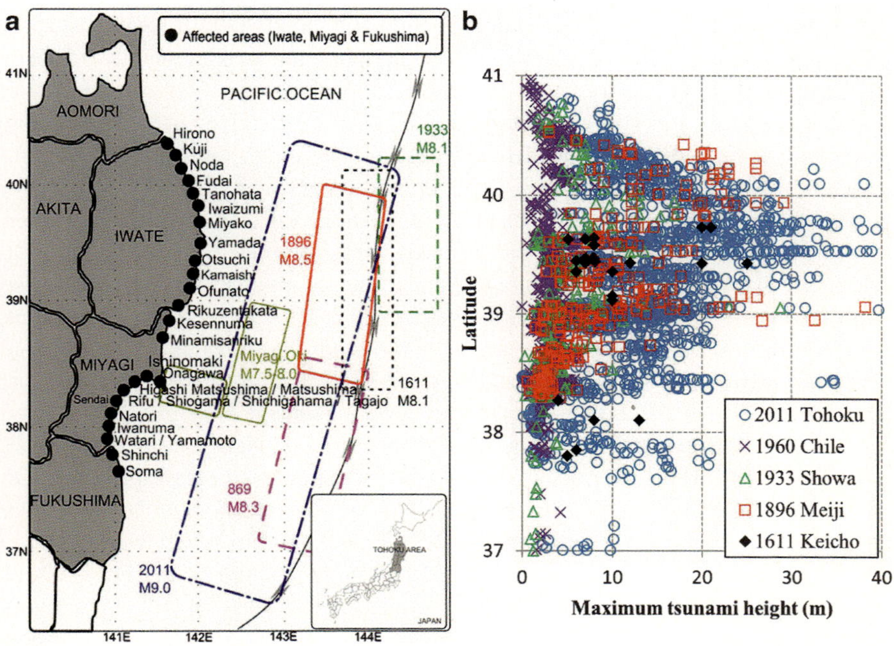

Fig. 17.1 Historical tsunamis in the Tohoku region the areas that were affected by the 2011 Tohoku tsunami and the maximum tsunami height of historical tsunamis

A great tsunami attacked Tohoku region again in 1960 but this time had even no shake that people can felt at all. The greatest earthquake magnitude in the world (M9.5) occurred at Chile and generated tsunami that attacked Tohoku region after approximately 1 day. At that time, maximum tsunami was measured as 10.7 m with 142 deaths were reported. This tsunami advocated a development of tsunami countermeasures in Japan to become the most advance country in the world (Abe and Imamura 2010). Besides, many people believed that tsunami will only occur in Pacific coast and not occur in Japan Sea coast. As results, large tsunamis with maximum height around 15 m in 1983 (Japan Sea tsunami) and 1993 (SW Hokkaido or Okushiri tsunami) claimed about 100 and 200 lives respectively (Yamashita 2008b). Finally, the 2011 Tohoku tsunami repeated the tragedy again as the greatest measured runup height of 40.5 m (The 2011 Tohoku Earthquake Tsunami Joint Survey Group 2011) was generated by M9.0 which never be expecting in Japan with results 19,000 deaths. This situation was similar to what happened in the 2004 Indian Ocean tsunami that caused almost 200,000 deaths in Indonesia and large damage in Sri Lanka, India and Thailand. This was because such a great earthquake was never expected to have occurred in that region before.

17.3 Human Loss Due to Tsunamis in Japan and Indian Ocean

Relationship between maximum tsunami height and fatality ratio for each historical tsunami event was used to explain on how experience and awareness may influence on human loss issue. An example of fatality ratio of historical tsunamis in Japan adapted from Kawata (1997) and Suppasri et al. (2012a) are given to explain on how experience safe life from tsunamis (Fig. 17.2a). The oldest tsunami that tsunami casualty data was available was in 1771 Meiwa tsunami in Okinawa. The tsunami caused fatality ratio higher than 0.9 in some locations since the average maximum tsunami height less than 10 m and can be classified as zone (1). The great one in 1896 Meiji-Sanriku tsunami had an average tsunami height greater than 10 m with also high fatality ratio and is classified as zone (2). After 37 years from the 1896 event, 1933 Showa-Sanriku had smaller fatality ratio even they also experienced large tsunami height (zone (3)). 1944 Tonankai and 1946 Nankai tsunami occurred just after about 10 years with fatality ratio was greatly reduced. This can be classified as zone (5) where both maximum tsunami height and fatality ratio were small. The 1933 Okushiri tsunami occurred in a place that never expected again but with lesser tsunami height leading to zone (3). The fatality ratio for the 2011 Tohoku tsunami in Sanriku coast is similar to the 1933 Showa-Sanriku as they were attacked by great tsunami but experience may help decreasing the fatality ratio. For the 2011 Tohoku tsunami in Sendai Plain coast can be classified as zone (4) as comparatively smaller tsunami but caused somewhat high fatality ratio. This figure shows importance of experience in reducing human loss by tsunami.

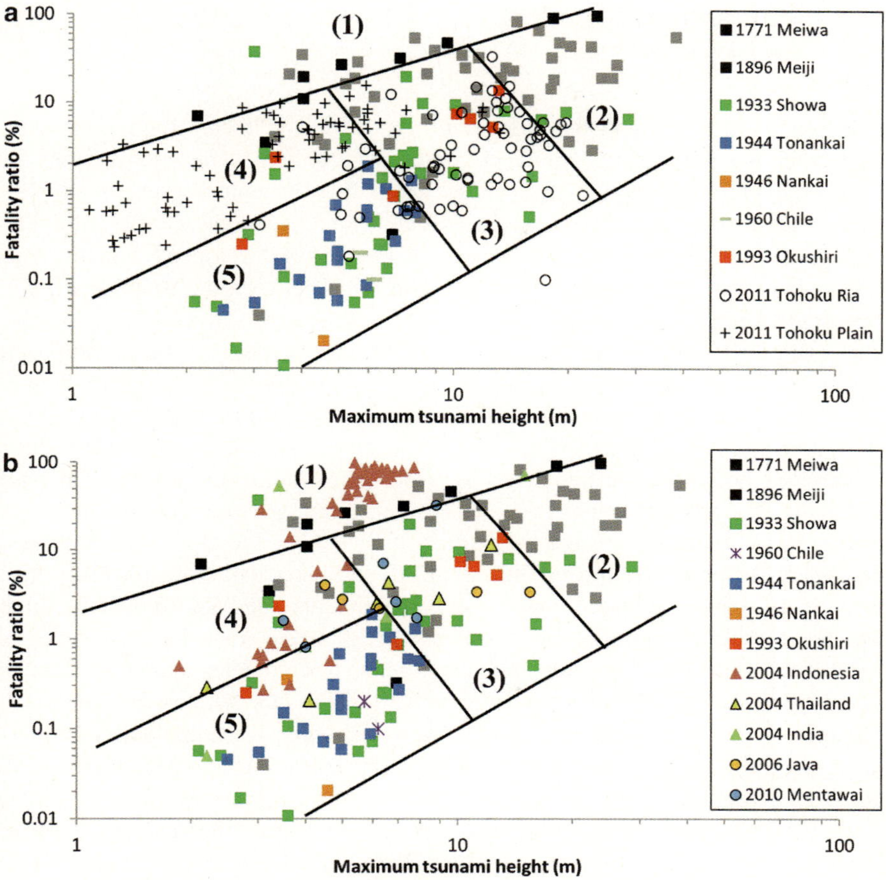

Fig. 17.2 (**a**) Tsunami fatality ratio in Tohoku region from historical tsunamis in Japan. (**b**) Tsunami fatality ratio in Tohoku region from historical tsunamis in Japan and Indian Ocean

Figure 17.2b shows a comparison including the Indian Ocean 2004 tsunami in some countries (triangle) and local tsunamis in Indonesia after the 2004 event (circle). It can be seen that fatality ratio of the 2004 tsunami in Banda Aceh, Indonesia is the highest (zone (1)) as coastal residence had no experience similar to the 1771 Meiwa tsunami in Japan. Other countries such as India and Thailand has comparatively smaller fatality ratio as they locate far from the tsunami source. Fatality ratio for Indonesian from the 2006 Java and 2010 Mentawai event (both are local tsunami) is reduced to zone (3) and (4) as they gain tsunami awareness from the 2004 event and with lower tsunami height.

17.4 Building Damage Due to Tsunamis in Japan and Indian Ocean

Building damage is also one important issue for the reconstruction process. Understanding of building damage characteristic will help as decision support for land use management and city planning against future tsunami. Tsunami fragility curves (Koshimura et al. 2009b) previously developed in countries affected by the 2004 Indian Ocean data were compared with the curves developed by the preliminary data of the 2011 Tohoku event. A comparison between the tsunami fragility curves for wooden houses is shown in Fig. 17.4. There are 150 wooden houses for the preliminary data of the 2011 event in Miyagi prefecture, Japan (Suppasri et al. 2012b). Tsunami fragility curves for wooden houses in Miyagi prefecture are plotted with those from other locations, such as Okushiri, Japan, (Koshimura et al. 2009a) and Sri Lanka (Murao and Nakazato 2010), and shown in Fig. 17.3.

Wooden houses in the latter two locations were rather old. For the Okushiri event, the area was suffered by the extensive fire during and after the tsunami attack which was not possible to differentiate in their study. Therefore, the combined damage by tsunami and fire in Okushiri leads to higher damage probability than those in Sri Lanka. Wooden houses in Okushiri, Japan, and Sri Lanka have high damage probability compared to those in Miyagi at the same inundation depth. For example, the probability for wooden houses to be completely damaged at a 2 m inundation depth in Okushiri is as high as 0.80, while this value in Sri Lanka and Miyagi is 0.15 and less than 0.05, respectively. However, if the inundation depth is increased from 3 to 4 m, the probability will increase from 0.85 to 0.95, 0.25 to 0.50 and 0.10 to 0.40 for Okushiri, Sri Lanka and Miyagi, respectively.

Even though most houses in Miyagi prefecture are constructed using wood, most of the houses are quite new and strong because of the Miyagi-oki earthquake. This

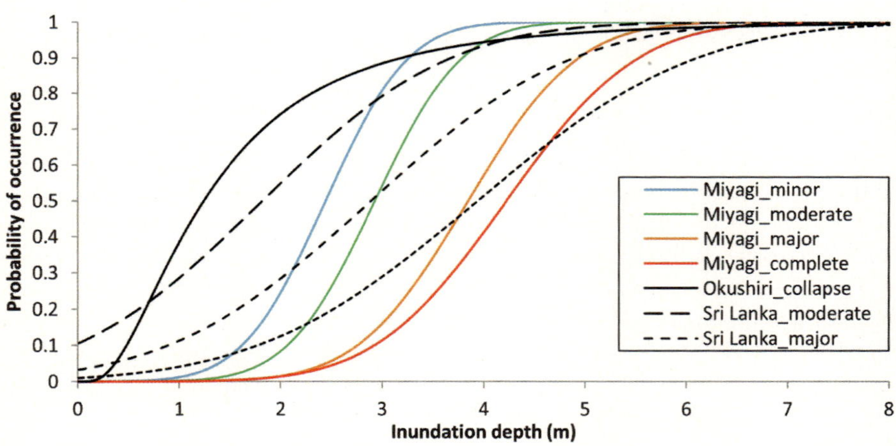

Fig. 17.3 Comparison between the tsunami fragility curves for the structural destruction of wooden houses in the Miyagi prefecture, Japan, and those at other locations

shows that 3–4 m is the critical inundation depth for Miyagi and the prefecture plan not to build a house in locations where expected inundation depth greater than 2 m. This agrees well with the full-scale experimental result on wooden walls in a water flume (Arikawa 2009). The wooden wall was broken when a tsunami depth of 2.5 m was applied.

17.4.1 Reconstruction and Adaptation

This section give information on successful reconstruction as an example from the 2004 Indian Ocean tsunami affected countries and what is going on and attempt to reduce impact for future tsunami in Japan.

17.4.1.1 The 2004 Indian Ocean Tsunami

Banda Aceh Province, Indonesia

The redevelopment of areas in the Aceh region was conducted through a community participatory method at the village level. For instance, a community would decide whether their area would be built with the employment of land consolidation or as it was. Along with these village-level reconstruction efforts, facilities and infrastructure for future tsunami mitigation were also prepared at a higher level. However, most of them were no use/no people live there. Most of the survivors just want to re-build their houses exactly where it was. Buildings for vertical evacuation are now ready for use in Banda Aceh (Fig. 17.4a). Nevertheless, the building was not used as expected during an evacuation during several earthquakes and tsunami warnings. Evacuation routes have been tested in national and international tsunami drills such as the IO-wave (Indian Ocean wave of 2009) in which all countries affected by the 2004 tsunami performed the drills. Recognizing that experiencing of the enormity of a tsunami is important, the media employed to spread the tsunami-related experiences of 2004 to the next generation come in various forms. A tsunami museum (Fig. 17.4b) is now established in Banda Aceh along with 85 tsunami poles indicating the height of the tsunami that hit the area. Other forms of memorial evidence such as ships, including a large diesel power plant vessel that was carried approximately 3 km inland, were left in place as a memorial to the next generation.

Galle District, Sri Lanka

Galle became a well-known district after the tsunami because noteworthy damage occurred along its entire coastline. In addition, severe damages to the Galle port, Galle bus terminal, fishery harbors such as Galle, Dodanduwa and Hikkaduwa, and

Fig. 17.4 (**a**) Evacuation building in the Banda Aceh region in 2009 and (**b**) Tsunami museum in the Banda Aceh region in 2009, (**c**) Tsunami memorial in Galle (27/4/2010) and (**d**) Patrol boat from the 2004 tsunami in Phang Nga (19/3/2009)

the International Cricket Stadium were major reconstruction challenges left to the community after the tsunami. However, the most noteworthy damage was that near Peraliya, where the tsunami surge overturned and submerged a Galle-bound train, killing over 1,000 people. Peraliya, the most devastated area in the Galle district, is now considered a memorial site for the Indian Ocean tsunami. A number of tsunami memorials were constructed in this area to sustain the memory of this event and to educate the younger generation who have never experienced a tsunami. Figure 17.4c shows a memorial built in this fashion, located at the site of the train accident that resulted in the worst human disaster in Sri Lankan history.

Phang Nga Province, Thailand

Nam Khem, a village that was most affected by the tsunami, now has many good strategies to promote tsunami recognition and awareness among residents during tsunami events. A school was chosen as the local evacuation shelter, and its location was moved higher above sea level and farther from the coast in response to lessons learned from the tsunami. The evacuation route is the main two-lane village road for vehicles. Two fishing boats that were carried 500 m inland, killing many residents and causing more housing damage, were left in place as a tsunami memorial. A memorial park was also constructed next to the beach for recreation, and a memorial gallery was erected inside the park with information on tsunamis, evacuation procedures and historical damage at Nam Khem. In Pakarang Cape, Phang Nga, an evacuation shelter near the beach was constructed in 2010 in accordance with shelter regulations stating that the distance to a safe zone must be greater than 1 km when there is no greenbelt to dissipate tsunami energy. In Khao Lak, the patrol boat that was in charge of protecting the royal family on their vacation was brought approximately 2 km inland by the tsunami (Fig. 17.4d). That boat was also kept in place as a symbol of the tsunami, and a small tsunami gallery was built nearby.

17.4.1.2 The 2011 Tohoku Tsunami

Tsunami Inundation Mark

Similar to other historical tsunamis in Japan such as 1960 Chile tsunami, tsunami inundation signs are now putting in the tsunami affected areas of the 2011 Tohoku tsunami in Miyagi prefecture. Figure 17.5a is designated evacuation building in Kesennuma city. It was survived by the tsunami that almost reached the third floor of the building. Figure 17.5b is a sign in Onagawa town's hospital that elevated on a 15 m high hill. Tsunami overtopped the hill and inundated the first floor of the hospital by about 2 m. Both buildings in Kesennuma and Onagawa shown here are located close to the sea. Although a pedestrian bridge in Natori city shown in Fig. 17.5c locates almost 2 km from sea but only less than 500 m from the Natori River. Tsunami inundated about 2 m in that area and some peoples survived staying over the pedestrian bridge. The last example is Sendai airport as shown in Fig. 17.5d. The airport is located approximately 1 km from sea and 3 m depth tsunami inundated the airport's first floor while most passengers, staffs and nearby residences survived staying at the second and third floor. These signs will help remaining tsunami hazard information of the great one in 2011 similar to those already have in the 2004 tsunami affected countries.

Damaged Structure as Tsunami Memorial

Unlike the 2004 tsunami affected countries, there were no tsunami memorials using damaged structure by a tsunami in Japan before. For historical tsunamis such as in 1896 Meiji-Sanriku or 1933 Showa-Sanriku, there were some rock plates, temples or shrines that recorded tsunami stories but never used damaged structure as a memorial. There are some power generation ship in Indonesia and fishery boats and patrol boat in Thailand that are now preserved as the 2004 tsunami memorials. This is a new adaptation in Japan that they found a need of using damaged structures to remind future generations about power of the great event that is willing not to be occurred so often.

Fig. 17.5 (**a**) Tsunami evacuation building in Kesennuma city (29/4/2012), (**b**) Onagawa town's hospital (16/3/2012), (**c**) Pedestrian bridge in Natori city (30/5/2012) and (**d**) Sendai airport (25/1/2012)

For the 2011 tsunami event in Japan, 70,000 pine trees in Rikuzen-Takata city were totally swept out remaining just only one tree which became a symbol of their hope for the reconstruction. Also in the same city, have a project called "Sakura Line 311" (Sakura Line 311 2011) to plant Sakura (Cherry blossom) tree along inundation limit of the 2011 tsunami. Figure 17.6a shows one large fishery ship that was moved more than 600 m inland by the tsunami in Kesennuma city. The city would like to keep the ship as a memorial but they are facing against strong objection from those who lost their families and houses because of that ship. This issue is now still under discussion and may take long time if the ship is to be preserved as a memorial. Figure 17.6b shows an example of overturned building in Onagawa town. Until now, overturned building is not common damage type by a tsunami as most of them were just collapses or washed away but not overturned. In fact, there were six overturned buildings in Onagawa and some in other locations. The town can only possible to keep three buildings as the one shown in Fig. 17.6b as a memorial (Onagawa town 2011) because of objections from owner of the buildings. Large fish oil tank in Ishinomaki city (Fig. 17.6c) was also another

Fig. 17.6 (**a**) Large fishery ship in Kesennuma city (29/4/2012), (**b**) Overturned building in Onagawa town (29/3/2011), (**c**) Large fish oil tank in Ishinomaki city (26/3/2012) and (**d**) Tsunami hazard map before the 2011 tsunami in Sendai city (18/1/2012)

example of massive structures that were damaged and destroyed by the tsunami. From the Fig. 17.6c, it looked like that the city intended to make a park surrounding the tank but finally the tank was removed with the reason that people living nearby that area felt scared about it. Figure 17.6d shows tsunami hazard map found nearby Sendai coast and prepared by Sendai city for the expected tsunami inundation zone before the 2011 event. In the figure, a printed paper with a message saying "please do not remove" was put by someone meaning that there was a need to keep these sign as a memorial or to show the lessons learned by the 2011 tsunami. However, the sign was finally removed.

Countermeasures for Future Tsunami

Examples on present reconstruction made in affected areas of the 2011 tsunami in Japan were shown in this section. Figure 17.7a shows a proposed model of seawall that is made of tsunami debris in Rikuzen-Takata city. Based on their explanation shown nearby the model, they said that this idea was successfully constructed with a 840 m length as river dike in Taiwan. Expressway that is located 5 km from sea in Sendai (Fig. 17.7b) help protecting areas behind it from tsunami by change and

Fig. 17.7 (a) Seawall making use of tsunami debris in Rikuzen-Takata city (29/4/2012), (b) Stair on expressway in Sendai city (25/1/2012), (c) Increasing land elevation in Natori city (3/8/2012) and (d) Hill of thousand years hope in Iwanuma city (3/8/2012)

some people survived by climbing this expressway. The city is therefore built stairs along this expressway so that people will be easier to get to the top of the expressway during an evacuation of future tsunami (Sendai city 2011). In Fig. 17.7c, Natori city has a plan to increase land elevation of tsunami affected areas by 3.9 m as the maximum inundation depth of the 2011 tsunami was 3.2 m (Natori city 2011). A model of this explanation of 3.9 m elevated land was built in the target area so that residence can be easily understand the image of the elevation. Iwanuma city has an idea to construct an artificial hill, the height of which varies from 10 to 15 m, using tsunami debris as a model shown in Fig. 17.7d. The purpose of the hill, called the "Hill of thousand years hope," is to create a memorial park and evacuation site for future tsunamis (Iwanuma city 2011). This hill will also help the new seawall to reduce a tsunami's energy before it reaches the populated area behind it.

17.5 Conclusions

M9.0-class earthquakes occurred twice in the last 7 years: in 2004 in the Indian Ocean and in 2011 off the Pacific coast of the Tohoku region in Japan. Both of these earthquakes resulted in huge tsunamis but had different impact scales. The 2004 tsunami was a near-field and far-field tsunami that propagated to all 15 countries located along the Indian Ocean coasts. This paper introduced examples of tsunami recovery from the 2004 Indian Ocean tsunami in Indonesia, Sri Lanka and Thailand. Because it was the first historically and instrumentally recorded event in the region, many tsunami activities have been performed since it occurred. Some of these include the construction of tsunami evacuation buildings, tsunami memorial parks and tsunami museums. On the other hand, the 2011 near-field tsunami caused devastating damage concentrated in Japan and had a comparatively small impact on other countries in Pacific coastal areas.

It is very clear that experience and evacuation helped reducing tsunami fatality ratio in coastal areas along the ria coasts of Miyagi and Iwate prefecture. It shows an importance of experience education for disaster prevention that people who live along the Sanriku coast have higher tsunami awareness and evacuation recognition. Overturned buildings in Onagawa could be one of examples for a tsunami memorial park to warn people in the next generation against the great tsunami which has return period much longer than human life. Selection of a proper tsunami evacuation should be reconsidered in some location as the experience from Minami-Sanriku town and Onagawa town that about half of tsunami evacuation shelter was inundated leading to great loss of live.

Even Japan is well known for the most advance in tsunami disaster prevention but they also suffered great loss from the 2011 Great East Japan tsunami. Similar to the 2004 Indian Ocean tsunami affected countries, Japan also have a plan to construct a memorial park or museum using the floating debris or damaged structure as a symbol to recall and advocate tsunami awareness for coastal residence in the area.

Acknowledgements We express our deep appreciation to the Tokio Marine & Nichido Fire Insurance Co., Ltd., the Willis Research Network and the Ministry of Education, Culture, Sports, Science and Technology (MEXT) for their financial support.

References

Abe I, Imamura F (2010) The 3rd International Tsunami Field Symposium (ITFS), Sanriku field trip guidebook, 11 pp

Arikawa T (2009) Structural behavior under impulsive tsunami loading. J Disaster Res 4(6):377–381

Iwanuma city (2011) Reconstruction plan home page. http://www.city.sendai.jp/fukko/1201143_2757.html (in Japanese)

Kawata Y (1997) Prediction of loss of human lives due to catastrophic earthquake disaster. Jpn Soc Nat Disaster Sci 16(1):3–13, In Japanese

Koshimura S, Matsuoka M, Kayaba S (2009a) Tsunami hazard and structural damage inferred from the numerical model, aerial photos and SAR imageries. In: Proceedings of the 7th international workshop on remote sensing for post disaster response. University of Texas, Texas, 22–23 Oct 2009, (CD-ROM)

Koshimura S, Namegaya Y, Yanagisawa H (2009b) Tsunami fragility – a new measure to assess tsunami damage. J Disaster Res 4(6):479–488

Murao O, Nakazato H (2010) Vulnerability functions for buildings based on damage survey data in Sri Lanka after the 2004 Indian Ocean tsunami. In: Proceedings of the 7th international conference on sustainable built environment, Kandy, 13–14 Dec 2010

National Oceanic and Atmospheric Administration (NOAA) (2011) Historical tsunami database at NGDC. www.ngdc.noaa.gov/hazard/tsu_db.shtml

Natori city (2011) Reconstruction plan home page. http://www.city.natori.miyagi.jp/news/node_14153 (in Japanese)

Onagawa town (2011) Reconstruction plan home page http://www.town.onagawa.miyagi.jp/hukkou/index.html (in Japanese)

Satake K, Tanioka Y (1999) Sources of tsunami and tsunamigenic earthquakes in subduction zones. Pure Appl Geophys 154:467–483

Sendai city (2011) Sakura line 311: http://sakura-line311.org/ (in Japanese). Reconstruction plan home page

Suppasri A, Koshimura S, Imai K, Mas E, Gokon H, Muhari A, Imamura F (2012a) Damage characteristic and field survey of the 2011 great east Japan tsunami in the Miyagi prefecture. Coast Eng J 54(1, Special Issue of 2011 Tohoku tsunami):1250005

Suppasri A, Mas E, Koshimura S et al (2012b) Developing tsunami fragility curves from the surveyed data of the 2011Great East Japan tsunami in Sendai and Ishinomaki plains. Coastal Eng J 54(1):1250008

Suppasri A, Shuto N, Imamura F et al (2012c) Lessons learned from the 2011 Great East Japan tsunami: performance of tsunami countermeasures, coastal buildings and tsunami evacuation in Japan. Pure Appl Geophys 170(6–8):993–1018. doi:10.1007/s00024-012-0511-7

The 2011 Tohoku Earthquake Tsunami Joint Survey Group (2011) Nationwide field survey of the 2011 off the pacific coast of Tohoku earthquake tsunami. J Jpn Soc Civ Eng Series B 67(1):63–66

Yamashita F (2008a) Tsunami and disaster prevention-Sanriku tsunami, Kokon-Shoin publishing 158 p, ISBN-13: 978-4772241175 (in Japanese)

Yamashita F (2008b) Tsunami tendenko: History of recent tsunamis in Japan, Shinnihon publishing 235 p, ISBN978-4-406-05114-9 (in Japanese)

Chapter 18
Tsunami-Deck: A New Concept of Tsunami Vertical Evacuation System

Abdul Muhari, Fumihiko Imamura, and Shunichi Koshimura

Abstract We introduce the concept of a new type of vertical evacuation shelter. It combines the function of pedestrian bridge and tsunami tower, which are then installed at intersection. The idea provides the ease of constructing and maintenance in addition to the loss of required land acquisition. We first evaluate the performance of pedestrian bridges during the 2011 tsunami in Japan by taking into account the exposure component in term of their position from the shoreline and the susceptibility component in term of the tsunami height around the bridge as the influencing parameters in determining damage probability. We developed fragility curves to show that pedestrian bridges lay on an area less than 500 m from the coastline (around 0.1 of the maximum inundation extent), or in the area where the tsunami flow depth are 1.5 of the height of the bridge's deck have more than 50 % probability to be damaged by the tsunami. If the function of pedestrian bridge will be expanded into 'Tsunami-deck', the information about tsunami behavior at the intersection therefore becomes important. For that reason, we performed a set of numerical experiment of tsunami flow at the intersection to determine conditions that will allow the installation of tsunami-deck with a height similar to the average height of the existing pedestrian bridge. By doing so, we attempted to determine preliminary placement criteria for Tsunami-deck in order to ensure the safety of evacuee. The results gave an opportunity to have more distributed vertical evacuation shelter in flat and densely populated areas.

Keywords Tsunami-deck • Pedestrian bridge • Sudden expansion • Hydraulic gradient

---

A. Muhari (✉) • F. Imamura • S. Koshimura
International Research Center for Disaster Science (IRIDeS), Tohoku University, Sendai, Miyagi, Japan
e-mail: aam@tsunami2.civil.tohoku.ac.jp

Y.A. Kontar et al. (eds.), *Tsunami Events and Lessons Learned: Environmental and Societal Significance*, Advances in Natural and Technological Hazards Research 35, DOI 10.1007/978-94-007-7269-4_18,

18.1 Introduction

In pre-disaster condition, difficult to find space for vertical evacuation structures is the common problem in a densely populated area. Utilizing the existing tall buildings especially in developing countries are sometimes risky because the structural strength is not adequate against the strong ground shaking before the tsunami (e.g. Muhari et al. 2010; Imamura et al. 2012). In densely populated cities, therefore, horizontal evacuation and utilizing the natural hills (if any) is the most common option for tsunami evacuation. However, the available evacuation time starts from the occurrence of the earthquake until the first arrival of the tsunami wave is very limited. Also, some physiological aspects play an important role in determining the evacuation behavior. Experience during the 2011 East Japan tsunami demonstrated that many people returned to the tsunami affected area to pick up their relatives using cars (Muhari et al. 2012). As an implication, massive traffic jams occurred along the roads that supposed to be used for evacuation, and thus delay the people to reach safe areas. To overcome the above situations, a new type of vertical evacuation structure namely Tsunami-deck is proposed. The underlying conception is the combination of two distinct structures with different conditions and experiences as evacuation shelter in Japan. The first is the pedestrian bridge. During the 2011 tsunami in Japan, many of them were survived and used for rapid evacuation by those who get trapped in a traffic jam around the intersection (Muhari et al. 2012). However, not so many people can be accommodated due to the limited space available on the bridge's deck. The second structure is tsunami tower, but applying the tsunami tower in a populated city is difficult because it requires large open spaces for the structures. We are then thinking to utilize both structures by eliminating their weaknesses and complement their strength to develop a flexible (easy to construct and to maintain) and solution-oriented evacuation structure of the above mention evacuation problems. The limited space in pedestrian bridge is trying to be solved by expanding the deck to occupy the whole area at intersection. This means that we will have an evacuation tower at the intersection (Fig. 18.1). By doing so, several advantages are obtained such as the applicability in densely populated areas, multipurpose in daily basis i.e. as pedestrian bridge, easy to construct and easy to maintenance, and the most important thing is that it gives an adequate temporary evacuation shelter for those who get trapped in a traffic jam during evacuation.

Since the basic idea is to extend the function of the pedestrian bridge, the limitations may come from the design height. The existing height of pedestrian bridge is around 4.5–5 m. Open structure with height taller than this might not be appropriate for the elderly in case of rapid evacuation. Therefore, criteria for its placement according to the tsunami characteristics at intersection are necessary.

We first evaluate the performance of pedestrian bridges during the 2011 tsunami in Japan. Evaluation is carried out by considering two parameters that influence the damage probability. First is the exposure in term of their position from the shoreline and the second is their susceptibility in term of the tsunami height around the bridge. Based on the developed fragility curves, we establish the placement

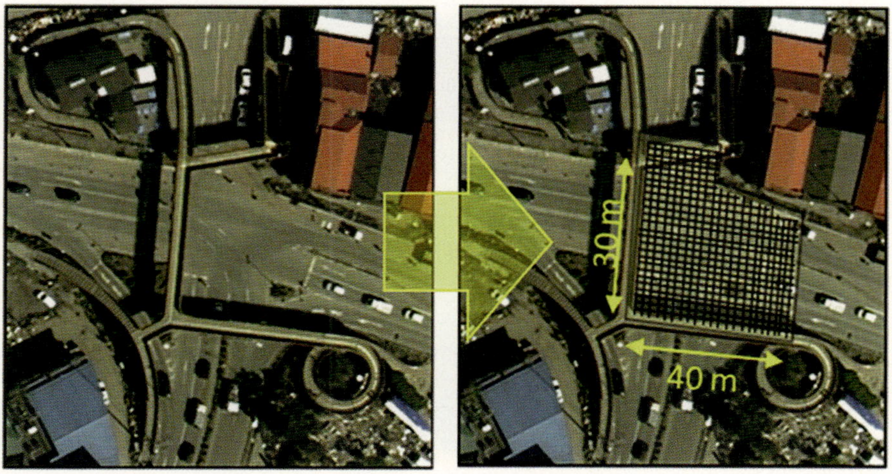

Fig. 18.1 An artist figure to describe the conception of tsunami-deck

criteria to ensure the safety of the evacuee if the tsunami-deck built to the same height as a pedestrian bridge. Starting from the concept of sudden expansion phenomena (e.g. Goto and Shuto 1983), we performed a set of numerical experiment to parameterized conditions that will allow the sudden drop of tsunami height due to the building configuration at intersection. Once the placement criteria for Tsunami-deck are established, further analysis from the structural point of view can be conducted.

18.2 Methodology

18.2.1 Fragility Curves for Pedestrian Bridge

We compiled 52 pedestrian bridges placed inside the tsunami affected areas through the visual inspection of satellite image, field survey data and utilizing the oblique photograph provided by the Geographic Survey Institute of Japan (2011). The damage levels of pedestrian bridges were then classified into three classes (see Table 18.1): (A) is heavily damaged, which described as a condition where bridges are swept away or damage on its deck, (B) is partially damage indicated by damage in the stairs or fence so it needs to be fixed before re-use, and (C) is a condition where bridges are only inundated and it can be directly used after the tsunami.

We are now analyzing the fragility of pedestrian bridge in term of their susceptibility related to tsunami flow depth around the bridge. In general, higher tsunami flow depth will increase the damage probability of the pedestrian bridge. Even though the damage might be caused by the impact of debris collision, in this study we use tsunami

Table 18.1 Damage criteria of pedestrian bridge

No	Damage classification	Indicator
1	Heavily damage (swept away)	A
2	Slight damage (i.e., stairs)	B
3	Survived (only inundated)	C

flow depth as the only observable parameter from field survey. Some of debris stranded on the bridge's stairs was visible from satellite imageries, but incomplete figures for all bridges might yield inconsistent interpretation. Therefore, debris impact is excluded from the analysis.

The cumulative damage probability is given through the widely used fragility equation (e.g. Koshimura et al. 2009) as follow,

$$P_i(x) = \Phi\left[\frac{\ln x\text{-}\mu'}{\sigma'}\right] \tag{18.1}$$

Here, P is the cumulative probability of specific damage classification (i), Φ is the standardized lognormal distribution function, x is the median of the observed tsunami flow depth, and is the mean and standard deviation of random variable of x respectively. In order to obtain the last two statistical parameters, we first calculate the inclination (μ_y) and intercept (σ_y through performing the least-square fitting plot of ln x and the inverse of lognormal distribution function Φ^{-1}. Next the mean and standard deviation were determined by ($-\mu_y/\sigma_y$) and ($1/\sigma_y$) respectively (e.g., Shoji and Moriyama, 2007).

For the fragility in term of the exposure, Eq. 18.1 is modified since the damage probability will be decreased along with the increment of the distance from the shoreline. The modification is given as follows,

$$1 - [P_i(x)] = 1 - \Phi\left[\frac{\ln x\text{-}\mu'}{\sigma'}\right] \tag{18.2}$$

By applying these equations, fragility curves for pedestrian bridges are developed and discussed in the next chapters.

18.2.2 Placement Criteria of Tsunami-Deck

In principle, the deck height should be higher than the maximum surrounding tsunami flow depth. If the buildings around the intersection withstand the tsunami, the so-called sudden expansion phenomenon where the flow depth is suddenly dropped due to the rapid change of flow direction toward different direction will be occurred depending on the building configuration at the intersection. This section will focus to elaborate the building configuration that will allow the sudden expansion phenomenon to occur.

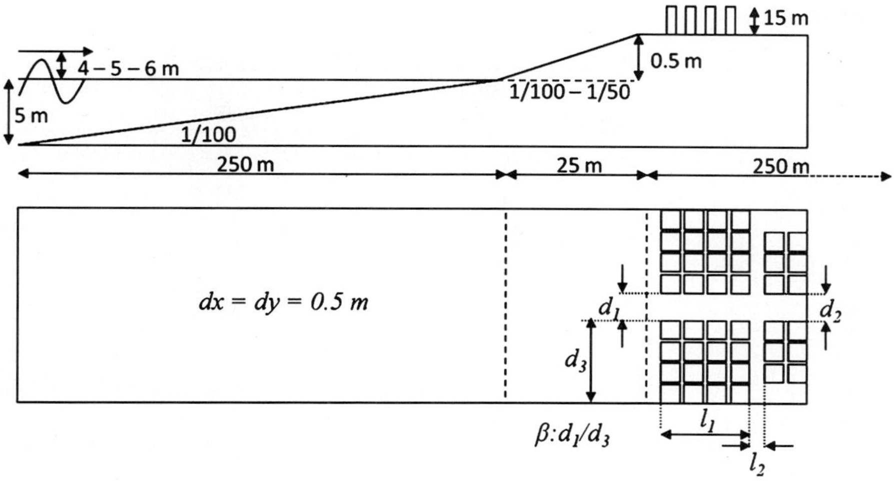

Fig. 18.2 Design of hypothetical town for numerical simulation of tsunami at intersection

Goto and Shuto (1983) analyzed factors that influence the expansion coefficient when tsunami flows along the lined obstacles. However, their works were limited in two parallel obstacles where there was no actual representation of the intersection. In this study, we perform numerical exercises using a hypothetical town that is set as shown in Fig. 18.2. Residential block prior the intersection occupied all areas, while the second is not. This design allows the flux at intersection to spread out through different direction, and minimizing the unnecessary reflecting flows from the sidewalls. In reality, this design can only able to represent areas where the width of residential block is much bigger than the width of the road between them.

A set of non-linear shallow water equation as given in Imamura (1996) is used to model the propagation and inundation of the tsunami, which is initialized by a sine wave in the ocean boundary of the domain. Goto and Shuto (1983), and Tsudaka et al. (2011) had demonstrated the reliability of 2D Shallow Water Equation (SWE) to represent flow passing through obstacles. They validated the result of numerical calculation using SWE with experimental data and obtained a good agreement. However, one should note that this approach has limitation on reproducing flood-water behind building (Tsudaka et al. 2011). Also, the absence of the momentum diffusion term in the traditional SWE might yield less accurate of modeled water level due to the lack of lateral diffusion at intersection.

We conducted the numerical exercises to assess the influence of four parameters in determining the flow characteristic at intersection. The initial sine wave – assumed as the tsunami – is set in different amplitude ranging from 2 to 6 m as shown in Table 18.2. The numerical exercises were conducted in flat (slope 0) and in sloping topography (1/100 and 1/50). In the following explanations, four parameters assessed in the numerical exercises are described. Symbols are referring to Fig. 18.2 and values of each numerical case are referring to Table 18.2.

Table 18.2 Scenarios for numerical simulation of tsunami at intersection

Case	Slope 0 d_2/d_1 (D)	l_1/l_2	l_2/l_1	β	Amplitude (A) (meter)	Slope 1/50 d_2/d_1 (D)	Amplitude (A) (meter)	Slope 1/100 d_2/d_1 (D)	Amplitude (A) (meter)
Case 1	1.20	5.9	0.17	0.11	2, 3, 4, 5, 6	1.20	5.00	1.20	5.00
Case 2	1.40	4.7	0.22	0.16	2, 3, 4, 5, 6	1.40	5.00	1.40	5.00
Case 3	1.60	3.5	0.27	0.21	2, 3, 4, 5, 6	1.60	5.00	1.60	5.00
Case 4	1.80	2.3	0.32	0.24	2, 3, 4, 5, 6	1.80	5.00	1.80	5.00
Case 5	2.00	–	–	0.28	2, 3, 4, 5, 6	2.00	5.00	2.00	5.00
Case 6	2.20	–	–	–	2, 3, 4, 5, 6	2.20	5.00	2.20	5.00

A. The influence of expansion ratio (*D*). This parameter is assessed by increasing the road width after the intersection (d_2) while the road width of the road before the intersection (d_1) is remaining.
B. The influence of road width parallel to the shoreline (l_2) is analyzed by increasing their width while the length of residential blocks prior the intersection (l_1) is constant.
C. The effect if some part of the houses along the road prior to the intersection were damaged by the tsunami is assessed by reducing the length of road prior to the intersection (l_1) and comparing it to the width of road that parallel to the shoreline (l_2).
D. Lastly, we estimate the effect of the ratio between the width of the road toward the intersection (d_1) and the width of residential blocks (d_3), named as beta (β) on influencing the sudden expansion at an intersection.

From the results of each scenario, a cross section at the middle of the road perpendicular to the shoreline is extracted and analyzed. The numerical exercises are conducted in a topographic domain with 0.5 m cell size. Selection of the grid size is according to an empirical relation of $dx/(g\ h_{max})^{1/2}\ T < 3\times10^{-3}$ to obtain $0.5 <$ predicted value/measured value < 2 proposed by Fujima (2012). The wave period of 0.5 h is selected for the sin wave as initial. This period is similar to the average observed period of the first wave during the 2011 Tohoku tsunami at Sendai buoys.

18.3 Results and Discussions

We classified the damage rank, cumulative frequency and cumulative damage probability in the 1-m range. In terms of their location; damage rank, cumulative frequency and cumulative damage probability were analyzed in every 250 m distance.

The regression analysis gave parameters shown in Table 18.3. There are used to determine the best fit of fragility curves in term of the exposure with the obtained

Table 18.3 Statistical parameters for fragility function obtained from regression analysis

Variable of fragility (x)	μ′	σ′	R^2
Exposure (distance from the shoreline)			
Damage classification (A)	4.12	1.53	0.74
Cummulative (A) + (B)	5.24	1.13	0.61
Susceptibility (tsunami inundation)			
Damage classification (A)	1.72	0.33	0.64
Cummulative (A) + (B)	1.20	0.42	0.81

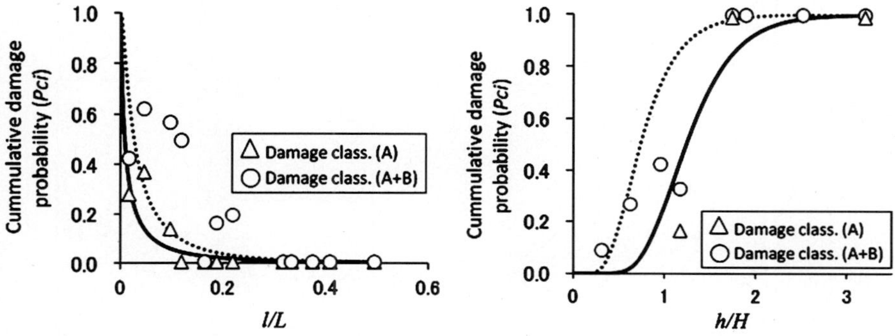

Fig. 18.3 Fragility curves for exposure (*left*) and susceptibility (*right*) parameters. *Solid line* indicates the fragility for criteria (A) and *dashed line* for cumulative (A) + (B)

coefficient correlations (R^2) as 0.74 for (A) and 0.61 for (A+B). In term of susceptibility, the coefficient correlations were obtained at 0.64 for (A) and 0.81 for (A+B).

The fragility curves are then developed by plotting the damage probability of the non-dimensional number for each fragility parameter (Fig. 18.3). The non-dimensional number for exposure is defined as the ratio between the locations of the bridge from the shore (l) with the maximum tsunami inundation distance (L), which is taken as 6 km. As for the fragility curve in term of susceptibility, the non-dimensional parameter is given by the ratio between the observed tsunami run-ups (h) to the height of the pedestrian bridge (H).

It can be seen that most of pedestrian bridges have more than 50 % probability to be damaged by tsunami if they were placed in the area up to 1.2 km from the coastline (or 0.2 of the assumed maximum inundation extent). Also, it has more than 50 % probability to be swept away by tsunami if the surrounding flow depths are 1.5 of the height of bridge's deck or around 6–7.5 m. This height of the tsunami was experienced by almost whole areas in the Sanriku coast within the distance of 0.5–1 km from the shore (Mori et al. 2011). Therefore, placement criteria for pedestrian bridge and tsunami-deck are necessary in order to fulfill the need to have evacuation facilities in the area described above.

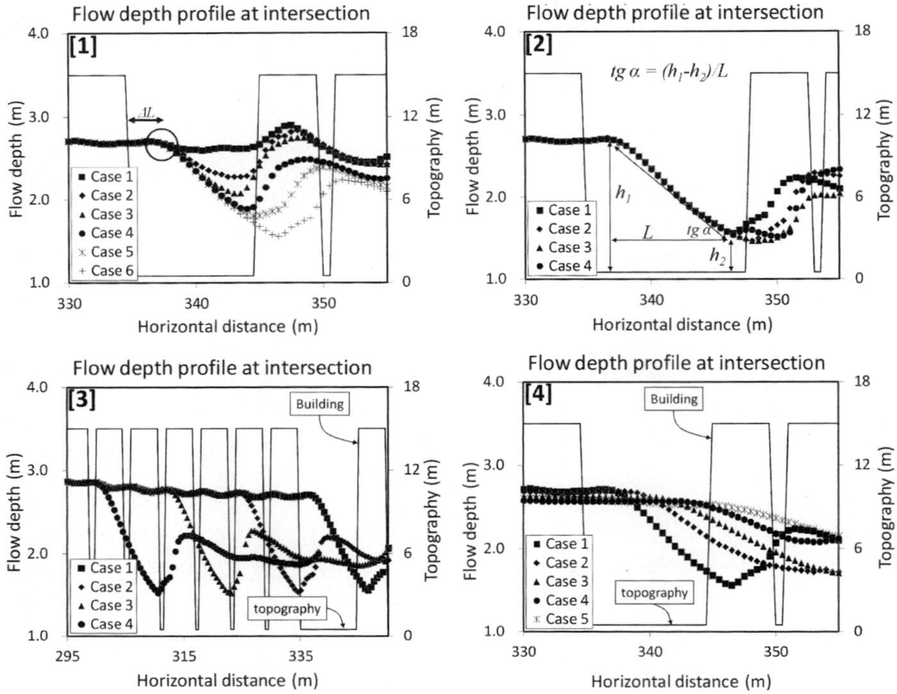

Fig. 18.4 The influence of the following parameters to the modeled water surface at intersection (refer to panel [*1*]–[*4*]): [*1*] expansion ratio (*D*), [2] road width parallel to the coastline (l_2), [3] the length of residential block (l_1) and [*4*] β (ratio between d_1 and d_2)

18.3.1 Placement Criteria for Tsunami-Deck

The dynamic features of tsunami flow at intersection might be varied depending on the geometric configuration of the junction and the surrounding building arrangement. We first analyzed the influence of expansion ratio (*D*) as given in Fig. 18.4, panel [1]. If (d_1) is relatively similar to (d_2), then the hydraulic gradient (tg α) become small (see description of tg α in Fig. 18.2, panel [2]). The tg α will start to be visible if the expansion ratio (*D*) 1.4, and saturated when *D* equal to 2.

In term of the Froude number (Fr), scenario we used in these numerical exercises brought the range of Fr as 1.13–1.34. This is higher than critical condition (Fr $>$ 1) because bore usually observed under super critical condition. By plotting the tg α with the expansion ratio (*D*) as given in Fig. 18.5 (A), we confirmed that higher Fr yields to larger hydraulic gradient at intersection. However, one should note that in a real situation, sudden expansion in high Froude number could only be observed if the surrounding building survived the tsunami.

The next analyses are taking into account the effect of road width parallel to the coastline (l_2), and the effect of road length prior to the intersection (l_1) on influencing the hydraulic gradient. Our results show that no significant effect of these factors.

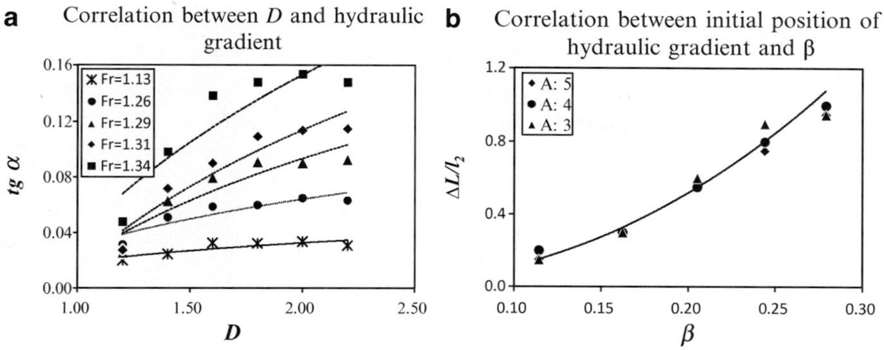

Fig. 18.5 Plotting correlation between: (**a**) expansion ratio (*D*) and hydraulic gradient (tg α), (**b**) correlation between the initial position of hydraulic gradient with the *β*

It means: first, the water mass that goes to roads parallel to the coastline will not significantly influence to the observed hydraulic gradient at the intersection. It just creates a slight difference on the initial position of hydraulic jump, which is not considered in this study (Fig. 18.4, panel [2]). Second, the length road prior to the intersection (l_1) is not influencing the hydraulic gradient. Their effect is limited only to slightly reduce the flow depth depending on the road length (Fig. 18.4, panel [3]). Therefore, one important thing to note is even if some houses along the road prior to intersection are damaged by the tsunami, the sudden expansion phenomena will still occur as long as the expansion ratio at intersection meets the requirement as discussed previously.

The influence of (β) is given in Fig. 18.4 panel [4]. It affects to determine the initial point of hydraulic gradient at intersection. Higher β will bring larger ΔL (see description of ΔL in Fig. 18.4 panel [1]). We drew the correlation between β with the non-dimensional parameter $\Delta L/l_2$. The last term indicates the ratio between initial points of hydraulic gradient from the rearmost cell of road prior to the intersection (ΔL) with the l_2. The result is imaged in Fig. 18.5 (B)–where 'A' denotes the amplitude of modeled waves–indicates that ΔL will overlap l_2 if $\beta > 0.25$ because the flow depth starts to decline when tsunami front already passed the intersection. It means there will be no hydraulic gradient observed at the intersection even if the expansion ratio meets the criteria as explained previously.

The results that describes of the influence of four parameters on determining the tsunami characteristics as we presented above only found in flat topography. We were not obtained similar results from sloping topography (1/100 and 1/50). Thus, this becomes a note that the dynamic features of tsunami flow at an intersection can only be used for consideration in determining the design height of the tsunami–deck if surrounding topography is relatively flat.

As the practical implications of the developed criteria; first, the tsunami-deck is applicable only if the predicted tsunami flow depth is lower than their height. Second, if the predicted maximum flow depth almost reaches the deck height,

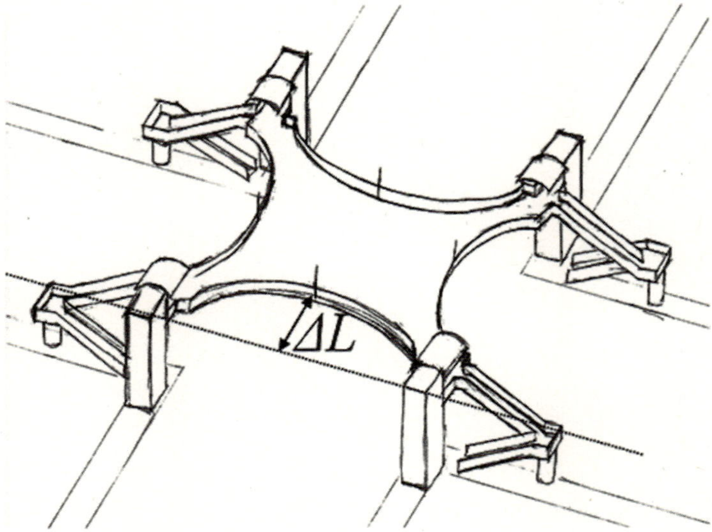

Fig. 18.6 An example of adjustment on the deck shape to accommodate the ΔL

then an adjustment should be made to the deck's design to ensure the flow passing through beneath the structure (Fig. 18.6). Also, it should be ensured that there is no source of large floating debris such as ships, boats and etc. that has possibility to strand on the tsunami-deck.

18.4 Conclusions

A new type of tsunami vertical evacuation structure is proposed to be applied in densely populated areas, especially in the developing countries. The easiness of their construction and maintenance as well as the flexibility to be put at an intersection is combined with larger accommodation space to create a reachable evacuation shelter once the traffic jam occurred during the evacuation.

As the basis for the analysis, fragility curves for pedestrian bridges are developed based on surveyed data in the tsunami affected areas along the east coast of Japan. The results indicate that pedestrian bridges placed in the area up to 1.2 km from the shore, or in the area where the estimated tsunami flow depths are 1.5 of the height of bridge's deck have a high probability to be damaged by the tsunami.

Since the above mention high risk's area might be crucial for the rapid evacuation, determining the placement criteria by using the sudden expansion phenomenon as the primary consideration solves the limitation due to the existing height. Intersections with the geometric configuration of the buildings that has (D) > 1.5 and (β) < 0.15 may be appropriate for tsunami deck.

The present study described only the hydrodynamic part in the introduction of tsunami-deck concept. It is highly acknowledged that the structural analysis and the potential of solid debris impact should be taken into account for the further study, which is now currently on going.

Acknowledgements We express our deep appreciation to JST-JICA project (Multidisciplinary hazard reduction from earthquakes and volcanoes in Indonesia) group 3, and the Ministry of Education, Culture, Sports, Science and Technology (MEXT) Japan for financial support through the study No.(22241042).

References

Fujima F (2012) Numerical simulation of the 2011 Tohoku tsunami in Miyako Bay, abstract for the Asian Oceania Geosciences Society (AOGS). Singapore, August 2012

Geospatial Information Authority of Japan (GSI) (2011) Information on Tohoku earthquake, available at http://www.gsi.go.jp/BOUSAI/h23_tohoku.html. Last access March 2012

Goto C, Shuto N (1983) In: Iida K, Iwasaki T (eds) Effect of large obstacle on tsunami inundation, Tsunamis–their science and engineering. Terra Scientific Publishing Company, Tokyo, pp 511–525

Imamura F (1996) Review of tsunami simulation with a finite difference method. In: Yeh H, Liu P, Synolakis C (eds) Long wave runup models. World Scientific, Singapore, pp 25–42

Imamura F, Muhari A, Erick M, Pradono MH, Post J, Sugimoto M (2012) Tsunami disaster mitigation by integrating comprehensive countermeasures in Padang city, Indonesia. J Disaster Res 7(1):48–64

Koshimura S, Oie T, Yanagisawa H, Imamura F (2009) Developing fragility function for tsunami damage estimation using numerical model and post-tsunami data from Banda Aceh, Indonesia. Coast Eng J 51(3):243–273

Mori N, Takahashi T, Yasuda T, Yanagisawa H (2011) Survey of 2011 Tohoku earthquake tsunami inundation and run-up. Geophys Res Lett 38:L00G14. doi:10.1029/2011GL049210

Muhari A, Imamura F, Natawidjaja DH, Post J, Latief H, Ismail FA (2010) Tsunami mitigation efforts with pTA in West Sumatra Province, Indonesia. J Earthq Tsunami 4(4):341–368

Muhari A, Koshimura S, Imamura F (2012) Performance evaluation of pedestrian bridge as vertical evacuation during the 2011 tsunami in Japan. J Natur Disaster Sci 34(1):79–90

Shoji G, Moriyama T (2007) Evaluation of the structural fragility of a bridge structure subjected to tsunami wave load. J Nat Disaster Sci 29(2):73–81

Tsudaka R, Shigihara Y, Fujima K (2011) Accuracy of tsunami numerical simulation with high-resolution topographic data. XXV IUGG General Assembly. International Association of Seismology and Physics of the Earth's Interior, Melbourne, 28 June–7 July, 2011

Chapter 19
Identifying Evacuees' Demand of Tsunami Shelters Using Agent Based Simulation

Erick Mas, Bruno Adriano, Shunichi Koshimura, Fumihiko Imamura, Julio H. Kuroiwa, Fumio Yamazaki, Carlos Zavala, and Miguel Estrada

Abstract In plain areas prone to tsunami, finding a way to shelter and escape from the inundation becomes a difficult task for residents. The 2011 Great East Japan Tsunami has shown that the horizontal evacuation using cars can compromise the safety of people. Another alternative is the vertical evacuation. In many cases, not only the capacity of these shelters plays an important role, but the spatial distribution and the evacuee preference for the nearest shelter. Such preference and location creates a conflict between capacity offer and demand. In this paper, we used an agent based model to simulate the evacuation of pedestrians and cars in La Punta, Peru. Twenty designated buildings for vertical evacuation are available for sheltering and escape from tsunami. The stochastic simulation of population spatial distribution and their refuge preferences revealed the over demand of some shelters. Finally, a capacity-demand map was created to share results with local authorities as a first step for future countermeasures in the district.

Keywords Agent based model • Evacuation simulation • Tsunami evacuation building • Tsunami simulation

E. Mas (✉) • B. Adriano • S. Koshimura • F. Imamura
International Research Institute of Disaster Science, Tohoku University, Sendai, Japan
e-mail: mas@irides.tohoku.ac.jp

J.H. Kuroiwa
Faculty of Civil Engineering, National University of Engineering, Lima, Peru

F. Yamazaki
Department of Urban Environment Systems, Chiba University, Chiba, Japan

C. Zavala • M. Estrada
Japan Peru Center for Earthquake Engineering Research and Disaster Mitigation, National University of Engineering, Lima, Peru

Y.A. Kontar et al. (eds.), *Tsunami Events and Lessons Learned: Environmental and Societal Significance*, Advances in Natural and Technological Hazards Research 35, DOI 10.1007/978-94-007-7269-4_19, © Springer Science+Business Media Dordrecht 2014

19.1 Introduction

Finding shelter during an emergency of tsunami for plain areas with a fast arrival time of waves is a challenging task for local people. The vertical evacuation to high buildings rising over the expected inundation depth in the area is one of the most suitable alternatives for this kind of urban districts. However, if the spatial distribution combined with the available capacity of these structures is not well displayed, conditions of over-demand and under-demand will be observed among them. In this paper, we conducted the numerical simulation of a future tsunami scenario using a predicted slip distribution model provided by Pulido et al. (2011), from the interseismic coupling model estimated by Chlieh et al. (2011). The case study area is La Punta district in Peru. Tsunami inundation is integrated with the agent based model of human evacuation. Using the stochastic simulation of several possible scenarios, the relation capacity-demand at each official Tsunami Evacuation Building in La Punta district in Peru was evaluated. The Capacity-Demand Index (CDI) is introduced as a way for mapping and identifying areas for future mitigation actions by local stakeholders.

19.2 Study Area

19.2.1 Historical Tsunamis Near La Punta: Peru

La Punta is part of the Constitutional Province of Callao in central Peru and one of the six districts in the First National Port city of Callao. It is a peninsula area and one of the smallest districts in Peru, with 4,370 inhabitants (INEI 2007) and a total land area of 0.75 km^2 (Fig. 19.1).

Major risk of earthquakes and tsunamis are present in this area, due to its low topography. La Punta was affected by several historical earthquakes and tsunamis, such as the July 9th, 1586, earthquake of magnitude 8.6 and a local tsunami height of around 5 m (Dorbath 1990), although some reports give a much larger value of 24 m (Berninghausen 1962). Another two earthquakes on October 20th and 21st, 1687, of magnitudes 8.0 and 8.4, respectively, struck this area. The first one had generated a 5–10 m local tsunami, while the second might have been located in southern areas (Dorbath 1990). One of the most memorable earthquakes in the Callao region is the great earthquake of October 28th, 1746, with a magnitude 8.0–8.6 which completely destroyed some central Peruvian coastal cities. A tsunami of 15–20 m in height resulted from this earthquake, it was reported half an hour after the ground shaking, and washed away Callao city with a 24 m run-up, killing 90 % of its population (Kuroiwa 2004). Two centuries later, the Peruvian central coast experienced more activity on May 24th, 1940, with a local earthquake and tsunami of 3 m in height. The October 3rd, 1974, event in front of Lima had a magnitude of 8.0 and a local tsunami height of 1.6 m (Langer and Spence 1995).

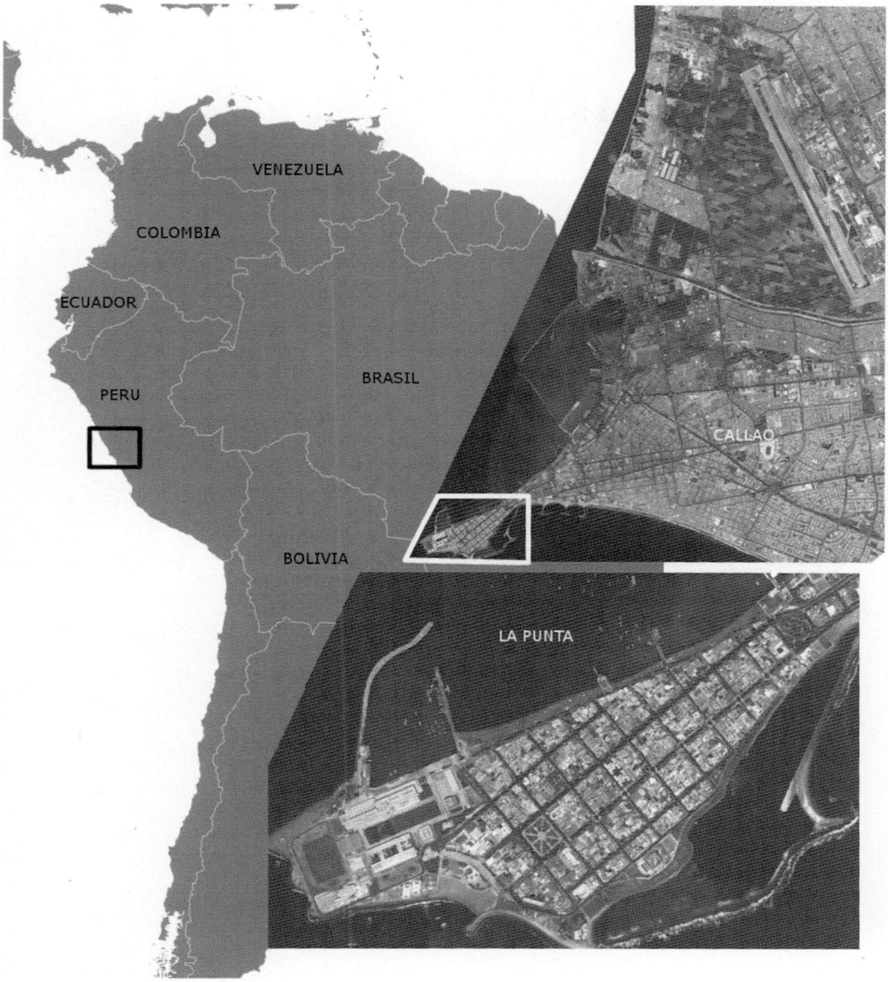

Fig. 19.1 Overview of the study area. Location of La Punta district, Peru

Since then, no large seismic activity has been reported in front of Callao area. A possible seismic gap might be located in this area, threatening La Punta and other coastal cities with a future large earthquake and tsunami.

19.2.2 Tsunami Evacuation Buildings

There are 19 official evacuation buildings in the district and a 20th building is located immediately outside of the district in the Callao province (Fig. 19.2). Previously it was considered in the evacuation plan, however after the 2007 Pisco Earthquake it suffered some structural damages and required retrofitting. Since then the district lost confidence on this building; however in 2009 it was successfully retrofitted.

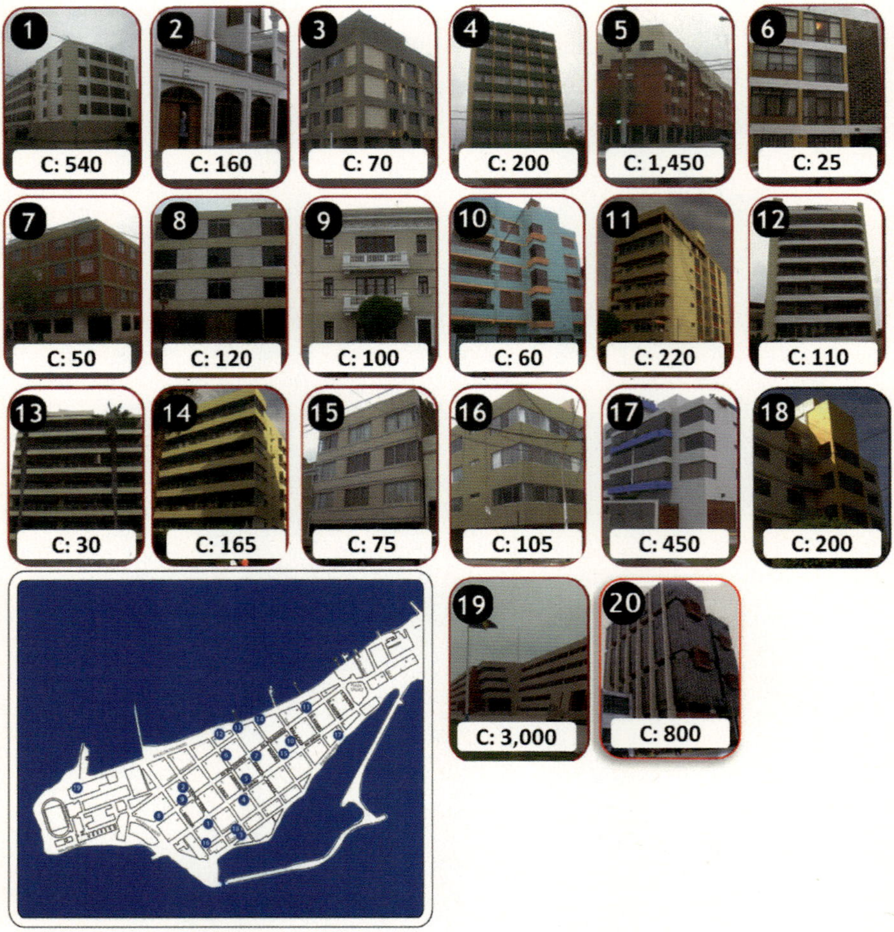

Fig. 19.2 Tsunami Evacuation Buildings (TEBs) in La Punta, Callao – Peru. The map shows the spatial distribution of TEBs and "C" is the capacity reported by the municipality

19.3 Tsunami Numerical Simulation

19.3.1 Tsunami Source

An instantaneous displacement of the sea surface identical to the vertical sea floor displacement is assumed in the model of tsunami source. The source of tsunami simulation is a result of the slip deficit rate with the interseismic period of 265 years since the 1746 historical earthquake in Peru (Fig. 19.3).

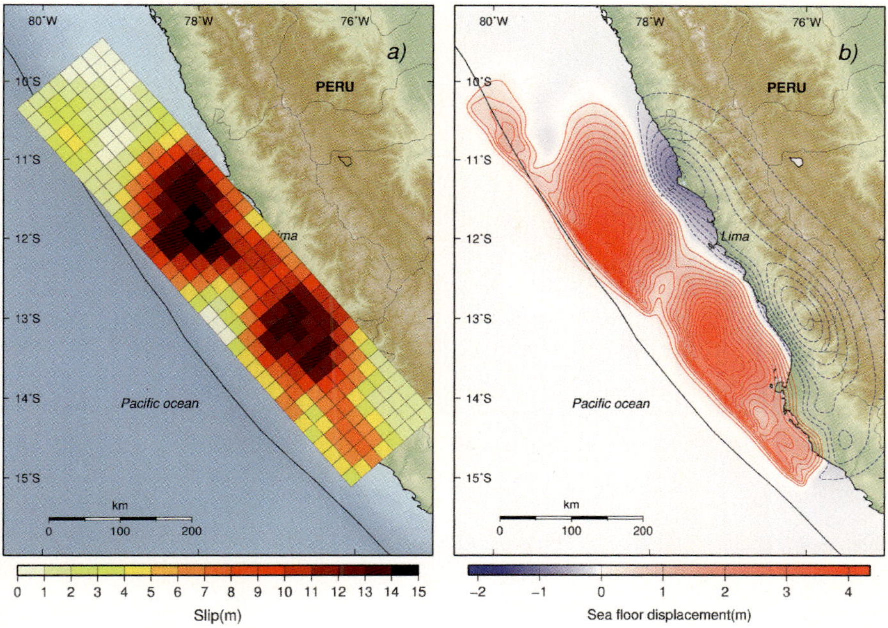

Fig. 19.3 Slip distribution from GPS data. There are 280 sub-faults of 20 × 20 km (Pulido et al. 2011)

19.3.2 Numerical Modeling

The Tohoku University's Numerical Analysis Model for Investigation of Near-field tsunamis (TUNAMI model) was used as the tool of tsunami modeling (Imamura 1996). A set of non-linear shallow water equations are discretized by the Staggered Leap-frog finite difference scheme. Bottom friction condition is in the form of Manning's formula constant in the whole domain.

19.4 Evacuation Simulation

We used the TUNAMI-EVAC1 (Mas et al. 2012, 2013b) to observe the tsunami inundation together with the resident evacuation behavior. The model was developed in NetLogo, a multi-agent programming language and modeling environment for simulating complex phenomena (Wilensky 2001). The model uses GIS data as spatial input and the tsunami numerical results to evaluate the probability of casualty at each step. Casualty estimation is a binary logistic model based on the experimental results of human instability reported in Takahashi et al. (1992). The population from the census data was categorized in four groups according to the age (Fig. 19.4).

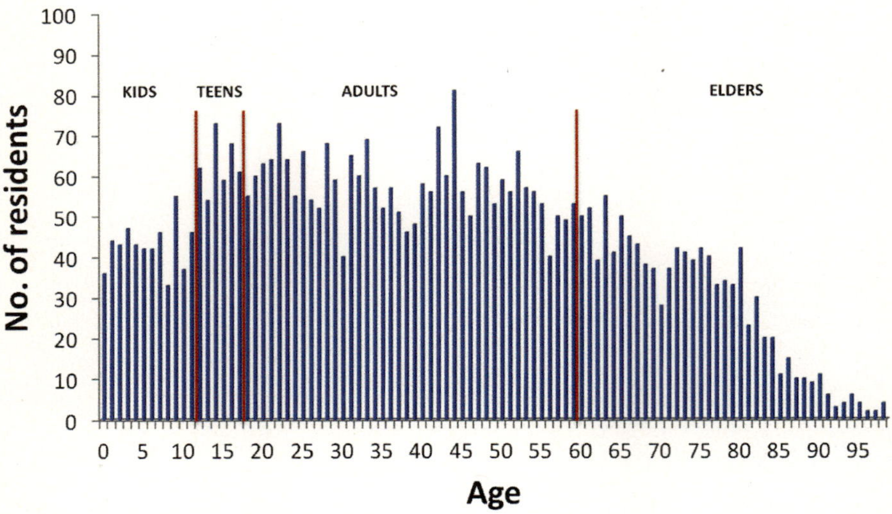

Fig. 19.4 Age distribution in La Punta and four groups of agent type for simulation

Table 19.1 Agent type number at each scenario and their maximum speed value allowed during simulation

Type	Vertical	Horizontal and vertical	Max. speed (m/s)
KIDS	514	514	1.06
TEENS	377	377	1.33
ADULTS	2,428	1,678	1.33
ELDERS	1,051	901	0.93
CARS	–	225	8.40

The main difference in all these groups or agent type is the maximum possible speed during evacuation. The speed varies as a half tail normal distribution of density in the agent field of view of 60° cone with 5 m distance for pedestrians and 10 m for cars. Table 19.1 shows agent types and their maximum allowed speed.

19.4.1 *Cases for Simulation*

For a better comprehension of the necessity of Tsunami Evacuation Buildings, a case for the only horizontal evacuation with no use of TEBs was conducted. Next, a Vertical Evacuation case and a combined case of Horizontal Evacuation and Vertical Evacuation were evaluated. These are the conditions for each case:

19.4.1.1 Horizontal Evacuation

Pedestrians and cars choose one of the possible two exits out of the district – two streets leading to the northeast of the district (Fig. 19.5 – triangles).

Fig. 19.5 Location of TEBs (*circles*) and Exits (*triangles*)

19.4.1.2 Vertical Evacuation

Twenty available TEBs are used for sheltering (Fig. 19.5 – circles). Evacuees – only pedestrians – choose the nearest shelter to their location regardless the capacity or condition at the shelter. To observe the over-demand of shelters, evacuees are allowed to enter the building without any restriction. Vehicles are not included in this case.

19.4.1.3 Horizontal and Vertical Evacuation

This is a combination of the two previous scenarios. Here, pedestrians and cars evacuate to any of all TEBs or Exits available in Fig. 19.5.

19.4.2 Start Time of Evacuation

In all cases the start time condition of pedestrians and cars follow the Tsunami Departure Curve method proposed in Mas et al. (2013a) (Fig. 19.6).

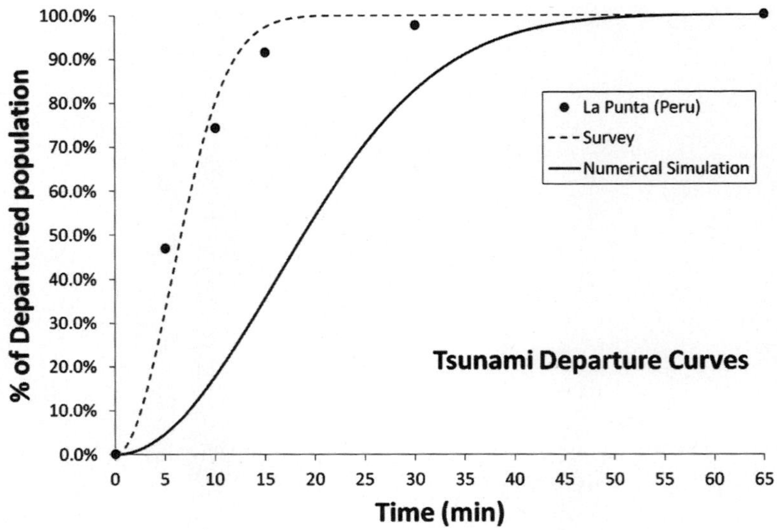

Fig. 19.6 Tsunami Departure Curves for the simulation cases. Behavior of start time is obtained through the stochastic simulation of random selected times bounded by two Rayleigh distributions. Left distribution obtained through stated preference surveys and the other distribution is based on the estimated arrival time of tsunami from numerical simulation

19.5 Results and Discussion

19.5.1 Casualty Estimation

Due to the many possibilities of spatial initial condition of residents, in the model, outcomes are averaged from a set of repetitions (250 runs) of different scenarios with agents loaded at random locations in the district (indoors or outdoors). For the result of these scenarios, it was observed that the Vertical case yields to fewer casualties (16 pers.) due to the short distance for a rapid evacuation and the less use of vehicles that contributed to stagnation points in the other two cases. The Horizontal Case shows the maximum average number of casualties (271 pers.), followed by the Horizontal and Vertical Case (153 pers.). As expected traffic congestion is observed along the exit roads and in particular near to the exits until the arrival of tsunami. Snapshots at 0, 10, 20, 30, 35, 40 min of the first case of simulation are shown in Fig. 19.7, while Fig. 19.8 shows the 40 min snapshots of the other two cases.

19.5.2 Shelter Demand

The number of agents sheltered at each TEB was counted and averaged from all repetitions; a comparison of demand and capacity is shown in Fig. 19.9 (next page).

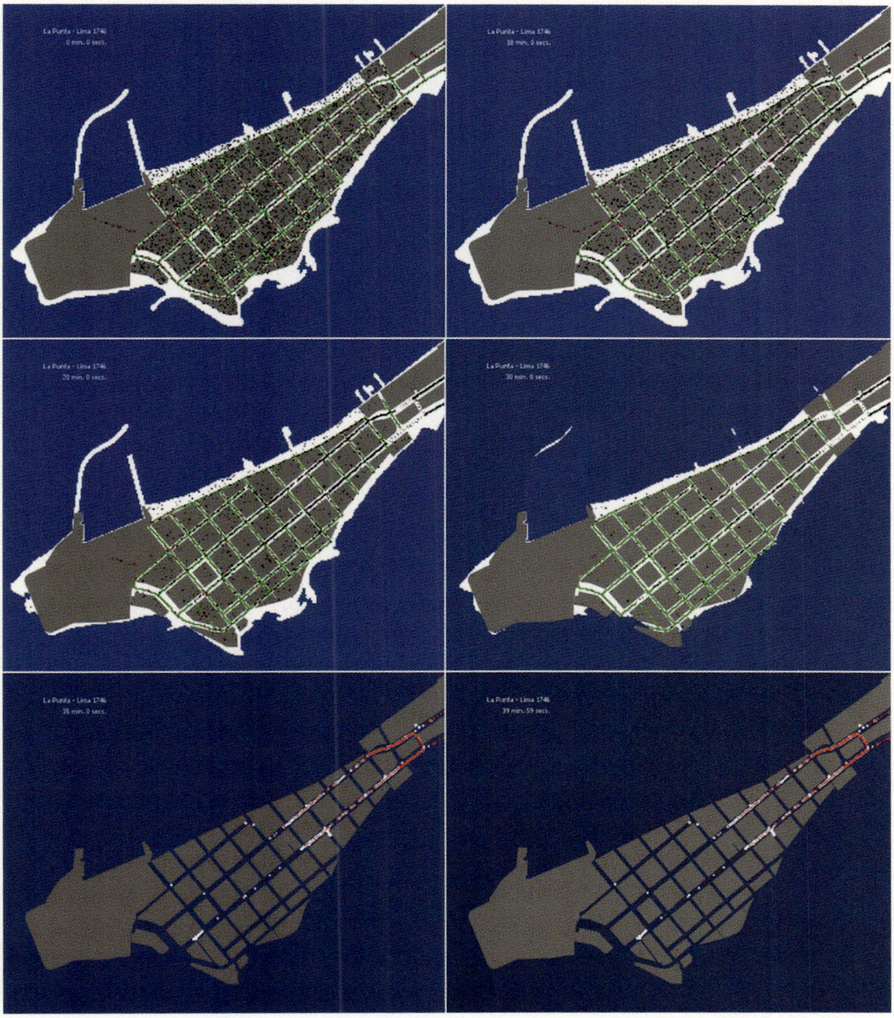

Fig. 19.7 Horizontal Evacuation Case. Casualties are observed at traffic congestion near the exit of the district. The necessity of a vertical evacuation is revealed here

Out of twenty (20) TEBs, 13 resulted with over-demands, while seven (7) were still with available space. It is necessary to ensure the safety of evacuees even at the over-demand shelters. As an example, building S11 with 25 persons of capacity may expect around 180 residents in the area looking for shelter. New areas for vertical evacuation should be implemented to support the rapid evacuation of residents in the area.

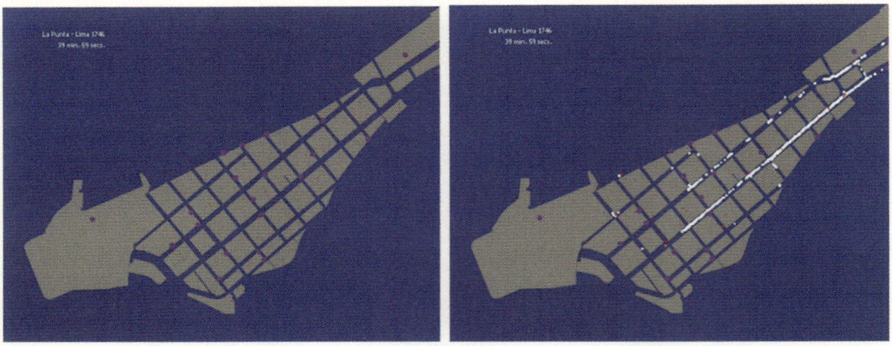

Fig. 19.8 Snapshot at 40 min simulation for the Vertical Evacuation case (*left*) and the Horizontal and Vertical Evacuation case (*right*). White dots show areas of stagnation.

Fig. 19.9 Capacity and Demand of Shelters in La Punta. Out of 20 TEBs, 13 might present over-demand and seven under-demands

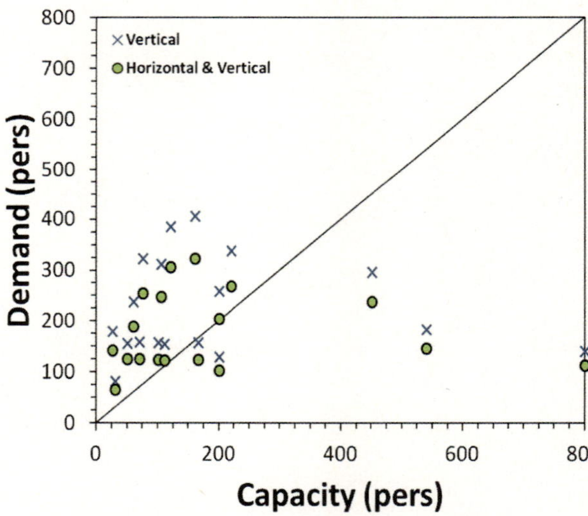

19.5.3 Mapping the Capacity-Demand

In order to facilitate the local authorities' decision of future countermeasures and alternatives for evacuation in La Punta, the resulting capacity-demand rate is mapped as observed in Fig. 19.10 (next page).

The relation capacity versus average demand of cases represents the Capacity-Demand Index (CDI) following the simple relation in Eq. (19.1):

$$\mathrm{CDI} = \frac{\mathrm{Demand}}{\mathrm{Capacity}} \tag{19.1}$$

Therefore, a value of 0.0–0.5 indicates that less than a half of the structure has been occupied; from 0.5 to 1.0 at most the total capacity of the structure is being used.

Fig. 19.10 Capacity-Demand mapping of TEBs. CDI > 1.0 (over demand) – CDI < 1.0 (under demand) (H&V case)

The important information comes for CDI values over 1.0, where over-demand conditions are probable. As a result, from the CDI Map it is possible to conclude that unfortunately the nearest buildings are comparatively of lower capacity with the ones located at the head and neck of the district. Due to the behavior of shelter selection (nearest), most of the structures closer to the beach area in the north coast present an over-demand while the south population apparently may fit in the available structures.

19.6 Conclusion

An evaluation of Tsunami Evacuation Building demand has been conducted through the simulation of agents in evacuation in La Punta. In this paper we have shown that the horizontal evacuation might lead to a high number of casualties due to the traffic congestion at the neck of the district. As estimated, the vertical evacuation would be a suitable solution for this area; however the spatial location of shelters complicates the evacuation process. Here, we identified the shelters with possible over-demand and their location according to the nearest shelter selecting behavior of evacuees. A CDI Map is proposed as a first step for future mitigation plans in the district. It is recommendable to allocate new areas for evacuation near to the north area of the district (close to the beach).

Acknowledgments We would like to express our deep appreciation to the Ministry of Education, Culture, Sports, Science and Technology (MEXT) and JST-JICA (Peru) for the financial support throughout the study. Also, a special thanks to the authorities and residents in La Punta for their participation and insights.

References

Berninghausen WH (1962) Tsunamis reported from the west coast of South America 1562–1960. Bull Seismol Soc Am 52(4):915–921

Chlieh M, Perfettini H, Tavera H, Avouac J-P, Remy D, Nocquet J-M, Rolandone F et al (2011) Interseismic coupling and seismic potential along the Central Andes subduction zone. J Geophys Res 116(B12):1–21. doi:10.1029/2010JB008166

Dorbath L (1990) Assessment of the size of large and great historical earthquakes in Peru. Bull Seismol Soc Am 80(3):551–576

Imamura F (1995) Review of tsunami simulation with a finite difference method. In: Long-wave runup models, World Scientific, p. 25–42

Imamura, F. (1996) Review of tsunami simulation with finite difference method. In: Yeh, H., Liu, P., Synolakis, C. (Eds.), Long-Wave Runup Models, World Scientific Publishing Co., Singapore. pp. 25–42

INEI (2007) Censos Nacionales 2007. XI de Poblacion y VI de vivienda @online

Kuroiwa J (2004) Disaster reduction: living in harmony with nature, 1st edn. Editorial NSG, Lima, Peru

Langer CJ, Spence W (1995) The 1974 Peru earthquake series. Bull Seismol Soc Am 85(3):665–687

Mas E, Suppasri A, Imamura F, Koshimura S (2012) Agent based simulation of the 2011 great East Japan earthquake tsunami evacuation. An integrated model of tsunami inundation and evacuation. J Nat Disaster Sci 34(1):41–57

Mas E, Imamura F, Koshimura S (2013a) Tsunami departure curves for evacuation simulation. A new approach to model the human behavior in tsunami evacuation. (under review)

Mas E, Adriano B, Koshimura S (2013b) An integrated simulation of tsunami hazard and human evacuation in La Punta, Peru. Journal of Disaster Research 8(2):285–295

Pulido N, Tavera H, Perfettini H, Chlieh M, Aguilar Z, Aoi S, Nakai S et al (2011) Estimation of slip scenarios for megathrust earthquakes: a case study for Peru. 4th IASPEI/IAEE International Symposium, Santa Barbara, pp 1–6

Takahashi S, Endoh K, Zen-ichiro M (1992) Experimental study on people's safety against overtopping waves on breakwaters. Port Harb Res Inst 31-4:3–31

Wilensky U (2001) Modeling nature's emergent patterns with multi-agent languages. EuroLogo, Linz, Austria

Chapter 20
Multiple Wave Arrivals Contribute to Damage and Tsunami Duration on the US West Coast

Aggeliki Barberopoulou, Mark Randall Legg, Edison Gica, and Geoff Legg

Abstract Tsunamis persist long after the triggering geophysical events diminish. The Tohoku, Japan tsunami of March 11, 2011 was an extreme event that continued to disturb the Pacific Ocean for many days following its initiation. Historically Japan was considered a source of low tsunami wave energy for the US West Coast. However, damage in California from the last great Japan tsunami was second to that suffered during the 1964 Alaska earthquake. Computer animations of the catastrophic Japan tsunami and other recent significant tsunamis combined with source wavelet cross-correlations help to identify multiple paths of tsunami wave energy refracted and reflected by complex bathymetry across the Pacific Ocean basin. Using recent large tsunamigenic earthquakes we demonstrate that the long duration and damage suffered in the far field during the great 2011 Tohoku Japan tsunami was a result of several factors. Shallow water waveguides and continental margins act as tsunami lenses and mirrors to direct the wave energy to diverse locations around the ocean basin; directionality affected by islands and seamounts, large reflections off of South America and Oceania (New Guinea region), bathymetric features far and near the area of impact and shelf geometry may delay and further amplify the main tsunami energy. This contribution of Ocean basin scatterers can be estimated a-priori and can help define impact zones vs. shadow zones and duration of events. This has direct implications on the prediction of tsunami impacts because the US West Coast often appears to receive maximum wave heights much later than first wave arrivals from far field tsunamis.

Keywords Tsunami • Tohoku

A. Barberopoulou (✉)
GNS Science, 1 Fairway Dr. Avalon, Lower Hutt 5010, New Zealand
e-mail: A.Barberopoulou@gns.cri.nz

M.R. Legg • G. Legg
Legg Geophysical, 16541 Gothard Street # 107, Huntington Beach, CA 92647, USA

E. Gica
NOAA Center for Tsunami Research, JISAO-University of Washington,
Seattle, WA 98105, USA

Y.A. Kontar et al. (eds.), *Tsunami Events and Lessons Learned: Environmental and Societal Significance*, Advances in Natural and Technological Hazards Research 35,
DOI 10.1007/978-94-007-7269-4_20,

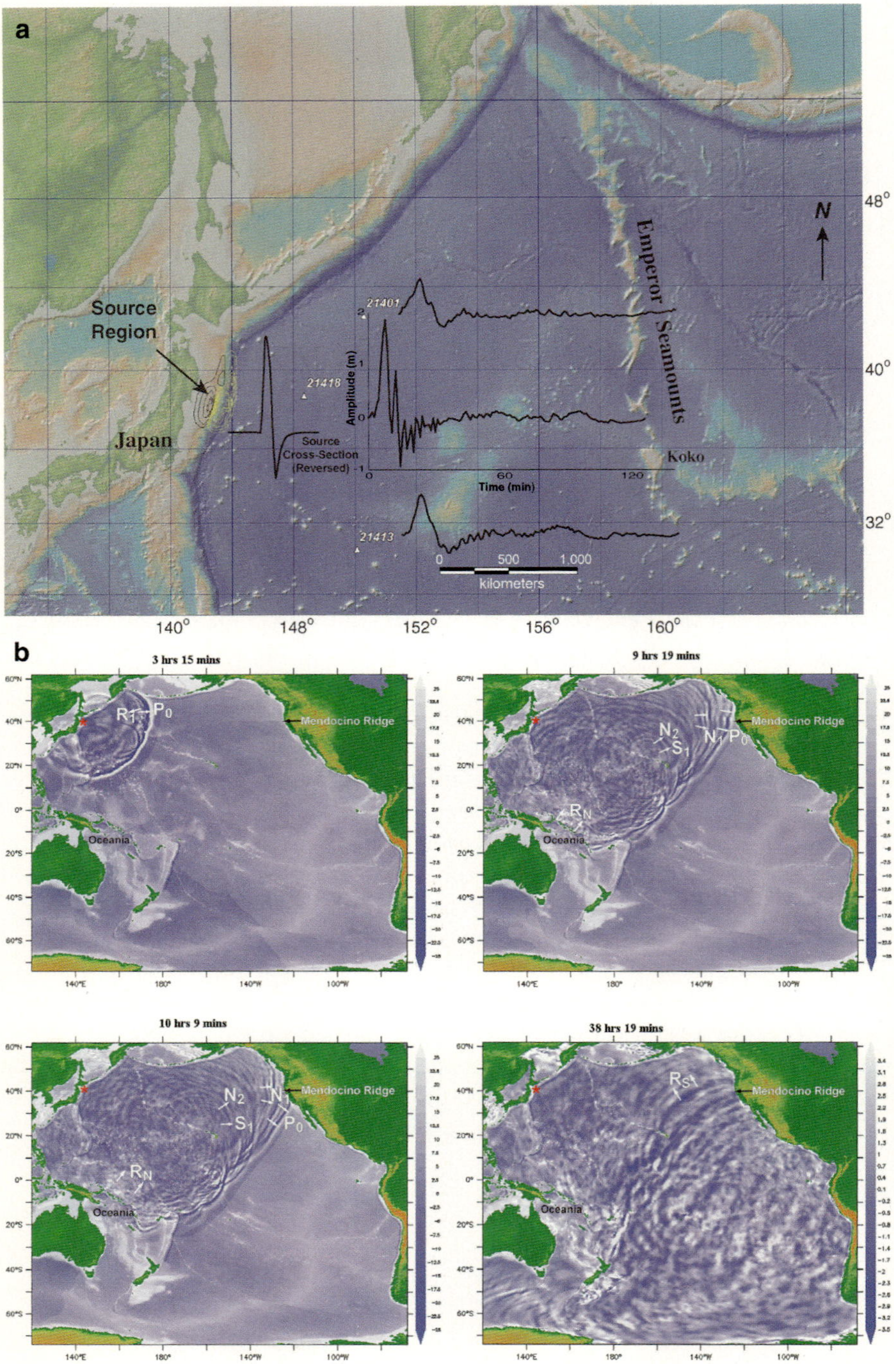

Fig. 20.1 (**a**). Tsunami source waveforms recorded on three near-source DART® buoys (*white triangles*) showing 128-min record of initial wave arrivals. A cross-section of the initial sea surface elevation based on the static dislocation model is shown reversed to represent the source time function expected at DART® buoy 21418 (cross-section along latitude of DART® 21418). (**b**) Snapshots of the modeled tsunami wave propagation during the March 2011 Tohoku tsunami in hours since earthquake origin. Red star indicates the earthquake epicenter. White arrows with labels indicate refracted/reflected waves off of prominent Pacific scatterers where

20.1 Introduction

On Friday, March 11, 2011 at 05:46:23 UTC a M9.0 earthquake occurred near the northeast coast of Honshu, Japan followed by numerous aftershocks, and a train of tsunami waves causing a level of destruction that has no parallel in recent history in Japan (Fig. 20.1). The great Tohoku earthquake now registers among the five most catastrophic events recorded during the past century. In particular, as of December, 2012 more than 2,300 earthquakes of magnitude 4.5 or greater have been registered in the epicentral region of the great earthquake of March 11, 2011.

The first tsunami waves were expected to arrive on the US West Coast approximately 9 h following the Tohoku earthquake (Fig. 20.1b previous page). However, damage was caused by later arrivals as previously noted during other tsunami events (Wilson et al. 2010). Local emergency personnel for most coastal areas ended their response procedures as extreme currents/surges subsided by morning of March 12, 2011 (PST) but wave action continued for the next few days. Unusually long water wave motions with distinct periods of 10–20 min were observed in many places along the coast of California. Even 4 days following the arrival of the first tsunami waves the ocean surface oscillations had not diminished to levels observed prior to the Japan event (first 35 h are shown in Fig. 20.2).

Similar duration was noticed on the coasts of Oregon and Washington but amplitudes of initial oscillations diminish faster than those along the California Coast (Fig. 20.2). To determine the cause of this prolonged duration and influence on the tsunami impact we analyzed and compared tide gauge records from the US West Coast, DART® buoy records from Japan and the US West Coast, and numerical simulations using MOST (Method of Splitting Tsunami; Figs. 20.3, 20.4a–d and 20.5a–c; Titov and González 1997; Titov and Synolakis 1998). The availability of near source recordings of this significant tsunami has provided the unique opportunity to calculate the tsunami source instead of an approximation based on numerical predictions. Computer animations of the Pacific-wide reach of the Tohoku, Japan tsunami, show complex refracted wave propagation in the mid-Pacific and multiple reflections off the coast of South America and the Australia-SE-Pacific Islands region. Bathymetric and topographic features of the Pacific Basin, source characteristics and the local basin physiography contribute to the extended wave activity in California over 20 h following the first waves as shown in Fig. 20.2 next page. Of particular concern are damaging waves and currents following late arrivals after many areas would consider the hazard to have passed. Although the idea of ocean

Fig. 20.1 (continued) semi-circular wavefronts are generated. Frame (*upper left*) shows strong wave reflection from the Emperor Seamounts (R1). Frame (*upper right*) shows divergent wavefronts approaching the US West Coast produced by refraction and diffraction by major seamounts including Koko (N1) and others along the Hawaiian Ridge (N2, S1). Frame (*lower left*) shows additional divergence of wavefronts propagating toward the US West Coast as well as significant reflected energy from Oceania. Re-entrants or cusps along wavefronts show focusing produced by elongate shallow water tsunami waveguides like the Hawaiian and Mendocino ridges. Frame (*lower right*) shows a fully-excited Pacific Ocean basin due to complex interference of tsunami energy wavefronts from reflections (Rs), refractions, and diffractions by major basin shoals and margins more than 38 h after tsunami generation

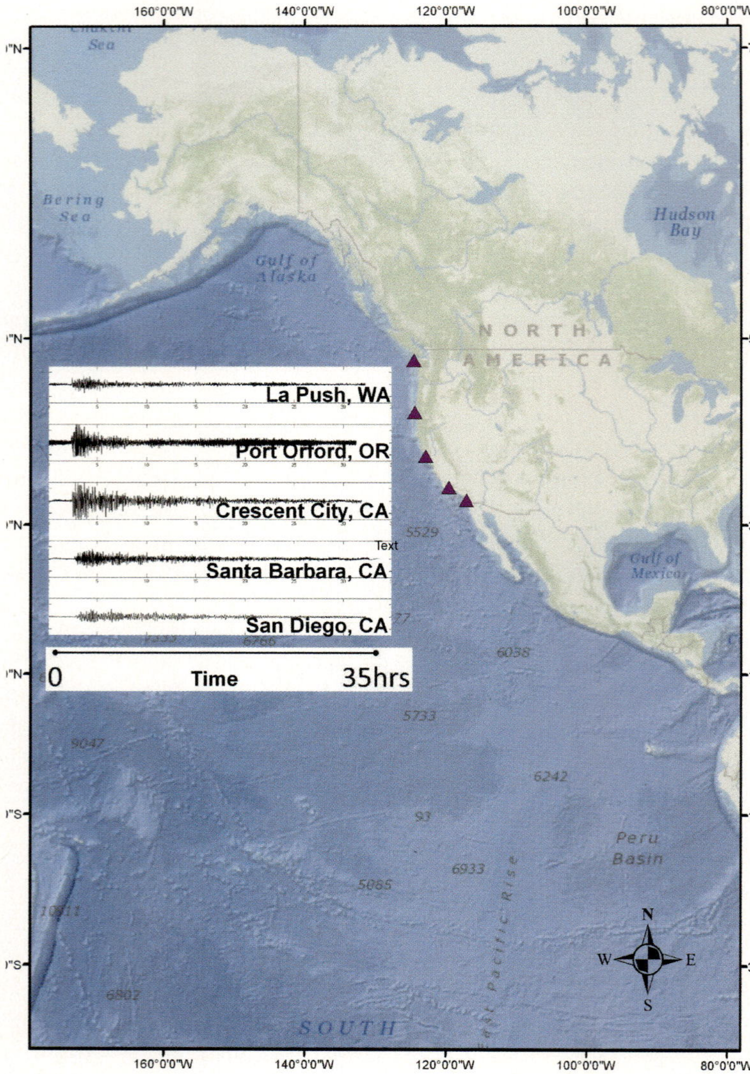

Fig. 20.2 Map of the US West Coast with five tide gauges from the Washington, Oregon and California coasts, showing prolonged wave activity beyond the first 24 h following the Tohoku tsunami. Each tide gauge record corresponds to 35 h of recording (x axis). Y axis shows amplitude in m (from −1.5 to 1.5 m)

basin scatterers is not new (i.e. Mofjeld et al. 2001) we adopt the cross-correlation method from seismology to show the reflections/refractions off of major island seamounts in records (i.e. "tsunami phases").

The significance of this is the ability to predict when strong energy will be redirected by ocean basin scatterers to locations that are vulnerable and which areas are in "shadow" zones therefore less impacted. Quantities such as the duration of a tsunami event affected by multiple wave arrivals can be predicted prior to first tsunami wave arrival and therefore be available to emergency management in time for precautionary measures to minimize damage and life loss.

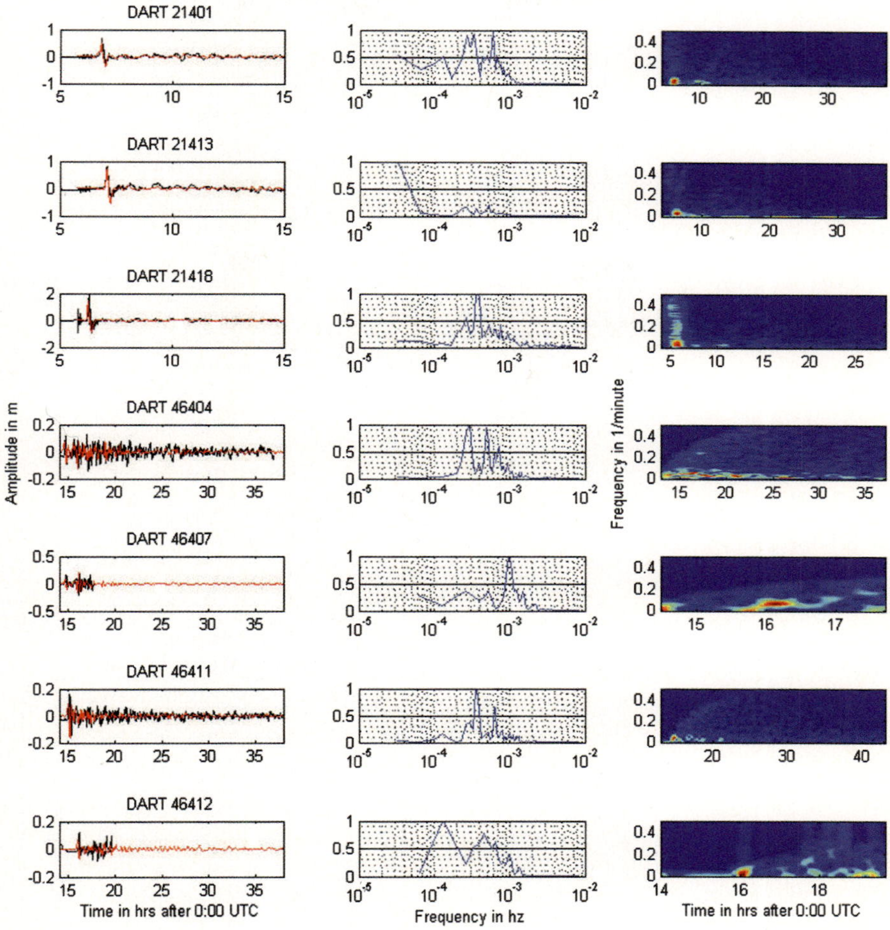

Fig. 20.3 Comparisons of DART® buoy data (*blue*) with simulations (*red*), spectra of DART® records (*middle column*) for three DART® buoys near Japan and four DART® buoys on the US West Coast, with corresponding spectrograms (*right column*). See Fig. 20.5 for DART® buoy locations

20.2 Observations

Historically, tsunamis generated off Japan were not large along the California coast (Wiegel 1976; Lander et al. 1993). Tsunami experts considered the approach angle of these tsunamis to be glancing or the complex basin-and-ridge physiography of the California Continental Borderland reduces tsunami amplitude by wave transformation and reflection at the margin of the broad coastal shelf (Van Dorn et al. 1979). However, damage along the California coastline from the March 11, 2011 tsunami is second only to that suffered during the 1964 Alaska earthquake.

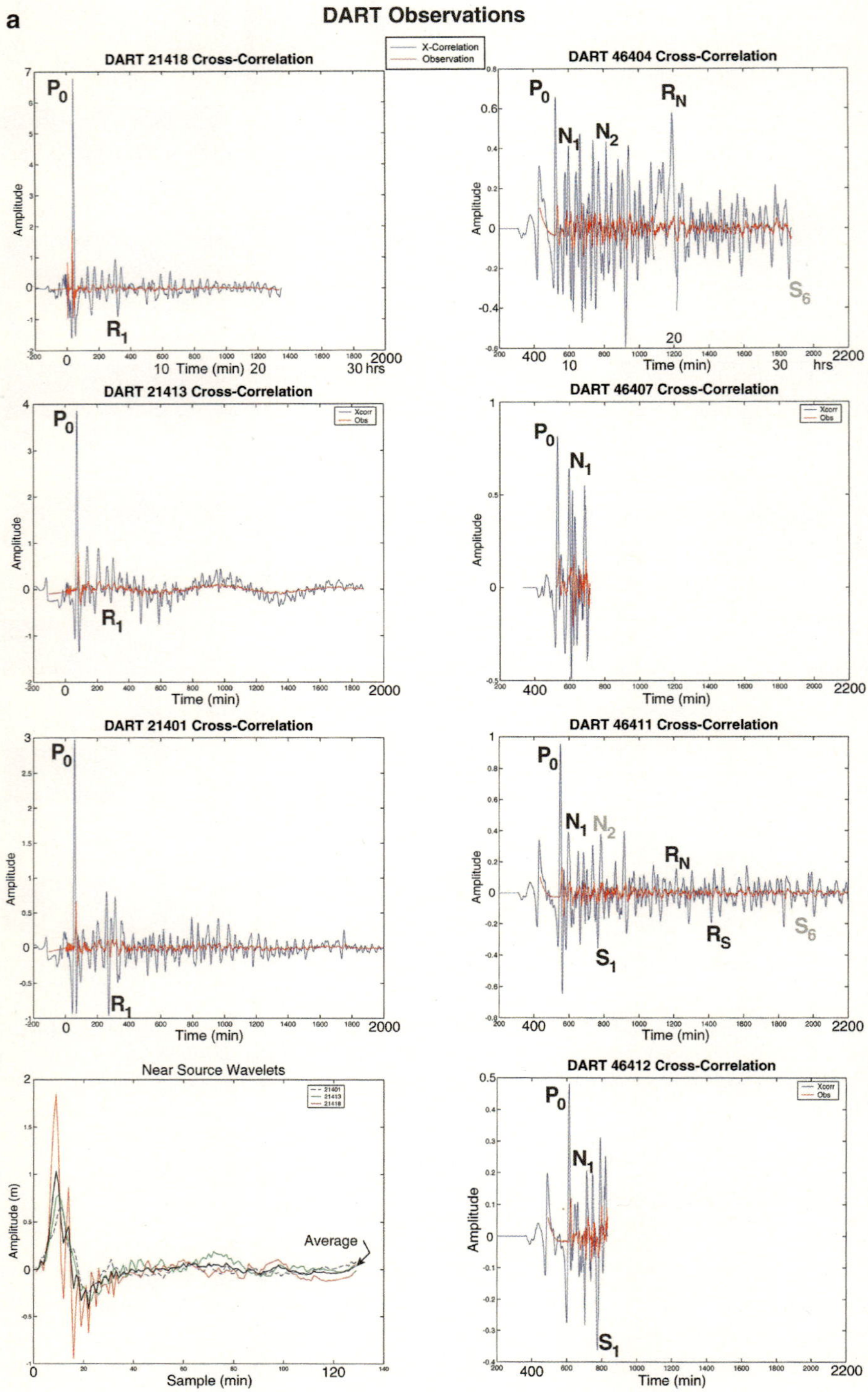

Fig. 20.4 (**a**). DART® records near source and in the farfield. Records (*red*) were cross-correlated with the average source wavelet (*blue*). P0 is the direct wave, *N picks* represent phases distinctive to coastlines of the Pacific Northwest of North America and *S picks* distinctive to Southern California. R phases are reflections from major features including Oceania (Rn) and South America (Rs).

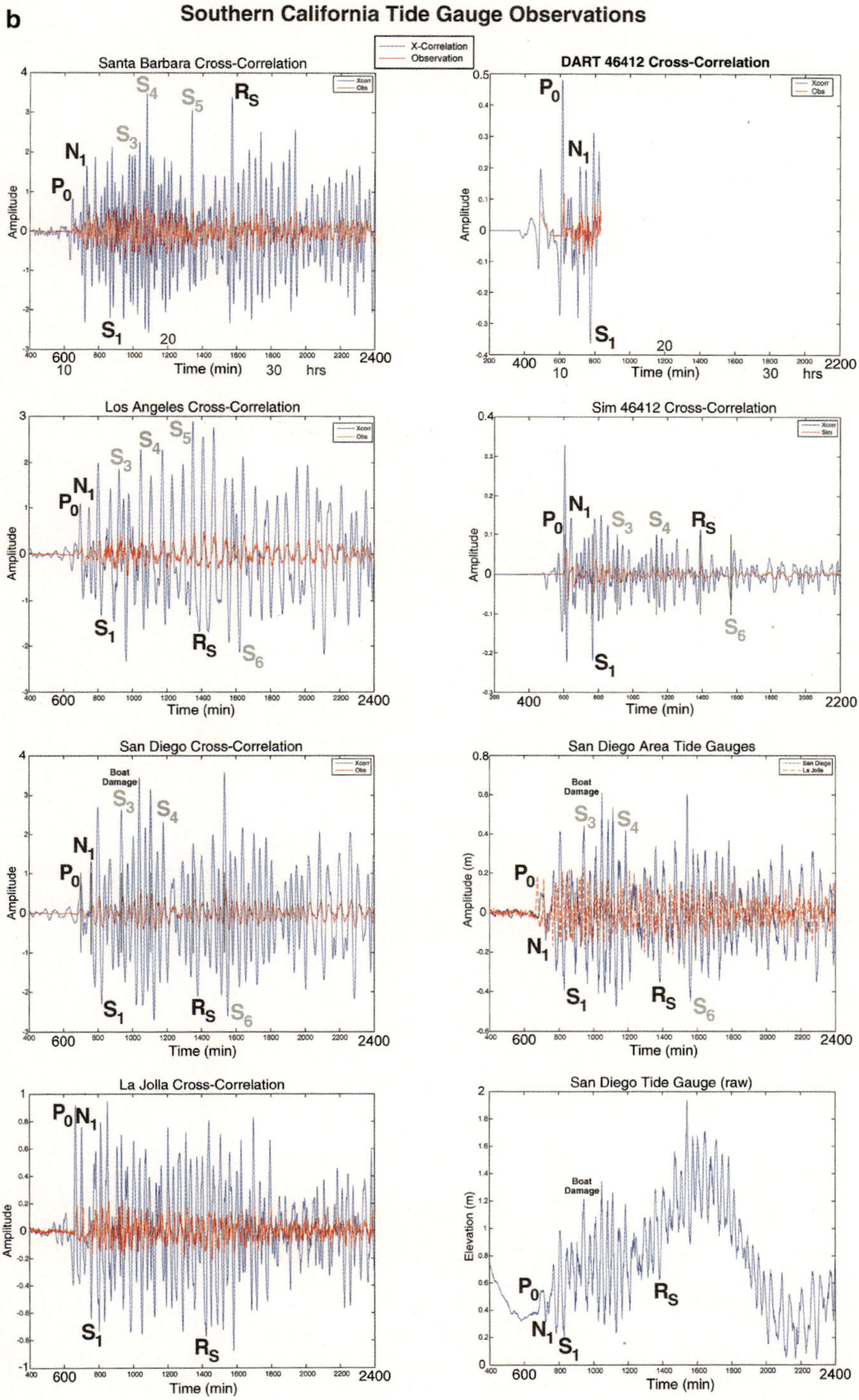

Fig. 20.4 (continued) (**b**). Tide gauge and DART® records (*red*) and cross-correlated (*blue*).

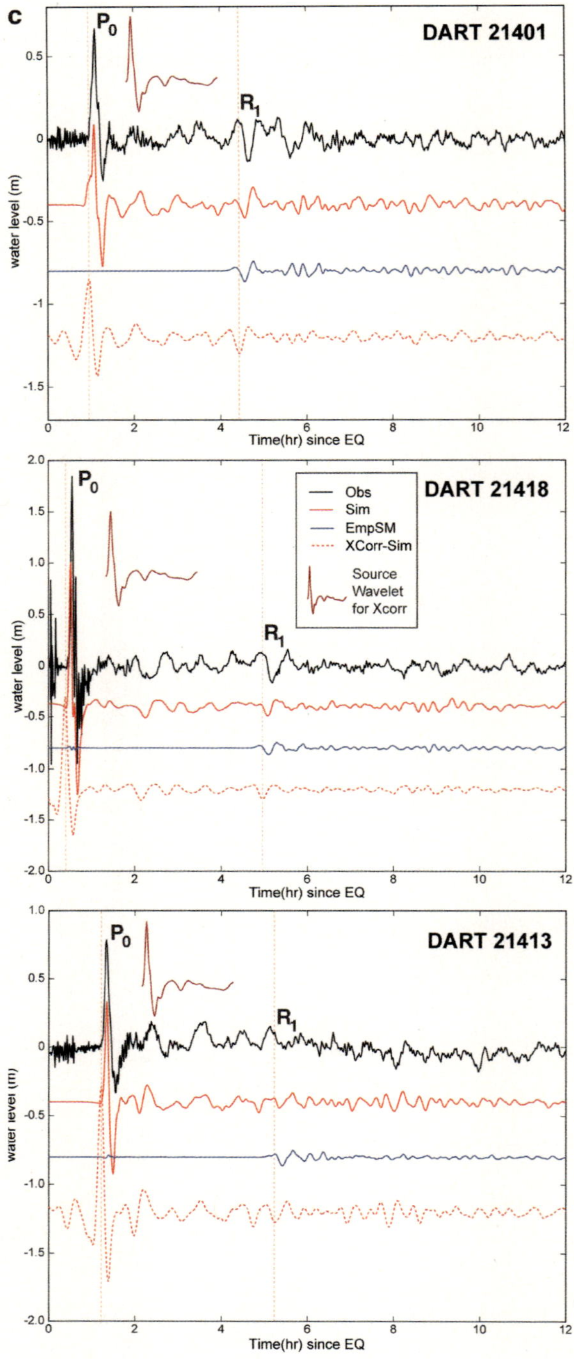

Fig. 20.4 (continued) (**c**). Near source DART® records compared to the simulated Emperor Seamount contribution (*blue*) show the strong reflected pulse (R1) from the Emperor Seamounts. Predicted arrival time of the reflected pulse is measured at the peak (trough) of the cross-correlation record. DART® 21413 is located farthest from the Emperor Seamounts and shows more complex reflected waveform due to interference from other wave arrivals scattered from closer features (Fig. 20.1a).

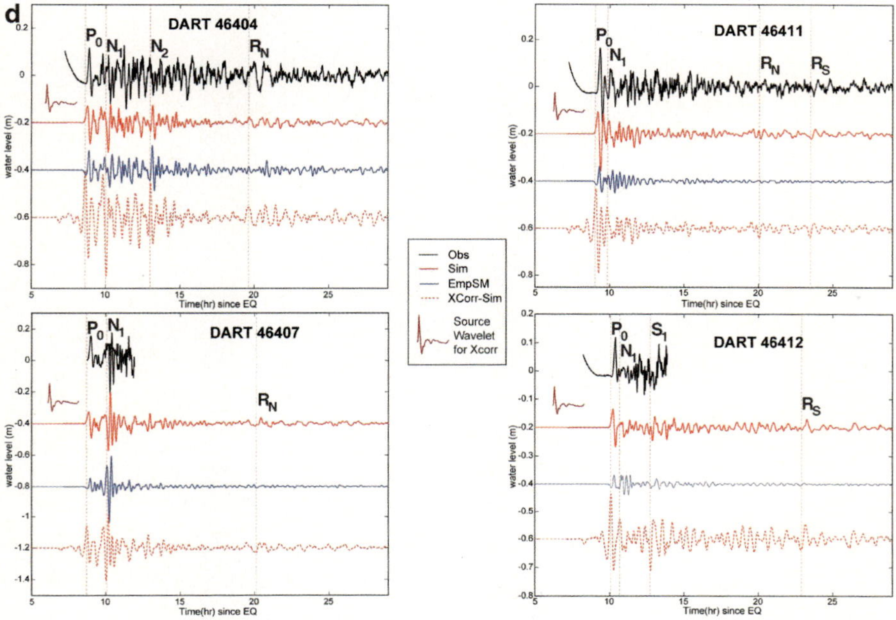

Fig. 20.4 (continued) (**d**). US West Coast DART® records compared to the simulated Emperor Seamount contribution (*blue*) show complex wave arrivals with greater amplitudes for later phases. The direct path from Japan to the US West Coast crosses the Emperor Seamounts and other shallow water waveguides further complicate the wavefield with dispersion and multipath effects

Following a weak first arrival in most coastal areas, California experienced stronger surges 4–8 h later. Wave action continued through the night with harbor personnel noticing distinct wave action with 10–20 min period for as long as 4 days after the event started. The highest tide of the day occurred during the night so that the inundation potential was greatest then. Based on tide gauge records it is uncertain when the tsunami event ends as persistent oscillations that exceed normal background levels in the ports are recorded through the 14th of March. Long duration tsunami wave activity along the California coast is not unusual. Four to five days of oscillations have been observed in San Diego, Santa Monica, Crescent City and parts of the Washington coast from large Pacific-wide tsunamis that occurred in 1946, 1960, and 1964 (Lander et al. 1993). The largest amplitudes were sometimes delayed for hours after the first arrivals at Crescent City, where significant damage and fatalities occurred, especially in the 1964 Alaska event. Likewise, the strongest currents observed in San Diego harbor occurred hours after the first arrivals in the 1960 Chile event. Low frequency oscillations persisted for a week (Miller et al. 1962). Late tsunami arrivals combined with a large south swell during the high tide caused flooding in Seal Beach/Sunset Beach, Orange County on the night following the February 27, 2010 Chile event. Currents reached 15 knots in

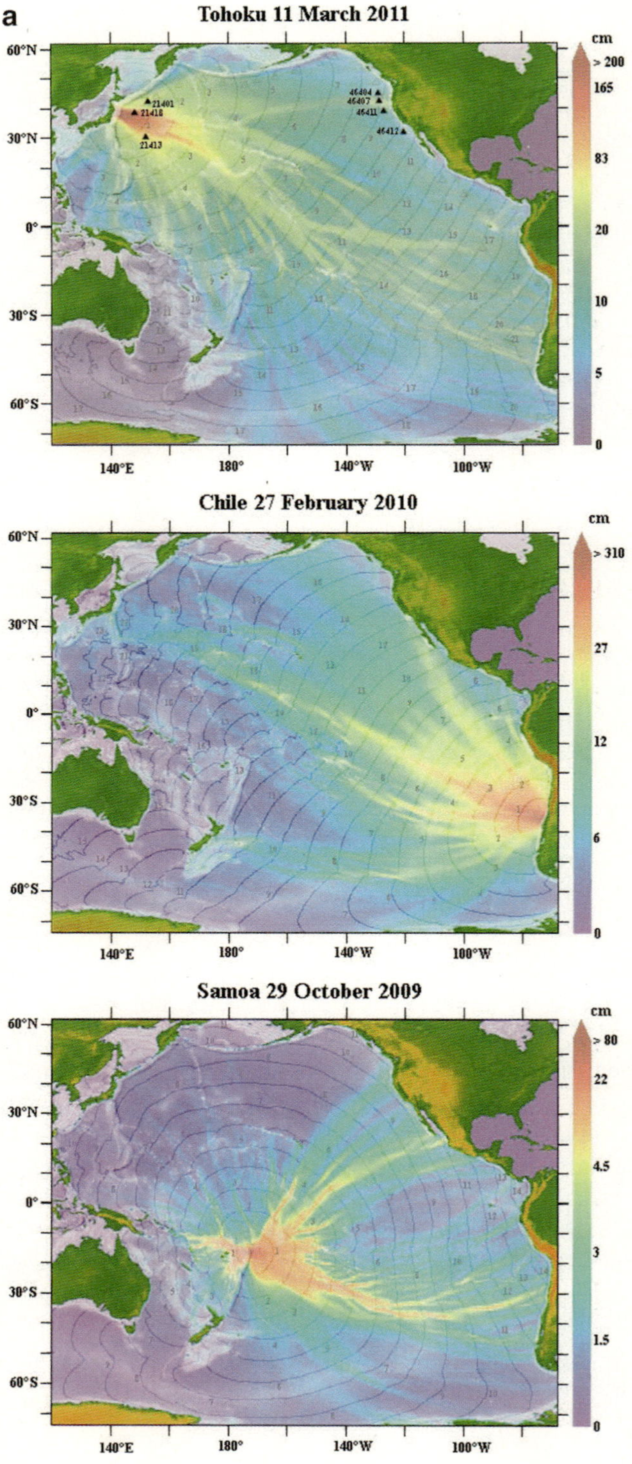

Fig. 20.5 (**a**). Energy plots of three recent large tsunamis. Black triangles are locations of DART®s that were used in this article. Contour lines represent arrival times (in hours) of first waves.

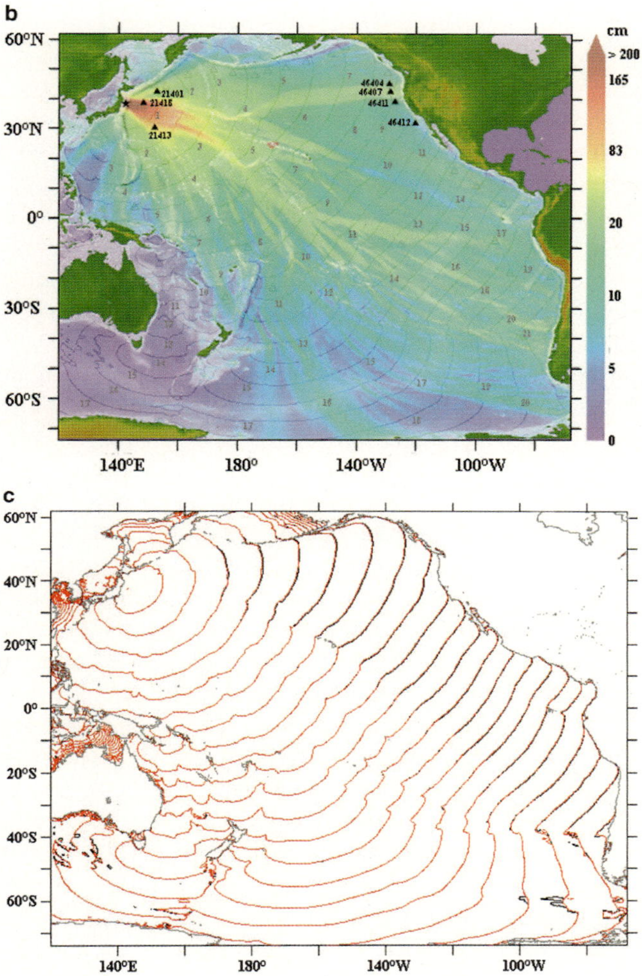

Fig. 20.5 (continued) (**b**). Energy plot of the Tohoku tsunami after the removal of the Emperor/Koko Seamount shows a stronger band of energy redirected to the east from the northern end of the Hawaiian Ridge, near the location of Koko Guyot, and focused along the Mendocino Ridge. (**c**) Arrival time (wave refraction diagram) for the Tohoku event with original bathymetry (*red contours*) and predicted arrivals after the removal of Emperor/Koko seamounts (*black contours*)

Southern California causing scouring and severe damage to docks and boats (Wilson et al. 2010). What appears to be unexpected for the Tohoku tsunami are the large amplitudes and strong currents that persisted for many days after the first arrivals in California from a source located to the west off Japan. In contrast, smaller magnitude events, such as the M8.3 Kuril Islands earthquake of November 2006 generated a tsunami that was hardly noticed on the coast of California but strong currents caused damage to the small boat basin of Crescent City that was estimated at 10 million USD (Dengler et al. 2009).

20.3 Review of Related Work

Previous investigations of tsunami hazards along the California coast considered that observed persistent wave activity or long "ringing" lasting several hours after the first tsunami wave arrival may be due to local resonances, derived from coastal configuration and offshore bathymetry. Raichlen (1972) showed that the ringing frequencies at widely spaced locations including Santa Monica, Los Angeles Harbor, and La Jolla were the same precluding local resonance as a dominant mechanism. More recently wavelet analysis indicated that local resonances may be the dominant mechanism in the delay of large waves observed on parts of the US West Coast (Tang et al. 2008).

Azimuth or approach angle of tsunami waves is recognized as important for creating large wave or current activity along distant coasts (Van Dorn et al. 1979). Directivity is based on the source orientation, where stronger wave energy is focused orthogonal to long subduction zone sources (Wiegel 1976; Miyoshi 1986). These raypaths may be evident in the tsunami energy diagrams (Fig. 20.5a–c) and show tendrils of focused energy propagation.

Multiple wave arrivals pump energy into bays and harbors that may result in extensive oscillations enhanced by local basin resonances (i.e. Rabinovich et al. 2011). Examination of the modeled wave propagation time history shows that the northwest Pacific appears to act as a significant source of scattered tsunami wave energy (Mofjeld et al. 2001). Dispersion along a tsunami waveguide and diffraction by distinct features within the waveguide, such as Koko Guyot, French Frigate Shoals, etc., may produce resonant wave trains depending on the geometry of the scatterers. The saddles and peaks in the amplitude spectra for the DART® records may demonstrate these resonances. Dispersion is evident as the progressively later arrivals of higher frequency energy travels slower in the shallower water near the axis of the waveguide (Marchuk 2009). In the case of Crescent City, Northern California, recent work demonstrates that local amplification is a result of redirected energy and local bathymetry (Kowalik et al. 2008; Horillo et al 2008).

20.4 Analysis

To determine the factors that contribute to the prolonged wave activity and enhanced impact of tsunamis along the West Coast of the United States, we run simulations of recent tsunamis. The numerical code used in the simulations is MOST (Titov and González 1997; Titov and Synolakis 1998). The code uses the depth-averaged non-linear shallow water wave equations in spherical coordinates with a numerical dispersive scheme that also takes into account the Coriolis parameter. Although our focus was on the Tohoku tsunami, two other tectonic tsunami sources were considered with $M > 8$ but widely different epicentral locations to demonstrate the effect of a broad range of bathymetric features on the radiation pattern and damage potential of tsunamis. We focused our investigation on strategically located DART® buoys and tide gauges to evaluate the extensive tsunami-induced oscillations of the Pacific ocean in the days following the 11 March 2011 Japan tsunami through the identification of multiple distinct energy pulses ("tsunami phases"). This was also complemented by a second simulation of the Tohoku tsunami over a modified bathymetry "missing" the Emperor and Koko seamounts to exhibit the contribution of this island chain to tsunami waves.

Seven DART®s were selected- three close to Japan (21401, 21413 & 21418; Fig. 20.1a) for near-source wave character and four adjacent to the US West Coast (46404, 46407, 46411, 46412; Fig. 20.3 next page). Tide gauge records from Southern California (Santa Barbara, Los Angeles, San Diego and La Jolla) were also used. Cross-correlation of a near-source wavelet with the tsunami record at distant sites is used to convert the tsunami pulses to zero-phase wavelets where the energy peaks at the exact arrival time of the energy pulse (Fig. 20.4a–d).

Because of the availability of near source records we can calculate the average true wavelet instead of a numerical approximation. This source wavelet cross-correlation process allows more accurate measurement of the arrival times of significant tsunami wave energy pulses. A better estimate of the "average" source wavelet, was obtained by averaging amplitudes for the three near source DART® buoy records. The source region contours were derived from the NOAA elastic dislocation model used in the MOST simulation and the bathymetry is derived from GeoMapApp (www.geomapapp.org). The source wavelet was windowed (128 samples = 128 min) to exclude significant reflected wave energy from more distant coasts or shoals, although it appears that it includes energy reflected from the coast of Japan or trench wall as can be seen by near source DART® records (Figs. 20.1a, 20.4a and 20.4c).

To identify the propagation path of the suspected tsunami phases, the tsunami animation was closely reviewed to identify distinct wave energy pulses as they propagate across the ocean basin and arrive at the time predicted by the cross-correlated record. Although the wavelet shape for different azimuths of propagation can change, for California and the US West Coast, the azimuths of tsunami energy propagation are within 45° of the orthogonal to the subduction zone source. The average wavelet includes near-source DART® records that fall within this range of azimuths.

US West Coast instruments show at least two large waves and strong oscillations that continue 15 h after the earthquake. Tide gauge and DART® records along the Southern California coast show complex long duration tsunami wave arrivals (Figs. 20.2, 20.3 and 20.4a–d). Animations also show multiple waves due to scattering of the propagating front as it encounters complex topography and bathymetry (Fig. 20.1b). Fourier spectra and spectrograms of near source DART® buoys compared to US West Coast instruments show dispersion and attenuation of the high frequencies present in the source signal (Fig. 20.3). Two distinct peaks are apparent in the near source DART®s one at 50 min period and one at approximately 65 min. Those two are almost identical on all source DART® buoys. The more complex source wavelet at DART® buoy 21418 shows the high frequency content of the source, which smooths out as the wave propagates farther as seen in the other two source wavelet records. The low frequency negative part (trough) of the source wavelet represents the subsidence that occurs on the western side of the source region toward the Honshu coast (Fig. 20.1a). Later arrivals in the DART® buoy records from the near source area likely represent the tsunami energy reflected back from nearby coasts or major seafloor shoals like the Emperor seamounts. The instruments from the US West Coast also show a peak at 50 min but they also show prominent peaks at shorter periods (33 min and 20 min approximately). The latter peak is apparent to what observers described during our field survey of Southern California. These peaks can be partially attributed to amplification due to the Mendocino escarpment (Kowalik et al. 2008; see energy finger on Fig. 20.5b pointing at 46407 instrument) but local effects may also be responsible. For example, Koko Seamount (Guyot) and an east-trending chain of seamounts create a significant shallow water tsunami waveguide that focuses energy toward the Mendocino Ridge and the California-Oregon border region (including Crescent City).

Based on the waveform analysis we can identify distinct arrivals at the far-field gauges. Some arrivals appear more prominently on the Pacific Northwest coast of the US (N_1, N_2 and R_N; Fig. 20.1b and 20.4a–d) and others appear distinctive to Southern California (N_1, S_1 and R_s). For example, N_1 and S_1 picks correspond to redirection of energy due to the Emperor Seamount Chain and Hawaiian Ridge and later arrivals correspond to reflection from South America. More prominent arrivals and the harmonic appearance in cross-correlated tide gauge records may indicate local amplification of the deep water tsunami pulses by local resonances and wave interactions. The bathymetry and topography of the Pacific Basin, as suggested by the deep ocean propagation of the Japan tsunami is partly responsible for the prolonged duration of oscillations observed along the US West Coast and for the delayed arrival of the first wave. Specifically, simulations of the Tohoku tsunami at 19 h following the earthquake appear to show shorter wavelength tsunami waves arriving from the southwest that may be responsible for the strong surges that damaged the San Diego Harbor Police docks and boats around 16:00PST (24:00 UTC). This coincides with phase S_1 corresponding to a refraction/diffraction from the French Frigate Shoals and the Hawaiian Ridge; waveguide dispersion delayed the arrival of shorter period energy. At approximately 23 h travel time, tsunami wave energy reflected from South America (R_S) reaches southern California. The distinct wavefront produced is the first of a series of strong reflections off of South America. After more than 30 h travel time, strong reflections off of the Chile/Peru coast reach Southern California (Figs. 20.1b and 20.4a–d).

Cross correlations were also performed of our simulations with the Ocean basin bathymetry after the removal of the Emperor Seamount Island chain at the seven DART® locations (Fig. 20.4c, d). Note that N_1 arrives shortly after the first arrival for DART® buoys 46411 and 46412, which are south of the Mendocino Ridge. The shallower water of the Gorda/Juan de Fuca plate north of the ridge delays this tsunami energy relative to the direct arrival – the energy is focused along the

Mendocino Ridge (waveguide effect) that amplifies the wave at 46407 (located near the Mendocino Ridge) and Crescent City. In contrast, 46404 located north of the ridge shows greater scattering and resonance whereas 46411 located south of the ridge shows less complexity. The Southern California 46412 is less affected by the Mendocino Ridge, but shows energy scattered from the Hawaiian Ridge a few hours after the initial arrival. Dispersion due to the waveguide effect is apparent, especially south of the Mendocino Ridge, like an edge wave or leaky mode propagation. This is most apparent on 46411.

A quick summary of how ocean basin scatterers can affect the distribution of energy can be seen from the modeling results of three recent large Pacific tsunamis (see Fig. 20.5a–c). Although most areas of higher wave energy appear to follow the propagation paths, there are important areas where the high-amplitude energy "stringers" are oblique to the wavefronts and wave-paths. Because of this mirror-like reflection, the northwest coast of North America was protected, including British Columbia and Alaska from significant wave energy generated by the Samoa event of September 29, 2009. In contrast, energy is focused toward Southern California by the Chile source and focused toward Mendocino and Crescent City for the Japan source. Although a comparison with records of the Feb 27, 2010 Chile tsunami on the US West Coast would have been ideal, only DART® buoy 46407 recorded the event. Tide gauge records of the Chile tsunami were also problematic and for this reason are not included in the paper. The September 29, 2009 simulation indicates another prominent refractor/wave-guide tsunami lens in the mid-Pacific equatorial region. This feature may have diverted energy from the Samoa tsunami to the east toward southern Mexico and Central America. The removal of the Emperor/Koko seamounts results in a wider impacted area on the US West Coast for the Tohoku event (Fig. 20.5b). The Emperor Seamounts appear to have blocked significant energy from reaching Alaska and British Columbia. Simulations also indicate a slight advance (early arrival) in the wavefronts, with Hawaii and the Mendocino Ridge (Escarpment) having the greatest effect on tsunami propagation toward the US West Coast (Fig. 20.5c).

20.5 Discussion

Directivity and angle of approach of tsunami waves are mechanisms that tend to focus tsunami energy across the Pacific, at South America from Japan and vice versa. For the California Coast, these mechanisms tend to create more glancing angles of approach from Japan sources. Because of non-uniform water depths across the Pacific Ocean basin, refraction effects from large-scale seamounts or island chains create waveguides that focus energy to distant shores (Figs. 20.1b and 20.5a, b). Wave scattering by major bathymetric features within the Pacific Ocean Basin may be represented by a finite number of focused "raypaths" that result in distinct wave arrival pulses or phases similar to those identified in earthquake seismology (Fig. 20.4a–d; Richter 1958).

Multiple wave arrivals may explain the large wave activity observed at Crescent City, Northern California, although waveguide focusing due to Koko Guyot and the Mendocino Ridge is also an important effect (phase N_1; Fig. 20.1b). The redirection of energy shown in Fig. 20.5a is from the Koko Guyot and Mendocino Ridge. Comparing DART® records offshore Oregon (46404) versus northern California (46411) south of Mendocino show higher amplitudes and more complex wave arrivals than north of the Mendocino Ridge. The Hawaiian Ridge in the northwest Pacific is oriented parallel to the direction of tsunami wave propagation from the Japan source, providing a long low-velocity (shallow-water) waveguide that redirects significant energy (phases N_2 and S_1) eastward towards northern California at a steeper angle relative to the coast (Fig. 20.1b). In contrast, the Emperor Seamount Chain is oriented at a high angle to the tsunami propagation and appears to block or reflect some tsunami energy directed toward the Pacific Northwest. Farther south the northwest-trending ridges and basins of the California Continental Borderland tend to dissipate or scatter tsunami wave energy arriving from the west. For azimuths from the north or south, sub-parallel to the dominant physiographic trends, the elongate shallow ridges and banks would act as waveguides to focus or refract/diffract tsunami wave energy along the coast. The latter effect is considered responsible for the more severe currents and related damage in San Diego Bay and small boat harbors of southern California from Alaska-Aleutian and South America source regions. Simulations for key scenarios can be available prior to an event and can indicate shadow zones due to main ocean basin scatterers as well impact zones. Multiple arrivals can also be predicted and can help in identifying the duration of the event for emergency planning and define the evacuation window.

Although the simulations do not reproduce observations well at all DART® buoy locations, phase arrivals from cross-correlation results appear promising. The wave propagation is a complex mix of refracted, diffracted (scattered), and reflected wave energy from numerous seafloor features and continental margins around the Pacific Ocean. So, the identification of valid tsunami phases is not a trivial exercise and for this reason the cross-correlation technique has to be investigated further as a tool for separating complex waveforms into simpler waveforms/contributions of the encounter of the tsunami waves with ocean bathymetric features.

20.6 Conclusions

The wavefield generated during large tsunami events, endures widespread scattering during refraction and reflection of the wave energy during its propagation. Potentially damaging waves from delayed scattered wave energy pulses are most likely for the very large Pacific-wide tsunamis generated during megathrust subduction earthquakes (M ~ 9+). The recent occurrence of major ocean wide-tsunamis around the globe has provided us with important data to help predict consequences that may result in future disasters (Titov et al. 2005; Geist et al. 2006). We suggest that there exist within an ocean basin a limited number of large scatterers capable of producing significant pulses in the wave field which may be recognized at distant recording stations. The idea of basin scatterers is not new but identifying tsunami phases using cross-correlation (widely used technique in seismology) is a new approach and animations of tsunami simulations are a crucial tool to assist in the identification of coherent wavefronts associated with specific scatterers in the tsunami wavefield.

Using records of the near source tsunami energy, distinct tsunami wave pulses (phases) were identified that represent specific reflected and refracted raypaths including point diffractors along major island or seamount chains. For example, based on the computed propagation of the March 2011 Tohoku tsunami across the Pacific Ocean, tsunami wave energy reflected from South America reached southern California after 23 h travel time from its initiation. The favorable southeast approach of reflected waves toward San Diego and other southern California locations may have produced significant wave amplitudes and possible inundation during the high tide that night, as occurred in Sunset Beach, California, during the February 27, 2010, Chile tsunami. Refracted/diffracted tsunami wave energy from the Hawaiian Ridge and Emperor Seamount Chain in the middle and northeast Pacific Ocean basin directed significant tsunami energy from Japan toward southern California with a southwest approach angle during the March 11, 2011 event. Numerical modeling of the Tohoku tsunami using a modified Pacific Ocean basin without the Emperor/Koko seamount chain showed the contribution of the Emperor Seamount. This technique can be further strengthened with improved calibrated numerical models.

Strong currents that damaged boats and docks in San Diego more than 8 h after the initial wave arrivals are suspected to result from scattered tsunami energy pulses. Focusing of tsunami energy by distant mid-Pacific and local wave-guides combined with coastal basin resonance produces damaging and persistent tsunami arrivals. Because multiple wave arrivals provide long duration of forcing to bays and harbors the recognition of distinct phases in other tsunami simulations –using cross correlation- may allow better prediction of damage potential from late tsunami wave arrivals at coastal sites.

Emergency response plans would benefit greatly by this technique and may save additional lives by warning of such delayed inundation potential that has been unexpected in past tsunamis.

20.7 Data and Resources

DART® records and tide gauge data used in this study can be obtained from the NOAA website for tides and currents found in http://tidesandcurrents.noaa.gov/.

Acknowledgments We would like to acknowledge NOAA Center for Tsunami Research for determining the tsunami sources for the tsunami events mentioned in this manuscript.

This publication is contribution 3708 from NOAA/Pacific Marine Environmental Laboratory and funded by the Joint Institute for the Study of the Atmosphere and Ocean (JISAO) under NOAA Cooperative Agreement No. NA100AR4320148, Contribution # 1862.

References

Dengler L, Uslu B, Barberopoulou A, Yim SC, Kelley A (2009) Tsunami damage in Crescent City, California from the November 15, 2006 Kuril event. Pure Appl Geophys (Pageoph) 166(1):37–54

Geist EL, Titov VV, Synolakis CE (2006) Tsunami: wave of change. Sci Am 294:56–63

Horillo J, Knight W, Kowalik Z (2008) Kuril islands tsunami of November 2006: 2. Impact at Crescent City by local enhancement. J Geophys Res 113:C01021. doi:10.1029/2007JC004404

Kowalik Z, Horillo J, Knight W, Logan T (2008) The Kuril Islands tsunami of November 2006: part I: impact at Crescent City by distant scattering. J Geophys Res 113:C01020. doi:10.1029/2007JC004402

Lander JF. Lockridge PA, Kozuch MJ (1993) Tsunamis affecting the west coast of the United States 1806–1992, NGDC Key to geophysical record documentation No. 29, NOAA, NESDIS, NGDC

Marchuk AG (2009) Tsunami wave propagation along waveguides. Sci Tsunami Hazards 28 (5):283–302

Miller GR, Munk WH, Snodgrass FE (1962) Long-period waves over California's continental borderland, II, tsunamis. J Mar Res 20:31–41

Miyoshi H (1986) Angle of energy flux at the origin of two major tsunamis. J Oceanogr 42:69–74. doi:10.1007/BF02109193

Mofjeld HO, Titov VV, González FI, Newman JC (2001) Tsunami scattering provinces in the Pacific Ocean. Geophys Res Lett 28:335–337. doi:10.1029/2000GL011710

Rabinovich AB, Stroker K, Thomson RE, Davis E. (2011) DART®s and CORK: high-resolution observations of the 2004 Sumatra tsunami in the abyssal northeast Pacific. Geophys Res Lett L08502. doi:10.1029/2011GL047063

Raichlen F (1972) Tsunami-responses of San Pedro Bay and shelf, California. J Waterw Harb Coast Eng Div Proc ASCE 98(WW1):104–110

Richter, C. F. (1958). *Elementary Seismology*. Freeman, San Francisco/London.

Tang L et al (2008) Tsunami forecast analysis for the May 2006 Tonga tsunami. J Geophys Res 113:C12015. doi:10.1029/2008JC004922

Titov V, González FI (1997) Implementation and testing of the method of splitting tsunami (MOST) model, NOAA Tech. Memo. ERL PMEL-112, NTIS: PB98–122773, NOAA/Pacific Marine Environmental Laboratory, Seattle

Titov VV, Synolakis CE (1998) Numerical modeling of tidal wave runup. J Waterw Port Coast Ocean Eng 124:157–171

Titov VV, Rabinovich AB, Mofjeld HO, Thomson RE, González FI (2005) The global reach of the 26 December 2004 Sumatra tsunami. Science 309:2045–2048. doi:10.1126/science/1114576

Van Dorn WG (1979) Theoretical aspects of tsunamis along the San Diego coastline in Earthquakes and other perils San Diego region, San Diego Association of Geologists Field Trip Guidebook, edited by Abbott PL, Elliott WJ, pp 115–116

Wiegel RL (1976) Seismic risk and engineering decisions, developments. In: Lomnitz C, Rosenblueth E (eds) Geotechnical engineering, tsunamis. Elsevier, Amsterdam, pp 225–286, Vol. 15, chap. 7

Wilson RI, Dengler LA, Legg MR, Long K, Miller KM (2010) The 2010 Chilean tsunami on the California coastline. Seism Res Lett 81(3):545–546

Chapter 21
Improving Tsunami Resiliency: California's Tsunami Policy Working Group

Charles R. Real, Laurie Johnson, Lucile M. Jones, and Stephanie Ross

Abstract California has established a Tsunami Policy Working Group to facilitate development of policy recommendations for tsunami hazard mitigation. The Tsunami Policy Working Group brings together government and industry specialists from diverse fields including tsunami, seismic, and flood hazards, local and regional planning, structural engineering, natural hazard policy, and coastal engineering. The group is acting on findings from two parallel efforts: The USGS SAFRR Tsunami Scenario project, a comprehensive impact analysis of a large credible tsunami originating from an M 9.1 earthquake in the Aleutian Islands Subduction Zone striking California's coastline, and the State's Tsunami Preparedness and Hazard Mitigation Program. The unique dual-track approach provides a comprehensive assessment of vulnerability and risk within which the policy group can identify gaps and issues in current tsunami hazard mitigation and risk reduction, make recommendations that will help eliminate these impediments, and provide advice that will assist development and implementation of effective tsunami hazard risk communication products to improve community resiliency.

Keywords Mitigation • Public policy • Tsunami hazards

C.R. Real (✉)
California Geological Survey, 801 K Street, MS 12-31 Sacramento, CA 95814-3531, USA
e-mail: creal@conservation.ca.gov

L. Johnson
Laurie Johnson Consulting | Research, 23 Portola Avenue, San Rafael, CA 94903, USA

L.M. Jones
US Geological Survey, 525 S Wilson Ave, Pasadena, CA 91106, USA

S. Ross
US Geological Survey, 345 Middlefield Road, MS-999, Menlo Park, CA 94025, USA

Y.A. Kontar et al. (eds.), *Tsunami Events and Lessons Learned: Environmental and Societal Significance*, Advances in Natural and Technological Hazards Research 35, DOI 10.1007/978-94-007-7269-4_21,

21.1 Introduction

The disastrous Great Sumatra Earthquake and Tsunami of 2004 elevated international awareness of tsunami hazards, triggering a reevaluation of tsunami risk and hazard mitigation efforts among coastal communities around the world. Followed by tsunamis in Samoa (2009) and Chile (2010), and Japan's Tohoku-Oki Earthquake and Tsunami in 2011, human casualties over the past two decades now approach one-quarter million, with financial losses in the hundreds of billions (US Dollars) from a natural hazard previously thought to be rare. Japan's tsunami caused millions of dollars of damage to California's harbors and is the largest tsunami to strike the state since the 1964 Great Alaska Earthquake and Tsunami (Wilson et al. 2012). Nearly 2 years after the Tohoku tragedy, the west coast of the U.S., Alaska, and Canada have begun dealing with tsunami debris just now reaching shore.

Located on the tectonically active Pacific Plate margin capable of generating large tsunamis, California is reassessing tsunami hazard, vulnerability, and options to reduce future losses. These efforts predated the Tohoku earthquake, but were strengthened based on the lessons learned in the wake of Japan's devastating event. This paper describes a unique process currently underway to improve estimates of California's tsunami hazard and risk, and is helping to identify and resolve weaknesses in current hazard mitigation efforts in order to strengthen the resilience of coastal communities. The cornerstone of this effort is formation of a State/Federal twin-project partnership linked by a unifying policy working group that is collectively quantifying the tsunami threat: facilitating discovery of principal issues, gaps and roadblocks preventing effective tsunami hazard mitigation, and is providing the necessary focus for group analysis by experts to develop practical, workable solutions.

21.2 Federal Tsunami Program

21.2.1 National Tsunami Hazard Mitigation Program (NTHMP)

The NTHMP is a partnership between the National Oceanic and Atmospheric Administration (NOAA), the Federal Emergency Management Agency (FEMA), the United States Geological Survey (USGS), the National Science Foundation, and Coastal States, Territories and Commonwealths, with a mission to reduce future tsunami losses by improving hazard assessments, providing effective notifications and warnings, and facilitating mitigation and preparedness at the local level. The NTHMP provides an overarching framework within which states can assist coastal communities in reducing the impact of tsunamis. Until recently the program has focused on short-term needs for preparedness and emergency management, supporting development of tsunami inundation maps for evacuation planning, providing assistance to local emergency services for response planning, funding community warning systems and placement of evacuation signs, and improving public awareness through education. Following the 2004 Indonesian tsunami, NTHMP began expanding its long-term hazard mitigation plans to consider land-use and construction strategies (NTHMP 2009). Success within the national program depends on sustained federal/state partnership and continued support through the federal Tsunami Warning and Education Act of 2006 (WSSPC 2011). California is among the first NTHMP state partners to initiate projects in this new direction.

21.2.2 Science Applications for Risk Reduction (SAFRR)

The U.S. Geological Survey initiated the Multi-Hazard Demonstration Project (MHDP) with a strategy to help improve resiliency to natural hazards by application of science to community decision-making and emergency response (Jones et al. 2007). Based on stakeholder inputs from workshops, a program approach was created that consists of multi-year, multi-disciplinary scientific development efforts to construct plausible scenarios for various natural hazard events; events large enough to evaluate the adequacy of current preparedness and hazard mitigation efforts. Two major scenarios resulted: ShakeOut, a large earthquake impacting the southern California metropolitan area (Jones et al. 2008), and ArkStorm, a massive statewide winter storm (Porter et al. 2011). The MHDP has since expanded nationwide to become SAFRR, with the current focus a tsunami scenario with a large credible tsunami striking the west coast of the U.S. Like its predecessors, The SAFRR Tsunami Scenario is increasing community awareness and resiliency to tsunami hazard and risk by applying advanced science to characterize the hazard, and conducting an assessment of direct and indirect damage and economic impacts to California and the nation.

The tsunami scenario is based on a magnitude 9.1 M-thrust earthquake occurring along the Aleutian Islands Subduction Zone, which presents the greatest distant tsunami threat to southern and central California. While local landslides could produce larger tsunamis, the distant events are more frequent and allow for warnings to be part of the scenario. Characterization of the tsunami source was underway when the Tohoku earthquake occurred and was modified based on that event. Goals are to improve tsunami modeling practices, include advanced modeling of water currents and inundation for the scenario event, to spawn further research on characterizing Alaska tsunami sources, and to better understand the impacts to coastal infrastructure and environment, and local, regional and national economies. Local emergency managers are enlisted to develop effective ways to communicate and respond to the threat, enhance decision-making when responding

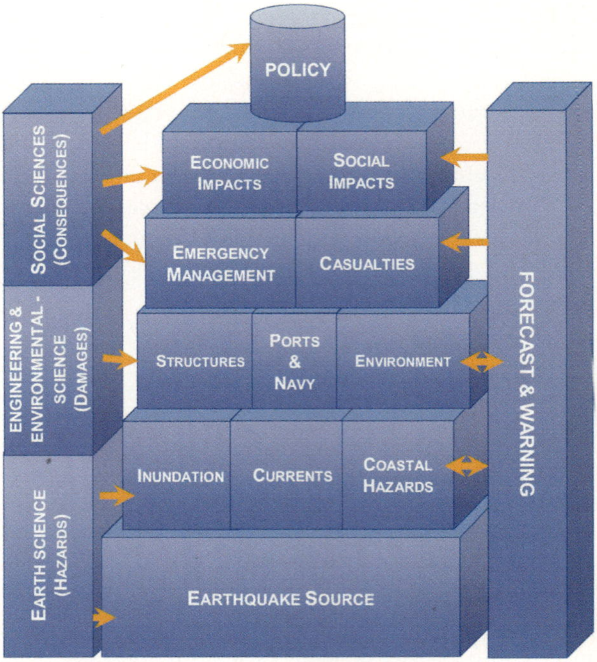

Fig. 21.1 The SAFRR Tsunami Scenario project involves an in-depth assessment of impact along Coastal California of a large hypothetical tsunami originating from an M 9.1 M-thrust earthquake in the Aleutian Islands Subduction Zone. Modeling the tsunami source, inundation, and consequent damage, both direct and indirect, is a multidisciplinary effort involving scientists, engineers, and dozens of stakeholders from government and industry

to tsunami warning center alerts, build the repository of lessons learned and evaluate mitigation options, and provide a tool for teaching preparedness.

Project participants include researchers and officials from government, private, and academic sectors. Ten working groups, with group leaders comprising the project steering committee, are established covering: earthquake source, tsunami geologic field studies to determine the history of past tsunamis, tsunami modeling, engineering and physical impacts, environmental impacts, ecological impacts, emergency management and education, social vulnerability, economic and business impact, policy, and impacts to Naval operations. Each group is composed of specialists in the respective field, who are providing direction to collaborators and oversee research and evaluations that identify issues for policy consideration Fig. 21.1.

A principal focus is the tsunami's impact on ports and harbor operations, particularly the ports of Long Beach and Los Angeles, including estimating the economic consequences of service disruptions. During the process, project staff engage with the community through extensive outreach activities, including presentations, education forums, workshops/panels, consultations, and the media. The results of The SAFRR Tsunami Scenario project are estimates of losses, physical, environmental and economic, that can be used to support policy decisions. Participation of local government officials is a key component of the scenario project, and helps to ensure transfer of findings.

21.3 State Tsunami Program

21.3.1 California Tsunami Preparedness and Hazard Mitigation Program (CTPHMP)

For more than a decade, the California Governor's Office of Emergency Services (Cal OES) and the California Geological Survey (CGS) have partnered with the NOAA under the NTHMP's grant program to help coastal communities better prepare for the impact of tsunamis and assist in the process of becoming "Tsunami Ready". Cal OES administers the program and coordinates activities through the California Tsunami Steering Committee, comprised of representatives from the four regional National Weather Service offices, all relevant state agencies, and each of the 20 coastal and Bay Area counties. As technical advisor to Cal OES, the CGS is California's science representative on the NTHMP's Coordinating Committee, a consortium of federal and state agencies funded by NOAA to mitigate tsunami hazards at a "grass roots" level. CGS staff co-chairs the NTHMP Mapping and Modeling Subcommittee, responsible for setting standards for tsunami source identification, numerical modeling, and, inundation and hazard mapping on a national level.

As a NTHMP partner, the CTPHMP works with NOAA's West Coast/Alaska Tsunami Warning Center to develop and implement emergency response activities/ plans when a *Tsunami Alert* is issued. In partnership with the University of Southern California's Tsunami Research Center, a series of tsunami inundation maps for developed portions of the California coastline were released to assist emergency managers in development of community evacuation plans (State of California 2009). This program also assists communities prepare for tsunamis generated locally offshore where there is insufficient time for a *Tsunami Alert* to be useful. In addition to assisting local emergency managers, pilot projects funded by NTHMP and FEMA are underway to support local tsunami hazard mitigation decision-making via land-use and construction, and to develop and design products for use in the construction of maritime facilities and response operations. The communities of Crescent City and Huntington Beach were chosen for study because their populations, economies, and exposure to tsunami hazard sharply contrast. This provides a broader spectrum of potential issues regarding their requirements and capacity to adopt and implement tsunami hazard mitigation measures. Maps depicting tsunami hazard for 100-, 500-, 1,000-, and 2,500-year return periods have been prepared using newly developed probabilistic methods (Thio 2010), which have been used to assess the vulnerability of the state's coastal transportation infrastructure. The objective is to provide the local planning and building departments with a more detailed up-to-date perspective of tsunami hazards, and collaboratively look at hazard mitigation options for reducing impact in a framework of acceptable risk. Products are being prepared that express hazard in terms of engineering parameters such as flow depth, current velocity, and momentum flux as a prototype dataset supporting building code provisions under development by the

American Society of Civil Engineers Subcommittee on Tsunami Loads and Effects (ASCE 2012). These products will help form the basis for tsunami design provisions in future editions of the California Building Code, as the statewide assessment of seismic hazard did more than a decade ago (Petersen et al. 1996). Products tailored for the maritime community that are based on detailed hydrodynamic modeling of harbor response are also under development to assist with identification of vulnerable port and harbor infrastructure (Wilson et al. 2012). Other maritime products include maps designating offshore safety zones for evacuation planning and response.

21.3.2 Tsunami Policy Working Group

The California Tsunami Policy Working Group (CTPWG) is a voluntary advisory body operating under the California Natural Resources Agency (CNRA), Department of Conservation, and is composed of experts in earthquakes, tsunamis, flooding, structural and coastal engineering and natural hazard policy from government, industry, and non-profit natural hazard risk-reduction organizations. The working group serves a dual purpose as an advisor to State programs addressing tsunami hazards and as a consumer of insights from The SAFRR Tsunami Scenario: raising awareness and facilitating transfer of policy concepts to other coastal states in the nation. CTPWG's role is to identify, evaluate and make recommendations to resolve issues that are preventing full and effective implementation of tsunami hazard mitigation and risk reduction throughout California's coastal communities.

Parallel work on The SAFRR Tsunami Scenario and State pilot projects provides a unique opportunity to simultaneously evaluate tsunami impacts with representative stakeholder groups in order to identify issues and gaps in preparedness, while evaluating alternative solutions, exploring issues and barriers to their implementation, and developing products and public policy strategies to facilitate tsunami hazard mitigation and risk reduction. The work is accomplished by having representatives of the State tsunami program placed on The SAFRR Tsunami Scenario Steering Committee representing *Geologic Field Studies* and *Emergency Management and Education*, and having a representative from SAFRR and CTPHMP Co-chair the CTPWG. This arrangement greatly facilitates transfer of project findings from The SAFRR Tsunami Scenario project and the State tsunami program to the CTPWG, which links both programs to policy makers at local, state, and federal levels Fig. 21.2.

Membership in the CTPWG is selected to represent entities responsible for coastal development, insurance, local and regional planning, public works, foreign and domestic disaster preparedness, recovery and seismic policy (see Acknowledgements). Among those selected are representatives of the two cities chosen for the State's tsunami pilot studies. Their participation in working group deliberations provides an opportunity to bring forth local implementation issues in a

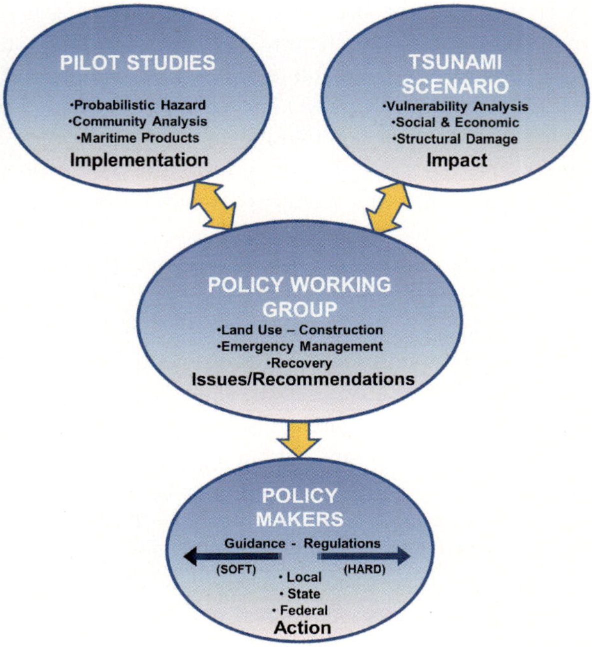

Fig. 21.2 The California Tsunami Policy Working Group plays a critical role assessing current tsunami hazard mitigation in California, identifying gaps and roadblocks and developing policy recommendations to enhance resiliency based on input from The SAFRR Tsunami Scenario project and the State Pilot Studies

forum conducive to multidisciplinary resolution, and an opportunity to incorporate a local perspective while formulating recommendations.

Furthermore, the composition of CTPWG provides linkages to agencies that not only provide valuable input to the group's mission, but can benefit from its findings and recommendations, helping to spread knowledge that will facilitate and expedite dissemination and utilization of tsunami hazard mitigation strategies. The principal product of the CTPWG is a report consisting of a general overview of the tsunami threat to California and shortcomings in coastal communities' preparedness and mitigation to lessen the impact, followed by a series of issues, and a list of specific recommendations to resolve each issue that have been carefully evaluated and prioritized based on viability and potential effectiveness in reducing future losses. CTPWG has a liaison from the Department of Conservation's Office of Government and Environmental Affairs, through which the document will be reviewed for actionable content of potential importance to CNRA and other agencies involved with coastal affairs.

The seven basic principles outlined in *Designing for Tsunamis* (NTHMP 2001) and detailed land-use strategies presented in the *Hawaii Coastal Hazard Mitigation Guidebook* (Hwang 2005) have been useful guides for CTPWG when assessing a community's preparedness. CTPWG obtains insights on the degree to which loss-reduction measures are implemented from the community *Action Plan* component of Local Hazard Mitigation Plans (GPO 2000), *zoning ordinances* that are implemented under Local Coastal Programs (PRC 2010), whether a community is certified *Tsunami Ready* (NOAA 2012), and whether a community participates in the National Flood Insurance Program. These programs, along with California's regulatory zoning for natural hazards, provide a strong policy framework for communities to reduce future losses from natural hazards, within which tsunami hazard mitigation in particular can be strengthened (Real 2010). Until recently, California's coastal communities almost universally considered tsunamis to be a rare event that ranks low in priority for risk reduction as evidenced by these documents. Following the recent tsunami disasters this perception is changing.

When developing recommendations to improve tsunami hazard mitigation, CTPWG considers application to land development, emergency management, and recovery for each of the following land-use categories: coastal infrastructure, coastal industry, maritime sector, and general development (all other types). Issues that are identified in The SAFRR Tsunami Scenario and the State pilot projects are sent to the CTPWG for analysis, where once categorized, they are prioritized based on importance/value in reducing loss, sufficiency of the existing knowledge base to resolve the problem, and the anticipated difficulty implementing a solution. The group meets quarterly to discuss issues and work on ways to resolve them. During the interim, group interaction continues through a collaborative website, e-mail, and conference calls. The website keeps members aware of new information, allowing for the generation and review of documents by all. Issues and recommendations can be shared and refined prior to meetings where they form the basis of discussion. Similar collaborative project websites have been established for The SAFRR Tsunami Scenario project and State Pilot projects, which are accessible to members of the CTPWG.

Policy recommendations generated by the group go through a screening process for viability based on their potential to evoke community resistance (conflict) according to three program categories: voluntary programs (least conflict), encouraged programs (moderate conflict potential), and mandatory (greatest potential for conflict). Each program category has several advantages and disadvantages that weigh in on viability, and can form a basis for considering alternative solutions during working group deliberations. For example, mandates within tsunami hazard zones that restrict land use too late in the development process, when property owners have accumulated investment-backed expectations, can incite sufficient public resistance to defeat the proposal. Experience in Hawaii demonstrates that such strategies should be operative as early as possible, before land is subdivided (Hwang 2005).

21.4 Conclusions

The CTPWG believes that significant reduction in future tsunami losses can be achieved through a balanced program of voluntary, incentive-based, and regulatory hazard mitigation programs that will increase the resiliency of California's coastal communities to tsunami hazard. The dual-track approach of The SAFRR Tsunami Scenario project analysis of impact and vulnerability combined with local implementation State Pilot projects serves to not only facilitate discovery of weaknesses in tsunami hazard resiliency, but documentation from these projects will also serve to support the California Tsunami Policy Working Group recommendations and related policy initiatives that they may spawn. The final Issues/Recommendations report will be a useful resource for federal, state, and local action.

Acknowledgements Members of the California Tsunami Policy Working Group are: James Barnts – Director, Crescent City Public Works, Mary Beth Broeren – Planning Manager, Huntington Beach Department of Planning, Lesley Ewing – Senior Coastal Engineer, California Coastal Commission, Laurie Johnson (*Co-Chair*) – Urban Planner, Laurie Johnson Consulting + Research, Ray Lenaburg – Chief, Risk Analysis Branch, FEMA Region IX, Robert Olson – Emergency Management/Public Policy Consultant, Robert Olson Associates, Inc., Ricardo Pineta – Chief, Floodplain Management Branch, DWR, Charles Real (*Co-Chair*) – Manager, Seismic Hazard Zonation, California Geological Survey, Rune Storesund – Consulting Engineer, Storesund Consulting/ASCE local Chapter President, Patti Sutch – Executive Director, Western States Seismic Policy Council, Ken Topping – Urban Planner, Topping Associates International/ Lecturer, California Polytechnic State University, Brian Tucker – Seismologist, President, Geohazards International. Operation of the working group is supported by a National Tsunami Hazard Mitigation Program grant from the National Ocean and Atmospheric Administration.

We thank Melanie Perron and Lauren Prehoda from the California Department of Conservation, Office of Government and Environmental Affairs for their assistance. We also thank Rick Wilson and Tim McCrink, California Geological Survey, and Anne Wein and John Haines, U.S. Geological Survey for their valuable review and comments.

References

ASCE (2012) ASCE 7 Subcommittee on tsunami loads and effects. http://www.asce.org/CommitteeDetail.aspx?committeeId=000001022422. Accessed 10 Sept 2012

GPO (2000) Disaster Mitigation Act. http://www.gpo.gov/fdsys/pkg/PLAW-106publ390/pdf/PLAW-106publ390.pdf. Accessed 10 Sept 2010

Hwang D (2005) Hawaii coastal hazard mitigation guidebook. Hagadone Printing Company, Hong Kong

Jones L, Bernknopf R, Cannon S, Cox DA, Gaydos L, Keeley J, Kohler M, Lee H, Ponti D, Ross S, Schwarzbach S, Shulters M, Ward AW, Wein A (2007) Increasing resiliency to natural hazards – a strategic plan for the multi-hazards demonstration project in Southern California. U.-S. Geological Survey Open-File Report 2007-1255

Jones LM, Bernknopf R, Cox D, Goltz J, Hudnut K, Mileti D, Perry S, Ponti D, Porter K, Reichle M, Seligson H, Shoaf K, Treiman J, Wein A (2008) The shakeout scenario – effects of a potential M7.8 earthquake on the San Andreas Fault in Southern California. U.S. Geological Survey Open-File Report 2008-1150, CGS Preliminary Report 25

NOAA (2012) Tsunami ready. http://www.tsunamiready.noaa.gov/. Accessed 10 Sept 2012

NTHMP (2001) Designing for tsunamis – seven principals for planning and designing for tsunami hazards. National Ocean and Atmospheric Administration

NTHMP (2009) National Tsunami Hazard Mitigation Program 2009 – 2013 Strategic Plan. National Ocean and Atmospheric Administration

Petersen MD, Bryant WA, Cramer CH, Cao T, Reichle MS, Frankel AD, Lienkaemper JJ, McCrory PA, Schwartz, DP (1996) Probabilistic seismic hazard assessment for the State of California. DMG Open-File Report 96-08, U.S. Geological Survey Open-File Report 96-706

Porter K, Wein A, Alpers C, Baez A, Barnard P, Carter J, Corsi A, Costner J, Cox D, Das T, Dettinger M, Done J, Eadie C, Eymann M, Ferris J, Gunturi P, Hughes M, Jarrett R, Johnson L,

Dam Le-Griffen H, Mitchell D, Morman S, Neiman P, Olsen A, Perry S, Plumlee G, Ralph M, Reynolds D, Rose A, Schaefer K, Serakos J, Siembieda W, Tock J, Strong D, Sue Wing I, Tang A, Thomas P, Topping K, Wills C; Jones L, Chief Scientist, Cox D, Project Manager (2011) Overview of the ARkStorm scenario: U.S. Geological Survey Open-File Report 2010-1312 and appendices

PRC (2010) California Coastal Act. http://www.coastal.ca.gov/coastact.pdf. Accessed 10 Sept 2012

Real C (2010) California's natural hazard zonation policies for land-use planning and development. J Disaster Res 5(5):567–574

State of California (2009) Tsunami inundation map for emergency planning – series. http://www.conservation.ca.gov/cgs/geologic_hazards/Tsunami/Inundation_Maps/Pages/Statewide_Maps.aspx. Accessed 10 Sept 2012

Thio HK (2010) Probabilistic tsunami hazard in California. Pacific Earthquake Engineering Research Center http://peer.berkeley.edu/publications/peer_reports/reports_2010/web_PEER2010_108_THIOetal.pdf. Accessed 10 Sept 2012

Wilson RI, Admire AR, Borrero JC, Dengler, LA, Legg, MR, Lynett P, Miller KM, Ritchie A, Sterling K, McCrink TP, Whitmore PM (2012) Observations and impacts from the 2010 Chilean and 2011 Japanese tsunami in California (USA). Pure and Applied Geophysics. http://dx.doi.org/10.1007/s00024-012-0527-z. Accessed 12 Sept 2012

WSSPC (2011) Tsunami hazard mitigation and preparedness – a perspective from state and territory tsunami programs in the high tsunami risk Pacific region. Western States Seismic Policy Council Report 2011-01. http://www.wsspc.org/resources/PublicationFiles/WSSPC_Tsunami_Report_2011-01.pdf. Accessed 12 Sept 2012

Chapter 22
Preliminary 2013 Solomon Islands Earthquake and Tsunami Data Report and Historical Retrospective

Paula Dunbar, George Mungov, Laura Kong, Heather McCullough, and Erica Harris

Abstract On 6 February 2013, at 01:12:27 UTC, a magnitude *Mw* 8.0 earthquake occurred at 10.738° S, 165.138° E, depth 29 km, in the Santa Cruz Islands region, Solomon Islands. The earthquake generated a tsunami that was observed throughout the entire Pacific Ocean. Maximum wave heights up to 3 m were observed on Nendo Island, located approximately 75 km east of the epicenter. On the islands of Malo, Nendo, Nibanga Noi, and Tomotu Niabona, the tsunami caused 11 deaths, 14 injuries, and damaged or destroyed more than 700 houses. The National Environmental Satellite, Data, and Information Service (NESDIS) National Geophysical Data Center (NGDC) and collocated World Data Service (WDS) for Geophysics maintains the global tsunami archive, providing integrated access to tsunami event observational, instrumental, and socio-economic impact data. This archive incorporates the historical tsunami database, imagery, raw and processed U.S. coastal tide gauge, and tsunameter data including Deep-ocean Assessment and Reporting of Tsunami (DART®) relevant to a tsunami event. The historical tsunami database includes information on the tsunami source, maximum wave heights, and socio-economic effects such as deaths and damage. NGDC works in collaboration with the UNESCO/IOC – NOAA International Tsunami Information Center (ITIC) to collect post-tsunami event information. As of 20 February 2013, NGDC has collected maximum tsunami amplitudes

P. Dunbar (✉) • H. McCullough
NOAA, NESDIS, National Geophysical Data Center, Boulder, CO 80305-3328, USA
e-mail: paula.dunbar@noaa.gov

G. Mungov • E. Harris
NOAA, NESDIS, National Geophysical Data Center, Boulder, CO 80305-3328, USA

CIRES, University of Colorado at Boulder 216 UCB, Boulder, CO 80309-0216, USA

L. Kong
UNESCO/IOC – NOAA, International Tsunami Information Center,
Honolulu, HI 96813, USA

Y.A. Kontar et al. (eds.), *Tsunami Events and Lessons Learned: Environmental and Societal Significance*, Advances in Natural and Technological Hazards Research 35,
DOI 10.1007/978-94-007-7269-4_22,

from 62 tide gauge observations, eight tsunameter and bottom pressure recorder (BPR) station observations, and 25 field survey and eyewitness reports for this event. Preliminary processing of tide gauge and tsunameter time series data revealed that the tsunami lasted between 6 and 9 h after arrival at the tsunameters located in the western Pacific and at the Pacific Island tide gauges. This paper presents a review of historical tsunamis in the Solomon Islands region and describes the data collected and processed for the 6 February 2013 earthquake and tsunami.

Keywords 2013 Solmon Islands Tsunami • Damage • Earthquake • Runup • Tide gauge • Tsunami • Tsunameter

22.1 Introduction

On 6 February 2013, at 01:12:27 UTC, a magnitude *Mw* 8.0 earthquake occurred at 10.738° S, 165.138° E, depth 29 km, in the Santa Cruz Islands region, Solomon Islands (USGS 2013). The earthquake generated a tsunami that was observed throughout the entire Pacific Ocean and caused deaths and damage locally with wave heights up to 3 m.

According to the USGS, during the month leading up to the 6 February earthquake, there were dozens of earthquakes in the epicentral region – in the preceding 7 days, 7 of these were larger than magnitude 6 (USGS 2013). Since the *Mw* 8.0 event, there have been numerous aftershocks. An *Mw* 6.8 aftershock on 8 February 2013 at 11:12 UTC damaged the wharf in Lata and approximately 4 h later an *Mw* 7.0 aftershock at 15:26 UTC caused additional damage to the coast and led to the evacuation of the Lata Hospital (NDMO Report 4 2013). The *Mw* 7.0 aftershock also generated a 9 cm tsunami at Lata Wharf, Nendo Island.

The Solomon Islands are a southwest Pacific archipelago of islands situated north of Australia. They are located on the eastern margin of the Australia tectonic plate, one of the most seismically active areas of the world due to high rates of convergence between the Australia and Pacific plates. In the region of the 6 February 2013 earthquake, the Australia plate converges with and subducts beneath the Pacific plate, moving towards the east-northeast at a rate of approximately 94 mm/year (USGS 2013).

The National Environmental Satellite, Data, and Information Service (NESDIS) National Geophysical Data Center (NGDC) and collocated World Data Service for Geophysics (WDS) maintains the global tsunami archive, providing integrated access to tsunami event observational, instrumental, and socio-economic impact data. This archive incorporates the historical tsunami database, imagery, raw and processed U.S. coastal tide gauge and tsunameter data, including Deep-ocean Assessment and Reporting of Tsunamis (DART®) relevant to a tsunami event. The historical tsunami database includes information on the tsunami source, maximum wave heights, and socio-economic effects such as deaths and damage. NGDC works in collaboration with the United Nations Educational, Scientific and Cultural

Organization (UNESCO)/Intergovernmental Oceanographic Commission (IOC) – National Oceanic and Atmospheric Administration (NOAA) International Tsunami Information Center (ITIC) to collect post-tsunami event information. In this paper, we review the regional history of tsunamis, describe the status of data associated with the 6 February 2013 tsunami event, and assess the processing and availability of both tide gauge and tsunameter data.

22.2 Historical Tsunamis in the Solomon Islands Region

A review of the NGDC historical tsunami database reveals that since 1926, 31 confirmed and 6 unconfirmed tsunamis were generated in the Solomon Islands, and five of these caused deaths locally (Table 22.1). Twelve of these tsunamis were observed greater than 1,000 km from the source (i.e., teletsunamis), but none of the 12 caused deaths outside of the local region. These results are similar to those found globally; 90 % of all deaths from tsunamis throughout history have occurred within 100 km of the source, or less than 1 h tsunami travel time from the source. The majority of the Solomon Islands tsunamis were generated by earthquakes (90 %) and the remainder resulted from volcanic eruptions (Timakula and Kavachi). The most devastating tsunami occurred in April 2007, when an *Mw* 8.1 earthquake located north of Guadalcanal generated a tsunami with 12 m runups (elevation at maximum inundation), that killed 52 people and destroyed several villages.

Earthquakes on the eastern margin of the Australia plate in the region south and east of the Solomon Islands have also generated fatal tsunamis. A New Caledonia earthquake in March 1875 generated a tsunami that washed away three settlements and caused 25 deaths on Lifou Island. An *Mw* 7.5 Vanuatu earthquake in November 1999 caused damage, 100 injuries, and five deaths on Pentecost Island, and generated a tsunami that contributed to an additional five deaths. On 29 September 2009, an *Mw* 8.1 earthquake south of Samoa generated a tsunami that killed 149 people in Samoa, 34 people in American Samoa, and nine people in Tonga.

Table 22.1 Fatal Solomon Islands tsunamis (Source: National Geophysical Data Center)

		Epicenter		Magnitude	Earthquake	Tsunami	Total
Date	Location	Latitude (°)	Longitude (°)	M_w	Deaths	Deaths	Deaths
3 Oct 1931	Solomon Sea	−10.93	161.02	7.8	0	50	50
30 Apr 1939	Solomon Sea	−9.30	159.23	7.9	0	12	12
10 Aug 1988	Solomon Sea	−10.26	160.90	7.6	0	1	1
1 Apr 2007	Solomon Sea	−8.46	157.04	8.1	54	52[a]	54
6 Feb 2013	Santa Cruz Islands	−10.74	165.14	8.0	0	11	11

[a]May include earthquake deaths

22.3 NGDC Tsunami Data for the February 2013 Solomon Islands Event

As of 20 February 2013, NGDC/WDS has collected maximum tsunami amplitudes from 62 tide gauge observations, eight tsunameter station observations, and 25 field survey and eyewitness reports. The tsunami was first observed 4 min following the main shock at the Lata Wharf, Nendo Island tide gauge with a wave amplitude of 1.01 m. Tide gauges in New Caledonia observed waves of 0.77 m (Hienghene), 0.57 m (Ouinne), and 0.51 m (Mare Island). All other tide gauge recordings were less than 0.5 m amplitude. A field survey reported 3 m waves on Nendo Island. The tide gauge and tsunameter observations were overlaid onto the calculated tsunami travel time map shown in Fig. 22.1.

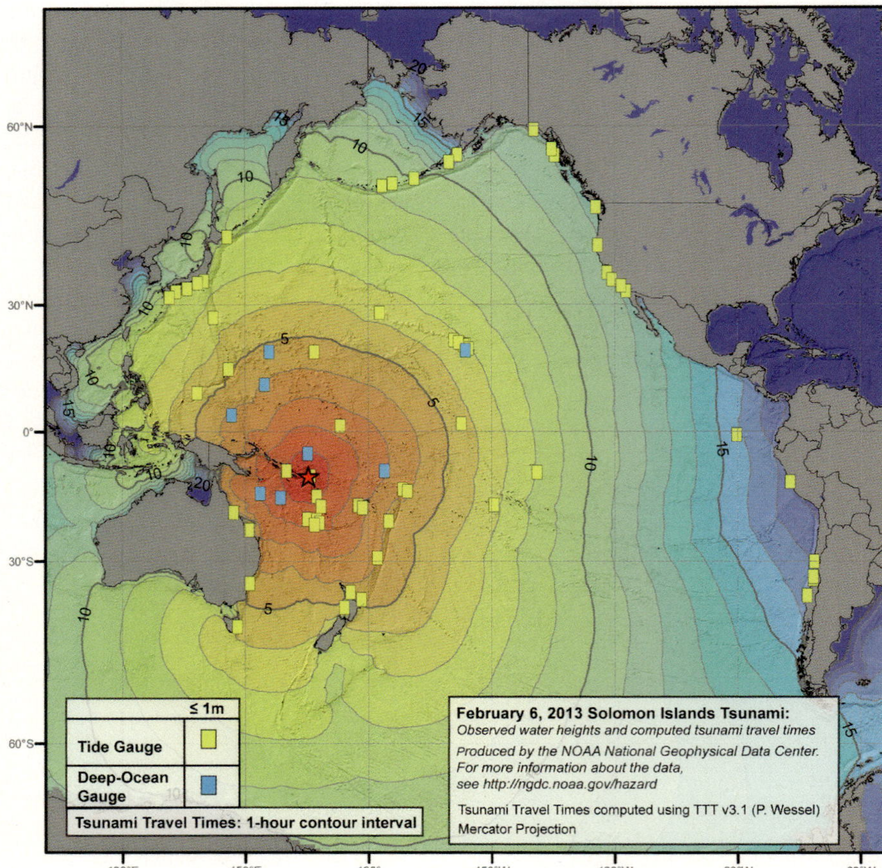

Fig. 22.1 Plot of the 6 February 2013 magnitude 8.0 Solomon Islands earthquake epicenter, locations of tide gauges and tsunameter stations that observed the tsunami wave, overlaid onto the calculated tsunami travel time map (contour intervals are in hours). A field survey reports 3 m runup heights on Nendo Island, located 75 km from the epicenter (Source: NGDC/World Data Service for Geophysics)

Table 22.2 Total population, deaths, total number of houses, and total number of houses damaged or destroyed for villages on Malo, Nendo, Nibanga Noi, and Tomotu Niabona Islands, Santa Cruz Islands, that were affected by the 6 February 2013 tsunami (Source: NDMO Reports 2, 6, and 8 2013)

Villages on Malo Is, Nendo Is, Nibanga Noi Is, and Tomotu Niabona Is	Village location		Total	Total	Total	Total houses	
	Latitude (°)	Longitude (°)	Population	Deaths	Houses	Damaged	Destroyed
Akamboi	−10.718	166.135	335		71	38	13
Baenga	−10.752	165.759	103		29	0	17
Bamoi	−10.787	166.070	123	1	29	20	4
Bania	−10.824	165.768	64	1	19	0	0
Bonate	−10.789	166.066	47		12	12	0
Gaito	−10.701	166.137	148		32	18	6
Lata Area 4	−10.722	165.800	225	1	51	18	18
Luava	−10.709	165.800	260		61	3	19
Lunelo	−10.791	166.062	2		1	1	0
Mali	−10.784	165.755	124		27	17	8
Malo	−10.698	165.798	597		147	80	35
Manoputi	−10.809	165.759	428		103	87	13
Masako	−10.751	165.964	6		1	0	1
Nabalue/Bemawa	−10.831	165.880	244		55	32	16
Nanggu	−10.785	166.048	206		83	24	12
Nea	−10.818	165.830	283	1	63	23	24
Nela	−10.725	165.787	214	4	58	7	46
Nemba	−10.786	165.754	453	1	83	24	30
Nonia	−10.864	165.865	238		54	9	0
Noole	−10.853	165.853	310		70	32	12
Venga	−10.736	165.777	302	1	85	0	46
Total			4,712	11	1,134	445	320

22.3.1 Deaths and Damage from the 6 February 2013 Earthquake and Tsunami

The February 2013 Solomon Islands tsunami caused 11 deaths, 14 injuries, and damaged or destroyed more than 700 houses in coastal villages on the islands of Malo, Nendo (Santa Cruz), Nibanga Noi (Lord Howe), and Tomotu Niabona in the Santa Cruz Islands region (Table 22.2, Fig. 22.2). Nendo Island, located approximately 75 km east of the epicenter, endured the majority of the tsunami effects (Fig. 22.3). It has a population of 5,000 and is the largest island in the westernmost Solomon Islands Province of Temotu. Reports confirmed that wave heights up to 3 m were measured at Nela Village on Nendo Island (NDMO Report 6 2013). In addition to the tsunami damage on the southern coast of Nendo Island, houses were damaged by the earthquake on the northern coast of the island (NDMO Report 8 2013).

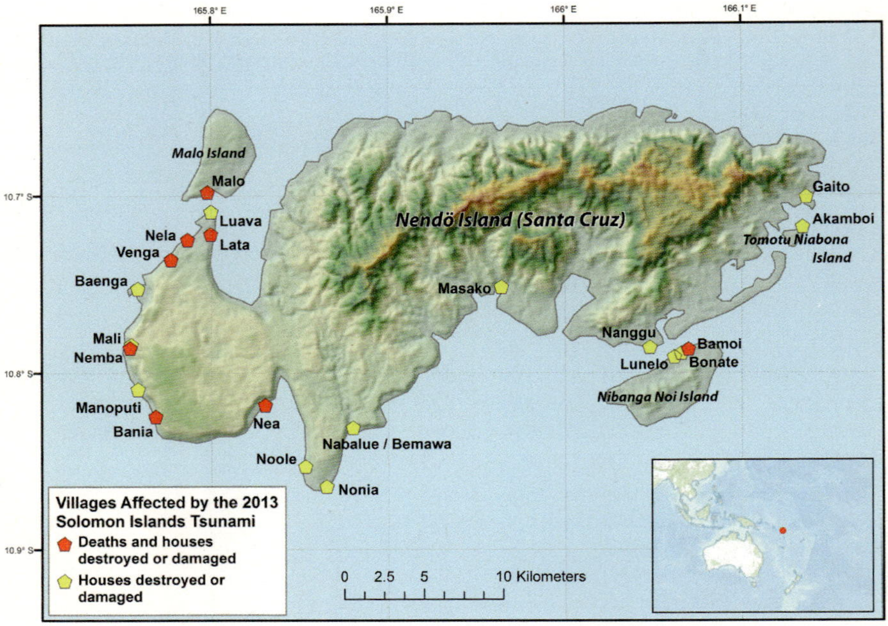

Fig. 22.2 Plot of villages on Malo, Nendo, Nibanga Noi, and Tomotu Niabona Islands, Santa Cruz Islands, that experienced deaths and damage from the 6 February 2013 tsunami. (Source: NDMO Reports 2, 6, and 8 2013). Background map is SRTM data (Source: Jarvis et al. 2008)

Fig. 22.3 Aerial photograph of west Nendo Island (Santa Cruz Island), showing damage to homes from the 6 February 2013 tsunami (Source: Solomon Islands National Disaster Management Office, Royal Australian Air Force)

Lata, the largest town on Nendo Island, reported wave damage to the terminal building and debris blocking the runway at the airport (NDMO Report 1 2013). Additional damage to infrastructure included the destruction of four buildings at the Venga Primary School, interference of water supplies to Lata and nearby villages, wreckage of pipes to entire communities, flooding of debris and salt water into water sources, and widespread contamination and destruction of wells (NDMO Report 2 2013). Aerial observations of the island indicated minimal damage to roads, while most of the major bridges, causeways, and culverts were intact (Baragamu and Aoraunishaka 2013).

Fenualoa Island, located approximately 50 km northeast of Nendo Island, saw no reports of loss of life, but flooding and damage did occur (Bone 2013). The only damage report outside of the epicentral region was in Vanuatu, where a small tsunami wave caused minor damage to a house and a church on the Maskelyne Islands, approximately 700 km from the epicenter (OCHA Report 2 2013).

22.4 Tsunameter, Including Deep-Ocean Assessment and Reporting of Tsunami (DART®), and Coastal Tide Gauge Data for the 6 February 2013 Tsunami

The tsunameter system is the primary component of the tsunami observational network, with each system sited in the deep-ocean near potential tsunami sources. Additional details about the DART® components, archive, station locations, and processing steps are described by Mungov et al. (2012) and Mofjeld (2009). Eight of the West Pacific Ocean tsunameters recorded the tsunami wave. Four of these stations are owned by the Australian Bureau of Meteorology (BOM); the data are transmitted by NOAA National Data Buoy Center (NDBC), and then processed and archived at NGDC. Processing of the tsunameter data reveals one large wave followed by a decreasing train of waves lasting 6-9 h (Fig. 22.4). Table 22.3 lists the station number, name and location.

Observations at tide gauge stations are the first confirmation of the tsunami at the coast. The U.S. owned and operated tide gauge stations are remotely located from the Solomon Islands. The closest U.S. station is Pago Pago, American Samoa, located almost 2,700 km from the source. Processing of these data does not reveal a tsunami signal on the U.S. West Coast. A tsunami signal is seen at the Pacific Islands stations (Fig. 22.5) and at some of the Alaska stations (Fig. 22.6). A first dominating wave is not observed in any of these records. The Alaska records show sequences similar to strong seiches that begin at the expected tsunami arrival time. The tsunami signal is only seen at Alaska stations that directly face the Pacific Ocean. Tables 22.4 and 22.5 list the station number, name, and location of the Pacific Island and Alaska tide gauges, respectively.

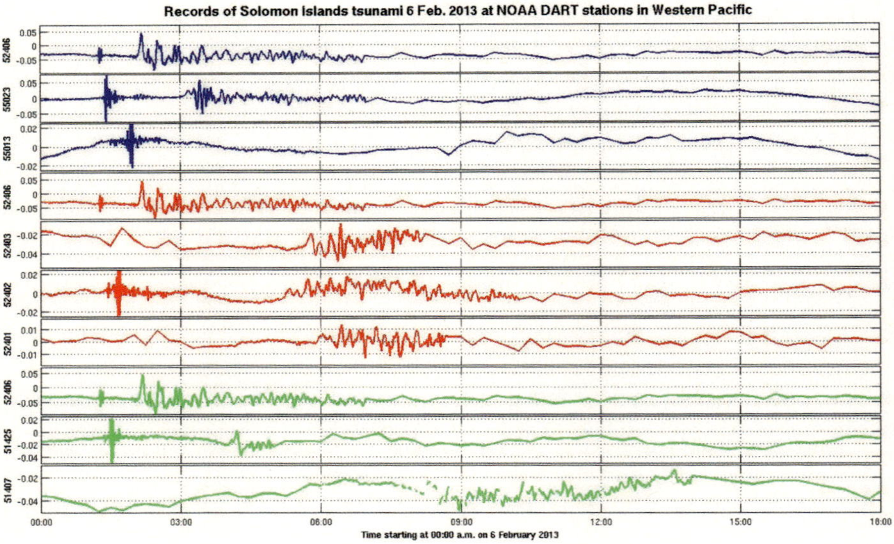

Fig. 22.4 Records of real-time data for the Solomon Islands 6 February 2013 tsunami at NOAA and Australian BOM tsunameter stations located in the Western Pacific. Tsunameters located south of the epicenter are plotted in blue, those in the north are in red, and those in the east are in green. Tsunameter station 52406, located closest to the epicenter, is repeated at the top in all of the plots

Table 22.3 West Pacific Ocean tsunameter stations

Station	NOAA and Australian BOM	Station location	
Number	Tsunameter station in West Pacific	Latitude (°)	Longitude (°)
51407	140 NM SE of Honolulu, Hawaii	19.6169	−156.5106
51425	370 NM NW of Apia, American Samoa	−9.5044	−176.2297
52401	West Pacific, 610 NM NE of Saipan	19.2547	155.7619
52402	790 NM NE of Kwajalein	11.8830	154.1100
52403	430 NM SW of Truk Lagoon	5.0000	145.5970
52406	450 NM NE of Guadalcanal	−5.2950	165.0120
55013	Tasman Sea	−46.6650	161.0100
55023	Coral Sea	−44.8530	164.7280

22.5 Data Access

NGDC and ITIC (http://www.ngdc.noaa.gov/hazard and http://itic.ioc-unesco.org/ respectively) are committed to delivering data and information that enables integration and supports multiple uses of data. Most of the data in the NGDC archive are within the public domain and may be used without restriction, although users should cite data sources. These data can be searched via web maps or text interfaces by attribute (date, location, magnitude, etc.) and displayed as tables, reports,

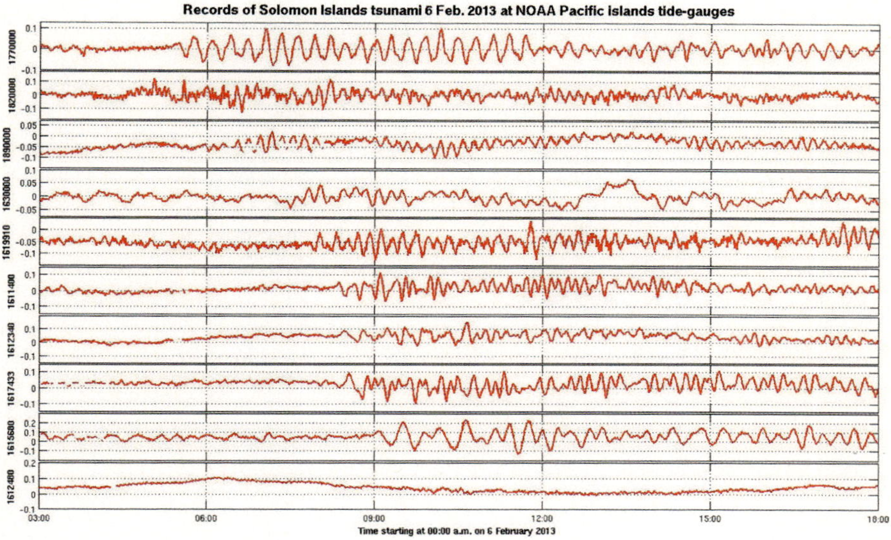

Fig. 22.5 Records of the Solomon Islands 6 February 2013 tsunami at NOAA Pacific Island tide gauge stations

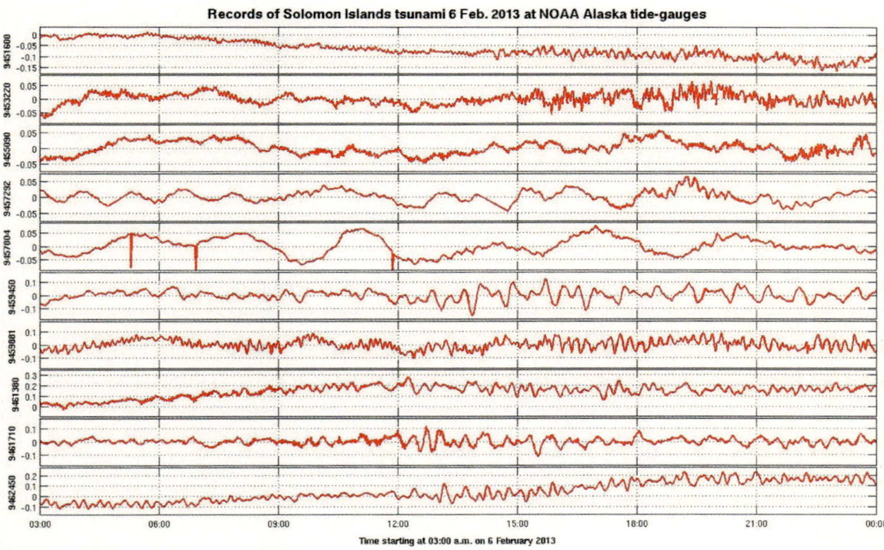

Fig. 22.6 Records of the Solomon Islands 6 February 2013 tsunami at NOAA Alaska tide gauge stations

Table 22.4 NOAA Pacific Island tide gauge stations

Tide-gauge	NOAA	Tide-gauge location	
Number	Pacific Islands tide-gauges	Latitude (°)	Longitude (°)
1611400	Nawiliwili, Hawaii	21.9544	−159.3561
1612340	Honolulu, Hawaii	21.3067	−157.8670
1612480	Makuoloe, Hawaii	21.4331	−157.7900
1615680	Kahului Harbor, Hawaii	20.8950	−156.4767
1617433	Kawaihae, Hawaii	20.0366	−155.8294
1619910	Sand Island, Midway Group	28.2117	−177.3600
1630000	Apra Harbor, Guam	13.4387	144.6539
1770000	Pago Pago, American Samoa	−14.2800	−170.6900
1820000	Kwajalein, Marshall Islands	8.7316	167.7362
1890000	Wake Islands	19.2900	166.6180

Table 22.5 NOAA Alaska tide gauge stations

Tide-gauge	NOAA	Tide-gauge location	
Number	Alaska tide-gauges	Latitude (°)	Longitude (°)
9451600	Sitka	57.0517	−135.3420
9453220	Yakutat (Yakutat Bay)	59.5485	−139.7334
9455090	Seward	60.1200	−149.4267
9457292	Kodiak	57.7317	−125.5120
9457804	Alitak	56.8983	−154.2470
9459450	Sand Point	55.3367	−160.5020
9459881	King Cove	55.0617	−162.3270
9461380	Adak Island	51.8633	−176.6320
9461710	Atka	52.2317	−174.1730
9462450	Nikolski	52.9406	−168.8713

interactive maps, and imagery. Additional details of the archive are described by Dunbar et al. (2008).

The 2013 Solomon Islands tide gauge and tsunameter time-series data are not yet integrated with the runup database, however NGDC disseminates summarized information for recent and significant tsunami events. Online access to processed and archived 15-s high-resolution records and real-time data from tsunameter, including DART® stations, are available for specific events. All additional data are available upon request to haz.info@ngdc.noaa.gov.

22.6 Conclusion

The 6 February 2013 Solomon Islands earthquake and tsunami is another example of a deadly local tsunami. Although it was not as devastating as the 2004 Indian Ocean or 2011 Japan tsunamis, it still reinforces the need to better understand

and prepare for this type of hazard. With these goals in mind, NGDC and ITIC endeavor to collect, preserve, and make available data and information for each tsunami event. By continually examining each aspect of this collaborative work, NOAA will lead the ongoing effort to increase our understanding of the past and our ability to forecast the impact of future tsunami events.

Acknowledgments The authors wish to acknowledge the partner agencies whose continued collaboration ensures the success of the United States' efforts to provide real-time tsunami warning to coastal communities during a tsunami event. NOAA's two Tsunami Warning Centers, the National Ocean Service, the National Data Buoy Center, the Pacific Marine Environmental Laboratory, the UNESCO/IOC – NOAA International Tsunami Information Center, and the National Geophysical Data Center each provide critical expertise. We also wish to acknowledge the Australian Bureau of Meteorology for their collaborative efforts in tsunameter networking in the southern oceans. Finally, we wish to acknowledge the Joint Solomon Islands National Disaster Management Office (NDMO) and National Emergency Operations Centre for providing their situation reports on the impacts of the 6 February 2013 earthquake and tsunami.

References

Baragamu G, Aoraunishaka P (2013) Temotu earthquake and tsunami aerial assessment report, 8 February 2013. National Disaster Management Office (NDMO) and National Disaster Council (NDC), Solomon Islands

Bone C (2013) Oceans watch helps out after Solomon Islands Tsunami. Sail-World.com. http://www.sail-world.com/USA/OceansWatch-helps-out-after-Solomon-Islands-Tsunami/106568. Accessed 15 Feb 2013

Dunbar PK, Stroker KJ, Brocko VR, Varner JD, McLean SJ, Taylor LA, Eakins BW, Carignan KS, Warnken RR (2008) Long-term tsunami data archive supports tsunami forecast, warning, research, and mitigation. Pure Appl Geophys 165:2275–2291. doi:10.1007/s00024-008-0419-4

Jarvis A, Reuter HI, Nelson A, Guevara E (2008) Hole-filled seamless SRTM data V4. International Centre for Tropical Agriculture (CIAT). http://srtm.csi.cgiar.org. Accessed 13 Feb 2013

Mofjeld HO (2009) Tsunami measurements. In: Robinson A, Bernard E (eds) The Sea, vol 15, Tsunamis. Harvard University Press, Cambridge, MA, pp 201–235

Mungov G, Eblé M, Bouchard R (2012) DART Tsunameter retrospective and real-time data: a reflection on 10 years of processing in support of tsunami research and operations. In historical and recent catastrophic tsunamis in the world: volume II. Tsunamis from 1755 to 2010. Pure Appl Geophys. doi:10.1007/s00024-012-0477-5

National Disaster Management Office (NDMO), Solomon Islands (2013) Temotu earthquake and tsunami, joint Solomon Islands National Disaster Management Office (NDMO) and National Emergency Operations Centre (NEOC) Situation Report Number 01. Accessed 6 Feb 2013

National Disaster Management Office (NDMO), Solomon Islands (2013) Temotu earthquake and tsunami, joint Solomon Islands National Disaster Management Office (NDMO) and National Emergency Operations Centre (NEOC) Situation Report Number 02. Accessed 8 Feb 2013

National Disaster Management Office (NDMO), Solomon Islands (2013) Temotu earthquake and tsunami, joint Solomon Islands National Disaster Management Office (NDMO) and National Emergency Operations Centre (NEOC) Situation Report Number 04. Accessed 9 Feb 2013

National Disaster Management Office (NDMO), Solomon Islands (2013) Temotu earthquake and tsunami, joint Solomon Islands National Disaster Management Office (NDMO) and National Emergency Operations Centre (NEOC) Situation Report Number 06. Accessed 13 Feb 2013

National Disaster Management Office (NDMO), Solomon Islands (2013) Temotu earthquake and tsunami, joint Solomon Islands National Disaster Management Office (NDMO) and National Emergency Operations Centre (NEOC) Situation Report Number 08. Accessed 16 Feb 2013

OCHA Regional Office for the Pacific (2013) Solomon Islands: earthquake and tsunami OCHA situation report No. 2. Accessed 7 Feb 2013

United States Geological Survey (USGS) (2013) Magnitude 8.0–76 km w of Lata, Solomon Islands. http://earthquake.usgs.gov/earthquakes/eventpage/usc000f1s0. Accessed 15 Feb 2013

Chapter 23
2004 Tsunami in Southern Thailand, Lessons Learned for the Thai Communities

Passakorn Pananont and Raykha Srisomboon

Abstract Unexpected devastation of the Thai communities along the Andaman coast lines by the 2004 Sumatra tsunami have changed the Thai perception on the earthquake and tsunami disaster forever. This paper tells about the economic and social impacts of the 2004 tsunami that affected the Thai society and the lessons learned from these tragic incident. Since then, a large amount of money and effort have been invested in the tsunami research and development, earthquake and tsunami monitoring and warning systems and a large number of educational outreach to young generations. The proper tsunami and earthquake mitigations of the central government collaborated with local governments integrated with the people's understandings of the nature of the earthquake and tsunami help increase the readiness of the Thai community for such disaster more than ever. Still, many issues remain to be resolved such as a lack of proper maintenance of the critical facilities for the earthquake and tsunami mitigation and inadequate research and development in earthquake and tsunami sciences in Thailand. It is very difficult to maintain the level of the awareness of the people about the great danger of the earthquake and tsunami and it is also a big challenge to increase the ability of the Thai society to have better critical scientific thinking about the natural disaster.

Keywords Tsunami mitigation • Earthquakes • Southern Thailand • Thai coastal communities

---

P. Pananont (✉)
Special Research Unit for Seismology and Tectonics of the Planet Earth (SEIS-SCOPE), Department of Earth Sciences, Faculty of Science, Kasetsart University, Bangkok, Thailand
e-mail: fscipkp@ku.ac.th

R. Srisomboon
Division of Commerce and Management, Mahidol University Kanchanaburi Campus, Kanchanaburi, Thailand

Program in Economics, School of Economics, University of the Thai Chamber of Commerce, Bangkok, Thailand

Y.A. Kontar et al. (eds.), *Tsunami Events and Lessons Learned: Environmental and Societal Significance*, Advances in Natural and Technological Hazards Research 35, DOI 10.1007/978-94-007-7269-4_23,

23.1 Introduction

The beaches along the southern Thailand are ones of the most popular tourist spots in the world that have attracted several million tourists annually. Tragically, on December 26th, 2004, an M_w 9.15 Sumatra earthquake (Chlieh et al. 2007) generated a devastating tsunami that caused over 220,000 casualties around the world and damaged large areas along the coastlines of Indonesia, Thailand, Malaysia, Sri Lanka, India, the Maldives and as far as Eastern Africa. In Thailand, it was the first time that the Thai society was aware of a real force of nature by the tsunami and it was the first time in Thai history that the country faced such a large scale of the natural disaster. Many villages near the coasts were completely destroyed and several other Thai communities along the Andaman coastlines were inundated by tsunami wave with a maximum run-up height of 5–12 m (NGI 2006). After this tragic incident, Thailand has gradually adapted itself to a resilient society that are well aware of the nature of earthquake and tsunami disaster. This paper intends to portray the overall picture of the current situation of the tsunami mitigation in Thailand, starting from the effects from the last 2004 tsunami in Thailand and the progress on the tsunami readiness of the country including several shortages that still exist.

23.2 Social and Economic Losses

After the disastrous incident, the Department of Disaster Prevention and Mitigation (DDPM), the Ministry of Interior of Thailand, reported that total of 407 villages in six provinces along the Thai Andaman coasts: Phuket, Phang Nga, Krabi, Ranong and Satun, were affected by this 2004 tsunami (DDPM 2005). Total numbers of 3,302 houses were completely destroyed and additional 1,504 houses were partial damages. Over 58,550 people (12,017 families) were suffered by this incident, including 5,395 casualties, 2,845 missings and 8,457 injuries. Among of the casualties, 1,953 deaths were identified as foreigners from 37 countries with the majority of the European countries such as German (454), Sweden (479), Finland (151), Switzerland (82), Norway (79) and the Great Britain (76). The identities of additional 1,481 bodies were still unidentified. This tsunami also generated over 1,221 orphans.

For an economic loss, as the tsunami hit the touristic Thai Andaman coastlines during the high season (the Christmas holiday) without a proper disaster mitigation, the loss was enormous for both lives and property damages. Moreover, a large portion of the economic loss came from losses of tourism and fishery industries and a loss of the coastal natural resources. The estimated total economic loss from government agencies corresponded well to those reported by private sectors. For example, the Office of National Economic and Social Development Board of Thailand, an organization responsible for directing the strategic plans and policies of the country, reported that the estimated economic loss was about 68,000 million baht (Kittiampon 2005). The average exchange rate between December 2004 and April 2005 varies from approximately 38.3 to 39.6 baht/dollar. At the same period, the Kasikorn Research Center of the Kasikorn Thai Bank (2004) reported that the estimated economic loss from this tsunami was 67,000 million baht and the Center for Economic and Business Forecasting of the University of the Thai Chamber of Commerce (UTCC 2005) estimated that the economic loss was 57,000 million baht. All groups commonly reported that over 30,000 million baht of the total losses were suffered by the tourism industries.

As the disaster occurred near the end of 2004, it affected the GDP of the country very little. It was predicted that the GDP of 2005 would decrease to 5.0–5.5 %. Although gaining very large amount of financial supports from several domestic and foreign institutions, the GDP of Thailand decreased from 6.3 % in 2004 down to only 4.2 % in 2005, mainly due to an increase of oil price, a current account deficit, the violence in the southern Thailand and the domestic politic instability in the country (Office of National Economic and Social Development Board 2011). These problems caused the tourism industry to recover much more slowly than expected, nevertheless both tourists and investors have gradually come back to Thailand since 2006 (Office of Tourism Development 2008).

23.3 Restoration of the Affected Areas

Although may communities were damaged by the 2004 tsunami, they recovered gradually by the supports of the governmental, intergovernmental and private sectors. Many destroyed houses were rebuilt by various financial supports and most of the survivals eventually returned to their new homes (Fig. 23.1). Several public lands near the beaches that were trespassed by several private businesses and the locals in the past were reclaimed by the government and were declared as public land. Some government properties located too close to the shore lines and destroyed by the tsunami have been no longer in service. For example, one of the TMD weather station at Ban Nam Khem, Phang Nga province, was adapted to be a site for installing a warning tower (Fig. 23.2).

Fig. 23.1 *Left*: Damages at Ban Nam Khem village in Phang Nga province by the 2004 Tsunami (photo taken in December 2004). *Right*: A rebuilt home supported by both government and private sectors (photo taken in August 2012)

Fig. 23.2 (*Left*) A notice (in Thai) of a reclaimed land by the government stating that this land belongs to a public at Ban Kuk Kak, Phang Nga province (photo taken in August 2012). (*Right*) A completely destroyed TMD Phang Nga weather station was modified to house a new warning tower near the beach at Ban Kuk Kak, Phang Nga province (photo taken in August 2012)

23.4 Research and Development for Tsunami Hazard Preparedness in Thailand

Since the Sumatra tsunami in 2004, a large number of research related to tsunami have been conducted in Thailand, by both international research collaborations and local researchers in most aspects, such as earth and environmental sciences (i.e. Hawkes et al 2007; Srivichai et al. 2007; Barbier 2008; Cochard et al. 2008; Jankaew et al. 2008; Fujino et al. 2009; Ruangrassamee and Saelem 2009; Srivihok et al. 2009; Rhodes et al. 2011; Lukaszewski et al. 2012; Malain et al. 2012; Prendergast et al. 2012), engineering (i.e. Ghobarah et al. 2006; Foytong 2007; Suppasri 2010; Srisutam and Wagner 2010 and Leone et al. 2011), economics (i.e. Henderson 2007; Lee et al. 2007; Asafu-Adjaye and Tapsuwan 2008; Becchetti and Castriota 2011), medical (i.e. Petju et al. 2007; Schuller-Götzburg and Suchanek 2007; Blanc et al. 2008) and social science (i.e. Kongsakon 2009; Falk 2010; Merli 2010). The results from these research work helped the country to gain better understanding of the nature of tsunami and their various impacts on the Thai communities and were integrated in the later tsunami mitigation of Thailand.

23.5 Government Initiatives for Tsunami and Earthquake Mitigations in Thailand

Since the 2004 tsunami incident, several developments for tsunami mitigations of Thailand have been conducted in various scales. For a national scale, the Thai government founded the National Disaster Warning Center (NDWC) in May 2005.

The NDWC is responsible for the coordinating and dissembling disaster information (covering geological, hydrological, meteorological and forest fire disasters) with the local governments throughout the country via various means of communications such as public radio, television and warning towers. The NDWC has the authority to order the local governments to evacuate the people from the potential disastrous areas, in contrast to other countries such as Japan and the United States where the local governments make their own decisions whether to issue an evacuation order or not. In addition to their routine work, the NDWC also conducts various education outreach activities regarding natural disasters such as earthquakes, tsunami, landslide, floods, and climate changes throughout the country.

23.6 Real Time Earthquake and Tsunami Monitoring and Warning Networks

In addition to founding the NDWC, the Thai government has spent several million dollars since 2005 to upgrade very few existing seismic monitoring stations and to setup a new real time earthquake and tsunami monitoring network in Thailand that are operated by the Thai Meteorological Department (TMD). Currently TMD operates 40 real time seismic stations (both short period and broadband seismometers) with the strong motion recorders and additional 26 strong motion recorders, in addition to existing seven analog seismic stations that are still in service. TMD also operates nine tidal gauges: four stations in the Andaman Sea and additional five stations in the Gulf of Thailand (Fig. 23.3). Theses tidal gauges are the last resource that can provide at least 15–20 min of the tsunami early warning time to the communities along the Thai Andaman coastlines to evacuate to the safe places.

Supporting by the NOAA, the first tsunami buoy was installed in the Andaman Sea to the west of the Andaman-Nicobar Islands, about 600 nautical miles to the west-northwest of Phuket in December 2006, nearly 2 years after the 2004 Sumatra earthquake. This buoy was in service until 2009. As the cost of servicing the buoy was as much as the cost of a new unit replacement, the Thai government decided to buy a new buoy which were installed at the same location in December 2009. In addition, The Thai government acquired two more buoys and installed them 248 km to the west of Phuket and 260 km to the west of the Surin Islands, near Thailand-Myanmar boarder in late 2010. Currently Thailand has capability to detect tsunami wave at both far away and near shore locations which should provide valuable early warning time for the locals to evacuate to shelters.

Besides upgrading the earthquake and tsunami monitoring systems, the Thai government also constructed 76 warning towers in the communities along the Andaman coasts. These warning towers can disseminate warning messages directly from the NDWC central command to the locals. Additionally, many local

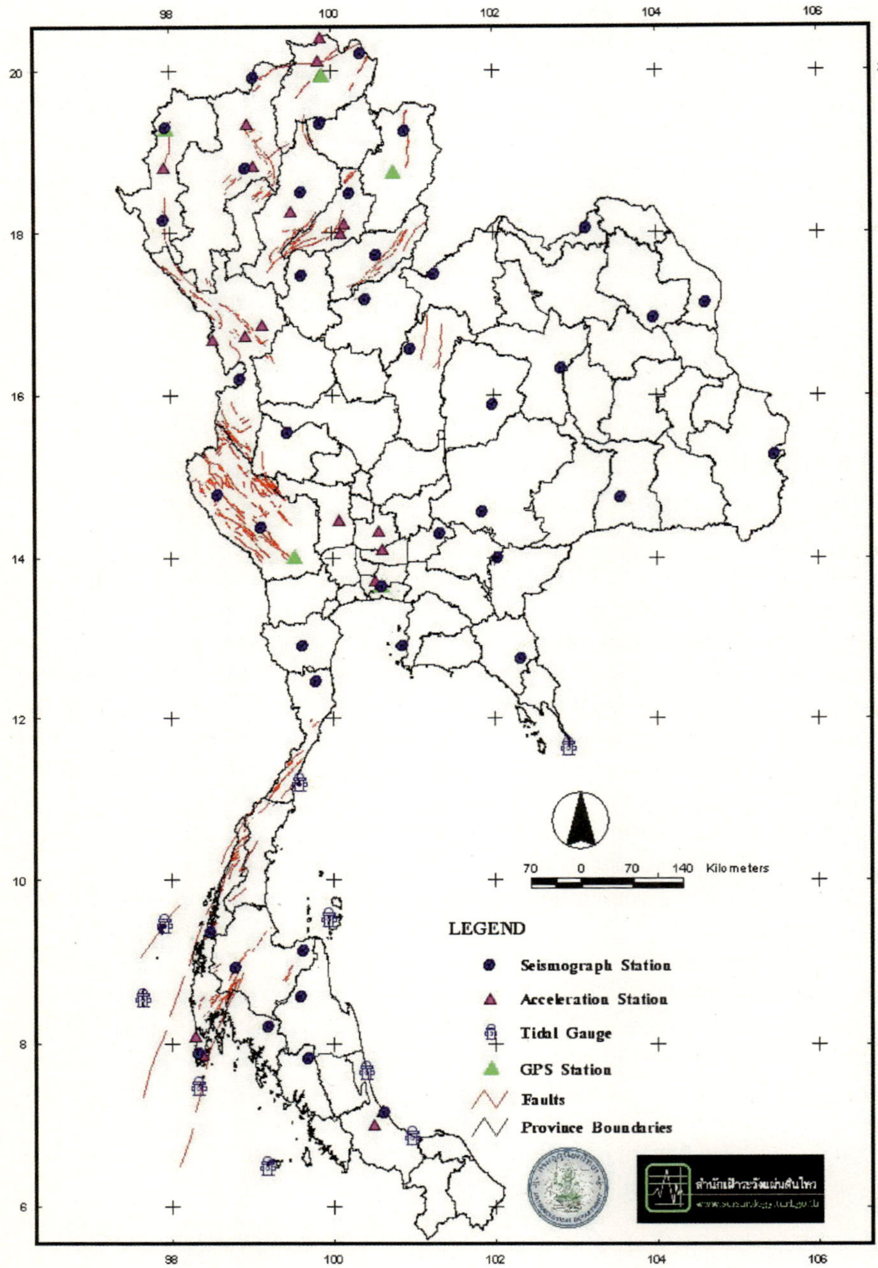

Fig. 23.3 Locations of a real time seismic monitoring system and tidal gauges operate by the TMD in Thailand

Fig. 23.4 Various types of the warning towers operated by the NDWC (*left*) and a warning tower built by the local government at Ban Suk Samran, Ranong province (*right*: photo taken in August 2012)

governments have constructed many more of their own warning towers that can choose to receive the warning message directly from the NDWC or can be operated locally (Fig. 23.4). Currently 129 warning towers are operated in six provinces along the Thai Andaman coastlines (NDWC 2012). These warning towers will help relay the warning messages to the locals living in the risk areas in order to receive important information regarding the incoming disaster and can prepare themselves accordingly.

23.7 Education Outreach and Mitigations

Before the December 26th, 2004, just a small fraction of the Thai populations know what a tsunami was. since then, enormous effort have been put into the Thai society to have people understand the nature of several natural disasters and how to prepare themselves when such disasters happen. Currently, most schools in Thailand add the curriculum of earthquake and tsunami including other natural disasters for their students as young as grade 6. Concurrently, due to a concern about the tsunami effect, the local communities along the Andaman coastlines have put several efforts for the tsunami mitigation, especially in the touristic towns. One of the most effective practices for the tsunami mitigation initiated by collaborations between the local and central governments is the tsunami evacuation drill. The central government, mainly by the DDPM, works together with the local governments that were affected by the tsunami in 2004 to conduct tsunami drills annually in order to increase the readiness of the authorities and the locals should the disaster have occurred. The drills were well received by the locals and the tourists in the risk areas.

Fig. 23.5 A tsunami evacuation map (*left*) and a warning post indicating the 2004 tsunami run-up height (5 m) at Ban Nam Khem village, Phang Nga province (*right*: photos taken in August 2012)

In addition to constructing additional warning towers and conducting the tsunami drills, more efforts have been put in. Many tsunami warning signs and evacuation maps based on a numerical modeling of tsunami wave height and the real observed tsunami run-up heights in 2004 were installed along the streets near the beaches or in public places in the communities along the coasts that were affected by the tsunami in 2004, especially in the touristic towns such as those in Phuket and Phang Nga provinces where majority of tourism activities take place (Fig. 23.5).

One of the factors that contributed a large casualties during the 2004 tsunami in Thailand was a lack of shelters that could protect tourists and the locals from the tsunami waves. As several areas along the Thai Andaman coastlines have very little topography, it can be very difficult for the victims to escape to the safe places within a short time. Figure 23.6 shows a topography of the Khao Lak beach in Phang Nga province, one of the most famous places for foreign tourists as an example. It can be seen that in several places, flat lands cover a large area for a long distance. In the 2004 tsunami, many vehicles on the Phetchakasem Highway, the main road to southern Thailand were wiped out of the road near Khao Lak area even if the road was located almost 1 km away from the beach. One of the best solutions for this situation is to construct temporary tsunami shelters that are high and strong enough to withstand a tsunami wave and are large enough to accommodate several evacuees. Several new

Fig. 23.6 Very little topography along the Khao Lak beach in Phang Nga province (the middle and lower half parts of the image. Without appropriate tsunami shelter, the tourists and the locals would have a difficult time escaping from the tsunami within a short period

tsunami shelters were built in coastal communities and tourist spots near the beaches so that people can escape to (Fig. 23.7). Unfortunately, due to the difficulty in the operation procedure and the vandalism problem, the local governments decide to distribute the provision and other important emergency items to the shelter only when the disaster strikes instead of storing them in the shelter. This drawback could cause a problem to provide the needed foods and emergency supplies to the evacuees if the disaster really happens due to the public chaos and the difficulty of the transportation that may exist during the incident. In addition, mosques in the coastline communities were adapted to be temporary tsunami shelters for the locals. Many schools near the beaches have also constructed multi-stories buildings, intended to be emergency temporary tsunami shelters.

As the tourism industries play an important role in the Thai revenue, one of the most essential tasks for the tourism industries along the Thai Andaman coasts is to make the tourists feel and be safe from such disaster. A good example of the tsunami mitigation for tourism industries can be seen in Phuket, one of the most touristic provinces in Thailand that attracts over six million tourists annually (Fig. 23.8). Currently, there are 20 warning towers installed along the beaches in Phuket. These warning towers can disseminate the warning messages directly from the NDWC in five languages. In addition, the local governments conduct a full scale tsunami evacuation drill twice a year and conduct the education outreach about the tsunami and other natural hazards to the local communities along the coasts throughout the year.

Fig. 23.7 Examples of temporary tsunami shelters in southern Thailand: (*Left*) A large building, over three stories high at Ban Nam Khem in Phang Nga province. (*Middle*) A mosque near the Pa Tong beach in Phuket province. (*Right*) A four-story school building at the Kamala beach, Phuket province (photos taken in August 2012)

Fig. 23.8 Very crowded tourism businesses along the Pa Tong beach, one of the most famous tourist spots in Phuket province that was affected by the tsunami in 2004 (photo taken in August 2012)

23.8 Shortcomings of Tsunami Mitigations in Thailand

Although Thailand has spent a large amount of money in order to increase the capability of the country to embrace tsunami in the most effective way, some persisting problems still exist. Most of the problems are due to the lack of continuation of the funding for proper maintenances of the facilities related to earthquake and tsunami monitoring and warning system. This includes both monitoring instruments, important tsunami warning signs and tsunami evacuation route

Fig. 23.9 A vandalized tsunami evacuation route map at the Kamala beach, Phuket province (*left*) and a faded tsunami warning sign at Suk Samran beach, Ranong province (photos taken in August 2012)

maps. Without adequate financial support for maintaining these facilities, they will deteriorate with time and eventually fail to operate which will increases the tsunami risk to the local communities (Fig. 23.9).

In addition to having a difficult time in keeping all disaster related facilities up and running at their full capacity, there is still another major problem: a shortage of tsunami experts in Thailand. Although Thailand has population over 60 million, there are only a handful of the earth scientists that conduct research in earthquake and tsunami sciences. This shortcoming makes the country lack of basic knowledge and important information for such disasters. For example, although many tsunami run-up models have been conducted along the Andaman communities of Thailand (i.e. Ioualalen et al. 2007, Yanagisawa et al. 2009), the lack of the high resolution bathymetry and topography near the shores makes the result less reliable. Therefore, there is a real and urgent need to construct a detailed bathymetry and topography map along the Thai Andaman coast lines so that these data can be used for more accurate tsunami inundation modeling. The better tsunami run-up models or more basic knowledge in earthquake and tsunami sciences will help the emergency response authorities to develop better tsunami mitigation and more effective evacuation plans for the locals which can reduce the loss of lives and properties in the risk areas.

Occasionally, several agencies from both governmental and local levels appear to have a conflicting or overlapping responsibility to manage the situation caused by the natural hazard. This problem is caused by the ambiguous duties and authorities that are authorized by the central government to different institutions. For example, at many times, the NDWC issues the earthquake public alert as well as the TMD even though the TMD operates the seismograph network of the country. It is often that several agencies such as the TMD, the NDWC, the Department of Mineral Resources (DMR) and even the DDMP issue the alert about the earthquake at the same time. The problem of overlapping and conflicting duty and authority among different agencies does exist for all natural disasters in Thailand which can make people very confused about whose information they should rely on. The Thai government needs to be clear about the chain command of duty and authority that each agency should have, otherwise it can create a confusion to the community during the disaster.

23.9 Excessive Disaster Concerns of the Thai Society

The awareness of the Thai society for the earthquake and tsunami have increased significantly since the 2004 tsunami incident, however, it can be seen that at times these awareness are too much that they have transformed to a panic. Two of the main reasons of this psychological problem are due to the overemphasizing of the earthquake disasters by the medias and the strong beliefs in religion and superstition including hoaxes of a large portion of the Thai population.

In recent years, numerous earthquakes around the world and in Thailand are reported by the medias through newspapers, televisions, radios and internet almost every day. The Thai population are exposed to this kind of reports all the time even though at many times, the magnitudes of the earthquakes reported are less than 2.5 (i.e. Thairath 2012a, b, c). These frequent reports increase the level of concern of the earthquakes for Thais because they are aware of only the number of the earthquakes reported, but are not aware of the magnitudes, the relationship between the number and the magnitude of the earthquakes or the tectonic origins of the earthquakes. These numerous and repeated reports make many Thais to visualize that the earthquakes have occurred much more often than what happened in the past. Moreover, now and then, there would be a famous person or a scientist who has never conducted any research in earthquake or tsunami giving an interview to the medias about a prediction of the earthquake or tsunami that would happen in Thailand. These people could be (but not limited to) a monk (Triphum 2011), a fortuneteller (Thairath 2011; Matichon 2012a) or someone who claims that his superstition can sense that the disasters are coming (Manager 2008). Unfortunately, even though these claims never or rarely came with scientific explanations and most of the time they were just hoaxes, they caught up many people and could disrupt the society terribly. Believing too much in superstition and religion can sometimes makes the Thai society have less scientific thinking than what they are capable of.

The disaster anxiety of the Thai population seems to have a cycle: The first cycle usually starts from all of the sudden, there will be a rumor originated from someone, whether famous (could be a nonscientist or a scientist that works very little in earthquake or disaster) or related to religion or superstition. This person will come to the public and give a prediction that the disaster is about to happen with (false) or without a scientific explanation. Then the Thai social will embrace this news overwhelmingly. The rumor about this prediction will then spread out to the public very rapidly by all means such as the internet or the public medias through televisions and newspapers.

This phenomena directly affects three groups of people. The first one is a group of people who live in the disaster prediction area. There will be chaos in these communities. Some people are too panicked about the disaster that they decide to relocate temporarily until the deadline of the disaster passes. The second group is the government agencies such as the NDWC (Apacnews 2011) who has to assure to the public regarding this false statement that this hoax is nearly impossible to happen and how much the agency is ready in case of the disaster should have really occurred. The third group of the people that are affected by the predictions are the real scientists that their research works are related to such predicted disaster. They will have to figure out how to calm down the public by approaching the medias with scientific explanations that why such prediction will not happen or is very unlikely to happen.

The second phase of the cycle is that the majority of the Thai population will be awaiting for the day when the disaster is predicted to occur. When the deadline passes and nothing bad happens, the public's panic and interest of this incident will gradually decrease and finally most of the people will forget about the prediction and live their normal lives until the next prediction comes out to public again.

These kinds of disaster hoaxes not only destroy the public serenity, but they also have a big negative impact on the local tourism industries. People trend to refrain themselves to travel to a place where the disaster is predicted to happen even though they are well aware that the predication is not true. A recent classic case was the prediction that there would be a large earthquake that could cause the Phuket Island to sink underneath the Andaman Sea on April 28th, 2012 (i.e. Dailynews 2012; Thaivisa 2012; UPI 2012). Even though this hoax was obviously very not true, many (Thai) tourists refrained themselves to visit Phuket during that time which caused a large damage to the local tourism industries.

Another infamous classic case that has been ongoing in Thailand for many years is the rumor about the breach of the Srinakarin Dam, the largest dam in Thailand that is located in Kanchanaburi province, western Thailand (i.e. Thairath 2009; Oknation 2011). Now and then there would be a rumor about the breaching of this dam by a large earthquake and the breached water would inundate the communities living along the river downstream. This rumor has caused the people who live downstream to panic and prevented the tourists from visiting the dam during that time which affects local tourism businesses. The public concern about this rumor is too intense that the Electricity Generating Authority of Thailand (EGAT), an agency who operates this dam, has to install a live webcam that can be accessed via an internet to ensure the public that the dam is still safe.

Surprisingly, Thai population might be one of very few groups in the world that are concerned or even panicked about the 2012 end of the world prophecies. There are many publications and news including interview programs about this issue on many medias throughout the year (i.e. Metteya 2009; DMC 2011; Matichon 2012b; Mthai 2012).

23.10 A Real Test for the Readiness in Tsunami Mitigation for Thailand

A real test for the Thai society to the tsunami mitigation came on April 11th, 2012 when the M_w 8.6 earthquake occurred in the Sumatra region (Duputel et al. 2012). The TMD issued a public announcement of this earthquake 13 min after the earthquake and the NDWC issued a tsunami warning and ordered the local authorities to evacuate the people in the tsunami risk areas 23 min after the earthquake. Although there were many confusions in the local communities at first, a routine tsunami evacuation drill helped people evacuate themselves to the safe places within a short time. Fortunately, as the major movements of the fault that generated this earthquake were mostly horizontal with minor vertical components, it created only small tsunami in Thailand. The measured tsunami wave heights were only 30 cm at Kho Miang in the Similan Islands, about 65 km west of Phang Nga shorelines and only 10 cm at Kho Taphao Noi, to the southeast of Phuket Island (NDWC 2012), therefore, the NDWC lifted the tsunami warning and the evacuation order at a later time. The Thai earth scientists had the opportunity to explain the reason why this large earthquake generated just a small tsunami to the public through the media shortly after that.

This incident showed how effective the tsunami mitigations in Thailand were. Most of the warning towers worked as designed although a few did not. It was evident that the mitigation worked effectively was that although most of the communications (such as telephone and internet) failed for nearly 2 h after the earthquake due to very heavily uses, the locals living at many places along the coasts could make a decision to evacuate themselves to safe places as a precaution. The local authorities received evacuation order from the NDWC in a short time to help coordinate the evacuation as effectively and as fast as possible. Most of the tourists along the beaches were well informed about this earthquake and tsunami warnings by both warning towers and mobile warning units operated by the local authorities. These mobile units helped relay a clear warning messages to the locals and make them be able to evacuate to safe places in time. All of these nearly perfect evacuation processes can happen because people understand the nature and the risk of tsunami by a proper education outreach and the local authorities have conducted tsunami evacuation drills routinely.

23.11 Conclusion

The losses of the Thai society due to the 2004 tsunami have made the Thailand become stronger and have better tsunami preparation and mitigation plans. Now it can be seen that Thailand is ready for such level of the great disaster. Thai citizens have learned to understand the nature of the disasters and have learned to prepare and protect themselves from such disasters. The government has invested a large amount of money in earthquake and tsunami monitoring and warning systems that can provide an early warning to the locals in time. Thailand is fortunate that most of the earthquakes capable of generating tsunami that can affects the southern Thailand are located relatively far from the country and most generated tsunami will take more than 1 h to reach Thai coastlines. Still, it is very important to maintain the readiness of the country and the awareness of the people for this type of a tragic disaster. The follows are the important aspects that the Thai government should pay attention to in order to maintain the readiness of the country for the tsunami hazard at high level:

- Provide enough financial support to maintain, upgrade and increase number of the critical facilities for earthquake and tsunami monitoring and warning network, including tsunami shelters, warning signs and evacuation maps, at best condition so that they can work effectively.
- Provide supports and increase the basic research and development of earthquake and tsunami sciences. This includes increasing number of earthquake and tsunami researchers and acquiring important database that can be use for tsunami hazard and risk evaluation.
- Maintain the awareness of the people for the danger of the earthquake and tsunami for both understanding their nature and how to prepare themselves for such disasters via educational outreach and drills.
- Decrease the conflicting/overlapping duties and authorities between the agencies and increase the cooperation and synergy among authorities for earthquake and tsunami hazard mitigation.
- Increase the ability of the society to have scientific thinking about the natural disaster rather than believing in superstition, religion, hoaxes or fortuneteller.

The most difficult task is how to maintain the readiness of the country at high level for both financially and a right mind setting of the society. It can be seen from many cases in the past that when the disasters have not occurred for a long time, people tend to let their guards down and will eventually suffer greatly when the disaster strikes. Although Thai population can be over panicked about the disasters at times, a proper and routine education outreach to the young generations to understand the nature of these disasters will eventually help decrease the panic on the disaster of the Thai society in the future.

References

Apacnews (2011) NDWC warn people not to panic of the disaster prediction at the end of year 2011 (in Thai). http://apacnews.net/news/?p=16332. Accessed 12 Oct 2012

Asafu-Adjaye J, Tapsuwan S (2008) A contingent valuation study of scuba diving benefits: case study in Mu Ko Similan Marine National Park, Thailand. Tourism Manag 29(6):1122–1130

Barbier EB (2008) In the wake of tsunami: lessons learned from the household decision to replant mangroves in Thailand. Resour Energ Econ 30:229–249

Becchetti L, Castriota S (2011) Does microfinance work as a recovery tool after disasters? Evidence from the 2004 tsunami. World Dev 39(6):898–912

Blanc V, Bohic J-Y, Bohic LL, Del Giudice P, Rodríguez-Nava V, Boiron P (2008) Nocardia spp. wound infection in a tsunami survivor from Thailand, 2004. J Mycolog Med 18(2):123–127
Chlieh M, Avouac JP, Hjorleifsdottir V, Song TRA, Ji C, Sieh K, Sladen A, Hebert H, Prawirodirdjo L, Bock Y, Galetzka J (2007) Coseismic slip and afterslip of the great Mw 9.15 Sumatra-Andaman earthquake of 2004. Bull Seismol Soc Am 97:152–173
Cochard R, Ranamukhaarachchi SL, Shivakoti GP, Shipin OV, Edwards PJ, Seeland KT (2008) The 2004 tsunami in Aceh and Southern Thailand: a review on coastal ecosystems, wave hazards and vulnerability. Perspect Plant Ecol Evol Systemat 10:3–40
Dailynews (2012) News reports on a sinking of Phuket (in Thai). http://www.dailynews.co.th/article/440/103856. Accessed 12 Oct 2012
Department of Disaster Prevention and Mitigation (DDPM), the Ministry of Interior, Thailand (2005) Tsunami victims' database. http://61.19.54.131/tsunami/. Accessed 23 Aug 2012
DMC (2011) The end of the world in 2012 part II (in Thai). http://www.dmc.tv/pages/top_of_week/tnt_think_around_all_2012_2_RIGHT.html. Accessed 12 Oct 2012
Duputel Z, Kanamori H, Tsai VC, Rivera L, Meng L, Ampuero JP, Stock JM (2012) The 2012 Sumatra great earthquake sequence. Earth Planet Sci Lett 351–352:247–257
Falk ML (2010) Recovery and Buddhist practices in the aftermath of the tsunami in southern Thailand. Religion 40(2):96–103
Foytong P (2007) Fragility of buildings damaged in the 26 December 2004 tsunami, Master thesis, Civil Engineering Department, Chulalongkorn University, Bangkok, Thailand
Fujino S, Naruse H, Matsumoto D, Jarupongsakul T, Sphawajruksakul A, Sakakura N (2009) Stratigraphic evidence for pre-2004 tsunamis in south-western Thailand. Mar Geol 262:25–28
Ghobarah A, Saatcioglu M, Nistor I (2006) The impact of the 26 December 2004 earthquake and tsunami on structures and infrastructure. Eng Struct 28(2):312–326
Hawkes AD, Bird M, Cowie S, Grundy-Warr C, Horton BP, Hwai ATS, Law L, Macgregor C, Nott J, Eong Ong J, Rigg J, Robinson R, Tan-Mullins M, Tiong Sa T, Yasin Z, Wan AL (2007) Sediments deposited by the 2004 Indian Ocean tsunami along the Malaysia-Thailand Peninsula. Mar Geol 242(1–3):169–190
Henderson JC (2007) Corporate social responsibility and tourism: hotel companies in Phuket, Thailand, after the Indian Ocean tsunami. Int J Hospit Manag 26(1):228–239
Ioualalen M, Asavanant J, Kaewbanjak N, Grilli ST, Kirby JT, Watts P (2007) Modeling of the 26th December 2004 Indian Ocean tsunami: case study of impact in Thailand. J Geophys Res doi:10.1029/2006JC003850
Jankaew K, Atwater BF, Sawai Y, Choowong M, Charoentitirat T, Martin ME, Prendergast A (2008) Medieval forewarning of the 2004 Indian Ocean tsunami in Thailand. Nature 455:1228–1231
Kasikorn Research Center (2004) Tsunamis and the impact to the Thai economy. http://www.kasikornresearch.com/EN/K-EconAnalysis/Pages/ViewSummary.aspx?docid=2269. Accessed 23 Aug 2012
Kittiampon A (2005) NESDB elucidated the effect of tsunami to Thai economy. Thai world affairs center. http://www.thaiworld.org/thn/thailand_monitor/answera.php?question_id=96. Accessed 23 Aug 2012
Kongsakon R (2009) P03-255 quality of life among the navies at six months after the tsunami in Phang-NGA naval base, Phang-NGA province, Thailand. Eur Psychiatry 24:S1254
Lee HY, Wu HC (2007) Contagion effect in financial markets after the South-East Asia Tsunami. Res Int Bus Finance 21(2):281–296
Leone F, Lavigne F, Paris R, Denain JC, Vinet F (2011) A spatial analysis of the December 26th, 24 tsunami-induced damages: lessons learned for a better risk assessment integrating buildings vulnerability. Appl Geogr 31(1):363–375
Lukaszewski Z, Karbowska B, Zembrzuski W, Siepak M (2012) Thallium in fractions of sediments formed during the 2004 tsunami in Thailand. Ecotoxicol Environ Saf 80:184–189

Malain D, Regan PH, Bradley DA, Matthews M, Al-Sulaiti HA, Santawamaitre T (2012) An evaluation of the natural radioactivity in Andaman beach sand samples of Thailand after the 2004 tsunami. Appl Radiat Isot 70(8):1467–1474

Manager (2008) Businesses in Kanchaburi burnt a dummy of Dr. Art-ong Jumsai Na Ayudhya for his disaster forecast (in Thai). http://www.manager.co.th/Local/ViewNews.aspx?NewsID=9510000116257. Accessed 12 Oct 2012

Matichon (2012a) Accurate disaster prediction by Sorajja Nual-yu (in Thai). http://www.matichon.co.th/news_detail.php?newsid=1334140742&grpid=01&catid=0202. Accessed 12 Oct 2012

Matichon (2012b) NASA refuse the fortuneteller prediction about the end of the world in 2012 (in Thai). http://www.matichon.co.th/news_detail.php?newsid=1325411930&grpid=00&catid=00. Accessed 12 Oct 2012

Merli C (2010) Context-bound Islamic theodicies: the tsunami as supernatural retribution vs. natural catastrophe in southern Thailand. Religion 40(2):104–111

Metteya (2009) Civilization's end or apocalypse or end time 2012? (in Thai). http://www.metteya.org/sriann/End-Time-NASA.html. Accessed 12 Oct 2012

Mthai (2012) Will the world in 2012? (in Thai). http://scoop.mthai.com/2012-end-of-the-world. Accessed 12 Oct 2012

Natural Disaster Warning Center (2012) Conclusion of procedure during the earthquake on 11 April 2012 (in Thai). http://www.ndwc.go.th/web/index.php?option=com_content&view=article&id=2695:2012-04-12-08-50-11&catid=41:earthquake&Itemid=219. Accessed 12 Oct 2012

Norwegian Geotechnical Institute (2006) Tsunami risk reduction measures with focus on land use and rehabilitation. Main report, NGI report 20051267-1

Office of National Economic and Social Development Board (2011) National Accounts of Thailand, new series (chain volume measures 1990–2010 editions)

Office of Tourism Development, Ministry of Tourism and Sport (2008) Tourism statistic in Thailand 1998–2007. http://www2.tat.or.th/stat/web/static_index.php. Accessed 23 Aug 2012

Oknation (2011) What will happen if the dam in Kanchaburi brereaches? (in Thai). http://www.oknation.net/blog/nonglek/2011/03/26/entry-1. Accessed 12 Oct 2012

Petju M, Suteerayongprasert A, Hassiri (2007) Importance of dental records for victim identification following the Indian Ocean tsunami disaster in Thailand. Public Health 121((4):251–257

Prendergast AL, Cupper ML, Jankaew K, Sawai Y (2012) Indian Ocean tsunami recurrence from optical dating of tsunami sand sheets in Thailand. Mar Geol 295–298:20–27

Rhodes BP, Kirby ME, Jankaew K, Choowong M (2011) Evidence for a mid-Holocene tsunami deposit along the Andaman coast of Thailand preserved in a mangrove environment. Mar Geol 282(3–4):255–267

Ruangrassamee A, Saelem N (2009) Effect of tsunamis generated in the Manila Trench on the Gulf of Thailand. J Asian Earth 36:56–66

Schuller-Götzburg P, Suchanek J (2007) Forensic odontologists successfully identify tsunami victims in Phuket, Thailand. Forensic Sci Int 171(2–3):204–207

Srisutam C, Wagner JF (2010) Reconstructing tsunami run-up from the characteristics of tsunami deposits on the Thai Andaman coast. Coast Eng 57:493–499

Srivichai M, Supharatid S, Imamuea F (2007) Simulation of inundation area and tsunami height for ban Bang Niang from the 2004 Indian Ocean tsunami, SOI Asia Special Symposium

Srivihok P, Das KD and Shanmugasundaram J (2009) Seismic data processing evaluation and tsunami modeling techniques for tsunami early warning system in the Indian Ocean and South East Asia, 3rd International Conference on Estuaries & Coasts (ICEC2009), Coastal Hazard II (Room 3, Sept. 14th 15:30–17:30)

Suppasri A (2010) Tsunami risk assessment of coastal population and building in Thailand, PhD thesis, Civil and Environmental Engineering, Tohoku University, Sendai, Japan

Thairath (2009) Proving a rumor of a breaching of the Srinakarin Dam (in Thai). http://www.thairath.co.th/column/pol/page1scoop/27383. Accessed 12 Oct 2012

Thairath (2011) Disaster, revolution, death with Sorajja Nual-yu prediction (in Thai). http://www.thairath.co.th/content/life/225479. Accessed 12 Oct 2012

Thairath (2012a) A report of an earthquake with magnitude 1.6 occurring on 19 September 2012 in Lampang province, Thailand (in Thai). http://www.thairath.co.th/content/region/292448. Accessed 12 Oct 2012

Thairath (2012b) A report of an earthquake with magnitude 2.3 occurring on 7 October 2012 in Myanmar (in Thai). http://www.thairath.co.th/content/region/296610. Accessed 12 Oct 2012

Thairath (2012c) A report of an earthquake with magnitude 2.2 occurring on 12 October 2012 in Myanmar (in Thai). http://www.thairath.co.th/content/region/297996. Accessed 12 Oct 2012

Thaivisa (2012) News reports on a sinking of Phuket. http://www.thaivisa.com/forum/topic/549462-thai-experts-confirm-phuket-will-never-sink-even-if-it-is-hit-by-a-strong-earthquake/. Accessed 12 Oct 2012

The Center for Economic and Business Forecasting of the University of the Thai Chamber of Commerce (2005) Economic evaluation of Thailand in 2005 after the Tsunami (preliminary). http://utcc2.utcc.ac.th/localuser/cebf/analysis.php. Accessed 23 Aug 2012

Triphum (2011) A warning of the earthquake and tsunami (in Thai). http://2012doomsday2555.blogspot.com/2011/04/1.html. Accessed 12 Oct 2012

UPI (2012) News reports on a sinking of Phuket. http://www.upi.com/Odd_News/2012/04/23/Phuket-not-sinking/UPI82731335222577/. Accessed 12 Oct 2012

Yanagisawa H, Koshimura S, Goto K, Miyagi T, Imamura F, Ruangrassamee A, Tanavud C (2009) The reduction effects of mangrove forest on a tsunami based on field surveys at Pakarang Cape, Thailand and numerical analysis. Estuar Coast Shelf Sci 81(1):27–37

Chapter 24
Understanding Challenges at the "Last-Mile" in Developing an Effective Risk Communication to Reduce People's Vulnerability in Context of Tsunami Early Warning and Evacuation

Neysa J. Setiadi

Abstract The paper presents a framework for assessing vulnerability in the context of tsunami early warning and evacuation at the local level. A case study in the city of Padang, Indonesia was given as a showcase on analysis of the vulnerability component of understanding and response to the early warning. Quantitative and qualitative analysis of household surveys as well as various semi-structured and non-structured interviews derive important parameters that should be considered in development an effective risk communication.

Keywords Early warning • Evacuation • Risk communication • Vulnerability

24.1 Introduction

Early warning system is one of the important measures to reduce vulnerability and risk by providing information that will enhance disaster preparedness of the people and elements at risk. Early warning is an important part of the disaster preparedness phase; it primarily does monitoring and provides information about the potential upcoming natural hazard events to the people at risk to response to. But solely delivering the information is not the goal of early warning itself, an early warning has to trigger appropriate risk reducing actions. In the recent decade, "people-centered" early warning systems have been promoted. No longer after the overwhelming Indian Ocean Tsunami in December 2004, the World Conference on Disaster Reduction 2005 was conducted in Hyogo, Japan, and called for the establishment of an effective and durable tsunami early warning system for the Indian ocean that also included a Special Session on "People-Centered Early Warning Systems". It emphasizes that in order to be effective, early warning systems must

N.J. Setiadi (✉)
United Nations University Institute for Environment and Human Security, Boon, Germany
e-mail: setiadi@ehs.unu.edu

Y.A. Kontar et al. (eds.), *Tsunami Events and Lessons Learned: Environmental and Societal Significance*, Advances in Natural and Technological Hazards Research 35,
DOI 10.1007/978-94-007-7269-4_24,

be embedded in, understandable by and relevant to the communities which they serve (UN/ISDR 2006). In addition to monitoring, forecasting catastrophic event, the element of response capability is added as an improved framework of early warning (Villagran de Leon et al. 2006).

UN/ISDR (2007) defines early warning as "the set of capacities needed to generate and disseminate timely and meaningful warning information to enable individuals, communities and organizations threatened by a hazard to prepare and to act appropriately and in sufficient time to reduce the possibility of harm or loss". This definition encompasses a complete warning chain that not only functions as an interface that delivers information on the natural hazard event, but also risk reducing actions that the people at risk perform as a response to the warning. Early warning should ensure clear messages that reach those at risk, and practiced and knowledgeable responses by risk managers and the public. In this regard the term "Last-Mile", which was firstly introduced by Shah (2006) to represent the people at risk with their configurations and challenges that needs to be understood in the context of earthquake risk mitigation, fits very well. Despite the fact that capability to respond was identified as one of the weakest elements of early warning systems, efforts and progress have been made in terms of consideration of risk assessment up to promoting response capability, i.e. people's reactions to early warning are taken into consideration in developing early warning systems (Dannenmann/ISDR, personal communication 30.03.2012).

Response to early warnings especially in case of major, sudden-onset hazard events, often involve mass evacuation activities. Here, the knowledge of social and spatial configuration of the people at risk (the "Last-Mile") as well as their capacity and behaviour in emergencies is very crucial. To be able to support specifically early warning and evacuation in the "Last-Mile", assessment of people's vulnerability should be defined in a way that it may show the gaps and needs of the people and guide relevant interventions to ensure that early warning is effective and timely evacuation is triggered. Vulnerability assessment should show where the bottlenecks are and why they are there.

A case study in the city of Padang, Indonesia, was conducted to develop and test a specific vulnerability assessment in the context of tsunami early warning and evacuation in scope of a joint research project "Last-Mile – Evacuation",[1] especially taking into consideration the social and institutional aspects. In this paper, we are going to focus only on aspects related with the risk communication. Based on lessons learned from this case study, some lessons learned and implications are derived.

[1] The "Last-Mile – Evacuation" Research Project intended to provide a detailed information system for tsunami early warning and evacuation planning, consisting of hazard and vulnerability information, as well as identification of potential bottlenecks in evacuation, for a coastal city of Padang, West Sumatra, Indonesia. UNU-EHS, under which the researcher conducted the study, was responsible for the Work Package 1000 "Socio-Economic Vulnerability Assessment" (Project Period: May 2007 – July 2010).

24.2 Conceptual Framework and Discussions

24.2.1 Overarching Vulnerability Assessment Framework

The term vulnerability is defined and conceptualized in various ways and is used in various disciplinary and practical contexts, as are its assessment methods (some extensive reviews can be found e.g. in Birkmann 2006; Cutter 1996; Cutter et al. 2009). The widely accepted definition of vulnerability is given by UN/ISDR (2004) as "*The conditions determined by physical, social, economic and environmental factors or processes, which increase the susceptibility of a community to the impact of hazards*; *for positive factors, which increase the ability of people to cope with hazards*". However, this definition is quite general and in order to develop vulnerability assessment framework and useful indicators for early warning interventions it needs to be further specified.

In the context of emergency management, Buckle (1998) argued that vulnerability has to be linked with the goal of effective delivery of services to the most appropriate target group and meeting needs and thereby the vulnerability needs to be categorized on basis of corresponding issues in the particular focus of emergency. Buckle proposed several meta-categories for vulnerability such as management capacity, access to services, cultural attitude etc., which should be linked with the dimensions of loss (lives, properties, social, psychological, etc.). In the context of this study, this argument also supports the idea of developing a specific vulnerability framework for specific context of tsunami hazard, and the corresponding early warning and evacuation measures to anticipate loss of lives.

One of the vulnerability concepts considered relevant for this study is BBC-Meta-Framework which was developed by Birkmann (2006), based on the works of Bogardi and Birkmann (2004) and Cardona (1999, 2001). It is one of the holistic approaches to vulnerability assessment. In this framework, vulnerability is composed exposure and susceptibility of various elements of concern, but also takes into consideration their coping capacity. This framework emphasizes the feedback-loop process of reducing risk by interventions to reduce vulnerability prior to hazard events rather than solely disaster (response) management. Additionally, Gebert (2011) suggested narrowing down the operationalization of this framework to assess and develop vulnerability indicators specific to disaster management phase or measures in question, e.g. disaster preparedness and early warning, in addition to assess the whole vulnerability as a baseline susceptibility, coping and adaptive capacity.

In the context of early warning and evacuation, this study proposes a working definition of vulnerability as "the conditions which increase the ability of people to respond to the warning and conduct appropriate evacuation to save their lives: dynamic exposure, access to early warning, evacuation readiness, evacuation capability". This focus on the response capability in the "Last-Mile" is based on the stages of early warning delivery at the local level up to people at risk taking action.

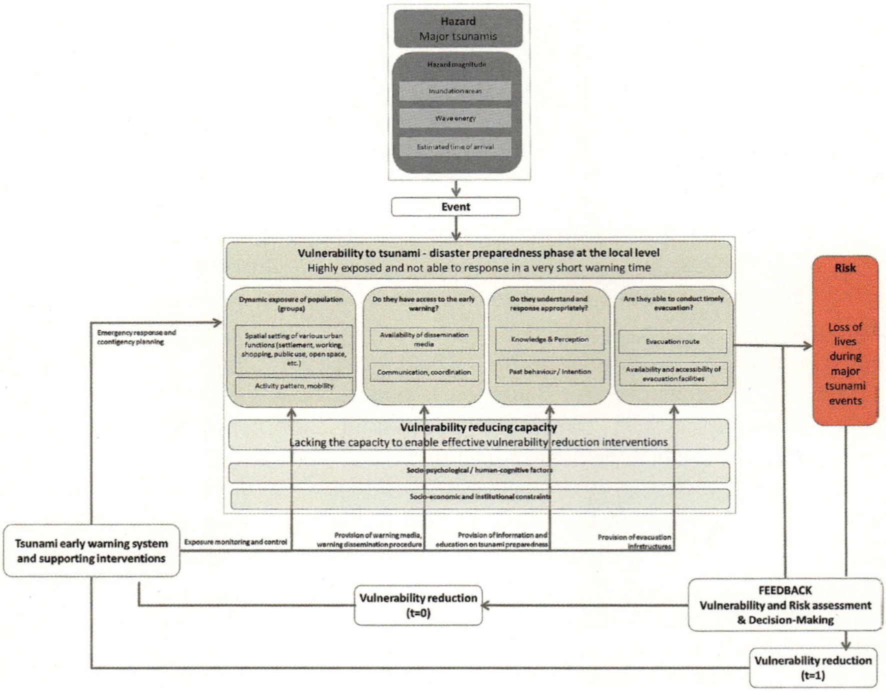

Fig. 24.1 Modified BBC-Framework for Vulnerability Assessment in the Context of Tsunami Early Warning and Evacuation. The framework shows various vulnerability components in this specific context (Source: Author, initial ideas are taken from Birkmann et al. 2008 and Gebert 2011)

Extending the basic ideas of the BBC-Framework,[2] the following framework was developed as a guidance to assess vulnerability and provide information basis for various interventions relevant for tsunami early warning and evacuation (Fig. 24.1).

The framework shows various vulnerability components in this specific context:

(i) Dynamic exposure of the people or population groups, this is influenced by the spatial setting of the urban areas in which the people are concentrated in different areas based on their daily activity pattern and locations of various city functions
(ii) Access to the warning, this shows the availability of various private and public media to disseminate early warning and relevant information and how these media are utilized
(iii) Understanding and response to the early warning, this relates mainly with evacuation behaviour of the people comprising perception and knowledge of the people on tsunami risk, early warning and evacuation measures, which will trigger timely decision to conduct evacuation and appropriate actions

[2] Initial vulnerability framework work was developed in the first phase of the "Last-Mile – Evacuation" research project jointly by Birkmann et al. (2008).

(iv) Ability to conduct evacuation, in addition to people's behaviour, the availability of sufficient evacuation infrastructures and facilities is very crucial and decisive for evacuation time

All of these components will shape the vulnerability of the people and, consequently, the risk of loss of lives that should be reduced by the effective tsunami early warning system and various supporting interventions. Additionally, the feedback loop diagram in the framework shows two vulnerability reduction options, prior to (t = 0) and during (t = 1) tsunami events. The vulnerability reduction during tsunami events assumes acceptance of the existing vulnerable conditions and coping by means of disaster (emergency) response. In contrast, vulnerability reduction prior to tsunami events suggests that the gaps and needs in various components are assessed and mitigated by means of various interventions. Such interventions will then require additional capacities of the people to support and implement them, which are represented by the additional component of vulnerability reducing capacity. Finally, each vulnerability component as well as intervention is shaped by the enabling condition which is institutional setting and capacity. Overall, this feedback loop system emphasizes the necessity of reducing vulnerability in a continuous and sustainable manner – in a long-term.

Operationalization of this framework into vulnerability indicators and development of the information basis is done separately for each component and require different methodologies as well as data sources. The vulnerability components of access to early warning and understanding and response to the early warning are directly linked with risk communication at the community level, while the other components of dynamic exposure and evacuation capability (infrastructures) are also linked with the spatial setting of the city and interventions in the urban spatial planning in the long-term. The following analysis is focusing only on the component understanding and response to the early warning.

24.2.2 Existing Studies on Response to Warning and Evacuation Behavior

Receiving the warning does not necessarily mean that people would understand what to do or comply with the instruction given. Studies on evacuation behavior, e.g. in case of hurricanes, by Riad and Norris (1998) found out that different communities interpreted the warning and the danger in a different manner, and the decision of evacuation participation is derived from both individual factors and social interactions. Other studies (e.g. Lachman et al. 1961; Baker 1991; Lindell et al. 2005; Tierney et al. 2001) also show the cases where warning and evacuation instructions were often not followed by evacuation and explored various influencing factors.

Evacuation behavior is a complex theme which involves not only physical (evacuation route and places) and institutional aspects (e.g. Standard Operational

Procedure, emergency plan), but also socio-psychological and socio-organizational aspects (Santos and Aguirre 2005). Bhatti (2001) suggests that early warning chain consists of five phases: receiving the warning, understanding the content of the warning, personalizing the warning and its sources, verifying through other sources, and taking reaction to warning. Mileti and O'Brien (1992) and Sorensen (2000) share similar phases and also suggest that these phases are influenced by individual characteristics (age, gender, education, etc.) as well as characteristics of the warning information (sources, repetition, etc.). Gregg et al. (2007) describes these response phases as "how the people perceive the risk and shape their behavior in every warning received". With regard to understanding the warning, knowledge generated from previous experience with the hazard events, formal and informal education, and own knowledge, may play a role (Shaw et al. 2004). However, knowledge is also often not directly transformed into action and not the only determining factor; e.g. in a study conducted by Indonesian Institute for Science (Hidayati et al. 2006) that measured household preparedness level to tsunami, it is shown although there were many households with a good level of knowledge about tsunami, there were very few with concrete emergency and preparedness plan.

Two theories related to individual or household behavior were selected to explain intended evacuation behavior in this study, namely "*Protection Motivation*" (Rogers 1983; Martin et al. 2007) and "*Planned Behaviour*" (Ajzen 1991; Paton et al. 2003). The first theory describes an intention that leads to a behavior or willingness to conduct a protective action as a product of four critical cognitives: perceived impact, perceived vulnerability, individual efficacy, and perceived efficacy of the particular action. It suggests that perceived costs of the action and assessment of other risk increasing actions influence the behavior. The second theory suggests that the expected result (or effectiveness) of the action and self-confidence on own capabilities to perform the action will influence the intention.

24.3 Brief Description of the Study Area

The city of Padang is located at the west coast of the island of Sumatra, Indonesia. It is the capital city of the province of West Sumatra. The city covers 694.96 km^2 of which about 10 % is built-area located mostly along the coast and in the low lying areas with a population of more than 800,000. The city has expanded and on its way to become a metropolitan city. However it is at the same time recognized as one of the most plausible areas to be affected by the next earthquake generated major tsunami (Borrero et al. 2006; McCloskey et al. 2010).

The city of Padang represents an urban context, where a tsunami early warning and disaster management at the local level is being established and the awareness of necessity of tsunami risk reduction is recently emerging based on tsunami experience of other areas (Indian Ocean Tsunami in December 2004 which affected severely the neighbouring province of Aceh). A tsunami hazard map for the city was developed based upon one plausible source scenario to identify potentially affected areas. Based on the analysis and modelling results available for the city, about 147,000 people in the night and 181,000 people during the day are exposed, two-thirds of them are women, children and elderly (Taubenböck et al. 2012). Those people (or more if considering additional people in the evacuation buffer zone) have to be evacuated within a short warning time. Based on the modelling, the estimated time of arrival is between 20 and 40 min after the strong earthquake occurs.

24.4 Methodology

Based on the literature study briefly described previously, the study argues that understanding of the early warning and appropriate response of the people will be triggered if people are in the position to identify, recognize the tsunami danger, recognize their own vulnerability and that there is something that they can do about it, and be confident of their own capability to conduct action (evacuation). Additionally, the intention to evacuate may vary in various social groups. Following this understanding some hypotheses were derived:

- Basic knowledge about tsunami enhances intention to evacuate
- Recognition of tsunami danger enhances intention to evacuate
- Recognition of own vulnerability and accordingly importance of preparedness enhances intention to evacuate
- Perception of own capabilities and efficacy of evacuation influences intention to evacuate
- Socio-economic characteristics of the households influence intention to evacuate

The main data source for analysis was the household survey covering 933 households which were conducted in April–May 2008 in some selected areas of the city of Padang. The results were compared with the household survey conducted by GTZ on the similar earthquake event in September 2009. Initially, descriptive analysis of the past earthquake events, especially in 2007, was done to get a picture of how various variables may influence evacuation decision. Subsequently, a set of variables for each hypothesis were selected from the household survey as independent variables to be tested with the intended evacuation response. Moreover, it is aimed to generate at the end a set of indicators for application in monitoring awareness activities in the community level; thus, applicability and relevance of the variables to awareness activities was also added as selection criteria. The process of the analysis and selection of variables for development of evacuation awareness indicators is delineated in Fig. 24.2. Some variables representing people's "objective" knowledge related to early warning and evacuation conduct is also measured, although for this focus no statistical analysis to test the hypothesis can be done, since no valid dependent variable is available (See Fig. 24.2).

Additionally, qualitative information, although at a small scale, was collected to provide context information, support the quantitative results, and capture parameters which were not measured directly in the household survey or did not show any statistical significance in the analysis. This was done by means of a focus group discussion with actors involved in disaster management in February 2008,

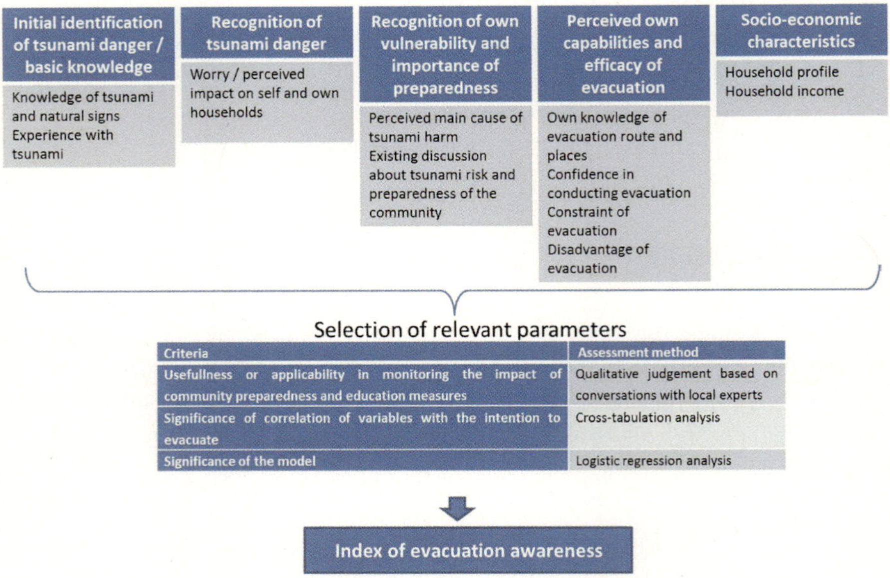

Fig. 24.2 Selection process of variables for index of evacuation awareness

semi-structured interviews in November 2008 with six purposively selected households with different intended evacuation behaviors, as well as non-structured conversations or interviews with various informants at the community level during field visits in 2008–2009. The analysis of this qualitative information was done using coding analysis of the transcribed interviews and discussions.

24.5 Results

24.5.1 Previous "Potential Tsunami" Experiences

Based on analysis of the household survey on the previous "potential tsunami" earthquake experience, especially in September 2007, it was found that there were only a small proportion of people evacuating after receiving the warning (35 %). The main reasons given by the respondents who did not evacuate were "did not trust the warning" and "wanted to wait and be sure about the tsunami occurrence".

Moreover, 42 % of the ones who evacuated did not immediately conduct evacuation but first gathered the household members, collected important items at home and/or securing the house before evacuated, etc. The analysis also found out that 98 % of the respondents performed evacuation in groups; most of them (75 %) evacuated with the family members. A simple correlation analysis was performed between the variable "did you evacuate" and some independent variables and resulting significant correlations as shown in Table 24.1 below.

Table 24.1 Correlation analysis between evacuation in the past and some selected variables (Source: Author, based on data analysis of UNU-EHS Household Survey in Padang 2008)

Independent variables	Kramer's V	Kendall tau b
Distance to the coast (further from the coast, less people evacuated)		−0.133
Household income (higher income, more people evacuated)		0.082
Gender (more female evacuated)	0.128	
Understanding the received warning message (understood better, more people evacuated)		0.122

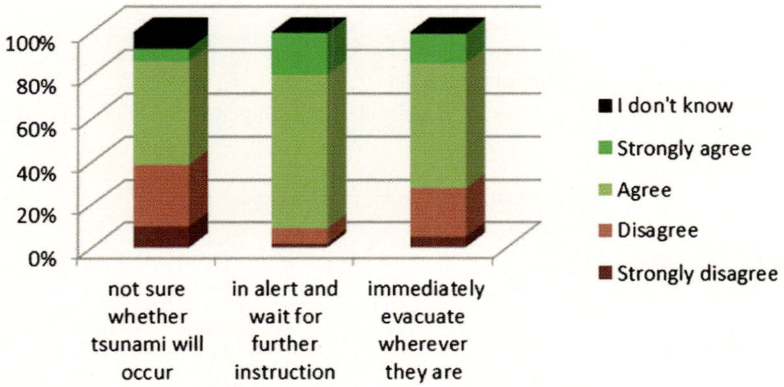

Fig. 24.3 Various interpretations of "potential tsunami" warning (Source: Author, based on data analysis of UNU-EHS Household Survey in Padang 2008)

These results also show that spatial setting, socio-economic characteristics and the format of the warning message are associated with the decision to evacuate. Additionally, the households that have elderly family members (with no children) showed a tendency of lower evacuation rate, although this has not shown a significant correlation due to small sample size for this particular group.

Noteworthy is especially the perceived clearness of the warning message. The official information given by the early warning center was "earthquake with potential tsunami" and followed by instruction by various sources to evacuate or stay in alert. Many people also received information from word of mouth, so that various interpretations were possible. Even for the official "potential tsunami", there have been different interpretations (See Fig. 24.3).

24.5.2 Intention to Evacuate

Although the evacuation rate was very low in the past experience, the intention for evacuating in case of future tsunami early warning showed higher rates (75 %). Several parameters were identified to be significantly influencing the intention to

Table 24.2 Selected parameters for evacuation awareness index and evacuation knowledge index (Source: Author, based on analysis of UNU-EHS Household Survey in Padang 2008)

Parameter	Weight
Basic knowledge	
What is tsunami	1
Natural indications accompanying potential tsunami occurrence	1
Recognition of tsunami danger	
Personal worries on the potential tsunami impacts	2
Recognition of vulnerability / importance of preparedness	
Cause of tsunami harm:not enough protection and preparedness	1
There is a lot of discussion on tsunami warning	2
Perceived own capabilities and evacuation efficacy	
Own knowledge of high building for evacuation	1
Own knowledge of higher ground for evacuation	1
Feel capable of reaching safe place on time and gathering family members and evacuating	2
Evacuation awareness index	11
Knowledge of recommended evacuation routes, location and signs	1
Knowledge of Tsunami Early Warning System	1
Participated in tsunami simulation or socialization (this variable was later extended to having household evacuation plan)	1
Evacuation knowledge index	3

evacuate as well as relevant for monitoring raising awareness activities (Table 24.2). The variables selected above were aggregated as the "evacuation awareness index" using simple additive methods with same weights. Additionally, several variables representing the "objective" knowledge of evacuation were also measured as evacuation knowledge index.

In order to validate the parameters, the evacuation awareness index that was measured previously was compared between the ones who evacuated in the strong earthquake events in September 2009 and the ones who did not. The results show that on average the ones who evacuated have higher evacuation awareness index, however, the dispersion seems to suggest that other explanatory parameters still need to be captured (Fig. 24.4).

In addition to the quantitative analysis of household surveys, there are several qualitative findings on the existing evacuation behavior as well as perception of the people on tsunami risk. It will be briefly discussed and demonstrated by some citations extracted from the qualitative data in the following.

24.5.3 Qualitative Findings on Evacuation Behaviour

Evacuation behaviors vary for different people depending on the available preparedness and arrangements at the household level and their knowledge of tsunami as well as evacuation procedure.

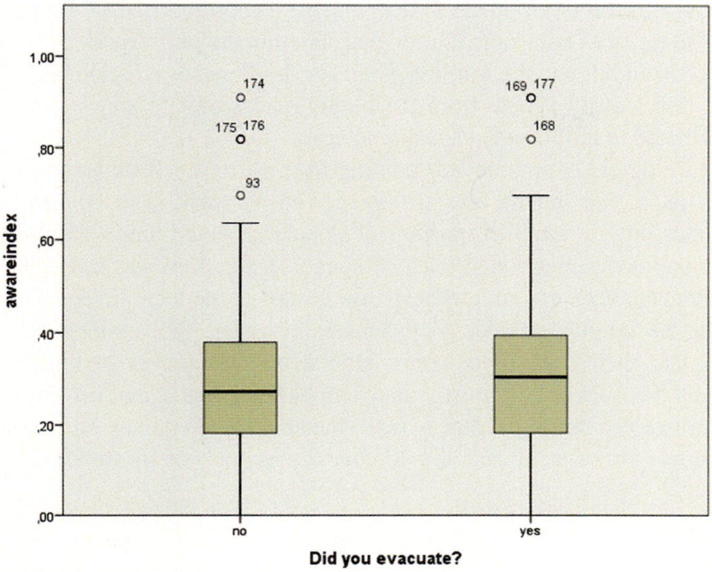

Fig. 24.4 Comparison of mean values and descriptive statistics of evacuation awareness index between respondents who evacuated and not during the earthquake event in 2009 (Source: Author, based on data analysis of GTZ Household Survey 2009)

> "we have prepared our family to evacuate if there is tsunami, if possible using the car, if not walk on foot. We will go to high buildings which are strong for vertical evacuation. We already have info on safe zone in Padang. We need better evacuation planning, I would pick up my children from school, but my older child has been instructed to go by himself to Gunung Pangilun (safe area), also my wife if we are in different locations".[3]

> "I would evacuate with my whole family, as far as possible not alone/ separately"[4]

> "we cannot avoid/forbid people to evacuate using their vehicles, we might recommend them to run (on foot) use bike, but we have to consider the worst-case scenario. The current SOP/ approach has not been tested and proven as effective"[5]

> "I was observing the beach as people were evacuating during the earthquake. I know the indications of tsunami from the television and books. If the water retrieves it means tsunami would occur, otherwise not. I will evacuate if there are tsunami indications and the earthquake is very strong"[6]

> "In the last experience, normally in the elementary schools, if an earthquake happened the parents picked up the children directly, so that schools handed over the children to them. (in context of potential arrangement that children's evacuation would be organized by the schools)"[7]

---

[3] Transcript P1 – 1:4 HH3.

[4] Transcript P 1- 1:4 HH4.

[5] Transcript P 6: – 6:2 BE.

[6] Transcript P 1- 1:9 HH4.

[7] Transcript P 9 – 9:1DP.

Moreover, the focus group discussion with the local actors in disaster management observed some other behaviors during evacuation in the past experience. The strong earthquake in addition to the warning given also triggered evacuation before warning issuance. Also, many people from the higher places or safe zones also evacuated and contributed to traffic and evacuation density.

The warning dissemination was coming from various private and public media, however, the public media was still perceived as weak. The household survey showed that only a small proportion of people received any warning by sirens (11.2 %) and some mentioned that the sound of the siren was not clearly heard. There have also been community response groups at the local level trained to help people in the neighbourhoods to evacuate; however, the number of the group members and their capacity is overwhelmed by the number of exposed people that should be supported. It was also emphasized that clear information from the authorities or warning centre is really needed in emergency situation in order for the people not to panic and able to conduct appropriate information.

24.5.4 Qualitative Findings on Tsunami Risk Perception and Risk Communication

It is still challenging for the people, especially the ones with lower socio-economic strength, to focus on reducing their risk to extreme natural hazard like tsunamis. They still deal daily with meeting the primary needs of their households. Different risk priority may often lead to responsive instead of preventive behavior. People would also expect "compensation" for their participation in preparedness activities, e.g. socialization or tsunami evacuation drills, and they also perceive such activities are economically costly as is conducting evacuation during strong earthquake events with uncertainty.

> "we heard a lot about tsunami from the media, but it is also like this in Padang, people just react or take action if there is an event, not much preparation before the event. If you would ask people in the coastal areas, they would think more about their daily life, what they are going to eat tomorrow – compared to worry about tsunamis. Why do the people not like to talk about natural disasters? We think it is in God's hands, what will happen happens"[8]

> ".. it is already difficult to deal with own livelihoods, let alone disaster preparedness socialization. Government may want to come to us, but the people would not participate, they still think about money to buy food".[9]

> "The people participated in government programs often times due to compensations, not because they are aware".[10]

[8] Transcript P 1- 1:6 HH2.

[9] Transcript 7 – 7:10 HB.

[10] Transcript P 2: – 2:2 KO.

People also relate their perception of tsunami risk and urgency to conduct preparedness action or evacuation is also related with their faith – or different interpretations of their faith. Some perceptions seem positive in responding to the ongoing preparedness efforts but others tend to lead to fatality behaviors.

> "some religious perceptions still influence the people's behavior: if we do good deeds, we will be protected. If we runaway, it means we are afraid/not faithful- this makes people ashamed to evacuate themselves immediately. We now try to cooperate with Ulama network but it is not easy"[11]

> "everybody (should) knows, being prepared is a command from God. That is indeed in God's hand, when disaster would happen, but we need to be in alert, do our best to save ourselves. Do not do suicide, we have to struggle".[12]

> "Sometimes people think that we are the ones who ask for tsunami disaster to come by preparing for it too much..people are scared..."[13]

Past experiences with strong earthquake with no tsunami occurrence may lessen the sense of urgency in responding to potential tsunami events. Additionally, long time experience of the coastal community, especially fishermen, with the sea shape their confidence of knowing better what could happen – or not happen.

> "In Padang there have been many earthquakes. Previously, people often escaped to the higher ground, but now people just stay at home or outside and pray that nothing would happen".[14]

> "after several earthquakes, people didn't want to stay in the coast, but now I think it is already normal as before again. The people living close to the coast are less fear than people living a bit further, we (people at the coast) know better about the sea from our experiences. I am not afraid, everything is in God's hand".[15]

> "the fishermen community feel that they know better about the sea and would not react to technology (TEWS) or only earthquake"[16]

With regard to the knowledge, as shown previously by the household survey analysis, not all people have been exposed to any socialization or have the same level of information needed for appropriate evacuation. Some respondents also mentioned they never had any socialization by the government or that the information given was still not clear.

> "please show us where to go to evacuate, there are already some maps, but they are not clear, not all people understand what the government announce about evacuation. Please conduct more socialization and provide more evacuation signs!"[17]

Various information channels at the local level like local radio, television, and mosques were expected. Additionally, the importance of community leaders in

---

[11] Transcript P 2 – 2:15 KO

[12] Transcript P 4 – 4:7 AT

[13] Transcript P 7 – 7:8 HB

[14] Transcript P 7 – 7:2 HB.

[15] Transcript P 1 – 1:11 HH4.

[16] Transcript P 2 – 2:16 KO.

[17] Transcript P1 – 1:1 HH2.

influencing people's decision is emphasized and their involvement in communicating risk or giving examples in appropriate behaviour is crucial.

> "one of the cultural values in the community is to pay respect to the community leaders. Many testimonies from the people mentioned that they evacuated if the leaders evacuated, we strongly suggest considering the capacity of community leaders and their knowledge of earthquake and tsunami".[18]

24.5.5 Capacities in Changing People's Perception and Behavior

Preparedness activities at the local level conducted, e.g. by the local NGOs and the Red Cross, have shown significant increase in level of awareness of the people. Activities are evident at the household level in pilot areas e.g. community-based response team building, socialization in mosque neighborhoods, in several pilot schools. However, the efforts have been localized in some pilot areas only and need institutional support from the city government, for example to provide legal basis and funding support for community efforts.

24.6 Discussions

Response to early warning and evacuation behavior comprises a combination of various different factors. As also concluded in other studies in other contexts, compliance with early warning and appropriate evacuation behavior is not to be taken for granted. This is also true for the case study of Padang city. An early warning has to be developed based on a good understanding of the context of the people at risk. In this matter, risk communication is the main key word. Raising awareness, changing perception and encouraging people to take preparedness actions has to take place way long before any (potential) tsunami events.

Clear and understandable information about the existing tsunami risk and existing early warning system is needed both prior to (through socialization, training and education activities) and during the potential tsunami event (through warning dissemination and clear guidance). This should encompass various options that people could do to prepare for potential events, such as family emergency plan, evacuation drills, as well as actions they are supposed to do in responding early warning. This has to match the information and instructions given accompanying an early warning of a potential tsunami event, so that people would tend to react based on their socialized knowledge.

[18] Transcript P10 – 10:1 KO.

Approach to the community is crucial here; especially considering that people have different risk priority and cultural values that shape their perception of tsunami risk and preparedness actions. Involvement of the local champions and leaders is necessary from the beginning. They are the ones who have to be empowered first. Socialization and training activities also have to take into consideration people's conditions, for example working time (in case of fishermen that have different rhythm of work) or possible compensation for loss of income due to such activities for certain groups with lower economic capacity.

In order to trigger appropriate actions, people also need to believe that such actions are do-able considering their capacity, thus, introduction should start from small actions such as family emergency plans, community initiatives to open evacuation path or widen the street voluntarily. But this has to be supported by concrete interventions at the city level by providing the necessary infrastructures and facilities, such as shelters, signs, and transportation assistance for special groups.

It is also noteworthy to see that past experiences with earthquakes may also undermine the urgency of evacuation in potential tsunami events. Socialization activities also need to take this into consideration and provide understanding on the necessity to conduct evacuation if instructed to do so in potential tsunami events, even though a tsunami may also not occur. However, in case of an extreme event with low frequency, the costs of preparedness and evacuation that have to be borne by the people also need to be reduced to that which is affordable in order to be sustainable.

Furthermore, the overall vulnerability framework helps to make clear that this component is interlinked with the other components and therefore, assessments as well as interventions have to take place on all components. Dynamic exposure will determine the constellation of exposed social groups with their behaviors and distributed household members in various locations that will determine their evacuation decision. Access to warning dissemination media will also have influence on the clearness of information received by those people upon which the evacuation decision and behavior will be made. And finally, positive response to the warning and intention to conduct evacuation has to be supported by sufficient infrastructures and facilities.

24.7 Conclusion

The study proposes a framework to assess people's vulnerability in the specific context of tsunami early warning and evacuation. In this paper, the parameters that should be considered in risk communication are identified: perception of tsunami risk and evacuation influenced by individual and cultural values, knowledge of evacuation, quality and clarity of information given prior to any potential event and in the warning message, arrangements and emergency plans at the household and community level.

Additionally, interventions related to risk communication also need to seek and promote the following capacity at the community level: existence of government programs for preparedness as well as champions, such as local NGOs and Red Cross that initiate efforts at the community level, acceptance of the community towards the programs and their active participation, capacity and involvement of community and religious leaders, institutional supports such as legal basis and funding scheme for preparedness activities at the community level, but finally, the need to reduce baseline socio-economic vulnerability of the people is inevitable.

Acknowledgement The author would like to thank the DFG/BMBF special Programme "Geotechnologies" – Early Warning Systems in Earth Management for funding the research (Sponsorship code: 03G0666A-E). The valuable contributions of the local experts and people in Padang, respected colleagues in Last-Mile Project and UNU-EHS especially P.D. Dr. Joern Birkmann and Niklas Gebert in providing their insights within the study are also acknowledged and highly appreciated.

References

Ajzen I (1991) The theory of planned behavior. Organ Behav Hum Decis Process 50:179–211

Baker EJ (1991) Hurricane evacuation behavior. Int J Mass Emerg Disasters 9(2):287–310

Bhatti A (2001) Risk perception, culture and communication: a South Asian experience. In 5th conference of the human reaction and risk perception to catastrophic events: a psychological and cultural perspective (disaster and social crisis research network. Session 6), Helsinki, 28 Aug-1 Sep. 2001

Birkmann J (2006) In: Birkmann J (ed) Measuring vulnerability to natural hazards – towards disaster resilient societies. United Nations University, Tokyo

Birkmann J, Setiadi N, Gebert N (2008) Socio-economic vulnerability assessment at the local level in context of tsunami early warning and evacuation planning in the City of Padang, West Sumatra. Paper for international conference on tsunami warning, Nusa Dua, 8–11 Nov. 2008

Bogardi JJ, Birkmann J (2004) In: Malzahn D, Plapp T (eds) Disasters and society – from hazard assessment to risk reduction. Logos Verlag, Berlin

Borrero JC, Sieh K, Chlieh M, Synolakis CE (2006) Tsunami inundation modeling for western Sumatra. Proc Natl Acad Sci 103(52):19673–19677

Buckle P (1998) Re-defining community and vulnerability in the context of emergency management. Aus J Emerg Manag 13(4):21–26

Cardona O (1999) Natural disaster management. In: Ingleton J (Hrsg). International decade for natural disaster reduction (IDNDR), Tudor Rose

Cardona O (2001) Estimación Holística del Riesgo Sísmico Utilizando Sistemas Dinámicos Complejos. Estimación Holística del Riesgo Sísmico Utilizando Sistemas Dinámicos Complejos

Cutter SL (1996) Vulnerability to environmental hazards. Prog Human Geogr 20(4):529–539

Cutter SL, Emrich CT, Webb JJ, Morath D (2009) Social vulnerability to climate variability hazard: a review of the literature. Final report to Oxfam America. Hazard and Vulnerability Research Institute, University of South Carolina, Columbia

Gebert N (2011) Risk and vulnerability assessment. Contributions to LIPI/DLR/UNU-EHS guideline for tsunami risk assessment in Indonesia: scientific proposal for practitioner and end users

Gregg C, Houghton B, Paton D, Johnston D, Swanson D, Yanagi B (2007) Tsunami warnings: understanding in Hawai'i. Nat Hazards 40:71–87

Hidayati D, Permana H, Pribadi K, Ismail F, Meyers K, Widayatun A et al (2006) Assessing and recognizing community preparedness in natural disasters in Indonesia. Kajian Kesiapsiagaan Masyarakat Dalam Mengantisipasi Bencana Gempa Bumi dan Tsunami di Indonesia. LIPI-UNESCO/ISDR, Indonesia

Lachman R, Tatsuaoka M, Bonk WJ (1961) Human behavior during the tsunami of May 1960. Science 133(3462):1405–1409

Lindell MK, Lu J-C, Prater CS (2005) Household decision making and evacuation in response to Hurricane Lili. Nat Hazard Rev 6(4):171–179

Martin IM, Bender H, Raish C (2007) What motivates individuals to protect themselves from risks: the case of wildland fires. Risk Anal 27(4):887–900

McCloskey J, Lange D, Tilmann F, Nalbant SS, Bell AF, Natawidjaja DH et al (2010) The September 2009 Padang earthquake. Nat Geosci 3(2):70–71

Mileti DS, O'Brien PW (1992) Warnings during disaster: normalizing communicated risk. Soc Probl 39(1):40–57

Paton D, Ronan K, Johnston D, Smith L, Johnston M (2003) Paper Number 166. Responding to earthquake hazards: promoting household resilience and preparedness

Riad J, Norris F (1998) Hurricane threat and evacuation intentions: an analysis of risk perception, preparedness, social influence, and resources. Disaster Research Center University of Delaware. Newark, Delaware

Rogers R (1983) Cognitive and physiological processes in fear appeals and attitude change: a revised theory of protection motivation. In: Cacioppo J, Petty R (eds) Social psychophysiology. Guilford Press, New York

Santos G, Aguirre BE (2005) In: Gaithersburg M, Peacock RD, Kuligowski ED (eds) Critical review of emergency evacuation simulation models. University of Delaware. Newark, Delaware, pp 27–52

Shah HC (2006) The last mile: earthquake risk mitigation assistance in developing countries. Philos Trans Royal Soc A 364:2183–2189

Shaw R, Hirohide KS, Kobayashi M (2004) Linking experience, education, perception and earthquake preparedness. Disaster Prev Manag 13(1):39–49

Sorensen JH (2000) Hazard warning systems: review of 20 years of progress. Nat Hazards Rev 1(2):119–125

Taubenböck H, Goseberg N, Lämmel G, Setiadi N, Schlurmann T, Nagel K, Siegert F, Birkmann J, Traub KP, Dech S, Keuck V, Lehmann F, Strunz G, Klüpfel H (2012) Risk reduction at the "Last-Mile": an attempt to turn science into action by the example of Padang, Indonesia. Natural Hazards (online first, doi:10.1007/s11069-012-0377-0)

Tierney KJ, Lindell MK, Perry RW (2001) Facing the unexpected. Joseph Henry Press, Washington, DC

UN/ISDR (2004) Living with risk. A global review of disaster reduction initiatives. United Nations, International Strategy for Disaster Reduction (UN/ISDR), Geneva

UN/ISDR (2006) Developing early warning systems: a checklist. EWC III Third International Conference on Early Warning From Concept to Action, 27–29 March 2006, Bonn, Germany. United Nations, International Strategy for Disaster Reduction

UN/ISDR (2007) Words into action: a guide for implementing the Hyogo Framework. Hyogo Framework for action 2005–2015: building the resilience of nations and communities to disasters

Villagran de Leon JC, Bogardi J, Dannenmann S, Basher R (2006) Early warning system in the context of disaster risk management. Entwicklung Ländlicher Raum 2:23–25

Chapter 25
The Great East Japan Earthquake and Tsunami Aftermath: Preliminary Assessment of Carbon Footprint of Housing Reconstruction

Cui Pan, Haibo Wang, Shaopeng Huang, and Haihua Zhang

Abstract The 9.0-magnitude Great East Japan Earthquake and the subsequent tsunami on March 11 of 2011 was the most devastating natural disaster ever to have hit Japan in its recent history. The aftermath of this catastrophic event is extensive, including a negative impact on the world-wide effort in mitigating global warming by carbon emission reduction. Not only has the Fukushima nuclear crisis dramatically slowed down nuclear power development as a low carbon alternative energy, but also the infrastructure reconstruction would come with a heavy carbon footprint. According to the statistic data released by the Japanese government, over 396,067 buildings were fully or partially destroyed, and more than 728,583 partially damaged in this disaster. The building reconstruction must consume large amounts of construction materials and energy, which would inevitably increase the greenhouse gas (GHG) emission. In this study, we identify housing construction, land use conversion to settlement, and earthquake/tsunami wreck clearance as three major GHG emission related processes of post disaster building reconstruction. We derive a first order estimate of the embodied GHG emission of these three processes based on official statistic data of damages, the quantification of major resources required for typical Japanese residential building types, and the IPCC Guidelines for National Greenhouse Gas Inventories. The result shows that the post Great East Japan Earthquake housing reconstruction alone has at least an emission of 26.3 million tons CO_2 equivalent, or about 2.1 % of the total greenhouse gas emission of Japan in 2010.

C. Pan • H. Wang
Xi'an Jiaotong University, Xi'an, China

S. Huang (✉)
Xi'an Jiaotong University, Xi'an, China

University of Michigan, Ann Arbor, MI 48109, USA
e-mail: shaopeng@mail.xjtu.edu.cn

H. Zhang
Kobe University Graduate School of Engineering, Kobe, Japan

Y.A. Kontar et al. (eds.), *Tsunami Events and Lessons Learned: Environmental and Societal Significance*, Advances in Natural and Technological Hazards Research 35,
DOI 10.1007/978-94-007-7269-4_25,

Keywords Carbon footprint • Energy consumption • Great East Japan Earthquake • Housing reconstruction

25.1 Introduction

The magnitude of 9.0 on the Richter scale earthquake off the Pacific coast of Japan, on 11 March, 2011 was the third most powerful earthquake ever recorded in the world, after the magnitude 9.5 Chile earthquake in 1960 and the magnitude 9.2 Alaska earthquake in 1964. The epicenter of this devastating earthquake, also known as the Great East Japan Earthquake, is approximately 70 km east of the Oshika Peninsula of Tōhoku and the hypocenter is about 32 km beneath the seafloor. This earthquake involved a magathrust fault where the Pacific Plate subducts below Eurasia Plate at a rate of 8–8.5 cm/year.

The fault plane of the Great East Japan Earthquake is estimated to be 440 × 220 km (Ide et al. 2011). Geophysical observations indicate that the distribution of coseismic fault slip exceeded 50 m in several places (Simons et al. 2011). The widespread displacement of the seafloor produced a destructive tsunami which swept the Sanriku coast of northern Japan and rolled through the entire Pacific Ocean. The enormous tsunami waves reached up to 40.5 m in the highest and penetrated up to 10 km inland (Brakenridge et al. 2013), causing tremendous casualties and massive losses in properties to Japan. The statistic data released by Japanese government confirms that a total of 15,873 people were killed and 2,744 remained missing nationwide as of November 21, 2012 due to the earthquake and tsunami (Japan National Police Agency 2012).

Building damages are among the most extensive phenomena in the disaster areas (Fig. 25.1). Across the 22 prefectures listed in the Japan National Police Agency damage report (Japan National Police Agency 2012), 129,627 buildings were totally collapsed, 266,440 buildings were half collapsed, and 728,583 buildings were partially damaged. The earthquake and tsunami also caused extensive and severe structural damage to roads, bridges, dikes, railways, and more significantly, the level 7 meltdowns of three reactors of the Fukushima Daiichi Nuclear Power Plant (Norio et al. 2011).

There have been a considerable amount of studies on the geological, social, environmental, and financial effects of the Great East Japan Earthquake (Kosaka et al. 2012; Morimoto 2012; Takahashi et al. 2012; Wang et al. 2012). But little attention has been paid to the consequence of this natural disaster on the worldwide effort to reduce greenhouse gas emissions as a measure of mitigation of and adaptation to global warming. The consequence is extensive: the Fukushima nuclear crisis has slowed down nuclear power development as a low carbon alternative energy; the recovery/replacement of the vast farmland ruined by tsunami, as well as the reconstruction of the infrastructure in the disaster areas would increase greenhouse gas emission. In this study, we provide a preliminary estimate of the carbon footprint of post disaster residential housing reconstruction.

Fig. 25.1 Photos of Rikuzentakata of Iwate prefecture in northern Japan before (*upper panel*) and immediately after (*lower panel*) the Great East Japan Earthquake. Photo credit: AP Photo/ Kyodo News

25.2 Review of Carbon Footprint Analysis

Carbon footprint is commonly used as a measure of the impact of a given product or process on the environment in terms of the amount of greenhouse gases produced (Wiedmann and Minx 2008). With global warming mitigation and adaptation high up on the scientific and political agenda, various approaches, ranging from online calculators to sophisticated life-cycle analysis methods and tools have been developed. These methods and tools can be classified into two main categories (Pandey et al. 2011; Wiedmann and Minx 2008): process based bottom-up analysis and input-output based top-down analysis.

The input-output based analysis has been widely used in the calculations of the embodied carbon emissions of various processes and in international trade (Wiedmann et al. 2007). An input-output based calculation of carbon footprint is usually presented in the form of inputs–outputs tables to establish carbon emission estimates in a comprehensive and robust way, taking into accounts all higher order impacts (Wiedmann and Minx 2008). For example, Suzuki and colleagues (Suzuki and Oka 1998; Suzuki et al. 1995) quantified the total amount of energy consumption and CO_2 emission during the construction, operation, maintenance, and renovation of residential and office buildings in Japan; Wiedmann (2009) compared energy footprints embodied in the trades from and to the UK in 2002; Liu et al. (2010) examined CO_2 emissions embodied in Japan-China trade.

A process-based analysis of carbon footprint uses traditional life cycle assessments to model all the activities related to a product or a service. For this method, all materials and energy consumption related to the activities during the life cycle of the product/process are considered. In reality, however, it is not always possible to reach the bottom or boundary in such a bottom-up approach. Very often only on-site, first-order, and some second-order impacts can be identified. Therefore, a process-based carbon footprint estimate is more conservative than is an input-output-based one. Process-based method has been applied in the assessment of carbon emissions on different scales. Börjesson and Gustavsson (2000) focused on carbon emissions of building material manufacture; Gustavsson and Sathre (2006) analyzed factors that contribute to the variation of energy and CO_2 balances of building material lifecycles of wooden and concrete-framed buildings; You et al. (2011) and Yan et al. (2010) investigated the carbon emissions in the life cycle of building construction in Hong Kong and mainland China, respectively.

The purpose of this study is to provide a first order assessment of the carbon footprint of housing reconstruction of the Great East Japan Earthquake and Tsunami (GEJET hereafter). We do not intend to make a comprehensive full life cycle analysis of building footprint. Instead, we focus our analysis on three major processes of the post-disaster housing reconstruction: clearance of earthquake/tsunami wrecks, possible conversion of certain amount of various land use types to settlement, and the actual housing construction. Methodologically, our analysis is first order process based. However, in estimating construction emissions, our calculation is based on the Japanese housing specific input-output based results of Suzuki and colleagues (Suzuki and Oka 1998; Suzuki et al. 1995).

25.3 Sources of CO_2 Emissions

Under the 2006 IPCC guidelines for national greenhouse gas (GHG) inventory, at the top level, sources and sinks are classified as four categories (IPCC 2006): (1) energy, (2) industrial processes and product uses (IPPU), (3) agriculture, forestry and other land use, and (4) waste. Housing constructions embraces all four categories of GHG emissions. However, a building life cycle analysis usually divides building construction into the following processes: land footprint, building construction, building use/operation, and building demolition and waste management (Gong et al. 2012; Rossi et al. 2012; Taborianski and Prado 2012).

There have been several researches on the carbon footprint of housing construction in different settings, adhering to the IPCC guidelines. For example, Cuéllar-Franca and Azapagic (2012) presents a full life cycle assessment for the three most common types of housing in the United Kingdom: detached, semi-detached and terraced. Their analysis includes house construction, use, and demolition after 50 years. Their result shows that the use stage has the largest contribution to most environmental impacts. In general, housing construction in the UK has an average footprint in the range of 309–455 t of CO_2 eq., around 90 % of which are from house

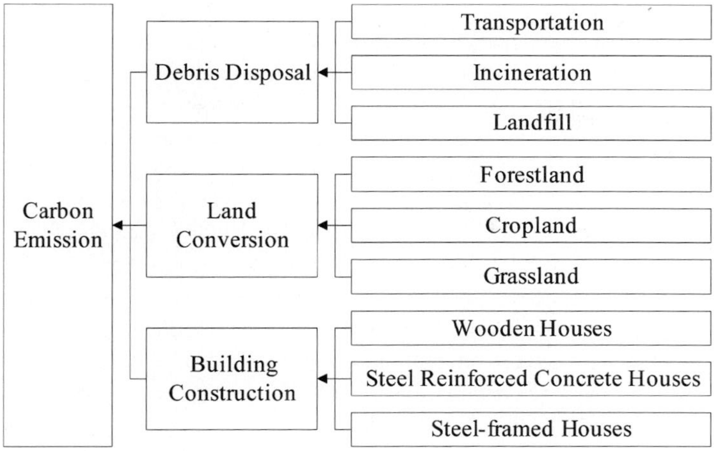

Fig. 25.2 Major sources of carbon emission in the post disaster reconstruction of residential buildings

use, 9 % from construction, and 1 % from the end-of-life waste management. The result of the UK case study is consistent with the analyses from several other regions (Gong et al. 2012; Rossi et al. 2012; Takahashi et al. 2012).

For a first order estimate of the GHG emission impact of the post GEJET disaster housing reconstruction, we do not include house use in our analysis, because house use is not special of the GEJET. In terms of GHG emissions, of particular importance to housing reconstruction after a disaster like the GEJET are the energy consumption for the debris removal and management, possible changes in the soil carbon stack due to land use and land cover change for housing construction, and the embodied GHG emission of the reconstruction of the ruined housing (Fig. 25.2).

One of the major relief and recovery activities after a destructive disaster is debris removal and management. It is also the first step of post disaster building reconstruction. Debris removal and management belong to the waste category in the 2006 IPCC guidelines (IPCC 2006). There are various disaster debris management options (Brown et al. 2011), such as temporary staging, recycling, open burning, and land reclamation. Different debris treatment options would result in different carbon emission intensities. In the case of the GEJET, wrecks are managed mainly through four approaches (Ministry of the Environment of Japan 2011): recycling, selling, incineration, and landfill. Recycling can reduce incineration and landfill, and is expected to play a key role in waste treatment. However, the overall effects of recycling on waste emission and environmental load are still not clear. For example, on the one hand the melting and solidification of incineration ash can help to reduce landfill consumption; on the other hand recycling process can emits flyash, CO_2, and other substances that load the environment (Nakamura and Kondo 2002). Therefore, the GHG emission from waste recycling is not covered in our calculation of debris disposal analysis. Selling of the wastes is also neglected in our first order assessment due to their relatively small share in the emission budget (Greenhouse Gas Inventory Office of Japan 2012). What is important to this analysis is the transportation of the earthquake/tsunami wrecks from the affected areas to treatment sites, which is energy and emission intensive. Transportation emission comes mainly from fuel combustion of vehicles traveling between the affected areas to solid waste landfills and waste incinerations.

Land use conversion from some previously non-residential areas to new residential areas is inevitable for the post GEJET housing reconstruction because some disaster affected areas may become unsuitable for residential building construction. In the 2006 IPCC guidelines, land use is classified into six categories, including forest, farmland, grassland, wetland, populated place, and others. Among the six categories of land use, forest is the most biomass intensive and carbon negative. Grassland, wetland, farmland, and others are less; and populated place is the least in carbon contents. A conversion among different land uses will change the soil carbon pool and result in carbon emission or removal (Greenhouse Gas Inventory Office of Japan 2012). The Japanese national GHG inventory data shows that wetland and others land types are unlikely to be converted to populated type. Therefore, new populated areas for the post GEJET reconstruction must be converted from forest, grassland, and farmland. The possible carbon emissions of the related land use conversion are evaluated in this study.

Building construction is the main source of carbon emission in the post disaster relief and recovery. Cole (1998) provides a detailed examination of the energy consumption and greenhouse emissions associated with building construction. The direct and indirect emission sources include manufacture of building materials, transportation of the building materials to construction sites, and on-site assembly of the houses. Consumption of manpower, water, fuels, electricity, and building materials will lead to carbon emission. Building material manufacture and transportation are two sub processes with most intensive carbon dioxide emission in this process (Gong et al. 2012; Rossi et al. 2012; Suzuki et al. 1995). Material manufacture includes all related emissions in factories and offices during manufacture process. Energy supply is another important source of CO_2 emission. Combustions of coal, gasoline, diesel, liquefied petroleum gas, coke and city gas all belong to this section. The fugitive emission of fuel in industrial facilities and the related end-use of fuel product are classified to IPPU. The manufacture of construction materials, such as cement, lime and glass production processes are also subclasses of IPPU. GHG emission from transportation comes from fuel combustion of vehicles in transporting the construction crew for the duration of their construction task, transporting building products from distribution centers to the building sites, and transporting construction equipment.

25.4 Data

25.4.1 Debris

The GEJET has generated an enormous amount of debris. Listed in Table 25.1 are the statistics of housing damage and related wastes based on the Japanese

Table 25.1 Statistics of the ruined residences and resultant debris of the Great East Japan Earthquake and Tsunami (Japan National Police 2012)

Prefecture	Total ruined residence (Door)	Half ruined residence (Door)	Partially damaged (Door)	Debris (kt)
Aomori	308	701	958	181
Iwate	19,199	5,043	8,784	4,755
Miyagi	85,331	151,768	224,124	15,691
Fukushima	21,034	72,110	162,491	1,726
Ibaraki	2,620	24,158	183,675	342
Chiba	799	10,024	52,026	82
Tochigi	260	2,109	72,143	n/a
Saitama	24	199	1,800	n/a
Tokyo	15	198	4,847	n/a
Yamagata	37	80	n/a	n/a
Kanagawa	n/a	39	445	n/a
Gunma	n/a	7	17,246	n/a
Hokkaido	n/a	4	7	n/a
Akita	n/a	n/a	3	n/a
Niigata	n/a	n/a	17	n/a
Shizuoka	n/a	n/a	13	n/a
Yamanashi	n/a	n/a	4	n/a
Total	129,627	266,440	728,583	22,777

Note: n/a indicates data not available or negligible

government released data (Japan National Police Agency 2012). There are guidelines for waste management after a large natural disaster (Kofoworola and Gheewala 2009; Nakamura and Kondo 2002; UNEP 2012). Different wastes are expected to be treated by different waste management methods and carry different GHG emission factors. However, the details about the categorical composition of the debris wastes generated by the GEJET are not available. We use the data of normal waste composition to estimate the amounts of the GEJET wastes applicable of different management methods. The 2012 National Greenhouse Gas Inventory Report of Japan (Greenhouse Gas Inventory Office of Japan 2012) shows that at the national level, 48 % of wastes are for recycle, 39 % for incineration, and 13 % for landfill. This proportion of waste management is used in our analysis of waste emissions.

For the GHG emissions from municipal solid waste incineration, Chen and Lin (2010) analyzed data from three incinerators in Taipei and find an average rate of 964 kg-CO_2-eq emission per ton of waste after the wastes were transported to the incinerators, similar to the result of Johnke (1999) which found that the incineration of 1 Mg municipal waste would release 0.7–1.2 Mg CO_2. For GHG emission from landfill, a conservative estimate of methane generation is about 50 Nm^3 per ton of waste disposed, or 16.836 t-CO_2-eq/t (Themelis and Ulloa 2007).

Although some portion of debris might be transported long distance for special management, most the wrecks are expected to be managed locally. Trucks are expected to be used in the transportation of the GEJET debris. Based on data from 2000 Japan Federation of Freight Industries, Takahashi (2005) estimates the

Table 25.2 The average distances from downtown to incinerators in Iwate and Miyagi

Prefecture	City	Distance from downtown (km)	Management capability (t/day)
Iwate	Kuji	2.0	2.5
	Miyako	1.6	150
	Kamaishi	2.7	160
	Ofunato	2.2	750
	Ichinoseki	11.6	300
Miyagi	Kesen-Numa	8.0	400
		3.8	12
		16.0	300
	Ishinomaki, Higashi-Matsushima and Onagawa	1.8	1589
	Ishinomaki	5.0	4
	Tagajyo, Shichigahama, Matsushima and Rifu	3.0	20
	Sendai	8.0	480
	Natori	5.7	190
	Iwanuma	5.3	195
	Watari	3.6	525
	Yamamoto	3.2	300

Data source: http://garekikouiki-data.env.go.jp, retrieved on December 20, 2012

amount of CO_2 emission per ton-kilometer of freight (business truck over 3 t) in Japan to be 174 g-CO_2/(t · km). In order to quantify the carbon emission of the earthquake/tsunami wreck clearance, we will need to estimate the transportation distance of the wrecks. We assumed the transportation distance to be 5 km based on the distances between downtown areas and incinerators in Iwate and Miyagi, so far the only two prefectures that have made such data publicly available on the internet (Table 25.2).

25.4.2 Land Use Conversion

Expansion of populated space has been a significant source of GHG emission in Japan over the recent years. The land use type categorized as populated place in the IPCC guidelines is categorized as settlement land in the 2012 National Greenhouse Gas Inventory Report of Japan. Overall, Japan has seen a steady decline in the area of land conversion to settlements over the past two decades, from an annual total of 43.8 kha in 1990 to 14.5 kha in 2009 (Greenhouse Gas Inventory Office of Japan 2012). There was a rebound in the land conversion in 2010 which recorded a total of 19.3 kha converted from forest, cropland, and grassland to settlements. According to the Report, the net emission of settlement increased in 2010 compared to 2009, mainly because forestland converted to settlement increased by 119 %. Post GEJET housing reconstruction will inevitably intensify the land conversion to settlement in the disaster affected areas. In this analysis, we assume that a size of about 20 % of the total reconstruction area of non settlement land will be converted to settlement with the same proportions of land conversions as shown in the 2012 national GHG inventory report. This assumption is based on the estimate that 10 % of the total and half destroyed houses will be relocated to newly converted land and the area of a residential lot is substantially greater than the floor area of a house (Sotoma et al. 2003).

25.4.3 Housing Construction

The GEJET related housing damages incurred in 17 prefectures, among which Miyagi, Fukushima, Iwate, Ibaraki, Chiba, and Anomori were hit most heavily. The main types of housing in Japan are multi-family steel reinforced concrete houses, wooden single family houses, and lightweight steel structure single-family steel framed houses. The average construction areas of these three housing types are 128.25, 150, and 146.5 m^2 for wooden houses, steel reinforced concrete houses, and steel framed houses, respectively (Gerilla et al. 2007; Suzuki et al. 1995). We combine the numbers of the total and half ruined residences (Table 25.1) and the proportion of each type of houses derived from the E-stat 2008 report to estimate the total post GEPET housing reconstruction areas (Table 25.3) (E-stat 2008). For a conservative assessment, we assume that only the total and half collapsed houses were to be reconstructed. In our calculation of GHG emission, we use CO_2 emission per construction area from Suzuki et al. (1995). In their study, an input-output life cycle analysis approach was applied to quantify the total energy consumptions and GHG emissions directly and indirectly related to the construction of various types of houses in Japan. CO_2 emissions resulting from the three types of housing construction are estimated to be 250, 850, and 400 kg-CO_2-eq/m ^{2}, respectively (Table 25.4).

25.5 Emission Calculation

25.5.1 Carbon Emission from Debris Disposal ($E_{Disposal}$)

In this study, the GHG emission of waste disposal covers transportation, incineration, and landfill. It is assessed by

$$E_{Disposal} = \sum M_i \cdot C_{Trans} + \sum M_i \cdot 39\% \cdot C_{Incin} + \sum M_i \cdot 13\% \cdot C_{Landfill} \quad (25.1)$$

where $E_{Disposal}$ is carbon emission generated from debris disposal; M_i is the mass of debris generated by the disaster in the given prefecture i; C_{Trans} is the unit volume of

Table 25.3 Estimates of the reconstruction areas of the three common types of housings of Japan

	Wooden houses		Steel reinforced concrete houses		Steel framed houses	
Prefecture	Proportion %	Reconstruction area (m^2)	Proportion %	Reconstruction area (m^2)	Proportion %	Reconstruction area (m^2)
Aomori	90	116,464	8	12,108	2	2,956
Iwate	86	2,673,771	10	363,630	3	106,544
Miyagi	67	20,373,324	6	2,133,891	6	2,084,100
Fukushima	79	9,437,117	15	2,095,740	5	682,280
Ibaraki	76	2,610,052	15	602,505	8	313,838
Chiba	60	832,830	33	535,739	7	110,990
Total	–	36,043,558	–	5,743,613	–	3,300,708

Table 25.4 Construction CO_2 emission of different residential houses

Housing type	Construction area (m^2)	CO_2 emission intensity (kg CO_2-eq/m^2) (Suzuki et al. 1995)	CO_2 emission (Mt)
Wooden houses	36,043,558	250	9.01
Steel enhanced concrete houses	5,743,612	850	4.88
Steel framed houses	3,300,708	400	1.32
Total	45,087,879	–	15.21

CO_2 emission from waste transportation, taken as 174 g-CO_2-eq/(t · km) (Takahashi 2005); C_{Incin} is the unit volume of CO_2 emission from waste incineration, 0.964 t-CO_2-eq/t (Ministry of the Environment of Japan 2011); $C_{Landfill}$ is the unit volume of CO_2-eq emission from landfill, 16.836 t-CO_2-eq/t (Themelis and Ulloa 2007); the 39 % and 13 % are the proportions of wastes for incineration and landfill respectively (Greenhouse Gas Inventory Office of Japan 2012).

25.5.2 *Land Use Emissions* ($E_{Landuse}$)

Calculation of carbon stock change from land use conversion requires the area of individual land types to be converted to settlement and the related GHG emission factors for land use conversion. Those data are not available at this stage. However, for the first order estimate in this study, we assumed that the post GEJET housing reconstruction will require a settlement land conversion of the size equivalent to 20 % of the reconstruction area, and the abandoned sites will remain within settlement area. The carbon emission from land conversion to settlement for post disaster housing reconstruction can be estimated by

$$E_{Landuse} = A_{conver} \cdot C_{Landconv} \tag{25.2}$$

where $E_{Landuse}$ is carbon emission from land use conversion for post GEJET housing reconstruction; A_{conver} is the total area of the land converted to settlement, equal to the floor plate of reconstructed buildings; $C_{landvonv}$ is the composited CO_2 equivalent emission factor associated with the land conversion to settlement, 183 t-CO_2-eq/ha (Greenhouse Gas Inventory Office of Japan 2012).

According to the National Greenhouse Gas Inventory Report of Japan, a total of 19.3 kha of land was converted to settlement in Japan in 2010: 10.8 kha from forest land, 7.2 kha from cropland, and 1.3 kha from grassland. The 19.3 kha land conversion to settlements was responsible for the release of GHGs of 3.53 Mt CO_2-eq. We use the 2010 statistics to define the parameter $C_{Landuse}$ in Eq. (25.2), implying that the composition of the land to be converted to settlement land for the post GEJET housing reconstruction would remain the same as the new housing construction in 2010.

Based on the data of ruined buildings (Table 25.1) and the composition of housing types in the disaster affected areas along with the average construction area of different dwellings, the total floor area of housing reconstruction can be estimated as

$$A_{\text{Floor}} = H_{count} \cdot \sum (c_k \cdot a_k) \tag{25.3}$$

where A_{Floor} is the total floor area of housing reconstruction buildings; H_{count} is the total number of the ruined buildings; k is the type of houses; c_k is the composition of housing of the main affected areas; a_k is the average construction area of different dwellings.

25.5.3 *Carbon Emission in Building Construction (E_{Const})*

Suzuki and colleagues assessed Japan specific housing construction footprint using the Input/Output top-down analysis method (Suzuki and Oka 1998; Suzuki et al. 1995). Based on their assessment of construction GHG emission factors, the post GEJET housing reconstruction can be estimated by

$$E_{Constr} = \sum (A_l \cdot C_{Constr,l}) \tag{25.4}$$

where E_{Constr} is the carbon emission of building construction; l refers to the housing types (e.g., wooden houses, steel enhanced concrete houses, and steel framed houses); A_l is the reconstruction area of l type housing; $C_{Constr,l}$ is the average CO_2 emission per unit construction area of l type housing (Table 25.4) (Suzuki et al. 1995).

25.5.4 *Total Carbon Emissions in the Residential Building Reconstruction (E_{Total})*

The total carbon emission in the residential building reconstruction is the sum of the three components discussed above:

$$E_{Total} = \sum \left(E_{Disposal} + E_{Landuse} + E_{Constr}\right) \tag{25.5}$$

25.6 Result and Discussions

The 9.0-magnitude Great East Japan Earthquake and the subsequent tsunami have an extensive effect on the world-wide carbon reduction effort to mitigate global warming. Our preliminary analysis of the three fundamental processes of the post

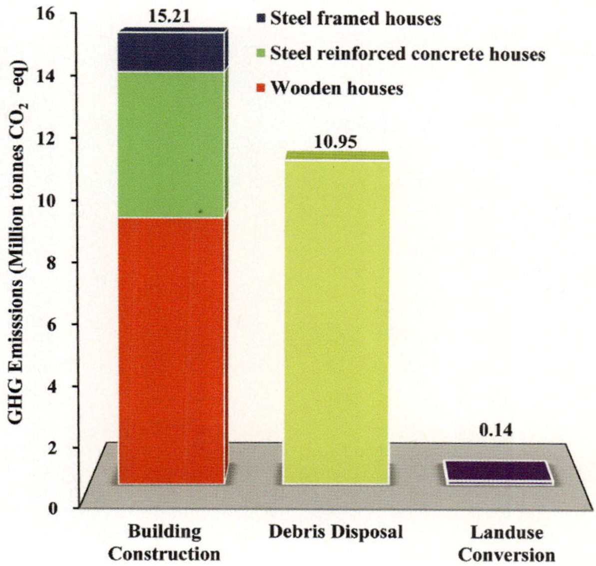

Fig. 25.3 The carbon footprint of the three major processes of the post Great East Japan Earthquake and Tsunami housing reconstruction

disaster reconstruction – debris disposal, land conversion to new settlement for housing relocation, and building construction – shows that the aftermath residential housing reconstruction has a total GHG emission of 26.30×10^6 t CO_2 equivalent (Fig. 25.3). The major source of the GHG emission in the post GEJET housing reconstruction is building construction which has an emission of 15.21 Mt CO_2-eq, accounting for 57.84 % of the total emission estimate. The GHG emission from debris disposal is 10.95 Mt CO_2-eq, a share of 41.62 % of the total. The CO_2 emission from land conversion for ruined housing relocation is estimated to be 0.14 Mt CO_2-eq, a small fraction of the total estimate. The GHG emissions of the reconstruction of the ruined wooden houses, steel reinforced concrete houses, and steel framed houses are 9.01 Mt CO_2-eq, 4.88 Mt CO_2-eq, and 1.32 Mt CO_2-eq respectively.

The Japanese government agreed to the target given by the Kyoto Protocol to reduce its greenhouse gas emissions by 6 % of 1990 levels within the Kyoto Protocol first compliance period from 2008 to 2012. Given the GEJET disaster, the likelihood of Japan meeting its Kyoto obligations has become slime. Our estimation of 26.30 Mt CO_2-eq housing reconstruction emission is equivalent to 2.18 % of the 1204.9 Mt CO_2-eq 1990 national emission level (Fig. 25.4). Even if this amount of emission would be spread over a couple of years, it is another burden that Japan can hardly overcome with respect to fulfilling its Kyoto Protocol commitment.

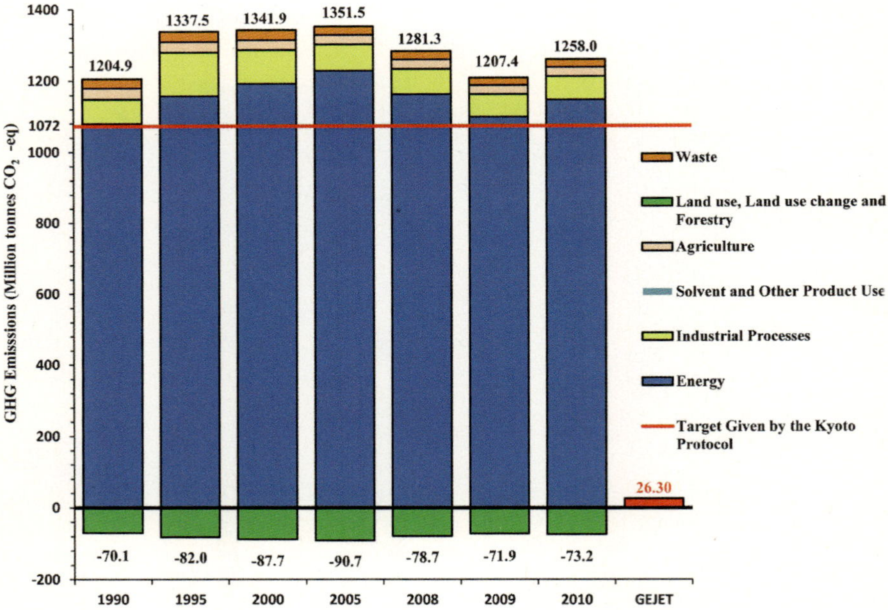

Fig. 25.4 Changes in Japan's national greenhouse gas emission since 1990 (Greenhouse Gas Inventory Office of Japan 2012)

Acknowledgements This study is supported in part by State Key Laboratory of Loess and Quaternary Geology through grant SKLLQG1006. The authors acknowledge the assistances of Ren Yongfei, Wang Jiayuan, Wang Xuezhi, and Peng Fen of the Geothermal and Environmental Research Laboratory of the Xi'an Jiaotong University.

References

Börjesson P, Gustavsson L (2000) Greenhouse gas balances in building construction: wood versus concrete from life-cycle and forest land-use perspectives. Energy Policy 28(9):575–588

Brakenridge GR, Syvitski JP, Overeem I, Higgins SA, Kettner AJ, Stewart-Moore JA, Westerhoff R (2013) Global mapping of storm surges and the assessment of coastal vulnerability. Natural Hazards 66(3):1295–1312

Brown C, Milke M, Seville E (2011) Disaster waste management: a review article. Waste Manag 31(6):1085–1098

Chen T-C, Lin C-F (2010) CO_2 emission from municipal solid waste incinerator: IPCC formula estimation and flue gas measurement. J Environ Eng Manage 20(1):9–17

Cole RJ (1998) Energy and greenhouse gas emissions associated with the construction of alternative structural systems. Build Environ 34(3):335–348

Cuéllar-Franca RM, Azapagic A (2012) Environmental impacts of the UK residential sector: life cycle assessment of houses. Build Environ 54:86–99

IPCC 2006, 2006 IPCC Guidelines for National Greenhouse Gas Inventories, Prepared by the National Greenhouse Gas Inventories Programme, Eggleston H.S., Buendia L., Miwa K., Ngara T. and Tanabe K. (eds). Published: IGES, Japan.

E-stat (2008) Dwellings by construction material (5 Groups)—Japan, Prefecture and 18 Major Cities, edited

Gerilla GP, Teknomo K, Hokao K (2007) An environmental assessment of wood and steel reinforced concrete housing construction. Build Environ 42(7):2778–2784

Gong X, Nie Z, Wang Z, Cui S, Gao F, Zuo T (2012) Life cycle energy consumption and carbon dioxide emission of residential building designs in Beijing. J Ind Ecol 16(4):576–587

Greenhouse Gas Inventory Office of Japan (2012), National greenhouse gas inventory report of JAPAN edited by Japanese Ministry of the Environment, National Institute for Environmental Studies, Tokyo, Japan

Gustavsson L, Sathre R (2006) Variability in energy and carbon dioxide balances of wood and concrete building materials. Build Environ 41(7):940–951

Ide S, Baltay A, Beroza GC (2011) Shallow dynamic overshoot and energetic deep rupture in the 2011 Mw 9.0 Tohoku-Oki earthquake. Science 332(6036):1426–1429

Japan National Police Agency (2012) Damage situation and police Countermeasures associated with 2011 Tohoku district—off the Pacific Ocean Earthquake. http://www.npa.go.jp/archive/keibi/biki/higaijokyo_e.pdf. Accessed 21 Nov 2012Rep

Johnke B (1999) Emissions from waste incineration. Background paper for Expert meeting on good practice in inventory preparation: emissions from waste. IPCC/OECD/IEA National Greenhouse Gas Inventories Programme, (Unpublished – TSU, Japan)

Kofoworola OF, Gheewala SH (2009) Estimation of construction waste generation and management in Thailand. Waste Manag 29(2):731–738

Kosaka K, Asami M, Kobashigawa N, Ohkubo K, Terada H, Kishida N, Akiba M (2012) Removal of radioactive iodine and cesium in water purification processes after an explosion at a nuclear power plant due to the Great East Japan Earthquake. Water Res 46(14):4397–4404

Liu X, Ishikawa M, Wang C, Dong Y, Liu W (2010) Analyses of CO_2 emissions embodied in Japan-China trade, energy policy security, prosperity and community – towards a common European energy policy? Spec Sect Regular Pap 38(3):1510–1518

Ministry of the Environment of Japan (2011) Guidelines (Master Plan) for disaster waste management after the Great East Japan Earthquake, edited by Ministry of the Environment of Japan

Morimoto A (2012) A preliminary proposal for urban and transportation planning in response to the Great East Japan Earthquake. IATSS Research 36(1):20–23

Nakamura S, Kondo Y (2002) Recycling, landfill consumption, and CO_2 emission: analysis by waste input–output model. J Mater Cycles Waste Manag 4(1):2–11

Norio O, Ye T, Kajitani Y, Shi P, Tatano H (2011) The 2011 Eastern Japan Great Earthquake disaster: overview and comments. Int J Disaster Risk Science 2(1):34–42

Pandey D, Agrawal M, Pandey JS (2011) Carbon footprint: current methods of estimation. Environ Monit Assess 178(1–4):135–160

Rossi B, Marique A-F, Reiter S (2012) Life-cycle assessment of residential buildings in three different European locations, case study. Build Environ 51:402–407

Simons M, Minson SE, Sladen A, Ortega F, Jiang J, Owen SE, Meng L, Ampuero JP, Wei S, Chu R (2011) The 2011 magnitude 9.0 Tohoku-Oki earthquake: mosaicking the megathrust from seconds to centuries. Science 332(6036):1421–1425

Sotoma M, Miyazaki H, Kyakuno T, Moriyama M (2003) Analysis of land use zoning regulations and green coverage ratio. J Asian Archit Build Eng 2(2):29–34

Suzuki M, Oka T (1998) Estimation of life cycle energy consumption and CO_2 emission of office buildings in Japan. Energy Build 22:38–41

Suzuki M, Oka T, Okada K (1995) The estimation of energy consumption and CO_2 emission due to housing construction in Japan. Energy Build 22(2):165–169

Taborianski VM, Prado RTA (2012) Methodology of CO_2 emission evaluation in the life cycle of office building façades. Environ Impact Assess Rev 33(1):41–47

Takahashi Y (2005) An evaluation study on the social experiment of modal shift to reduce carbon dioxide emission. J East Asia Soc Transp Stud 6:13

Takahashi T, Goto M, Yoshida H, Sumino H, Matsui H (2012) Infectious diseases after the 2011 Great East Japan Earthquake. J Exp Clin Med 4(1):20–23
Themelis NJ, Ulloa PA (2007) Methane generation in landfills. Renew Energy 32(7):1243–1257
UNEP (2012) Managing post-disaster debris: the Japan experience Report. Retrieved from www.unep.org/disastersandconflicts on December 20, 2012
Wang Z, Huang W, Zhao D, Pei S (2012) Mapping the tohoku forearc: implications for the mechanism of the 2011 East Japan Earthquake (Mw 9.0). Tectonophysics 524–525:147–154
Wiedmann T (2009) A first empirical comparison of energy Footprints embodied in trade – MRIO versus PLUM. Ecol Econ Methodol Adv Footprint Anal 68(7):1975–1990
Wiedmann T, Minx J (2008) A definition of 'Carbon Footprint'. In: Pertsova CC (ed) Ecological economics research trends. Nova, Hauppauge, pp 1–11
Wiedmann T, Lenzen M, Turner K, Barrett J (2007) Examining the global environmental impact of regional consumption activities – part 2: review of input-output models for the assessment of environmental impacts embodied in trade. Ecol Econ 61(1):15–26
Yan H, Shen Q, Fan LCH, Wang Y, Zhang L (2010) Greenhouse gas emissions in building construction: a case study of One Peking in Hong Kong. Build Environ 45(4):949–955
You F, Hu D, Zhang H, Guo Z, Zhao Y, Wang B, Yuan Y (2011) Carbon emissions in the life cycle of urban building system in China – a case study of residential buildings. Ecol Complexity Coupled Hum Nat Syst 8(2):201–212

Index

Y.A. Kontar et al. (eds.), *Tsunami Events and Lessons Learned: Environmental and Societal Significance*, Advances in Natural and Technological Hazards Research 35, DOI 10.1007/978-94-007-7269-4, © Springer Science+Business Media Dordrecht 2014

U

Printed by Printforce, the Netherlands